Current Topics in Microbiology 194 and Immunology

Editors

Mechanisms in B-Cell Neoplasia 1994

Edited by M. Potter and F. Melchers

With 152 Figures and 47 Tables

Springer-Verlag
Berlin Heidelberg New York
London Paris Tokyo
Hong Kong Barcelona
Budapest

MICHAEL POTTER, M.D.
Chief

Laboratory of Genetics
Bldg. 37, Rm. 2B04
National Cancer Institute
National Institutes of Health
Bethesda, MD 20892
USA

Professor Dr. FRITZ MELCHERS
Director

Institute for Immunology
Grenzacherstr. 487
CH-4005 Basel
Switzerland

Cover illustration: The cover design depicts typical highly differentiated plasmacytoma cells from an ascites Wright-Giemsa stained preparation. The figure shows differences in cell size and a large mitotic figure in a plasmacytoma cell. Continuous mitotic activity in differentiated cells is a characteristic of many mouse plasmacytomas.

Cover design: Harald Lopka, Ilvesheim

ISSN 0070-217X
ISBN-13:978-3-642-79277-9 e-ISBN-13:978-3-642-79275-5
DOI: 10.1007/978-3-642-79275-5

Softcover reprint of the hardcover 1st edition 1995

Production: PRODUserv Springer Produktions-Gesellschaft, Berlin
Typesetting: Camera-ready by authors
SPIN 10128460 27/3020-5 4 3 2 1 0 – Printed on acid-free paper.

Preface

The 12th workshop on MECHANISMS IN B-CELL NEOPLASIA was held in the Holiday Inn, Bethesda, MD, on April 18–20, 1994, with approximately 150 persons in attendance. The overall purpose of this series of workshops is to explore the workings underlying the process of neoplastic development in the B-cell lineage. This lineage has unique features as a model system of neoplastic development. First, its development is punctuated by a series of rearrangements and mutations in immunoglobulin genes, and these events allow in many situations a means for dating the time and site of origin of an oncogenic event that contributes to the development of a particular tumor. These physiological changes in genomic DNA may incur a transient instability in the functioning genome and open the door for mutational events such as chromosomal translocations. Second, there are many points along the extensive course of B-cell development from which progression towards autonomous growth may deviate and develop. This generates a considerable diversity of B-cell tumors. Third, punctuated development in the B lymphocyte lineage is also associated with a changing expression of new cell surface receptors and responses to exogenous growth and differentiation factors. The regulatory mechanisms that govern these changes may be vulnerable to mutation, and some of these mutations may give rise to the unusual growth behavior of a B-cell tumor. We shall highlight several areas of emphasis in this workshop.

Multiple Myeloma

Multiple Myeloma emerged as the dominant B-cell tumor under discussion in this workshop. This was tempered by a growing interest in this disease and by the great success of the IVth International Workshop on Myeloma held at

Rochester, MN, in October 1993 which Dr. ROBERT KYLE of the Mayo Clinic organized.

Multiple Myeloma (MM), once considered a rare disease, has become, unfortunately, increasingly more common. Several contributing factors, perhaps trivial, are the improvement in diagnostic procedures and the expansion of the high risk older age population (S. DEVESA: *Epidemiology and Biology of Multiple Myeloma*, Springer-Verlag, Heidelberg, 1991, p 3; D. A. REIDEL and L. M. POTTERN: Hematology/ Oncology Clinics of North America 6:225, 1992). Specific risk factors, however, have remained elusive. Possible genetic influences have been implied by the finding that the incidence of myeloma in the American black population has increased since 1970. In a continuing study at the Mayo Clinic, a most important relationship of MM to a benign abnormality in B-cell/plasma cell proliferation known as MGUS (Monoclonal Gammopathy of Undetermined Significance) has been revealed. MGUS is far more common than myeloma, and it is estimated in some studies that 3% of individuals over the age of 70 have a monoclonal expansion in their B-cell compartments (WITZIG et al.). At the Mayo Clinic 241 MGUS cases have been followed for 20 to 35 years, and it has been found that 24.5% of them progressed to become malignant MM or to develop amyloidosis (R. KYLE, Hematology/Oncology Clinics of North Merica 6:347, 1992). Thus, MGUS is a potential precursor of MM. With the availability of the more sensitive immunofixation procedure for diagnosing MGUS, we should learn much more about the prevalence and natural history of abnormal B-cell clones. A critical question for epidemiologists interested in MM is whether the true indicator of a trend or the identification of risk factors for MM lies in first understanding the prevalence of MGUS.

MM is semantically a paradoxical paradigm. All of the cells produce the same immunoglobulin (Ig) molecule and by this definition are part of the same B-cell clone. MM is also a tumorous process that involves multiple discontinuous bone marrow sites. The name, originally introduced in 1873 by J. VON RUSTIZKY, now raises important questions about how to explain the pathophysiological origin of separate marrow cavity tumors which can all be traced to B-lymphocytes that have the identical V-D-J-rearrangements, isotype switches and somatic hypermutations. In essence, how and when does this "myeloma" clone of B-cells spread to so many locations? Are these multiple bone marrow cavities individually seeded from a single extramedullary source (possibly some kind of transformed B lymphocyte) or did the process begin focally in one marrow cavity at the plasma cell stage and then selectively spread (metastasize) to new similar soil?

In 1974 it was reported that the myeloma idiotype could be found on peripheral blood lymphocytes using rabbit antisera (MELLSTEDT et al., Clin Exp Imm 17:371, 1974). Since this time there has been a continuous series of papers implicating lymphocytes as part of the myeloma clone.

Now, with the availability of more sensitive molecular clonal markers such as the ASO-RT-PCR, and DNA fingerprinting methodologies (see BILLADEAU et al. and BERGSAGEL et al.) coupled with rigorous cell sorting technologies, it becomes possible to extend the study of the relationship of B-lymphocytes to the plasma cells in MM to greater levels of sensitivity.

New questions and controversies have developed in appraising and identifying the lymphocytes with the myeloma idiotypes (see BERENSON et al., WITZIG et al., BILLADEAU et al., And BERGSAGEL et al.). The problem has taken on serious clinical implications precipitated by the increasing use of autologous bone marrow transplantation in the therapy of MM. In this procedure peripheral blood hematopoietic stem cells obtained during remission are used to reinfuse into the radiated recipient. It would be self defeating to be reinfusing myeloma stem cells in the form of $CD34^+$ (myeloma) lymphocytes. Thus, purging of these preparations of members of the myeloma clone coupled with specific selection of stem cells become critical components of the procedure. A major unresolved controversy now focuses on the stem cell antigen CD34 (see references cited above) which has been claimed to be expressed on lymphocytes carrying the myeloma idiotypes.

Plasma cells are not generally regarded as circulating cells, as they are very rarely if ever found in normal human peripheral blood. It is known that in advanced MM, neoplastic plasma cells do enter the blood and with sensitive selection techniques can be detected there (see WITZIG et al.). The presence of these plasma cells complicates the analysis of the peripheral blood lymphocyte compartment in MM.

Underlying these practical questions about the nature of the myeloma process is the yet unanswered question concerning when and where neoplastic transformation begins in MM. At present the answers are only hypothetical and fall into two categories. First, the process begins in one of the B lymphocyte compartments and undergoes some kind of clonal expansion as a lymphocyte. Two consequences of this hypothesis are (1) that transformed lymphocytes as well as plasma could then be components of the myeloma clone, and (2) the recirculating transformed lymphocytes could account for the dissemination of the transformed cells to multiple bone marrow cavities. The second type of hypothesis is that a later stage B-cell/plasma cell undergoes transformation in the bone marrow. One focal tumor then could expand and metastasize to other bone marrow cavities.

B-lymphocytes, particularly those that have emerged from the centroblast-centrocyte pathway in germinal centers of lymphoid tissues (Liu et al., Immunol Today 13:17, 1992), have been selected for increased binding affinity for antigens. These antigens presumably are exogenous and are largely introduced into the body by infections. Somatic hypermutation has been associated with the process of germinal center development. This is not the only mechanism for generating somatic hypermutations. Berenson et al. propose that human myeloma proteins bear the marks of somatic hypermutation. One implication of such a natural history is that the myeloma proteins should be monoclonal antibodies with high affinity to some relevant antigen. A paucity of supporting data for this exists (i.e., the demonstration that myeloma proteins have high affinity binding activity to exogenous antigens) but should be pursued as antigen binding activity of the myeloma protein could provide clues about the origin of the myeloma clone.

B-lymphocytes related to centroblast-centrocyte development in germinal centers are intriguing candidates as precursors for MM as these cells normally can develop into plasma cells and circulating memory B lymphocytes. If an immortalization event occured in cells related to centrocytes or their immediate derivatives, it could retain this potential as a neoplastic cell and hence give rise to both circulating lymphocytes and plasma cells.

B-blasts activated by different modes may comprise alternative precursors of plasma cell tumors. One candidate, not yet implicated, which has attracted special interest is the B1-a lymphocyte which expresses CD5. Plasma cells rarely express this antigen, and this makes it difficult to relate a neoplastic plasma cell to B1-a cells. Human B-CLL cells, however, express CD5 consistently. The interest in the $CD5^+$ B-cell relates to the nature of the associated immune responses. Responses to thymus independent type 2 and polysaccharide antigens (KENNY et al.), autoreactivity and polyreactivity are kinds of activities linked to $CD5^+$ B-cells. Progress in understanding CD5 on B-cells has emerged from biochemical and biological studies relating to its expression. BANDYOPADHAY et al. present evidence that activation of B-cells by sIgM crosslinking induces the accumulation of CD5 expression at the mRNA level. The interaction of expressed CD5 and its ligand CD72 results in the activation of B-cells. CHEN et al. postulate that CD5-CD72 binding could play a role in B-B interactions and that may be responsible for the self-renewing characteristic of B1-a cells and, further, that B-B interaction may be associated with reactivities to self antigens. Intriguingly, CHEN et al. have constructed a CD5 transgenic mouse and found that these mice develop a plasma cell hyperplasia. The question of heavy chain switching and somatic hypermutation in these cells remains unsettled. Antigen binding activities of mouse myeloma proteins have shown that between 5 and 10% bind T-independent antigens. Possibly, many of the others will more closely resemble so called "natural", polyreactive or autoreactive antibodies. The area of antigen binding activity of myeloma proteins has been neglected in the last few years. Perhaps further study will reveal interesting clues about whether a segment of the B-cell population is more prone to neoplastic development.

Regulation and Deregulation of Proliferation in Late B-Cells

The deregulation of *c-myc* is a pivotal step in the development of many diverse types of B-cell tumors. The transcription of this gene is deregulated consistently in bursal lymphomas in the chicken, plasmacytomas in the mouse, immunocytomas in the rat, endemic and sporadic Burkitt's lymphoma in humans and less consistently in a variety of human non-Hodgkin's lymphomas. Understanding the function of this gene and how its deregulation alters the behaviour of B lymphocytes is a fundamental problem, and we now begin to see how this may happen. *c-myc*, as might be anticipated, regulates complex functions such as mitotic cycling and apoptosis and is truly a regulatory gene. This has presented many technical problems. Beginning with the discovery of the *max* gene product that forms heterodimers with *myc* protein (BLACKWOOD and EISENMAN, Science 251: 1211, 1991), there has been a steady stream of progress towards an understanding of *myc* function. The myc/max heterodimers bind specifically to regulatory motifs in DNA. Thus, myc can function as a positive transcriptional regulator. The important specific targets are just beginning to emerge, e.g., ornithine decarboxylase (CLEVELAND et al.). Now a new series of activities has opened up with the observation that *myc* protein binds to other Helix-Loop-Helix zipper proteins TfII-I and YinYang1 (YY1). YY1 is a transcription factor that represses or activates the function of many genes (for list see SHRIVASTAVA and CALAME). The interaction of MYC protein with YY1 protein blocks the ability of YY1 to activate the transcription of a number of genes.

The transcription of *c-myc* can activate the apoptotic pathway leading to cell death in the WEHI 231 cell line (LEE et al.). In the BCL1 cell line antiidiotypic antibodies activate a cascade of events that leads to neither active cell growth or elimination by apoptosis, but rather a dramatic state. This apparently is mediated through a separate signal transduction pathway (SCHEUERMANN et al.). The dormant state may persist for 200 days. During the dormant state there is a slow balanced turnover and elimination of the cells. This model system may be relevant to the indolent behavior of some lymphomas and MGUS.

PHILLIP et al. (Mol Cell Biol 14:4032, 1994) have recently shown that the constitutive expression of *c-myc* results in the repression of cyclin D1. These results suggest that the *myc* protein interacts with another nuclear factor and that this heterodimer represses the expression of cyclin D1. Cyclin D1 is active during G1 and is repressed as cells go into S phase (PHILLIP et al.), thus repression of cyclin D1 may in effect may remove an essential block in the progression through the cell cycle. The chromosomal translocation t(11; 14) is consistently found in mantle zone lymphomas in humans (WILLIAMS et al.) and is associated with aberrant expression of Cyclin D1 by a mechanism that is not yet elucidated. Cyclin D1 is not expressed in normal B-cells and other Non Hodgkin Lymphomas (WILLIAMS et al.). Mice carrying the cyclin D1 transgene (*Eμ-cyclin*) apparently do not develop increased incidence of B-cell tumors. However, when these mice were mated to *Eμ-c-myc* transgenics, the formation of preB and B-cell lymphomas was accelerated in these double transgenics (HARRIS et al.). These studies suggest a cooperative effect from constitutive *c-myc* and cyclin D1 transcription.

The role of IL-6 as a critical factor in plasma cell tumor proliferation is still not resolved. It is clear that IL-6 is an essential requirement for the *in vitro* growth of mouse PCTs and human myelomas. IL-6 also has other properties acting as a maturation factor and increasing the production of Ig. Two signal transductions are activated by the binding of IL-6 with its receptors (CHENKIANG), the transient and immediate JAK-STAT and the delayed pathway that activates *ras* dependent mitogen activated protein kinases. The mechanism by which IL-6 stimulates proliferation is not well understood. An intriguing link between the surface receptor CD-40 (expressed on normal and neoplastic plasma cells) has been described by WESTENDORF et al., who show that myeloma lines can be stimulated by CD40L binding to produce IL-6 in an autocrine fashion.

Genomic Instability

Genomic instability is a semantic term that embraces a variety of interrelated phenomena. Tumor processes that exhibit a diversity of chromosomal breaks and rearrangements, such as MM and B-cell, are possibly global examples. The remarkable diversity of tumor cell phenotypes with established tumors may be another example of genome instability. Many experimental approaches to the problem of genomic instability, however, focus, on specific rather than global genetic regions.

Many of the consistently recurring chromosomal rearrangements, hallmarks of certain types of B-cell lymphomas, trace their origin to the VDJ rearrangements or to heavy chain isotype switching. Both processes are regarded as "physiological" in B-cells, but they must be nonetheless potentially dangerous as they expose DNA for breakage and are probably associated with the activiation of a variety of "recombinase" enzymes as well as housekeeping DNA repair machinery. The enzymatic participants in heavy chain switching are not yet defined, we are still learning about how and when they bring about recombinations. The key question in B-cell neoplasia, though, concerns why the other non-Ig participants in the illegal recombinations become the ostensible and selective target. Further, once identified each of these target genes, e.g., *c-myc*, *bcl-1*, *bcl-2*, *pvt-1* and more recently *bcl-6* (see DALLAFAVERA), now becomes a sepparate area relevant to the B-Cell neoplasia field.

The consistent rearrangement, e.g., t(8; 14), t(14; 18), etc., have been determined cytogenetically, but in the last few years PCR methodologies have been worked out for detecting each of these (e.g., see JANZ, MÜLLER, WILLIAMS), thus making it possible to apply this sensitive detection technique for exploring the natural history of lymphoma development. Remarkably, several of these rearrangements now have been identified in non-neoplastic tissues of in tissues sites where tumors are undergoing development (MÜLLER et al.). JANZ et al. present evidence that other highly repetitive genes are illegimately rearranged to *c-myc*.

The biochemical mode of action of the bcl-2 gene has become a focus of interest in the genomic instability field with the evidence that bcl-2 protects cell against various types of DNA damage, i.e., γ-irradiation, formation of hydroperoxidases. This has implicated bcl-2 as part of the antioxidant defense system. An overview of this complex area has been given by BORNKAMM and RICHTER. How these biochemical mechanisms contribute to the inhibition of apoptosis, if at all, remains to be investigated.

Acknowledgements

The meeting was sponsored by the National Cancer Institute, Division of Cancer Diagnosis, Biology and Centers, with the encouragement and enthusiam of Dr. ALAN RABSON. Generous as this support was, it was insufficient. Because of the economic crunch of the 1990's we nedded additional help. First, many of the participants used their own travel budgets to attend. We were most grateful for this cooperation, as it reflected the value of this meeting. Dr. BART BARLOGIE of the University of Arkansas Medical Center secured critical additional funding from a non-government source. Dr. GERALD MARTI obtained a contribution from the Food and Drug Administration. Scientific meetings, particularly those of a workshop nature, serve the important function of information exchange which often results in an ordering of a complicated set of facts. Trends are perceived and collaborative projects initiated. In an era in which so much new data is emerging, these syntheses are useful and particularly so in complex problem areas such as B-cell neoplasia where there are so many different model systems and fundamental components.

We are very grateful to Ms. VICTORIA ROGERS for her assistance in the preparation of this book.

Eric Humphries
1944–1993

ERIC HUMPHRIES was a regular and highly valued participant in the mecahnisms in B-cell Neoplasia workshops. The following was contributed by Dr. Jonathan Uhr.

"The death of Eric Humphries in December, 1993, left a void in many of his colleagues and friends. Eric was walking by himself in his beloved West Virginia woodlands when he apparently suffered a cardiac arrhytmia several weeks after replacement of a heart valve for rheumatic heart disease. He had been in extraordinarily good spirits and appeared to be convalescing uneventfully from the surgery.

The emotional impact of his death came not only from the loss of Eric as a scientific colleague but also because he emanated extraordinary enthusiasm for science and life in general accompanied by a special vitality and warmth. These qualities earned him an enormous affection and respect from all of us who knew him.

He received his graduate training with HOWARD TEMIN and his postdoctoral fellowship with ROBIN WEISS. Eric began his independent scientific career in the Department of Microbiology at the University of Texas Southwestern Medical School with very basic studies of the role of the Bursa of Fabricius in the development of normal and neoplastic B lymphocytes. He combined his expertise in virology with a growing understanding of immunology to unravel the biology of this interesting organ. He used it to elucidate the role that it played in B-cell differentiation and how oncogenic viruses could exploit it for their own purpose. Along this route he performed some of the pioneering studies on the oncogene REL and how these genes led to the development of B-cell neoplasia.

Eric was a scientist's scientist. He was creative, disinterested in trendy research and took great pride in the care and rigor with which his experimental data were accumulated and interpreted. There was no "fluff" or credibility gap in Eric's publications. He was an excellent and caring teacher and his students responded with loyalty and affection. He was a major contributor to the intellectual atmosphere of the Department of Microbiology here at the University of Texas Southwestern for 15 years.

Eric leaves his wife Carolyn (who has returned to our department) and his children."

Table of Contents

The "Myeloma" Clone

The Myeloma/Plasmacytoma Phenotype

Plasmacytoma Induction in Mice

Human and Mouse B-Cell Lymphomas

Growth Regulation: IL-6

The Late B-Cell

Growth Regulation – C-myc

Growth Regulation: Apoptosis, BCL-2, Dormancy

Growth Regulation: Cyclin D1, ABL

Genomic Instability: General Topics

Genomic Instability: Heavy Chain Switch Related Problems

List of Contributors

(Their addresses can be found at the beginning of their respective chapters.)

The "Myeloma" Clone

Detection of Peripheral Blood Myeloma Cells by Three-Color Flow Cytometry

T. E. Witzig, T. K. Kimlinger, and P. R. Greipp
Mayo Clinic, Rochester, MN 55905

Introduction

A variety of techniques have documented the presence of malignant cells in the peripheral blood (PB) of patients with multiple myeloma (MM) [1-5]. These cells are usually not detected on a white blood cell (WBC) differential performed by the routine clinical laboratory instruments or microscopes because they are in such small numbers or because they do not have plasma cell morphology. Cells that have the gene rearrangement but not the morphology of the plasma cell have commonly been referred to as precursor cells. These cells are important to characterize if we are to better understand the pathogenesis of MM and also learn which is the most important cell population to target.

There are several issues that remain to be addressed. The precursor cells must be studied to learn their morphology, immunophenotype, and what factor(s) lead to their differentiation to mature plasma cells. Although our ability to detect precursor cells has improved with newer immunological and molecular techniques, we still do not fully understand their clinical importance in MM. It must be realized that from a clinical standpoint, MM is a disease of monoclonal plasma cells. When MM is diagnosed the BM is replaced with monoclonal plasma cells that inhibit the normal BM function producing cytopenias that lead to infection and the need for blood transfusions. It is this plasma cell that secretes IL-1 beta which causes painful lytic bone lesions [6]. Treatment such as chemotherapy or radiation that lowers the number of tumor cells leads to clinical improvement; however, when these cells proliferate again the patient's condition deteriorates.

Monoclonal gammopathy of undetermined significance (MGUS) is the earliest form of detectable dysproteinemia and may be a useful condition to study to learn the clinical importance of precursor cells. It is found in approximately 3% of people over the age of 70 years of age. When BM examinations are performed on these patients, a population of monoclonal plasma cells (<10% by definition) can usually be detected indicating that the transformation from precursor cell to mature Ig-secreting plasma cell has already occurred. Despite the presence of monoclonal plasma cells, long-term (median, 22 years) studies on these patients indicate that only about 20% of these patients will ever get MM or amyloidosis [7]. This indicates that additional changes must occur to transform this dormant plasma cell clone into one that produces MM.

Methods

Our studies of the PB compartment have utilized a slide-based immunofluorescence peripheral blood labeling index (PBLI) technique [8]. This technique detects and quantitates circulating plasma cells by their morphology and monoclonal cIg light chain staining, and determines whether they are in S-phase by the bromodeoxyuridine labeling index.

In a previous study utilizing this technique we demonstrated that the circulating cells in patients with new or relapsed MM (ie active disease) had a higher labeling index than the cells in patients with inactive disease (MGUS or smoldering MM) [4]. A more recent report quantitated the circulating cells and, as expected, patients with active disease had higher numbers of circulating cells as compared to those patients with inactive disease [5]. In our experience, if the patient with circulating plasma cells responds to therapy the cells will decrease or disappear from the blood only to return at the time of relapse.

Recent studies have concentrated on comparing the PBLI technique with three-color flow cytometry (FC) on the same sample of PB. Our goal was to learn if the FC technique could detect small numbers of CD38+ 45- plasma cells seen on the slide technique and whether monoclonal cells would be detected in the CD38+ 45+ and 38- 45+ cell populations. PB was obtained from patients with MM and the mononuclear cells were split into two fractions- one was used for the standard PBLI technique and the other was used for three-color FC utilizing antibodies to CD38, CD45 and either cytoplasmic immunoglobulin (cIg) kappa or lambda light chain.

Results

In all three patients with documented monoclonal plasma cells by the immunofluorescence PBLI method we found 38+45- cells by FC. Three-color FC enabled us to also gate on the CD38+45+ and 38-45+ populations to determine whether there were monoclonal B-cells (by detection of an abnormal kappa/lambda ratio) in those populations. Figs 1-3 are examples of the patterns found. Fig 1 is an example of a case exhibiting clonal cells only in the CD38+45- compartment. Fig 2 demonstrates clonal cells both in the CD38+45- and 38+45+ compartments but not in the CD38-45+ population. Lastly, Fig 3 demonstrates clonal cells in all three compartments. In other cases studied to date (data not shown), we have only found clonal cells in the CD38+45+ and 38-45+ compartments in cases where there are also monoclonal cells in the CD38+45- compartment.

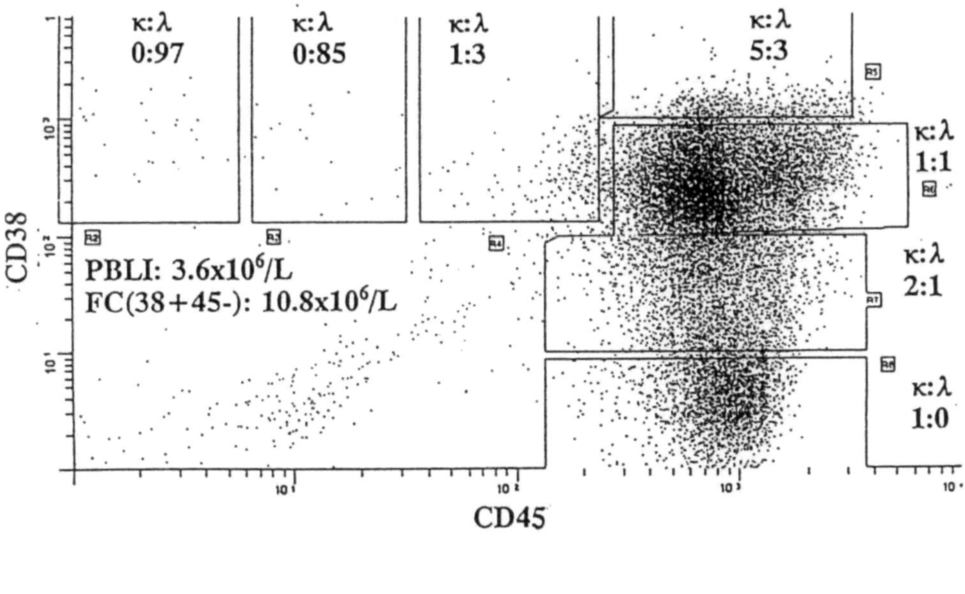

Fig 1: CD38/45/ kappa or lambda dot-plot from peripheral blood of a patient with 3.6 x 10^6/L monoclonal plasma cells by immunofluorescence slide technique. The flow cytometry technique calculated 10.8 x 10^6/L CD38+ 45- cells that expressed monoclonal lambda cytoplasmic light chain isotype. The CD38+ 45+ and 38- 45+ quadrants contained no monoclonal B-cells.

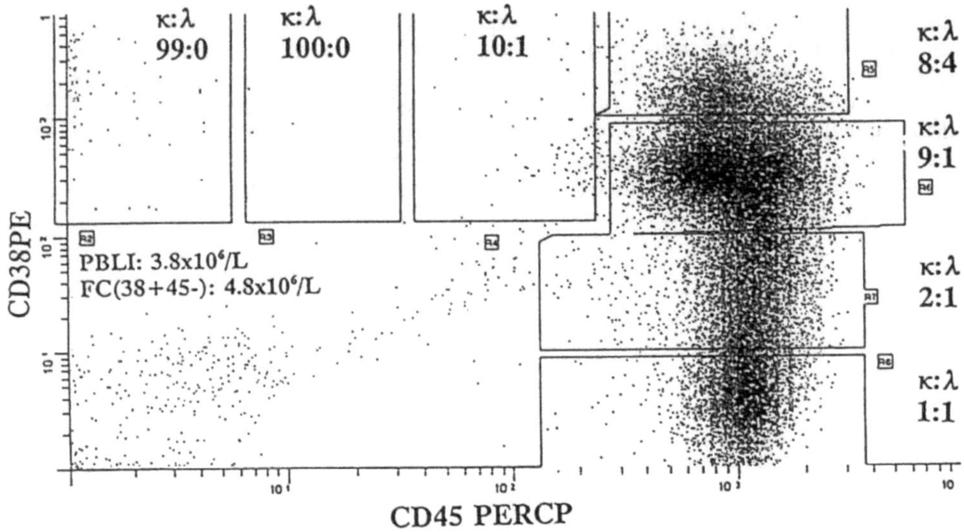

Fig 2: CD38/45/ kappa or lambda dot-plot from peripheral blood of a patient with 3.8 x 10^6/L monoclonal plasma cells by immunofluorescence slide technique. The flow cytometry technique calculated 4.8 x 10^6/L CD38+ 45- cells that expressed monoclonal kappa cytoplasmic light chain isotype. The CD38+ 45+ contained monoclonal cells but the 38- 45+ quadrant did not.

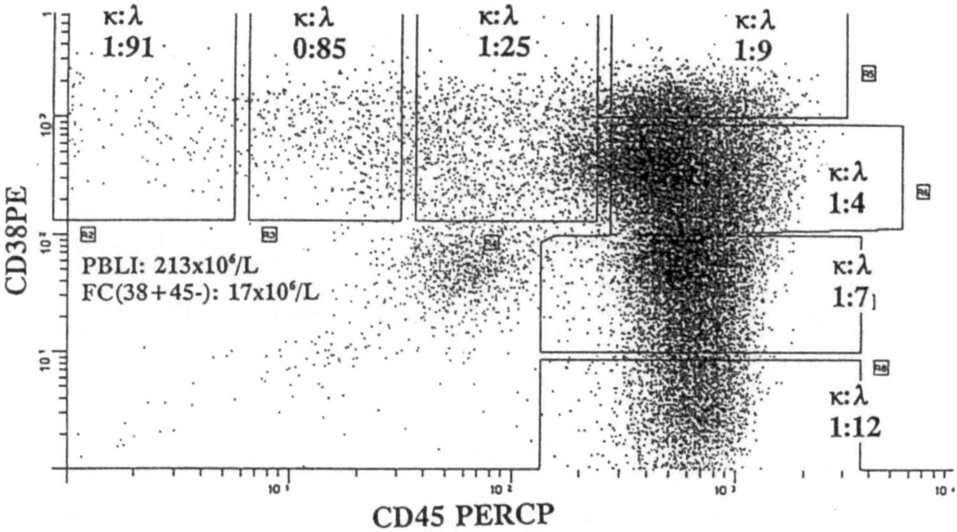

Fig 3: CD38/45/ kappa or lambda dot-plot from peripheral blood of a patient with 213 x 10^6/L monoclonal plasma cells by immunofluorescence slide technique. The flow cytometry technique calculated 17 x 10^6/L CD38+ 45- cells that expressed monoclonal lambda cytoplasmic light chain isotype. Both the CD38+ 45+ and 38- 45+ quadrants contained monoclonal cells by flow cytometry.

Summary and Future Directions

These results confirm that monoclonal cells frequently populate the PB of patients with active MM and that they can be easily detected by three-color FC. They also indicate that the clonal cells may have a variety of immunophenotypes, yet still belong to the malignant clone. Additional studies are needed to learn the molecular characteristics of the CD38+45+ and 38-45+ cells. In order to determine the importance of the CD38+45+ and 38-45+ cells at reproducing MM, these populations need to be isolated and placed in SCID mice or on to stromal cells to see if monoclonal plasma cell outgrowths can be obtained. Studies also need to be performed on PB cells from patients with MGUS to learn whether monoclonal plasma cells and precursor cells exist in the blood of those patients. These studies should provide us with useful information on the characteristics of precursor cells and allow better targeting of therapy for MM.

References

1. Berenson J, Wong R, Kim K, Brown N , Lichtenstein A (1987) Evidence for peripheral blood B-lymphocyte but not T lymphocyte involvement in multiple myeloma. Blood 70:1550-1553
2. Billadeau D, Quam L, Thomas W, Kay N, Greipp P, Kyle R, Oken M , Van Ness B (1992) Detection and quantitation of malignant cells in the peripheral blood of multiple myeloma patients. Blood 80:1818-1824
3. Fend F, Weyrer K, Drach J, Schwaiger A, Umlauft F , Grünewald K (1993) Immunoglobulin gene rearrangement in plasma cell dyscrasias: detection of small clonal cell populations in peripheral blood and bone marrow. Leuk Lymphoma 10:223-229
4. Witzig T, Gonchoroff N, Katzmann J, Therneau T, Kyle R , Greipp P (1988) Peripheral blood B-cell labeling indices are a measure of disease activity in patients with monoclonal gammopathies. J Clin Oncol 6:1041-1046
5. Witzig T, Dhodapkar M, Kyle R , Greipp P (1993) Quantitation of circulating peripheral blood plasma cells and their relationship to disease activity in patients with multiple myeloma. Cancer 72:108-113
6. Cozzolino F, Torcia M, Aldinucci D, Rubartelli A, Miliani A, Shaw A, Lansdorp P , DiGugielmo R (1989) Production of interleukin-1 by bone marrow myeloma cells. Blood 74:380-387
7. Kyle R (1993) "Benign" monoclonal gammopathy after 20-35 years of follow-up. Mayo Clin Proc 68:26-36
8. Witzig T, Gonchoroff N, Ahmann G, Katzmann J , Greipp P (1991) T-cell depletion using anti-CD2 coated magnetic beads simplifies the detection of peripheral blood plasma cells. J Immunol Methods 44:253-256

Detection of B-Cells Clonally Related to the Tumor Population in Multiple Myeloma and MGUS

D. Billadeau[1], P. Greipp[2], G. Ahmann[2], T. Witzig[2], and B. Van Ness[3]

[1]Laboratory Medicine and Pathology, The University of Minnesota, Minneapolis, MN, [2]The Mayo Clinic, Rochester, MN, and [3]The Institute of Human Genetics, The University of Minnesota

Introduction

Multiple myeloma is characterized by the clonal expansion of the plasma cells in the bone marrow (BM) and production of monoclonal immunoglobulin (Ig). We have used a sensitive allele-specific oligonucleotide (ASO)-PCR technique to detect cells clonally related to the malignant plasma cell population of monoclonal B cell gammopathies [1-3]. The PCR approach utilizes consensus primers to amplify the CDR3 region of a rearranged IgH allele from the tumor population, followed by sequence determination of the PCR product. This sequence is clonally unique due to novel combinations of V-D-J gene segments, random nucleotide insertions and deletions, and somatic mutations. Allele specific oligonucleotide (ASO) primers can then be synthesized and used to detect the tumor specific allele [1].

Although the malignant plasma cells are generally restricted to the bone marrow, clonally related cells are readily detected in the peripheral blood in myeloma [2] and some, but not all MGUS (monoclonal gammopathies of undetermined significance), even in the absence of observable clonal circulating plasma cells. Because of the increase use of peripheral blood stem cell harvests in a transplant approach for the treatment of myeloma, the identification and characterization of these related populations becomes a significant issue to resolve. The ASO-PCR technique offers a valuable tool in identifying cells clonally related to the plasma cell tumor in both unsorted stem cell harvests as well as selected populations (i.e. CD34+ populations) being used for transplant.

If B cells at various stages of differentiation are clonally related, it would be expected they share common Ig gene rearrangements. Because the ASO-PCR approach is specific for the tumor rearranged heavy chain allele, it can also provide a useful tool to identify expression of clonally related transcripts. RNA isolated from the BM of patients with IgG or IgA myeloma was reverse transcribed with μ, δ, γ, or α constant region primers, and the cDNA amplified using the tumor specific ASO. Based on the sequence of the PCR products, we confirm the presence of transcripts that are identical to the tumor IgH V-D-J CDR3 region, but associated with different isotypes, including μ. These results suggest the presence of B cells clonally related to the tumor at earlier stages of differentiation. Sorting of marrow or peripheral blood using CD38 and CD45 markers demonstrate the clonally related cells can be found in different cell compartments. Moreover, sequence comparisons between μ–δ expressing transcripts and the clonally related, more differentiated γ or α transcripts show a remarkable lack of accumulating somatic mutations.

The Use of ASO-PCR to Detect Tumor Associated DNA in the Peripheral Blood of Myeloma and MGUS Patients

As previously described [3], a pool of V-region, family-specific primers to the conserved regions of framework 1, and a J(H) consensus primer are used to amplify the CDR3 region of the rearranged immunoglobulin heavy chain allele from the predominant tumor population in the marrow of a newly diagnosed myeloma patient. The resulting PCR product can be sequenced to generate an allele-specific oligonucleotide primer. This primer can then be used in a subsequent PCR to specifically detect tumor associated DNA [1,2] or RNA [3].

We have used ASO-PCR to detect and quantitate circulating cells clonally related to the plasma cell malignancy in the marrow [2]. As shown in Table 1, we were able to detect tumor associated DNA from 13 of 14 myeloma peripheral blood samples. We found that the peripheral blood content did not correlate with the stage of the disease, nor marrow tumor burden. However, it is important to note that from the PCR analysis we

Table 1. Detection of cells clonally related to the marrow plasma cell population[a]

Sample	Stage of Disease	BM(%)[b]	PB(%)[c]
MM1	2A	15	3
MM2	3A	15	<.0001
MM3	3A	68	0.001
MM4	3B	22	3
MM5	1A	38	0.21
MM6	1A	3	0.0014
MM7	2A	3	0.0075
MM8	3B	36	0.03
MM9	3A	1.3	1.6
MM10	1A	90	33
MM11	2B	13	0.11
MM12	3A	33	0.10
MM13	3A	25	0.23
MM14	3B	36	0.25
MGUS1	-	10	0.20
MGUS2	-	16	0.02
MGUS3	-	10	<0.001
MGUS4	-	3	<0.03

a Methodology and details of detection given in reference 2

b Densitometric determination of DNA blot hybridization of rearranged J(H) locus

c Densitometric determination of ASO-PCR product; ND=not done cannot distinguish malignant plasma cells that escape from the marrow from possible circulating precursor cells that may continually contribute to the malignant population in the marrow. In some cases we found significant levels of circulating tumor cells, despite institutional reports indicating no circulating clonal plasma cells [2]. This suggested to us that there may be cells clonally related to the tumor population, but morphologically and developmentally distinct.

We have also examined a limited number of peripheral blood samples from MGUS patients. In one case we detected tumor associated DNA in the peripheral blood (Fig. 1), and estimated the cells represented about 0.2% of the total circulating mononuclear cells. Interestingly, immunohistological analysis showed no evidence of circulating clonal plasma cells, but a distinct clonal lymphoblastoid population, representing about 0.2% of the total circulating mononuclear cells, was observed. This suggests that even in some MGUS patients there may also be developmentally distinct populations clonally related to the expanded plasma cell population in the marrow.

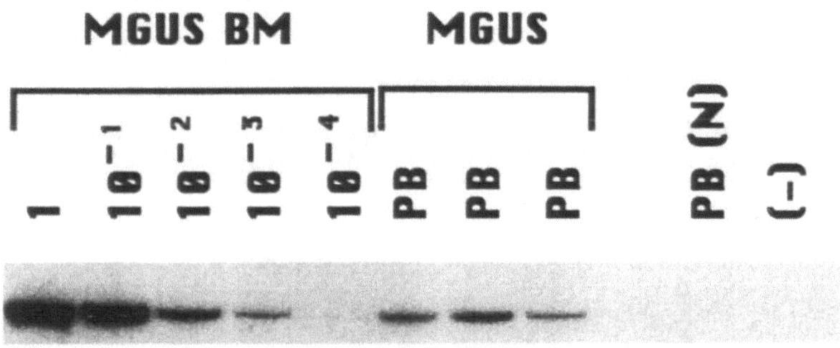

Fig. 1. ASO-PCR detection of tumor associated DNA in the marrow and peripheral blood of a MGUS patient. The marrow (BM) DNA was serially diluted into normal peripheral blood DNA to establish a titration of target DNA. PCR amplification of marrow and peripheral blood DNA (in triplicate) was performed, using α^{32}P-labeled primers to incorporate radioactivity. Products were eletrophoresed on a polyacrylamide gel, which was then exposed to x-ray film.

ASO-PCR analysis of peripheral stem cell harvests for transplant

Because of the increasing use of peripheral blood stem cell harvests in transplant therapies for myeloma, detection of tumor associated cell populations could impact on strategies for stem cell selections. One patient analysis is shown in Fig. 2, including analysis of the peripheral blood at diagnosis, at peripheral stem cell harvest, and after sorting for the CD34+ stem cell populations. This figure highlights the utility of the ASO-PCR analysis in assessing tumor contamination. Tumor-associated PCR products were readily detectable in both the marrow and peripheral blood. A significant reduction in the tumor-specific PCR product was noted at time of harvest, and there was significant additional reduction in CD34+ sorted stem cell populations. We believe the faint, but detectable PCR signal observed in the sorted sample likely represents a very low level of residual disease contaminating the sorted population; however, we cannot rule out the possibility that the signal derives from CD34+ cells that are clonally related to the malignant plasma cell.

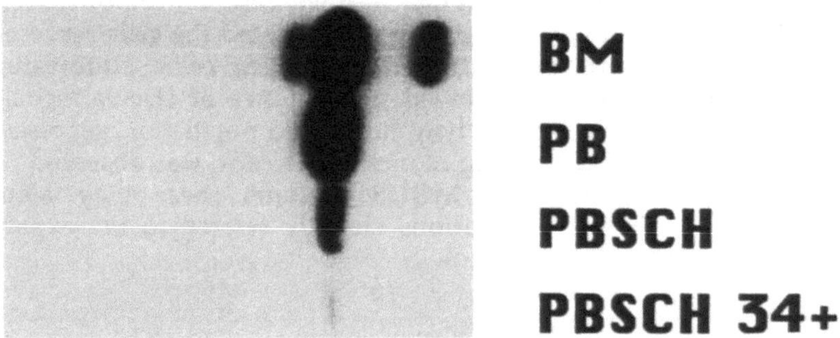

Fig. 2. ASO-PCR detection of tumor related DNA in the bone marrow (BM), peripheral blood (PB), peripheral blood stem cell harvest (PBSCH), and PBSCH CD34+ sorted cells.

Detection of Clonally Related but Developmentally Distinct RNA Species by ASO-RT-PCR in Myeloma Patients

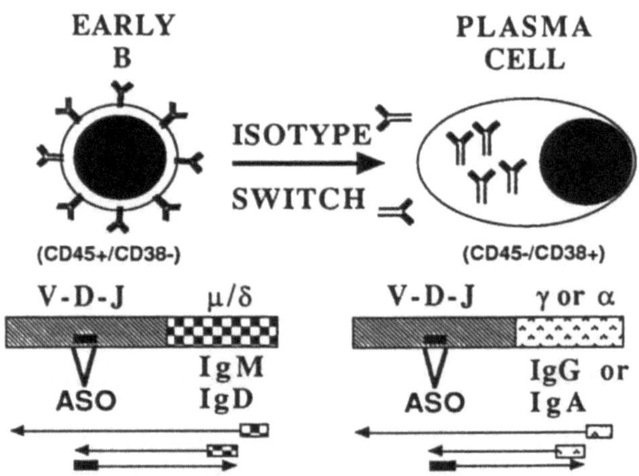

Fig. 3. The detection of clonally related, developmentally distinct B cells. Early B cells express V-D-J-Cμ and CD45. Clonally derived plasma cells express the same V-D-J associated with γ or α isotype and CD38. Using selected primers, isotype specific cDNA can be synthesized. The tumor specific ASO, together with an internal isotype-specific primer can then be used in a PCR reaction to detect products clonally related to the plasma cell tumor.

Although myeloma is characterized by the clonal proliferation of a plasma cell, there is continued debate over the presence of earlier lineage cells that are clonally related to the malignancy. Most myeloma cells express either Cγ or Cα heavy chain immunoglobulin isotypes. We reasoned that if B cells earlier in development were present, they may express pre-switch isotypes (Cμ and Cδ) and show clonal identity with the plasma cell by virtue of the unique V-D-J sequences. Tumor specific ASO-PCR was used to address this possibility.

Fig. 3 schematically represents the distinguishing features of pre- and post- switch B cells. The PCR strategies to detect B cells clonally related to the plasma cell malignancy, but developmentally distinct, are also indicated. Briefly, primers were used to reverse transcribe specific isotypes, and the cDNA generated was amplified using the tumor specific ASO-PCR. Using this approach, we [3] and others [4,5] have demonstrated the existence of pre-switch isotype species that are clonally related to the myeloma tumor. The sequence of the PCR products obtained from two representative patient samples is shown in Fig. 4. These sequences confirm the clonal relationship of the PCR products. The first line shown for each sample represents the IgH CDR3 sequence corresponding to the predominant plasma cell malignancy (γ or α). In each case the sequence of other isotypes (notably, pre-switch μ and δ sequences) showed almost complete identity. This demonstrates a significant sequence stability after switch recombination, and suggests the clonal proliferation was not under continuous antigen selection.

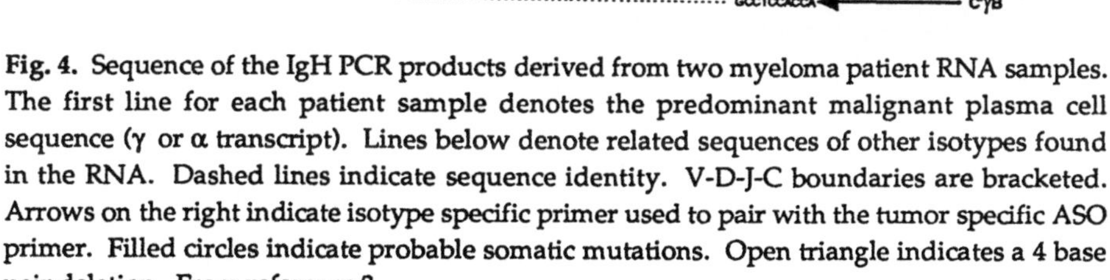

Fig. 4. Sequence of the IgH PCR products derived from two myeloma patient RNA samples. The first line for each patient sample denotes the predominant malignant plasma cell sequence (γ or α transcript). Lines below denote related sequences of other isotypes found in the RNA. Dashed lines indicate sequence identity. V-D-J-C boundaries are bracketed. Arrows on the right indicate isotype specific primer used to pair with the tumor specific ASO primer. Filled circles indicate probable somatic mutations. Open triangle indicates a 4 base pair deletion. From reference 3.

As shown in Fig. 3, some antigenic CD surface markers are differentially expressed during B cell development. Notably, CD45 is expressed early in B cell development, through B cell activation, but is lost on plasma cells. In contrast, CD38 is expressed early in B cell development, is lost during B cell differentiation, and is expressed at high levels on plasma cells. The majority of myeloma plasma cells are CD38+/CD45-. Thus, separation of cells based on the expression of these two markers can yield distinct B cell populations.

Cells from myeloma patients were sorted based on these two markers and examined using the isotype restricted ASO-RT-PCR. Fig. 5, panel A, shows the FACS profile of unsorted cells, as well as the identification of clonally related, but isotypically distinct PCR products. Panels B and C show that the isotype-specific PCR products are differentially detected in two sorted populations. The CD38+/CD45- cells are shown to uniquely amplify with the γ-specific ASO-PCR, whereas the clonally related, but isotypically distinct PCR products are associated with the CD38-/CD45+ sorted cells. These results confirm that developmentally distinct populations are present that are clonally related to the malignant plasma cell.

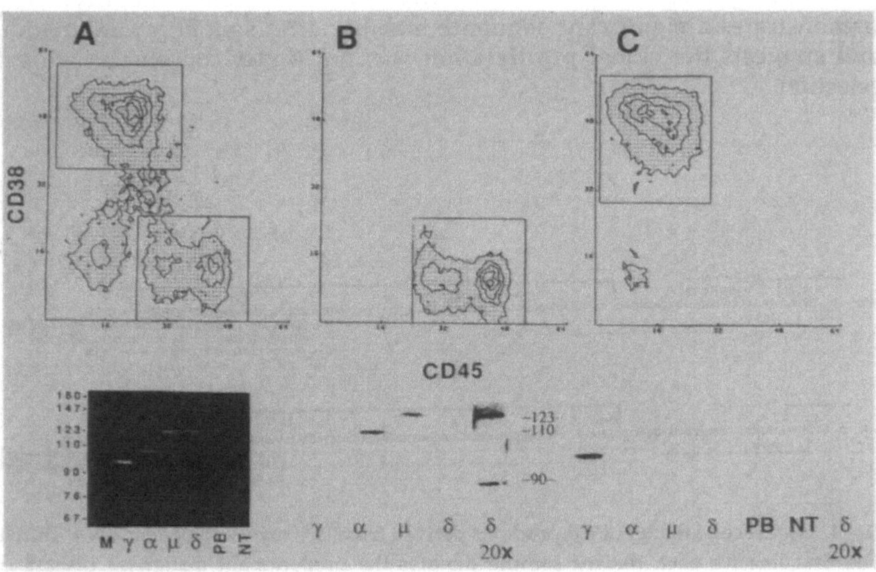

Fig. 5. RT-ASO-PCR identification of clonally related but isotype distinct RNA molecules in CD45/38 cell-sorted poulations. (A) Presorted cells from a myeloma patient. Below the FACS plot is shown the PCR products detected using isotype-specific primers. (B) CD45+/CD38- sorted cells. Isotype specific, clonally related ASO-RT-PCR products detected in this sorted population include μ, δ, α, but not γ. (C) CD38+/CD45- sorted cells. Only the γ isotype RNA is amplified with ASO-RT-PCR. From reference 3.

Discussion

A number of reports have suggested the existence of earlier lineage B cells that may represent myeloma precursors [6-10]. Our analysis has focused primarily on the peripheral blood cells of myeloma and MGUS patients. We find that we can detect circulating cells clonally related to the plasma cell malignancy in most myeloma patients, and in some, but not all, MGUS patients. Of particular interest was the observation that our estimates of cell numbers in the peripheral blood frequently exceeded estimates of the observed plasma cell content in the blood. This suggested that there may be clonally related populations that are morphologically, phenotypically and developmentally distinct. With the availability of a tumor-specific PCR assay we could find evidence for μ transcripts that shared sequence identity with the γ transcripts of the malignant plasma cell. Thus, we found evidence for pre-switch transcripts, that presumably were derived from an immature B cell population clonally related to the plasma cell tumor. Others have obtained very similar results [4,5]. Moreover, FACS sorting showed that tumor related pre-switch transcripts were derived from a cell compartment distinct from the plasma cell. The question that remains is whether these cells represent myeloma precursors.

If a myeloma precursor population exists that serves to continually populate the marrow with plasma cells, it would seem likely that the cell would have already progressed through heavy and light chain rearrangements; otherwise clonal rearrangements with identical CDR3 sequences would be required to obtain the clonal plasma cell populations found in the marrow. However, because the plasma cell tumor generally expresses a predominant isotype, a precursor model involving the μ-expressing, clonally related cells would require that these cells undergo a dominant, potentially programmed, switch recombination. Indeed, programmed switching of heavy chain immunoglobulin genes has been documented *in vitro* [11].

Additional studies will be required to characterize developmentally distinct cell populations clonally related to the plasma cell malignancy. One of the primary concerns is whether these cells truly influence the disease process. Subsequent studies could address this by sorting these clonally related populations and examining the differentiative and proliferative capacity of these cells either in culture or in SCID mouse models.

References

1. Billadeau D, Blackstadt M, Greipp P, Kyle R, Oken MM, Kay N, Van Ness B. (1991) Analysis of B-lymphoid malignancies using allele-specific polymerase chain reaction: A technique for sequential quantitation of residual disease. Blood 78:3021
2. Billadeau D, Quam L, Thomas W, Kay N, Greipp P, Kyle R, Oken M, Van Ness B (1992) Detection and quantitation of malignant cells in the peripheral blood of multiple myeloma patients. Blood 80:1818
3. Billadeau D, Ahmann G, Greipp P, Van Ness B (1993) The bone marrow of multiple myeloma patients contains B cell populations at different stages of differentiation that are clonally related to the malignant plasma cell. J Exp Med 178:1023

4. Carradini P, Boccadoro M, Voena C, Pileri A (1993) Evidence for a bone marrow B cell transcribing malignant plasma cell VDJ joined to C-mu sequence in IgG and IgA secreting multiple myeloma . J Exp Med 178:1091
5. Bakkus, MH, Van Reit I, Van Camp B, Theilemans K. (1994) Evidence that the clonogenic cell in multiple myeloma originates from a pre-switched but somatically mutated B cell. Br J Haematol 87:68
6. Kubagawa H, Vogler LB,Capra JD, Conrad ME, Lawton AR, Cooper MD. (1979) Studies on the clonal origin of multiple myeloma. Use of individually specific (idiotype) antibodies to trace the oncogenic event to its earliest point of expression in B-cell differentiation. J Exp Med 150:792
7. Cassel A, Leibovitz N, Hornstein L, Quitt M, Aghai E. (1990) Evidence for the existence of circulating monoclonal B-lymphocytes in multiple myeloma patients. Exp Hematol 18:1171
8. Caligaris-Cappio F. Bergui L,Tesio L, Pizzolo G, Malavasi F. Chilosi M, Campana D, vanCamp B, Janossy G. (1985) Identification of malignant plasma cell precursors in the bone marrow of multiple myeloma. J Clin Invest 76:1243
9. Epstein J, Barlogie B, Katzmann J, Alexanian R (1988) Phenotypic heterogeneity in aneuploid multiple myeloma indicates pre-B cell involvement. Blood 71:861
10. Kiyotaki M, Cooper MD, Bertoli LF, Kearney JF, Kubagawa H (1987) Monoclonal anti-id antibodies react with varying proportions of human B lineage cells. J Immunol 138:1401
11. Shockett P, Stavnezer J. (1991) Effect of cytokines on switching to IgA and alpha germline transcripts in the B lymphoma I.29 mu. Transforming growth factor-beta activates transcription of the unrearranged C alpha gene. J Immunol 147:4374

The Blood B-Cells and Bone Marrow Plasma Cells in Patients with Multiple Myeloma Share Identical IgH Rearrangements

P. L. Bergsagel, A. Masellis Smith*, A. R. Belch*, and L. M. Pilarski*
NCI-Navy Medical Oncology Branch, National Cancer Institute, Bethesda, MD 20889-5015
and *Departments of Immunology and Oncology, University of Alberta, Canada

Abstract

Previous reports have described the phenotypic and functional properties of monotypic late stage B cells in the blood of patients with multiple myeloma and have speculated that these B cells represent a malignant circulating component of myeloma. Here we show that blood B cells have IgH rearrangements identical to those expressed by the bone marrow plasma cells by using Ig Fingerprint and Allele-Specific Oligomer (ASO) polymerase chain reaction (PCR) methods. DNA from purified blood B cells and bone marrow plasma cells taken at the same time, and blood B cells taken at subsequent patient visits was amplified using consensus IgH primers, or ASO primers. In 10/16 patients, a single IgH rearrangement was amplified from the bone marrow plasma cells. In all 10 of those patients the same clonotypic rearrangement was amplified from the purified blood B cells. The relationship of these clonal blood B cells to the malignant bone marrow plasma cells remains undetermined

Introduction

Multiple myeloma (MM) is characterized by massive accumulations of plasma cells in the bone marrow (BM), high concentrations of monoclonal immunoglobulin (MIg) in the blood, and multiple lytic bone lesions. It is almost uniformly lethal with an average survival of 3 years post-diagnosis [1-4]. Although most patients initially respond to therapy, their disease eventually becomes refractory to further treatment [4-7]. Nearly all therapies reduce the number of plasma cells in the BM, decrease MIg levels in blood, and render patients transiently asymptomatic.

Funded by The National Cancer Institute of Canada with funds from the Canadian Cancer Society and by the Alberta Cancer Board Research Initiatives Program

However there is no correlation between the degree of plasma cell kill as measured by a decrement in serum MIg and patient survival [1,2]. This suggests that a cell other than the plasma cell is the ultimate determinant of outcome. Although the plasma cells in MM undergo occasional cell division, it is unclear whether they represent a self-renewing population. The view that plasma cells constitute the primary source of malignant growth and spread has been challenged by the demonstration of a monoclonal population of late stage blood B cells with DNA aneuploidy and multi-drug resistance. Thus MM may represent a heterogeneous population of continuously differentiating malignant B cells that give rise to BM plasma cells.

The work reported here is based on the premise that MM plasma cells lack generative potential and are the final differentiation stage of an earlier precursor that persists despite chemotherapy and continuously populates the BM with plasma cells. Blood B cells in MM are a heterogeneous set of monoclonal B cells that are multi-drug resistant, express CD34, have generative potential in vitro and exhibit extensive DNA hyperdiploidy, an attribute of malignancy [11-13]. These presumptively malignant B cells persist in "remission" BM, and in peripheral stem cells used for autologous transplantation, and may survive cytoreductive treatment. A fundamental issue regarding these B cells is the extent to which they include cells with clonotypic IgH rearrangements, an unequivocal indicator of clonal relationship with the malignant bone marrow plasma cells in MM. Although the presence of the clonotypic rearrangement does not necessarily denote malignant status, any circulating malignant cells should have the clonotypic rearrangement.

The Clonotypic IgH Rearrangement is Present in the Blood B Cells in MM

MM patients have 6-80% $CD19^{+}20^{+}$ B cells in peripheral blood mononuclear cells (PBMC) [13,14] with comparable mean values before and after chemotherapy, and increased numbers in relapse or progression. The total number of B cells in blood at diagnosis is 2 fold higher and the % B PBMC is 4 fold higher than in normal donors. By Southern blot the PBMC and BM PC share the same Ig heavy chain gene J_H rearrangement [14]. Analysis of Ig light chain mRNA by northern blot revealed substantial amounts of light chain mRNA restricted to or as predicted from the light chain type of the MIg [11,15]. PBMC MM B cells have cytoplasmic Ig [11,16]. B cells were monoclonal as defined by restricted Ig light chain expression, with heterogeneity in the intensity of staining suggesting progressive B cell differentiation within the $CD19^{+}$ population [16].

To determine the number and type of IgH rearrangements expressed within a population of blood B cells, they were purified by cell sorting based on their CD19 expression. Since all of these sorted B cells express cytoplasmic Ig, all must by definition have a rearranged IgH locus. Hemi-nested PCR was used to compare IgH rearrangements in sorted populations of blood B cells with sorted plasma cells from the bone marrow (BM). PCR using consensus oligonucleotides to the IgH variable region framework 2 (FR2) and the IgH J segment (JH) genes amplifies rearranged heavy chain genes, but not germline heavy chain genes (because germline FR2 and JH are too distant to be amplified). Because of different D gene lengths, and N region diversity, the length of the VDJ rearrangements amplified varies within a range of about 48 nucleotides, changing in increments of three nucleotides (because the majority of rearrangements preserve an intact open reading frame). When electrophoresed on a high resolution sequencing gel, the rearrangements present in a polyclonal B cell population thus appear as a ladder of ~16 bands, spanning 48 nucleotides. For a monoclonal B cell population, such as purified BM plasma cells from a patient with myeloma, only a single rearrangement should be amplified (an Ig fingerprint). In 10 patients we amplified a unique rearrangement from their BM plasma cells, and we examined their blood $CD19^+$ cells by Ig fingerprinting. The $CD19^+$ B cells were subdivided into a subset with high forward/side scatter properties and one with small scatter, and into $CD34^+$ and $CD34^-$ subsets.

The clonotypic rearrangement was detected in the blood B cells of all 10 patients. In 5/10 patients only a single rearrangement was amplified by CDR3 PCR from the blood B cells or in B cell subsets defined by physical or phenotypic properties, indicating clonotypic involvement. The rearrangement was determined to be the clonotypic rearrangement based on the fact that it was identical in length to the rearrangement amplified from the BM. In one case the rearrangement from the blood was also sequenced and shown to be identical without evidence of any mutations. In all cases where only a single rearrangement was amplified by CDR3 PCR, ASO-PCR also amplified a rearrangement of the expected size. Different patient's blood B cells amplified at the same time generated clonotypic rearrangements of different lengths, providing an internal control. These clonotypic rearrangements were unique to cells within the B lineage as T cells from the same patients lacked IgH rearrangements, providing a negative control for technical artifacts. In 5/10 patients many rearrangements were amplified by CDR3 PCR. For these patients, ASO primers were prepared and ASO-PCR using the FR2 and ASO oligos confirmed the presence of clonotypic rearrangements in the $CD19^+$ cells, but not the T cells. The results of the CDR3 and ASO PCR in these 10 patients are summarized in Table 1.

Table 1. Detection of clonal cells in CD19+ subsets by CDR3 and ASO PCR.

Patient	Small	Large	CD34-	CD34+
1			+	-
2			+	+
3			-	+
4		+	+	+
5	-	+	-	+
6	+	+		
7	-	+		
8	+	+		
9	+	+		
10	-	+		

Blood B cells in myeloma are distinct from plasma cells

The presence of clonotypic sequences among $CD19^+$ B cells is most reasonably interpreted as an indication that B cells at a differentiation stage earlier than that of the plasma cells are included within the malignant clone. However an alternate interpretation is that plasma cells have entered the blood from the bone marrow, although plasma cells are not usually detectable in the blood of myeloma patients. A considerable quantity of data distinguishes blood B cells from BM plasma cells [12-17] which is summarized in Table 2.

Table 2. Blood B cells are phenotypically and functionally distinct from plasma cells

	Blood B Cells	Plasma Cells
Clonotypic	Yes	Yes
B markers	CD20+24+	CD20-24-
Adhesive receptors	+++	Few
Motility	Motile (20%)	Sessile
P-gp	+++	-/lo
cIg	Lo/Med	High
CD34	++/+++	-/+
Morphology	Lymph/Myeloid	Plasmacytoid
DNA aneuploid	Yes	Yes

Although blood B cells and plasma cells share the characteristics of DNA aneuploidy and the expression of clonotypic sequences, they differ in biologically important properties such as the expression of phenotypic markers associated with earlier stages of B cell differentiation. Blood B cells, but not plasma cells, express CD20 and CD24 and have the morphology of lympho/myeloid cells. Unlike plasma cells, blood B cells in myeloma express adhesion receptors associated with motility/invasion (α2ß1 and α6ß1 integrins, CD11b) as well as having highly motile behavior. Blood B cells express CD34, a marker also expressed on hematopoietic stem cells, and associated with adhesive and possible invasive behavior. Plasma cells have very low or no expression of CD34. As expected based on their phenotypic differentiation stage, blood B cells have detectable but low-moderate cytoplasmic Ig (cIg) while plasma cells have high expression of cIg. Finally, while plasma cells are killed by cytotoxic drugs and appear to lack a functional multidrug transporter, blood B cells in MM have high functional expression of the multidrug transporter and appear to resist toxic drugs as indicated by their persistence after chemotherapy.

Blood B cells in MM express CD34 but have important differences from CD34+ hematopoietic stem cells

The expression of CD34 by blood B cells in MM was confirmed using a large panel of anti-CD34 antibodies (in collaboration with Dr. P. Landsdorp) and they express mRNA encoding CD34 as defined by reverse transcriptase PCR. The B cells analyzed are present in blood in much greater numbers than are normal stem cells, even after peripheral blood mobilization procedures employed clinically. Approximately 10-20% of total PBMC express a moderate density of CD34 in MM PBMC, and all of these coexpress B cell markers. When these $CD34^+$ B cells in blood were purified and analyzed for clonotypic rearrangements by ASO-PCR, they were found to be clonally related to the malignant plasma cells in 4/5 patients tested. Methods used to purify $CD34^+$ hematopoietic stem cells for transplant will, by definition, exclude most of these cells since they recover less than 1% of total PBMC. The extent to which purified hematopoietic stem cells from MM patients are contaminated with some of the CD34+ blood B cells remains to be established. CD34+ stem cells express a higher density of CD34 than do most myeloma B cells, and the flow cytometry sort gates that collect cells with low scatter properties, or avidity of binding required for column purification is likely to exclude most, but perhaps not all the presumptively malignant B cells described here.

Blood B cells in MM are motile while plasma cells are sessile

The cell type responsible for spread of MM within and outside the bone marrow must be able to extravasate and locomote throughout blood and tissue. Adhesion receptors involved in migration are only weakly or not expressed on B lineage cells in BM including CD19+ B and pre-plasma cells. BM B/plasma cells in MM express adhesive receptors involved in anchored behavior or homotypic aggregation including VLA-4, CD44, CD31 (PECAM), CD56 (NCAM) and some VLA-5 (fibronectin receptor). In contrast, PBMC B cells express L-selectin, VLA-2 and VLA-6 (receptors for collagen and laminin) and CD11b. Migration to the bone marrow requires extravasation, with CD11b mediating transendothelial migration [15], and penetration of the basement membrane requiring binding to extracellular matrix (ECM) [17]. Only a subset of the small PBMC B cells in MM express CD11b or VLA integrin receptors for ECM but nearly all the large B cell have abundant expression, suggesting they are programmed for migratory behavior. A subset of small PBMC B cells, all large PBMC B cells and most plasma cells express a receptor for hyaluronan-mediated motility (RHAMM) [16,18-20], required for B cell motility. The absence of RHAMM on normal B cells, and its prominent expression of MM B/plasma cells suggests a role in the malignant process. RHAMM has been shown to behave as an oncogene converting transfected cells from normal to metastatic growth (E. A. Turley, pers.comm).

Using video timelapse microscopy, among purified MM blood B cells, many undergo active and vigorous cell deformation to form protrusions, and approximately 20% are highly motile on hyaluronan with rapid migration speeds of about 16µm/minute. By comparison, normal B cells are not at all motile, and hairy cell leukemia B cells locomote at about 3µm/minute [20]. In distinct contrast, plasma cells from bone marrow are sessile with no cell deformation or motility. The motility of the blood B cells is inhibited by antibodies to RHAMM. Among locomoting MM B cells, but not on sessile RHAMM+ plasma cells, RHAMM co-localizes with the proto-oncogene ras. This suggests ras is intimately involved in RHAMM- mediated motility and probable invasive spread in MM. These studies further support the idea that the potential for invasive spread resides among the blood B cells not the bone marrow plasma cells, and suggest that they represent malignant members of the myeloma clone.

Biological Significance of Clonotypic Blood B Cells in MM

Historically there have been repeated reports in myeloma of circulating B cells with apparent expression of monoclonal immunoglobulin (Ig) or with highly abnormal properties, and speculations that multiple myeloma (MM) is perpetuated by a B cell at an earlier stage of differentiation than bone marrow (BM) plasma cells. The idea that the malignant clone in MM includes B cells that escape chemotherapy and which underlie relapse is thus conceptually attractive. We have described a large population of abnormal B cells in the blood of patients with MM with properties that might be expected of malignant progenitor cells. Important questions thus arise: 1) Do these blood B cells express the clonotypic Ig rearrangements found in BM plasma cells? 2) Are the clonotypic blood B cells distinct from plasma cells? 3) Are clonotypic blood B cells self renewing? Using Ig fingerprint analysis and allele-specific oligomer (ASO) PCR we analyzed IgH VDJ rearrangements among PBMC sorted for $CD19^+$ B cells, as compared to VDJ sequences from sorted BM plasma cells. In all 10 patients clonotypic sequences were amplified from blood B cells and for some, only the clonotypic band was amplified, conclusively showing that clonotypic B cells circulate in the blood. $CD34^-$ blood B cells always included cells with the clonotypic rearrangement, with little heterogeneity. Only the clonotypic rearrangement was amplified from the $CD34^+$ blood B cells in 2 patients, ASO-PCR amplified the clonotypic rearrangement from 2 more patients, and for one patient, no clonotypic rearrangements have yet been amplified from the $CD34^+$ subset although there were clonotypic cells in the $CD34^-$ subset.

Blood B cells are phenotypically and functionally distinct from BM plasma cells and from normal resting B cells. MM blood B cells express adhesion receptors needed for invasive spread, which are mostly lacking from BM plasma cells and absent from normal B cells. MM circulating B cells are motile while BM plasma cells are sessile, as are normal B cells. Circulating MM B cells express functional P-glycoprotein (P-gp), the multidrug transporter, while BM plasma cells lack any detectable functional export pump. MM blood B cells, but not plasma cells or normal B cells, express a moderate to high density of CD34, confirmed by the staining with 18 anti-CD34 mAbs, and by the detection of CD34 mRNA from sorted B cells. CD34 is low or absent from BM plasma cells in MM. Blood B cells have a 10-100 fold lower intensity of cytoplasmic Ig than do plasma cells, are heterogeneous in their expression of cIg suggesting a population of differentiating B cells, and morphologically do not resemble plasma cells.

The clonotypic and highly abnormal circulating B cells in MM may represent a malignant component of myeloma, or could represent the persistence of original clone of B cells (MGUS?) that provided the target for malignant transformation. If myeloma arises from a chronically stimulated antigen-specific B cell clone, the chronic stimulus is likely to persist in blood and/or bone marrow even after malignant transformation immortalizes one member of that clone. The cells described here could include a subset that is a historical "fossil footprint" of the early events in myeloma. The population of blood B cells in MM may also represent, or include, components of the malignant clone upon which dormancy has been successfully imposed coexisting with clonal relatives that have escaped the mechanisms maintaining dormancy. Table 3 lists the possible interpretations, which should not be viewed as mutually exclusive.

Table 3. Clonotypic B cells in the Blood of MM Patients

1. Non-malignant relatives of the MM clone
2. Migratory progeny of a BM-localized stem cell
3. Migratory progeny of a stem cell outside the BM
4. The malignant stem cell

The probable malignant status of the MM blood B cells is indicated by their extreme DNA aneuploidy suggestive of replicative abnormality. In nearly all patients, blood B cells exhibited extensive DNA hyperdiploidy with on average 8-10% excess DNA as compared to T cells. $CD34^+$ $P\text{-}gp^+$ blood B cells had the greatest degree of DNA hyperdiploidy. Normal B cells have no detectable DNA aneuploidy. The significant motility of circulating B cells as compared to either MM plasma cells or normal B cells also supports their designation as a malignant component of the myeloma clone responsible for traffic and spread of the cancer. Clonotypic blood B cells may include the malignant stem cell in MM, or alternatively may be the migratory progeny of a clonal progenitor resident in or outside the BM. Spread from an original lesion to distant BM sites must be accomplished by migration of malignant cells through blood. Our evidence suggests a possible role for clonotypic blood B cells as drug resistant mediators of malignant spread and a reasonable target for chemotherapy. Their origin and ability for self-renewal is as yet unknown.

References

1. Bergsagel DE (1990) Treatment of plasma cell myeloma. Ann Rev Med 30:431-3
2. Palmer M, Belch A, Hanson J, Brox L (1989) Reassessment of the relationship between M-protein decrement and survival in multiple myeloma. Br J Cancer 59:110-2
3. Belch A, Shelley W, Bergsagel D, Wilson K, Klimo P, Willan A (1988) A randomized trial of maintenance versus no maintenance melphalan and prednisone in responding myeloma patients. Br J Cancer 57:94-9
4. Griepp PR (1992) Advances in the diagnosis and management of myeloma. Sem Hematol 29:24-45
5. Barlogie B, Gahrton G (1991) Bone arrow transplantation in multiple myeloma. Bone Marrow Transplant 7:71-9
6. Harousseau JL, Milpied N, Laporte JP, et al. (1992) Double-intensive therapy in high-risk multiple myeloma. Blood 79:2827-33
7. Barlogie B, Epstein J, Selvanayagam P, Alexanian R (1989) Plasma cell myeloma - new biological insights and advances in therapy. Blood 73:865-79
8. Sonneveld P, Durie BGM, Lokhorst HM, et al. (1992) Modulation of multidrug-resistant multiple myeloma by cyclosporin. Lancet 340:255-9
9. Salmon SE, Dalton WS, Grogan TM, et al. (1991) Multidrug-resistant myeloma: laboratory and clinical effects of verapamil as a chemosensitizer. Blood 78:44-50
10. Dalton WS, Grogan TM, Meltzer PS, et al. (1989) Drug-resistance in multiple myeloma and non-Hodgkin's lymphoma: detection of p-glycoprotein and potential circumvention by addition of verapamil to chemotherapy. J Clin Oncol 7:415-24
11. Pilarski LM, Mant MM, Belch AR (1993) Circulating, monoclonal, multi-drug resistant B cells may comprise the malignant stem cells population in multiple myeloma. In: Dammaco F (ed) Frontiers in Immunology and Cancer Research. (in press)
12. Pilarski LM, Belch AJ (1994) Circulating monoclonal B cells expressing p-glycoprotein may be a reservoir of multidrug resistant disease in multiple myeloma. Blood 83:724-36

13. Pilarski LM, Jensen GS (1992) Monoclonal circulating B cells in multiple myeloma: A continuously differentiating possibly invasive population as defined by expression of CD45 isoforms and adhesion molecules. Hematol Oncol Clin North Am 6:297-322
14. Jensen GS, Mant MJ, Belch AJ, Berensen JR, Ruether BA, Pilarski LM (1991) Selective expression of CD45 isoforms defines CALLA+ monoclonal B lineage cells in peripheral blood from myeloma patients as late stage B cells. Blood 78:711-9
15. Jensen GS, Belch AR, Kherani F, Mant MJ, Ruether BA, Pilarski LM (1992) Restricted expression of immunoglobulin light chain mRNA and of the adhesion molecule CD11b on circulating monoclonal B lineage cells in peripheral blood of myeloma patients. Scand J Immunol 36:843-53
16. Pilarski LM, Masellis Smith A, Belch AR, Yang B, Savani RC, Turley EA (1994) RHAMM, a receptor for hyaluronan-mediated motility, on normal lymphocytes, thymocytes and in B cell malignancy: An oncogene in B cell malignancy? Leuk Lymphoma (in press)
17. Jensen GS, Belch AR, Mant MJ, Ruether BA, Yacyshyn BR, Pilarski LM (1992) Expression of multiple beta-1 integrins on circulating monoclonal B cells in patients with multiple myeloma. Am J Hematol 43:29-36
18. Hardwick C, Hoare K, Owens R, et al. (1992) Molecular cloning of a novel hyaluronan receptor that mediates tumor motility. J Cell Biol 117:1343-50
19. Turley EA, Austin L, Vandligt K, Clary C (1991) Hyaluronan and a cell associated hyaluronan binding protein regulate the locomotion of ras-transformed cells. J Cell Biol 112:1041-7
20. Turley EA, Belch AR, Poppema S, Pilarski LM (1993) Expression and function of a receptor for hyaluronan-mediated motility (RHAMM) on normal and malignant B lymphocytes. Blood 81:446-53

Multiple Myeloma Clones are Derived from a Cell Late in B Lymphoid Development

J.R. Berenson[1], R.A. Vescio[1], C.H. Hong[1], J. Cao[1], A. Kim[1], C.C. Lee[1], G. Schiller[1], R.J. Berenson[2], and A. K. Lichtenstein[1].
[1]Divisions of Hematology/Oncology, D.V.A. West Los Angeles and UCLA School of Medicine, Jonsson Comprehensive Cancer Center, Wilshire and Sawtelle Blvds., Los Angeles, CA 90073.
[2]CellPro Incorporated, 22322 20th Avenue SE, Suite 100 Bothell, Washington 98021

Abstract

We have previously demonstrated that the immunoglobulin (Ig) heavy chain variable region (VH) sequences expressed by the malignant clone in multiple myeloma (MM) contain a high degree of somatic mutation without clonal diversity. This sequence can be used to identify all members of the malignant clone in this B cell malignancy. We sequenced the variable regions expressed by patients with MM and generated primers from the complementarity determining region (CDR) sequences specific for each patient's tumor. Using these primers, we performed PCR amplification on highly purified subpopulations of cells separated by expression of CD10, CD34 and CD38. The results of these experiments demonstrate: 1) there is a small fraction of CD10-expressing tumor cells in MM patients, 2) CD34-bearing malignant cells do not exist in MM, and 3) although the vast amount of tumor is in the CD38-expressing cells, a small amount of tumor is in the CD38-negative population. We also used these primers to determine whether pre-class switch (i.e., Cμ-expressing lymphocytes) clonal cells exist in these patients. After PCR amplification with CDR1 and Cμ primers, colony hybridization was performed using both framework 3 (FR3) and CDR3 probes. Out of >200 FR3-hybridizing colonies, ≤ 5 colonies also hybridized with the CDR3 probe. Colonies which hybridized with both these probes were sequenced, and none of these sequences matched even closely the CDR3 expressed by the malignant clone. These results make the existence of a pre-class switch malignant cell unlikely in MM. Overall, these results suggest that the malignant clone in MM derives from a cell late in B lymphocyte development.

Introduction

Previous studies in our laboratory have shown that the Ig VH sequences expressed by the malignant clone in MM bone marrow (BM) contain a high degree of somatic mutation without clonal diversity or change throughout the course of the patient's disease (1). The combination of highly mutated genes without sequence diversity makes the expressed VH segment an ideal tumor marker. We utilized this gene to identify all malignant cells in MM to gain further insight into the spectrum of cell types which are part of this tumor. We developed primers complementary to unique areas of the VH gene expressed by the malignant clone in MM patients. These primers then were used to determine both the stages of B cell differentiation and types of cell surface markers present on the malignant cells in MM.

Data has accumulated to support malignant involvement of nonplasma cell types in MM (2-6). Evidence in support of possible pluripotential stem cell involvement in MM include the expression of nonplasma cell markers on the aneuploid cell population in myeloma and the high incidence of acute nonlymphoblastic leukemia (ANLL) in longterm MM survivors (2,3). The possible involvement of these immature cells in the pathogenesis of MM has become of increasing importance with the development of therapeutic approaches involving myeloablative chemotherapy combined with autologous bone marrow support. Although initial response rates to these treatment regimens was encouraging, relapse rates in patients were relatively high due in part to the reinfusion of contaminating autograft tumor cells (for review see (7)). Peripheral blood (PB) progenitor cell transplantation is being used to minimize tumor reinfusion. This technique may not substantially reduce the autograft tumor load since the total cell number infused is greater and circulating tumor cells have been detected in the PB (8,9). Highly purified hematopoietic progenitor cells bearing the early hematopoietic cell antigen CD34 are capable of engrafting patients following myeloablative chemotherapy in patients with advanced breast cancer (10). A similar CD34-selected transplantation in MM would only be advantageous if this antigen was not expressed on any of the tumor cells. Normally, CD34 antigen expression is gradually lost during hematopoietic cell maturation, but does persist on the earliest B-lymphocyte precursor cells (11). Although the CD34 antigen is not normally found on the surface of plasma cells, the existence of a clonally less differentiated CD34-expressing B-lymphocyte or even stem cell precursor or aberrant expression of this antigen on MM cells may occur. Thus, we determined whether any malignant cells in MM bear the CD34 antigen by using PCR amplification with VH gene primers on CD34 cells from MM BM.

A preclass-switched malignant population in MM has also been suggested from several studies. First, idiotype expressed by the malignant clone was also found on the surface of IgM-bearing lymphocytes (4). Second, PCR amplification of Cμ-expressing cells using Ig gene primers derived from CDR3 sequences expressed by the MM clone shared sequence identity (5,6). However, the high degree of somatic mutation without clonal diversity (1) would suggest that the malignant cell of origin starts late in B cell development. Thus, we employed a colony hybridization technique using patient-specific CDR1 and Cμ primers on MM samples in order to determine whether malignant Cμ-expressing cells exist in MM.

The expression of CD38 has been described as present on all normal plasma cells and all malignant cells in myeloma (12,13). Thus, if the expression of CD38-bearing cells in MM include all of the malignant population, it may be possible to use antibodies directed against this antigen for ex vivo purging and in vivo therapeutic manipulations. We determined whether cells lacking CD38 expression are part of the MM clone using PCR amplification with patient specific Ig primers.

The CD10 antigen was initially derived from antisera developed against acute lymphoblastic leukemia cells (15), and initial studies suggested its expression was limited to immature lymphoid cells (16). However, more recent studies suggest that this antigen is also expressed on germinal center B cells (17) and other hematopoietic (18) and nonhematopoietic cell types (19). The expression of this marker on the malignant cells in MM has been reported to occur in approximately one-fourth of cases (for review see (20)). These patients have been reported to have a shorter surivival than patients lacking CD10-expressing malignant cells (20). However, no studies have been performed to determine whether this antigen is present on a subset of malignant cells in a larger proportion of MM patients using more sensitive tumor detection techniques. We used PCR with VH primers on highly purified CD10-expressing MM bone marrow cells to search for the presence of a CD10-bearing malignant subpopulation.

Materials and Methods

Patient material. BM and PB were obtained from patients with multiple myeloma after informed consent and in accordance with the Human Subjects Review Boards of the DVA and UCLA Medical Center. The mononuclear layer was obtained after density centrifugation and washed after exposure to RBC lysis buffer.

Myeloma VH gene determination. Total RNA was extracted from BM mononuclear cells using the guanidine isothiocyanate method. cDNA was made using antisense primer specific for either Cγ or Cα and M-MLV. A consensus primer to the leader sequence of one of the six VH gene families was then added to the cDNA mixture along with Taq polymerase. Thirty-two cycles of PCR amplification were performed as this allowed for detection of the monoclonal VH gene product in most myeloma patients and yet did not yield a product in normal patients (1). Six reactions were performed for each patient using a sense primer from each of the VH gene families, electrophoresed through a ethidium bromide impregnated agarose gel and assayed by exposure to UV light. PCR product was detectable in only one of the six reactions for each patient.

Myeloma variable gene sequencing: The remainder of the PCR product was purified and excised from a 1.6% low-melt agarose gel. The extracted DNA was treated with T4 polynucleotide kinase and then blunt end-ligated into Bluescript II SK^- vector (Stratagene, La Jolla, CA). One-fifth of the ligation mixture was then used to transform E. Coli strain XL1-Blue which were then plated onto ampicillin-impregnated agar plates. Colonies containing appropriately sized inserts were selected after mini-prep purification and Pvu II (Stratagene) plasmid digestion. Dideoxy chain termination sequencing was performed on a minimum of three clones obtained from at least two separate PCR reactions using the Sequenase II kit (US Biochemicals, Cleveland, OH) with ^{35}S α-dATP.

Purification of CD34+ bone marrow cells: BM mononuclear cells were resuspended and incubated for 25 minutes on ice with a biotinylated 12.8 antibody (murine anti-CD34 from CellPro, Bothell WA). The cells were passed through the CEPRATE LC34-BIOTIN system (CellPro) (10) and adsorbed cells were collected and cryopreserved until needed. On the day of sorting, the CD34+ cells were resuspended after washing and incubated with an antibody to a separate epitope on CD34, HPCA-2 PE (Becton Dickinson, Mountain View, CA), on ice for 30 minutes. The cells were washed and sorted on a FACStar Plus (Becton Dickinson). Cells were sorted in the normal-C mode after gating to exclude debris and cell doublets at a rate of 1500-2000 cells/sec and the CD34+ cells were collected for DNA extraction.

Purification of CD10+ BM cells: BM mononuclear cells were washed, resuspended, and incubated with both anti-CD10-FITC (W8E7 clone-mouse IgG2a kappa, Becton Dickinson) and anti-CD10-PE (SS2/36-mouse IgG1 kappa), DAKO, Glostrup, Denmark). Mouse IgG1 PE, mouse IgG2a-FITC, and FITC were used as negative controls. FAC sorting was done as outlined above for CD34 cell purification, and cell fractions differing in CD10 expression collected for PCR analysis.

Purification of CD38+ PB and BM cells: BM and PB mononuclear cells were washed, resuspended and incubated with two antibodies which recognize different epitopes of the CD38 antigen (anti-CD38-PE, Becton Dickinson and antiCD38-FITC, AMAC, Inc., Westbrook, ME). Mouse IgG1 PE and FITC were used as negative controls. FAC sorting was done as outlined above for CD34 cell purification, and different cell fractions collected for VH gene studies.

PCR conditions: FAC-sorted or unsorted BM or PB mononuclear cell DNA was extracted and quantitated using the DNA DipStick kit (Invitrogen, San Diego CA). Oligonucleotide primers were prepared (Operon, Alameda CA) complementary to the most unique sequences in the CDRs for each patient and also to the β-actin gene (Stratagene). PCR amplification was performed on 0.1 μg of DNA. Thirty μl of the PCR product was then electrophoresed through an ethidium bromide-treated agarose gel and the quantity was assayed by UV light exposure.

Colony Hybridization with Cμ: M-MLV was used to make cDNA using bone marrow RNA and a Cμ oligonucleotide primer. PCR amplification was performed using a patient specific 5' CDR1 primer and Taq polymerase, and the appropriately sized band was removed after electrophoresis on an agarose gel. The PCR products were purified and ligated as described above, and E. coli bacteria were transformed. White colonies were selected, transferred to nitrocellulose filters and lysed to obtain DNA. The filters were first hybridized with a ^{32}P-labeled FR3 probe which identified colonies containing the specific VH gene expressed by the malignant clone. Next, these same filters were hybridized with a labeled probe derived from the CDR3 sequence expressed by the patient's malignant clone. Finally, all colonies which hybridized both to the FR3 and CDR3 probes were sequenced, and compared to the MM Cγ sequence.

Results

Patient characteristics: Each of the patients studied had Stage II or III MM by the criteria proposed by Durie and Salmon (21), and a large amount of serum monoclonal Ig protein. BM and PB for these studies were obtained from patients at the time of diagnosis or relapse.

Myeloma VH gene sequencing and oligonucleotide primer/probe design: BM RNA was subjected to RT PCR with a Cγ or Cα primer plus an upstream VH gene oligonucleotide primer. Amplified PCR product was obtained from only one of the six VH reaction mixtures in each case, which represented the V gene family used by the MM clone. This PCR product was then sequenced after insertion into a plasmid. At least three plasmid clones were sequenced and all of the sequences were identical in each patient. The MM sequence was then compared to published germline VH, D and JH gene sequences to determine the location of the CDR regions and amount of somatic mutation. Patient specific oligonucleotide primers and probes were then designed complementary to the most unique regions of the CDRs.

Are CD34-expressing malignant cells present in MM?

CD34+ cell purification: BM mononuclear cells from five MM patients were exposed to a biotinylated anti-CD34 antibody 12.8 (22) and passed through the CEPRATE immunoadsorption column. When the adsorbed cells were assayed for the CD34 antigen by a second antibody which binds to a different epitope (HPCA-2-PE), the percentage of CD34-positive cells after this initial purification varied from 50-92%. The adsorbed cell yield ranged from 0.5-1.5% of the initial sample. The adsorbed cell CD34+ purity and yield from these MM patients is similar to that obtained from normal BM specimens done in our lab (Berenson JR and Vescio RA, unpublished observations) and by others (10), and compares well to the expected percentage of CD34+ cells in normal bone marrow which is 0.5-2% (22). These CEPRATE-enriched cells were further purified by sorting the cells through a FACStar Plus after labelling with the HPCA-2PE antibody. The positively staining cells were collected, re-analyzed, and the CD34+ cells were >99.99%.

Sensitivity of the PCR assay to detect MM tumor cells: Using CDR1 and CDR3 primers specific for each patient, PCR amplification was performed on previously quantitated BM diluted with placental DNA to maintain a final concentration of 0.1 μg of DNA per reaction in order to determine the sensitivity of our assay to detect clonal cells (Figure 1). PCR product was still detectable in each patient after dilution of the BM DNA to $1:10^3$ in two cases and $1:10^4$ in three cases. After adjusting the results for the percentage of plasma cells in the initial sample, the sensitivity of the PCR assay to detect malignant cells in the five patients studied ranged from 1:1,600 to 1:16,600 cells per reaction.

Assay for clonality in the CD34+ population: As a control experiment, the PCR was used to amplify 0.1 μg of DNA extracted from the purifed CD34+ cell population, placenta and unseparated BM with primers specific for the β-actin gene, (present as a single copy gene in all cells). All reactions were done concurrently, and when the PCR products were run on an agarose gel, the intensity of the PCR product was identical (Fig. 1). This implied that the DNA was of adequate quantity and quality in each sample. When the same PCR reactions were performed using the patient specific oligonucleotide primers on a second aliquot of CD34+ DNA, (0.1 μg), no detectable product was obtained on any of the five patients' samples (Fig. 1).

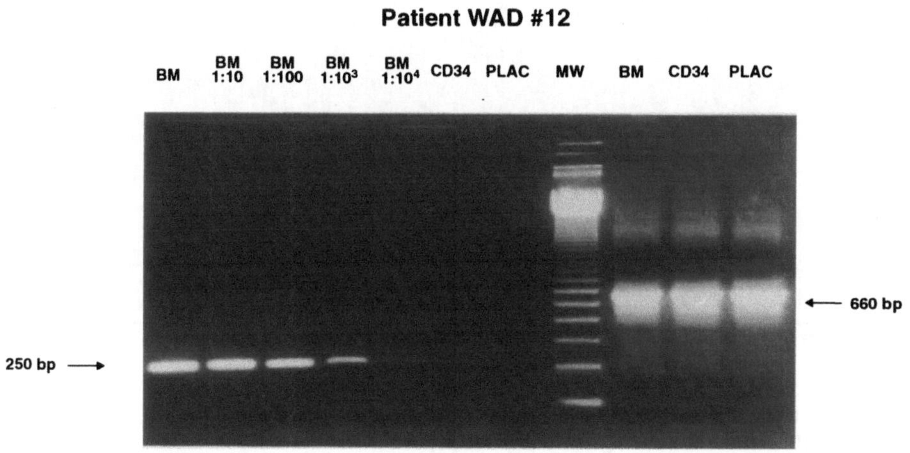

Fig. 1. Amplified PCR product (250 bp band) was detected using CDR1 and CDR3 primers on serially diluted unsorted BM from WAD#12 but not in the CD34-sorted or placental DNA. PCR product with β-actin primers is equivalent (660 bp band).

All reactions were done in duplicate, and concurrently with amplification of an equivalent quantity of serial diluted BM DNA and placental DNA. There was never any PCR product detectable when placental, normal or different MM patient's BM DNA was amplified with the patient's specific CDR1 and CDR3 primers demonstrating the specificity of these primers. Since our assay was capable of detecting one tumor cell in 1,600 to 16,600 normal cells and these PCR reactions were done in duplicate, our ability to detect tumor ranged from one MM cell in 3,200 to 33,200 nonmalignant BM cells. These results make the existence of a CD34-bearing malignant cell highly unlikely in MM.

Is CD38 Expressed on All Malignant Cells in Multiple Myeloma?

CD38+ cell purification: PB and BM mononuclear cells obtained from six MM patients were double FAC-sorted cells using the CD38-FITC and CD38-PE antibodies into the brightest, 90%, 70%, 50%, 30%, and 10% dimmest staining fractions. The cells in the positively staining fraction in each of the six cases analyzed (range, 1-7% of total cells sorted) were separated on the basis that their fluorescent intensity was higher than any cells present in the isotype control populations.

PCR amplification of CD38-sorted fractions with CDR primers: First, the PCR was used to amplify 0.1μg of DNA extracted from the CD38-sorted and unfractionated BM and PB fractions with primers specific for the β-actin gene, and showed similar amounts of intact DNA. When the same PCR reactions were performed using the patient specific CDR primers on a second aliquot of DNA (0.1 μg) extracted from these same CD38-separated and unseparated fractions, detectable product was obtained in each sample although the intensity of the band was markedly higher in the $CD38^{bright}$ fraction than in any of the other sorted fractions (Fig. 2). Thus, although the vast majority of malignant cells in MM bear CD38, there exists a small clonal subpopulation which does not bear CD38 on its cell surface.

MW UNS $^{+}$5% $^{-}$90% $^{-}$70% $^{-}$50% $^{-}$30% $^{-}$10% H_2O MW

Fig. 2. Detection of amplified Ig PCR product on unseparated (UNS), $CD38^{+}$ ($^{+}$5) and $CD38^{-}$ (90% to 10% dimmest staining fractions) PB populations on WAD#26.

Is there a CD10-Expressing Subppopulation in Patients With MM?

CD10+ cell purification: BM mononuclear cells obtained from six patients with MM were double FAC-sorted cells using the CD10-FITC and CD10-PE antibodies into the 1% brightest, 90% dimmest, and 10% dimmest staining fractions. The 1% brightest staining cells demonstrated fluorescent intensity above the background isotype controls.

PCR amplification of CD10-sorted fractions with CDR primers: β-actin primers yielded similar PCR product on these samples. Next, PCR was performed on 0.1 ug of DNA from these CD10-sorted and unsorted samples using the patient specific CDR primers, and detectable product was obtained in each sample although the intensity of the band was markedly higher in cells lacking CD10 expression (both the $CD10^{90\%dimmest}$ and $CD10^{10\%dimmest}$ fractions) than the CD10-bearing population (Fig. 3). These results imply that there is a small population of CD10-bearing cells in most if not all patients with myeloma.

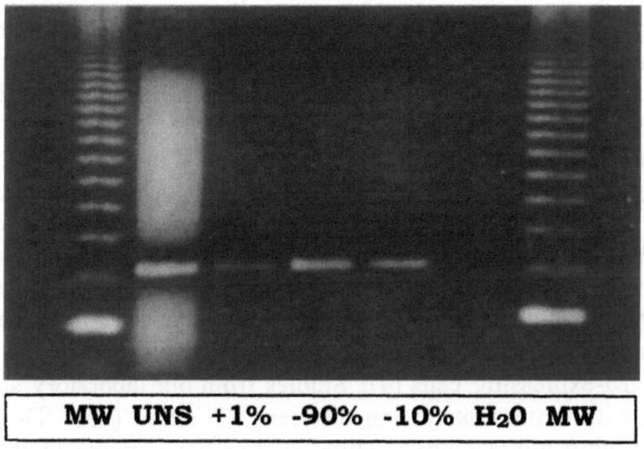

Fig. 3. Detection of amplified Ig PCR product on unseparated (UNS), CD10+ ($^{+}$1) and CD10- (90% and 10% dimmest staining fractions) BM populations on WAD#12.

Are Pre-Class Switch Cells Part of the MM Clone?

Despite attempts to amplify BM RNA using the patient specific CDR1 and Cμ primers in six MM patients, the PCR produced product in only one case. In the other five cases, BM DNA with this primer and the Cγ or Cα primer expressed by the malignant clone yielded PCR product. However, using PB DNA, we could obtain appropriately sized products in all six cases studied with CDR1 and Cμ primers. Sequencing of these PCR-amplified Cμ-containing products from PB in all six cases and BM in the one case (≥5 colonies/case) revealed sequences which were identical to the FR3 expressed by the malignant clone but none of these colonies even closely matched the MM CDR3. Because these initial studies were limited to small numbers of Cμ-expressing colonies containing CDR3s which did not match the malignant clone's sequence, we attempted to increase our chances of finding Cμ-expressing cells which express the MM's VH sequence by performing colony hybridization. Over 200 colonies were isolated from three of these cases which showed evidence of hybridization with a probe complementary to the FR3 sequence. These colonies represented pre-class switch (i.e., Cμ-containing) clones which expressed the same VH germline gene as the one used by the malignant post-class switch (i.e., Cγ-expressing) MM clone. When these filters were rehybridized with the patient specific CDR3 probe, only a small number of positive colonies were identified (4 in WAD#1, 5 in WAD#23, and 1 for WAD#25). These positive colonies were then sequenced. Although each sequence showed homology in the FR3 region and often in the 3' end of the CDR3 sequence (where the CDR3 probe was designed from), the remainder of the CDR3 sequences were markedly different. Thus, it is unlikely that a pre-class switch malignant MM clone exists.

Discussion

Characterization of the types of cells that make up the malignant clone in MM is of both biological and clinical importance. Attempts to both ex vivo purge autologous cells to be used to support hematopoietic recovery following myeloablative chemotherapy and targetting the tumor cells for in vivo therapeutics depend upon the accurate determination of cell types and surface markers on the MM cells. Our studies suggest that the cell of origin begins late in B lymphoid development as evidenced by the absence on any malignant MM cell of CD34 expression and lack of any malignant Cμ-expressing cells. Previous studies have suggested stem cell involvement in MM based on the expression of other hematopoietic lineage markers on malignant plasma cells. However, recent studies by Terstappen show that normal plasma cells may bear some of these nonlymphoid markers including myeloid antigens on their cell surface (12). In addition, the aberrant expression and non-lineage specificity of many of these "lineage" markers have been found in other hematologic malignancies (for review see (23)),(24). Consequently, the presence of these antigens on a malignant cell does not prove the existence of a malignant less differentiated B-cell or stem cell component. The high rate of ANLL development in MM can be explained by the prolonged exposure of these patients to alkylating agents (3). Our inability to detect the early CD34 marker on malignant cells in MM should also allow collection of a tumor-free autograft (i.e., positively selected for this antigen) which is capable of engfrafting MM patients after myeloablative chemotherapy.

Previous studies suggesting that pre-class switch cells were part of the MM clone were based on idiotypes on IgM-expressing cells (4). Studies from our laboratory show that a single anti-idiotype may react with tumors expressing different VH genes (25). Thus, these reagents are not specific in identifying tumor clones. Other studies have shown PCR amplification of small segments of CDR3 and JH segments identical to the MM clone's sequence using CDR3 and Cμ primers (5,6). These results may be explained by the relative nonspecificity of the regions PCR amplified and compared between the malignant and Cμ-expressing clones' sequences. We have demonstrated that the CDR3 sequences in MM mostly contain germline sequences (26). Thus, PCR product using these relatively nonspecific CDR3 primers on MM BM may result from cells present in the patient's nonmalignant B lymphocyte population.

In these studies, we have clearly shown that a small fraction of tumor exists in cells lacking CD38 expression. Kawano et al. suggested that all of the malignant cells in MM express high levels of CD38 (13). However, the identification of malignant cells in his study relied on phenotypic analysis; and, as a result, would be unable to identify rare MM cells which may lack CD38 expression. However, our findings suggest that the vast majority of tumor resides in the CD38-positive fraction. This has several important therapeutic implications. First, although this cell surface marker has also been described on other types of hematopoietic cells, it is absent from the true pluripotent stem cell (14). Thus, if the vast majority of tumor is in the CD38-positive fraction, it is possible to negatively select for largely tumor-free hematopoeitic progenitors which are capable of engrafting patients after high-dose chemotherapy. Since we have also demonstrated the lack of CD34 on any MM cell, it should be possible to positively select for CD34 which will both reduce both total cell numbers (1% of the starting cell number) and purge tumor cells. The few contaminating tumor cells remaining in the CD34-selected fraction can be removed by negative selection for CD38. Second, in vivo approaches using anti-CD38 antibodies are being tested (27). Their success will depend upon the ability of these antibodies to bind ubiquitously to the tumor population. However, our results suggest that a small MM fraction requires targetting by another approach.

Although previous studies suggested that the CD10 antigen could only be found in a small fraction of MM cases, our data demonstrates that most if not all MM patients contain tumor cells that expresses this antigen. Since a small subset of normal plasma cells have been found to also express CD10 (28), its expression is certainly not limited to plasma cells of malignant origin. Normal germinal center cells also contain populations bearing this surface marker (17), and its presence on a subpopulation of malignant cells in MM may suggest that this CD10-expressing fraction and the disease itself originate in the germinal centers as suggested by other investigators (20). In support of this, we have previously demonstrated a great amount of somatic mutation which is antigenically driven without clonal diversity in the malignant cells in MM which are characteristics of cells from germinal centers.

References

1. Vescio RA, Cao J, Hong CH, Newman R, Lichtenstein AK, Berenson JR (1994) Somatic hypermutation of VH genes in multiple myeloma is unaccompanied by intraclonal diversity. Blood 82 (suppl. 1):259a
2. Epstein JH, Xiano-Yan H (1990) Markers of multiple hematopoietic-cell lineages in multiple myeloma. N Eng J Med 322:664-668
3. Cuzick J, Erskine S, Edelman D, Galton DAG (1987) A comparison of the incidence of the myelodysplastic syndrome and acute myeloid leukemia following melphalan and cyclophosphamide treatment for myelomatosis. Br J Cancer 55:523-529
4. Kubagawa H, Vogler LB, Capra JD, Conrad ME, Lawton AR, Cooper MD (1979) Use of individually specific (idiotype) antibodies to trace the oncogenic event to its earliest point in B-cell differentiation. J Exp Med 150:792-807
5. Billadeau D, Ahmann G, Greipp P, Van Ness B (1993) The bone marrow of multiple myeloma patients contains B cell populations at different stages of differentiation that are clonally related to the malignant plasma cell. J Exp Med 178:1023-1030
6. Corradini P, Boccadoro M, Voena C, Pileri A (1993) Evidence for a bone marrow B cell transcribing malignant plasma cell VDJ joined to C mu sequence in immunoglobulin (IgG)- and IgA-secreting multiple meylomas. J Exp Med 178:1091-1096
7. Vesole DH , Jagannath S, Glenn L, Barlogie B (1993) Autotransplantation in multiple myeloma. Hematology/Oncology Clinics of North America 7:613-630
8. Berenson JR, Wong R, Kim K, Brown N, Lichtenstein A (1987) Evidence for perippheral blood B lymphocyte but not T lymphocyte involvement in multiple myeloma. Blood 70:1550-1553
9. Billadeau D, Quam L, Thomas W, Greipp P, Kyle R, Oken MM, Van Ness B (1992) Detection and quantitation of malignant cells in the peripheral blood of multiple myeloma patients. Blood 80:1818-1824
10. Shpall E, Jones RB, Franklin RB, et al. (1994) Transplantation of enriched CD34-positive autologous marrow into breast cancer patients following high-dose chemotherapy: Influence of CD34-positive periphral-blood progenitors and growth factors on engraftment. J Clin Onco 12:28-36
11. Scmitt C, Eaves CJ, Lansdorp PPM (1991) Expression of CD34 on B cell precursors. Clin Exp Immunol 85:168-173
12. Terstappen LWMM, Johnsen S, Segers-Nolten IMJ, Loken MR (1990) Identification and characterization of plasma cells in normal human bone marrow by high-resolution flow cytometry. Blood 76:1739-1747
13. Kawano MM, Huang N, Harada H, Harada Y, Sakai A, Tanaka H, Iwato K, Kuramoto A (1993) Identification of immature and mature myeloma cells in the bone marrow of human myelomas. Blood 82:564-570
14. Sato N, Sawada K, Koizumi K, et al. (1993) In vitro expansion of human peripheral blood CD34+ cells. Blood 82:3600-3609
15. Brown G, Hogg N, Greaves M (1975) Candidate leukemia-specific antigen in man. Nature 258:454-456
16. Greaves MF, Hairi F, Newman RA, Sutherland DR, Ritter MA, Ritz J (1983) Selective expression of the common acute lymphoblastic leukemia (gp100) antigen on immature lymphoid cells and their malignant counterparts. Blood 61:628-639
17. Hsu SM, Jaffe ES (1984) Phenotypic expression of B-lymphocytes. 1. Identification with monoclonal antibodies in normal lymphoid tissues. Am J Pathol 114:387-395
18. Cossman J, Neckers LM, Leonard WJ, Greene WC (1983) Polymorphonuclear neutrophils express the common acute lymphoblastic leukemia antigen. J Exp Med 157:1064-1069
19. Metzgar RS, Borowitz MJ, Jones NH, Dowell BL (1981) Distribution of common acute lymphoblastic leukemia antigen in non-hematopoietic tissues. J Exp Med 154:1249-1254
20. Warburton P, Joshua DE, Gibson J, Brown RD (1989) CD10-(CALLA)-positive lymphocytes in myeloma: Evidence that they are a malignant precursor population and are of germinal centre origin. Leukemia and Lymphoma 1:11-20
21. Durie BGM, Salmon SE (1975) A clinical staging system for multiple myeloma. Cancer 36:842-852
22. Andrews RG, Singer JW, Bernstein ID (1986) Monoclonal antibody 12.8 recognizes a 115-Kd molecule on both unipotent and multipotent hematopoietic colony-forming cells and their precursors. Blood 67:842-845
23. Drexler HG, Thiel E, Ludwig W-D (1993) Acute myeloid leukemias expressing lymphoid-associated antigens: Diagnostic incidence and prognostic significance Leukemia 7:489-498
24. Can't-Rajnoldi A, Putti C, Saitta M, et al. (1991) Co-expression of myeloid antigens in childhood acute lymphoblastic leukaemia: relationship with stage of differentiation and clinical significance. Brit J Haematol 79:40-43
25. Berenson JR, Hart S, Cao J, et al. (1991) Expression of shared idiotypes and VH gene families by patients with monoclonal gammopathies. in the Third International Workshop on Myeloma. pp 31-32
26. Kunkel LA, Vescio R, Cao J, Schiller GJ, Lichtenstein AK, Berenson JR (1993) The CDR3 regions of multiple myeloma patients reveal characteristics seen in fetal lymphoid development. Blood 82 (suppl 1):259a
27. Stevenson FK, Bell AJ, Cusack R, et al. (1991) Preliminary studies for an immunotherapeutic approach to the treatment of human myeloma using chimeric anti-CD38 antibody. Blood 77:1071-1079
28. Caligaris-Cappio F, Bergui F, Tesio L, et al. (1985) Identification of malignant plasma cell precursors in the bone marrow of multiple myeloma. J Clin Invest 76:1243-1251

The Myeloma/ Plasmacytoma Phenotype

Multiple Myeloma - Recent Developments in Molecular and Cellular Biology

B. Barlogie, M.D., R. Hoover, M.D and J. Epstein, D.Sc.
Division of Hematology-Oncology and Arkansas Cancer Research Center, University of Arkansas for Medical Sciences, Little Rock, Arkansas

Supported in part by CA 55819 and CA 59340 from the National Institutes of Health, Bethesda, Maryland

INTRODUCTION

Multiple myeloma (MM) belongs to a group of monoclonal B cell malignancies affecting mainly the elderly, more males than females and causing symptoms both directly by way of tumor mass effects and indirectly by secretion of a variety of cytokines by both tumor and normal cells in response to tumor cell cytokines[1].

Molecular studies of the immunoglobulin genes revealed somatic hypermutations of heavy chain genes suggesting that the clonogenic tumor cell in MM is a post-germinal center memory B cell or plasmablast[2]. These observations do not rule out an oncogenic event at an earlier precursor cell level, although this is rather unlikely. MM shares somatic hypermutations with hairy cell leukemia which, however, displays intraclonal variations[3].

Genetic abnormalities are common in MM, as evidenced by nuclear DNA abnormalities in about 80% of patients examined[4]. This feature is shared by clonal plasma cells of subjects with benign monoclonal gammopathy or monoclonal gammopathy of undetermined significance (MGUS), which is considered a pre-MM lesion with a propensity for progression to overt MM of approximately 2% per year[5].

Unlike a presumed circulating plasmablastic precursor compartment with high proliferative activity[6], the predominant mature-appearing MM plasma cells are hypoproliferative with a median labelling index of 1% at diagnosis[7]. As a result, informative karyotype results are available in only 30% of cases[1]. Growth factor combinations may increase the mitotic yield to the 50% range[8]. Available cytogenetic data indicate highly complex anomalies, both numeric and structural in nature. Among 100 abnormal karyotypes recently analyzed, trisomies 9,11 and 15 occurred jointly in 26%; t (11;14) as in mantle zone lymphoma were noted in about 40%, and another 40% displayed breakpoints at 14q 32[9]. No significant differences were noted in cytogenetic anomalies present among younger and older MM patients. Prognostically, those with normal diploid karyotypes experienced a more favorable clinical outcome reflecting, in all likelihood, a lower proliferative potential of tumor cells so that mainly normal hemopoietic metaphases were observed.

The low mitotic yield and complexity of karyotypic changes both have been a hindrance to a directed approach to the study of oncogenes and suppressor genes. A survey of genes recognized to be important in other B cell malignancies revealed abnormalities of c-MYC, BCL-1, BCL-2, N-ras, and H-ras oncogenes as well as of Rb and p53 suppressor genes[1,4]. In contrast to murine plasmacytoma, MYC translocation to Ig genes are exceedingly rare in human MM[10]. However, MM cells, despite their hypoproliferative status, display high levels of c-MYC RNA and protein[11]. As reported by Hoover in a contribution to this workshop, both human MM cell lines and fresh clinical bone marrow samples displayed a 3.1 kb MYC message, probably as a result of P_0 promoter utilization instead of transcription from the commonly employed P_1 and P_2 promoters[12,13]. Extending cytogenetic observations of monosomy 13 as well as del 13q, fluorescence in situ hybridization (FISH) analysis revealed Rb deletion in over 50% of almost 100 patients evaluated (Y. Gazitt, personal communication)[14]. These data resemble results obtained in chronic lymphocytic leukemia[15]. Rearrangements of BCL-1 and BCL-2 were noted in 15% to 20% each[1]. BCL-2 protein expression is typical of the normal plasma cell stage and is also observed in the majority of MM plasma cells[16]. Using single strand conformational polymorphism (SSCP) technology, mutations of N-ras and p53 occur in up to 30% of usually advanced MM[for review, see 1].

The above molecular abnormalities do not shed light on the mechanism of myelomagenesis. It is interesting, however, to speculate that the presence of BCL-2 as a survival gene, also inhibiting cell cycle progression, opposes apoptotic and/or proliferative signals provided by c-MYC and possibly cyclin D, the product of the BCL-1 gene. A similar consideration applies to abnormalities of p53 and Rb, no longer exerting cell cycle arrest and apoptosis-inducing functions either through altered interactions with other gene products and/or through derepression of interleukin (IL-6), which has been touted as the major growth factor for MM[see 1]. Recent data, however, demonstrated IL-6 expression and production by preplasmacytic MM cells19[17]. Together with the observation that exogenous IL-6 rescues human MM cell lines from DEX-induced apoptosis (via downregulation of constitutive IL-6 expression), these data suggest that IL-6 may play a role in MM cell survival and differentiation rather than proliferation[18]. Unlike DEX, 2-chlorodeoxyadenosine and HMBA induce MM cell apoptosis through downregulation of BCL-2, an effect not observed in the case of DEX exposure[19].

Much progress can be anticipated from a systematic study of all of the aforementioned oncogenes and suppressor genes and their collective effects on growth-promoting and - inhibitory signals. In the case of IL-6, its biological activity is modulated, in addition, by circulating soluble IL-6R as well as soluble gp-130, the signal-transducing molecule for IL-6, LIF, oncostatin-M and CNTF.

The availability of stromal cell culture systems[20] and the severe combining immunodeficiency (SCID) mouse model[21,22] has enriched the armamentarium available to the MM biologist to study the growth- regulatory functions of the above mentioned gene products and the ever increasing list of cytokines expressed and produced by MM cells and by the bone marrow microenvironment, possibly in response to the entrapment of circulating MM cells through cell adhesion and extracellular matrix molecules[23]. Advancements in the understanding of cytokines involved in osteoblast and osteoclast growth, differentiation and function have stimulated new areas of research, especially since osteoclasts are derived from hemopoietic progenitors and osteoblasts from stromal cells, i.e. important components for both normal hemopoiesis and myeloma pathophysiology.

Potentially fruitful areas for therapeutic intervention pertain to cytotoxic agents and biologicals capable of inducing apoptosis in MM cells. Likewise, approaches aimed at disrupting cytokine pathways are being explored. In this regard, interferons displaying relatively limited antitumor effect clinically, are of interest as they seem to activate IL-6 transcription and inhibit IL-6 R and gp-130 (M. Schwabe, personal communication). Trials with anti-IL-6 monoclonal antibody have proven effective mainly in alleviating the manifestations of MM[24] and a related plasma cell dyscrasia, Castleman's disease[25] IL-6-chimeric molecules, such as IL-6-pseudomonas exotoxin, are being further developed for clinical use[26]. The observation of triggering IL-6 production through the CD40 pathway may open up new interventional avenues[27]. MM cell lines commonly express CD28, alone or in combination with B7; these molecules play crucial roles in T cell activation[28]. As in murine plasmacytoma, there is an expansion of T cells expressing an immunoglobulin isotype-concordant FcR which is shed. Upon binding to the surface Ig of MM cells, c-MYC and Ig are downregulated, and MM cell kill ensues. In the case of IgG MM, FcR-γ III (CD16)-expressing $CD8^+$ T cells are expanded[29]. Serum concentrations of s-CD16 are inversely correlated with tumor mass[30]. Efforts are underway in our laboratories to transduce CD16 into $CD34^+$ hemopoietic stem cells, thus providing sustained release of a tumor-controlling cytokine, once marked tumor cytoreduction has been achieved with myeloablative therapy and autologous transplant[31].

Systematic pursuit of studies delineated above should advance our understanding of MM pathophysiology and thereby open novel avenues for therapeutic intervention.

REFERENCES:

1. Barlogie, B., Alexanian, R., and Jagannath, S. (1992): Plasma cell dyscrasias. J. Am. Med. Assoc., 268:2946-2951.
2. Bakkus, M.H.C., Heirman, C., Van Riet, I., Van Camp, B., and Thielemans, K. (1992): Evidence that multiple myeloma Ig heavy chain VDJ genes contain somatic mutations but show no intraclonal variation. Blood, 80:2326-2335.
3. Wayner, S.D., Martinelli, V., and Luzzatto, L. (1994): Similar patterns of V_k gene usage but different degrees of somatic mutation in hairy cell leukemia, prolymphocytic leukemia, Waldenstrom's macroglobulinemia, and myeloma. Blood, 83:3647-3653.
4. Barlogie, B., Epstein, J., Selvanayagam, P., and Alexanian, R. (1989): Plasma cell myeloma - New biologic insights and advances in therapy. Blood, 73:865-879.
5. Kyle, R.A. (1993): "Benign" monoclonal gammopathy--after 20 to 35 years of follow-up. Mayo Clinic Proceedings, 68:26-36.
6. MacLennan, I.C.M., Chan, E.Y.T. (1991): The origin of bone marrow plasma cells. In: Epidemiology and Biology of Multiple Myeloma, edited by G.I. Obrams, and M. Potter, pp. 129-135. Springer-Verlag, Berlin.
7. Greipp, P.R., Katzmann, J.A., O'Fallon, W.M., and Kyle, R.A. (1988): Value of beta-2-microglobulin level and plasma cell labeling indices as prognostic factors in patients with newly diagnosed myeloma. Blood, 72:219-223.
8. Facon, T., Lai, J.L., Nataf, E., Preudhomme, C., Zandecki, M., Hammad, M., Wattel, E., Jouet, J.P., and Bauter, F. (1993): Improved cytogenetic analysis of bone marrow plasma cells after cytokine stimulation in multiple myeloma: a report on 46 patients. Brit. J. Haematol. 84:743-745.
9. Barlogie, B. (1994): Plasma cell myeloma. *In*: Hematology, edited by E. Beutler, M.A. Lichtman, B.S. Coller, and T.J. Kipps. McGraw-Hill, New York, in press.
10. Selvanayagam, P., Blick, M., Narni, F., Van Tuinen, P., Ledbetter, D., Alexanian, R., Saunders, G., and Barlogie, B. (1988): Alteration and abnormal expression of the *c-myc* oncogene in human multiple myeloma. Blood, 71:30-35.
11. Greil, R., Fasching, B., Loidl, P., Huber, H. (1991): Expression of the *c-myc* proto-oncogene in multiple myeloma and chronic lymphocytic leukemia: An in situ analysis. Blood, 78:180-191.
12. Travis, P., Kaushal, V., Baltz, B., Ivey, A., and Hoover, R. (1993): Abnormal *c-myc* transcript size in 70% of patients with multiple myeloma (MM). Blood, 82:261a.
13. Hoover, R.G., Kaushal, V., Lary, C., Travis, P., and Sneed, T. (1994): *C-myc* transcription is initiated from Po in 70% of patients with multiple myeloma. *In*: Mechanisms of B Cell Neoplasia, 1994, edited by M. Potter and F. Melchers. Editiones Roche, Basel, Switzerland, in press.
14. Dao, D.D., Sawyer, J.R., Epstein, J., Hoover, R.G., Barlogie, B., and Tricot, G. (1994): Deletion of the retinoblastoma gene in multiple myeloma. Leukemia, in press.
15. Stilgenbaure, S., Dohner, H., Bulgay-Morschel, M., Weitz, S., Bentz, M., and Lichter, P. (1993): High frequency of monoallelic retinoblastoma gene deletion in B-cell chronic lymphocytic leukemia shown by interphase cytogenetics. Blood, 81:2118-2124.
16. Pettersson, M., Jernberg-Wiklund, H., Larsson, L.G., Sundstrom, C., Givol, I., Tsujimoto, Y., and Nilsson, K. (1992): Expression of the *bcl-2* gene in human multiple myeloma cell lines and normal plasma cells. Blood, 79:495-502.
17. Hata, H., Xiao, H., Petrucci, M.T., Woodliff, J., Chang, R., and Epstein, J. (1992): Interleukin-6 gene expression in multiple myeloma: a characteristic of immature tumor cells. Blood, 81:3357-3364.
18. Hardin, J., MacLeod, S., Grigorieva, I., Chang R., Xiao, H., Barlogie, B., and Epstein, J. (1994): Dexamethasone controls myeloma through IL-6 deprivation. Blood, in press.
19. Siegel, D., Zhang, X., Niesvizky, R., Busquets, X., and Michaeli, J. (1994): Modulation of *bcl-2* protein expression in spontaneous and induced program cell death (apoptosis) in multiple myeloma. Proc. Am. Assoc. Clin. Res., 35:7.
20. Huang, Y., Richardson, J.A., Tong, A.W., Zhang, B., Stone, M.J., and Vitetta, E.S. (1993): Disseminated growth of a human multiple myeloma cell line in mice with severe combined immunodeficiency disease. Ca. Res., 53:1392-1397.
21. Caligaris-Cappio, F., Bergui, L., Gregoretti, M.G., Gaidano, G., Gaboli, M., Schena, M., Zallone, A.Z., and Marchisio, P.C. (1991): Role of bone marrow stromal cells in the growth of human multiple myeloma. Blood, 77:2688-2693.
22. Feo-Zuppardi, F.J., Taylor, C.W., Iwato, K., Lopez, M.H., Grogan, T.M., Odeleye, A., Hersh, E.M., and Salmon, S.E. (1992): Long-term engraftment of fresh human myeloma cells in SCID mice. Blood, 80:2843-2850.

23. Caligaris-Cappio, F., Gregoretti, M.G., Ghis, P., and Bergui, L. (1992): In vitro growth of human multiple myeloma: implications for biology and therapy. Hematol. Oncol. Clin. N. Amer. 6:257-271.
24. Portier, M., Zhang, X.G., Caron, E., Lu, Z.Y., Bataille, R., and Klein, B. (1993): α-Interferon in multiple myeloma: inhibition of interleukin-6 (IL-6)-dependent myeloma cell growth and downregulation of IL-6-receptor expression in vitro. Blood, 81:3076-3082.
25. Beck, J.T., Hsu, S.M., Wijdenes, J., Bataille, R., Klein, B., Vesole, D.H., Hayden, K., Jagannath, S., and Barlogie, B. (1994): Alleviation of systemic manifestations of Castleman's disease by monoclonal anti-IL-6 antibody therapy. N. Engl. J. Med., 330:602-605.
26. Kreitman, R.J., Siegall, C.B., Fitzgerald, D.J.P., Epstein, J., Barlogie, B., and Pastan, I. (1992): Interleukin-6 fused to a mutant form of *Pseudomonas* exotoxin kills malignant cells from patients with multiple myeloma. Blood, 1775-1780.
27. Westendorf, R., Tschumper, R.C., and Jelinek, D.F. (1994): Differential effects of CD40 stimulation in normal and malignant human plasma cells. Proc. FASEB, April:A1012.
28. Guba, S.C., and Kornbluth, J. (1994): Potential role of CD28-B7 autostimulation in the proliferation of myeloma plasma cells. Proc. Am. Assoc. Ca. Res., 35:518.
29. Hoover, R.G., Jary, C., Page, R., Travis, P., Owens, R., Flick, J., Kornbluth, J., and Barlogie, B. (1994): Autoregulatory circuits in myeloma: tumor cell cytotoxicity mediated by soluble CD16. J. Clin. Invest., in press.
30. Mathiot, C., Teillaud, J.L., Asselain, B., El Malak, M., Mosseri, V., Euller-Zeigler, L., Daragon, A., Grosbois, B., Michaux, J., Facon, T., Bernard, J., Duclos, B., Monconduit, M., and Fridman, W. (1993): Correlation between serum soluble CD16 (sCD16) levels and disease stage in patients with multiple myeloma. J. Clin. Immunol., 13:41-48.
31. Munshi, N.C., Ding, L., Srivastava, A., Hoover, R., and Barlogie, B. (1993): Soluble CD-16 from AAV mediated CD-16 gene transfer to human cell lines is lytic to a multiple myeloma cell line. Blood, 82:264a.

Potential Role of CD28-B7 Interactions in the Growth of Myeloma Plasma Cells

J. Kornbluth
Departments of Medicine and Microbiology and Immunology, Arkansas Cancer Research Center, University of Arkansas for Medical Sciences and McClellan VA Medical Center, Little Rock, Arkansas, 72205

Introduction

Multiple myeloma is a malignancy of plasma cells. Patients with myeloma have associated immune deficiencies, including decreased numbers of B cells in the peripheral blood and corresponding decreased humoral responses. Myeloma patients also show significant alterations in T cell function. There are decreased numbers of CD4+ T cells while CD8+ T cells are normal to slightly increased [1,2].

The process of T cell activation requires both an antigen-specific signal delivered through the T cell receptor complex and a second, co-stimulatory signal delivered by CD28 binding to its ligand on antigen-presenting cells. This co-stimulatory signal potentiates and sustains proliferation of activated T cells by synergizing with T cell receptor-mediated signals at the transcriptional and post-transcriptional level [3]. Moreover, in the absence of this second signal, antigen-exposed T cells become anergic [4]. The CD28 receptor is a 44 kD homodimer expressed constitutively on the majority of CD4+ T cells and approximately half of CD8+ T cells [5,6]. CD28 has also been found on normal plasma cells, plasmacytomas and myeloma cells [7]. The ligand for CD28 is B7, which is expressed by activated B cells, dendritic cells and macrophages. Two additional members of the B7 family have recently been described. B7-2, which was discovered using B7 knockout mice, is expressed by unstimulated antigen-presenting cells. B7-3, like the original B7 (now designated B7-1), is expressed only by activated antigen-presenting cells [8-10]. These B7 molecules also bind, with very high affinity, to CTLA-4, which is expressed by activated T cells [11]. Whether CTLA-4 has a unique role in sustaining T cell responses remains to be determined.

B7 expression can be induced on T cells by activation, indicating that activated T cells can coexpress both CD28/CTLA-4 and the counter-receptor B7. Azuma et al have demonstrated that B7+ T cell clones can stimulate a primary allogeneic mixed lymphocyte response, indicating that the B7 expressed by these cells is functional. In addition, anti-B7 antibodies could partially block this response [12].

These findings suggest that activated T cells could potentially be autostimulated by self-self CD28-B7 interactions.

These data suggest several possible roles for the CD28 pathway in multiple myeloma. The decreased numbers of CD4+ T cells and B cells in the peripheral blood of myeloma patients, essentially resulting in a deficit in CD28 and B7 bearing cells, might contribute to the T cell anergy seen in these patients. The expression of CD28 on human plasma cells and myeloma cells has not been assessed functionally. It is possible that, as with activated T cells, CD28 on myeloma cells may be involved in autostimulation and thereby contribute to the proliferation of these tumor cells. In this paper, the expression of CD28 and B7 was examined in an extensive panel of human myeloma cell lines and freshly obtained bone marrow tumor cells from myeloma patients. The effect of antibodies to CD28 and B7 on the proliferation of myeloma cell lines in vitro was also examined.

Results

Bone marrow derived myeloma tumor cells from patients express CD28
Ficoll-Hypaque separated bone marrow cells from nine myeloma patients were examined by flow cytometry for expression of B7 and CD28. The B7 monoclonal antibody used, BB1, was generously provided by Dr. Edward Clark, University of Washington, Seattle [13]. This antibody reacts with B7-1 and B7-3. Anti-CD28 antibody 9.3 was kindly provided by Bristol-Myers Squibb [5]. Tumor cells within the bone marrow population were first identified by high CD38 expression, low to intermediate CD45 cell surface expression and light scatter properties of the cells [14]. These were then evaluated for expression of B7 and CD28. Table 1 shows the results from nine patients. The numbers in the table represent the percentage of B7+ and CD28+ myeloma plasma cells.

Table 1. CD28 and B7 expression by myeloma tumor cells

Patient #	% B7+ cells	% CD28+ cells
1	0	38
2	0	88
3	0	4
4	0	31
5	0	69
6	0	81
7	0	31
8	0	80
9	0	0

As can be seen from Table 1, a proportion of myeloma plasma cells from 7 of 9 patients expressed CD28. The percentage of CD28+ myeloma cells within each patient's tumor cell population was quite variable, ranging from 31 to 88%. Tumor cells from two patients were essentially CD28 negative. However, myeloma plasma cells from all patients were uniformly B7 negative. The lack of expression of B7 on myeloma plasma cells from patients is consistent with their inability to induce sustained anti-tumor T cell responses.

CD28 and B7 expression by myeloma cell lines

A panel of 14 myeloma cell lines were examined by flow cytometry for their cell surface expression of CD28 and B7 using antibodies 9.3 and BB1, respectively. Three of the cell lines, U266, 8226 and ARH-77, are long established myeloma cell lines. The remaining cell lines were generated within the last five years from patients followed at the Arkansas Cancer Research Center. The flow cytometry results are shown in Table 2. The cell lines fall into four categories with regard to their expression of CD28 and B7. Category 1, represented by two of the 14 cell lines, ARD and 8226, is the phenotype of the majority of the patient bone marrow samples: B7-, CD28+. Category 2, into which the largest number of cell lines fall (6 of 14), has the reciprocal phenotype: B7+, CD28-. A third pattern of expression is seen with 5 of the cell lines in which there is co-expression of both CD28 and B7. In these cell lines, the vast majority of the cells are B7+ (85-100% positive); however, CD28 expression is uniformly lower. Moreover, for all of these cell lines, there are discrete populations of CD28+ and CD28- cells within the clone. Only one cell line (DIN) lacked expression of both B7 and CD28.

Table 2. CD28 and B7 expression by myeloma cell lines

Cell line	Category	% B7+ cells	% CD28+ cells
ARD	1	0	100
8226	1	0	90
EST	2	100	0
CLO	2	100	0
COL	2	90	0
RAM	2	73	0
U266	2	70	0
ENZ	2	48	0
ARH-77	3	100	62
LES	3	85	59
ARK	3	100	48
MIT	3	100	22
MER	3	100	33
DIN	4	0	0

In a limited study, some of the myeloma cell lines which appeared to be B7 negative by their lack of reactivity with antibody BB1 were stained with the CTLA-4Ig fusion protein, kindly provided by Bristol-Myers Squibb [11]. CTLA-4Ig binds to B7-1, B7-2 and B7-3, while antibody BB1 recognizes only B7-1 and B7-3. One of the BB1- cell lines, 8226, stained strongly with CTLA-4Ig, indicating that although it lacks B7-1 and B7-3, it is likely to be B7-2+ (Fig. 1). Studies are now in progress to examine patient bone marrow myeloma cells for expression of B7-2 with this reagent.

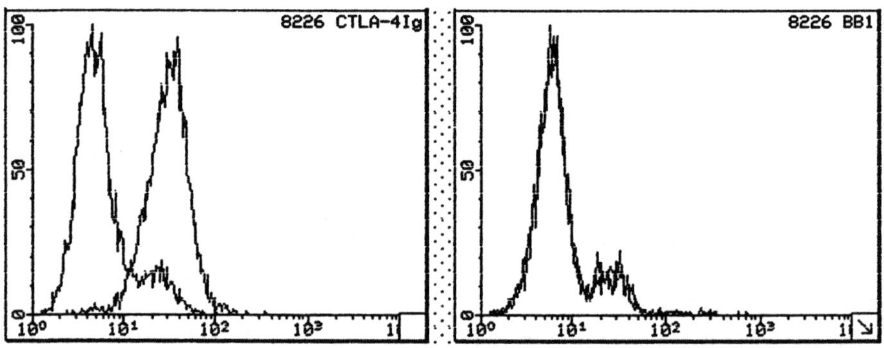

Fig. 1. Expression of B7 by myeloma cell line 8226. Cells were incubated with CTLA-4Ig, followed by FITC-labeled goat anti-human Ig Fc gamma (left panel) or BB1, followed by FITC-labeled goat anti-mouse Ig (right panel). The x-axis represents fluorescence intensity (log scale) while the y-axis denotes the proportion of positive cells. The left peak on each panel is the control staining.

Effect of anti-B7 and anti-CD28 antibodies on the proliferation of myeloma cell lines

The co-expression of B7 and CD28 on a number of myeloma cell lines suggested the possibility that CD28-B7 autostimulation might contribute to the growth of these tumor cells. A series of experiments were performed in which myeloma cells were cultured in the absence and presence of anti-CD28 antibody 9.3 or anti-B7 antibody BB1. The effect of anti-HLA-DR specific monoclonal antibody 210 [14] was also assessed since all of the cell lines tested were HLA-DR+. Results of a representative experiment are shown in Fig. 2. All antibodies were used at 5 ug/ml; cells were plated in triplicate microtiter wells per condition at 100,000 cells per well. After 24 hours, wells were pulsed for an additional 18 hours with tritiated thymidine and harvested. Results are expressed as the percentage of the control (no antibody) response for each cell line. Anti-CD28 antibody 9.3 inhibited the proliferation of MER by 50%, but had no effect on the proliferation of the two other B7+ CD28+ cell lines tested. Anti-B7 antibody BB1 partially inhibited the proliferation of all three cell lines, with inhibition ranging from 30-

45%. These results suggest that CD28-B7 interactions may contribute to the growth of some myeloma cells.

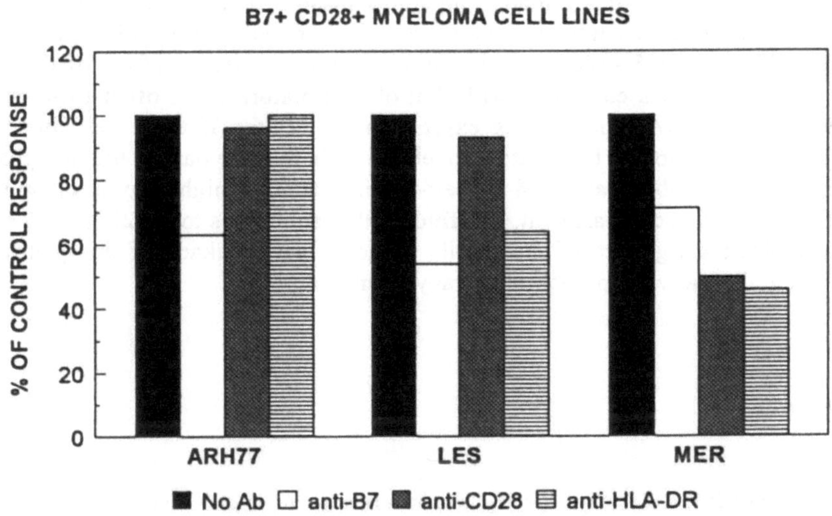

Fig. 2. Effect of CD28 and B7 antibodies on the proliferation of CD28+ B7+ myeloma cell lines. Tritiated thymidine incorporation was measured by scintillation counting. The graph represents proliferation as a percentage of control.

Discussion

The results of this study indicate that the majority of mature, myeloma plasma cells from patients (CD38+, CD45-, +/-) express CD28 but not B7-1 or B7-3. The lack of expression of B7 by these cells is consistent with their inability to induce sustained T cell responses, and may, in fact, play a role in the T cell anergy often seen in myeloma patients. The lack of co-expression of CD28 and B7 on these mature cells is also consistent with their weak proliferative capacity. It is not clear what role CD28 plays in myeloma plasma cells. In resting T cells, CD28 receptor engagement alone has no effect on proliferation. However, as a second signal or in primed T cells, CD28 ligation triggers tyrosine phosphorylation of specific substrates, ultimately leading to enhanced cytokine synthesis due to both enhanced transcription as well as stabilization of cytokine mRNA [3]. One can speculate that CD28 engagement may have a similar effect in myeloma cells, leading to enhanced production of immunoglobulin or cytokines such as IL-6 and TNF-α. This is a testable hypothesis, and such experiments are underway using several of the CD28+ myeloma cell lines.

The presence of B7 on the majority of myeloma cell lines but not on patient plasma cells may indicate that the state of activation or level of maturity of the cell lines differs from that of the mature myeloma plasma cells in the bone marrow. It is clear that most cell lines reflect a less differentiated, more proliferative cell than the CD38+ CD45 - plasma cells within the bone marrow. Many of the cell lines have phenotypes consistent with that of less mature cells, often expressing CD19 and CD45(RA) and lacking expression of CD56. If these cell lines do represent a population of less mature myeloma cells that are part of the malignant clone, it is conceivable that CD28-B7 self-self interactions might play a role in the proliferation of this compartment. The finding that antibodies to CD28 and B7 can partially block the growth of these cells suggests that blockade of this pathway may be an alternative approach to therapy in myeloma.

References

1. Hoover RG, Kornbluth J (1992) Immunoregulation of murine and human myeloma. Hematology/Oncology Clinics of North America 6:407-424.
2. San Miguel JF et al (1992) Lymphoid subsets and prognostic factors in multiple myeloma. Br J Haematol 80:305-309.
3. Linsley PS, Ledbetter PS (1993) The role of the CD28 receptor during T cell responses to antigen. Ann Rev Immunol 11:191-212.
4. Harding FA, McArthur JG, Gross JA, Raulet DH, Allison JP (1992) CD28-mediated signalling costimulates murine T cells and prevents induction of anergy in T cell clones. Nature 356:607-609.
5. Hansen JA, Martin PJ, Nowinski RC (1980) Monoclonal antibodies identifying a novel T cell antigen of human lymphocytes. Immunogenetics 10:247-260.
6. Hara T, Fu SM, Hansen JA (1985) A new activation pathway used by a major T cell population via a disulfide-bonded dimer of a 44 kilodalton polypeptide (9.3 antigen). J Exp Med 161:1513-1524.
7. Kozbor D, Moretta A, Messner HA, Moretta L, Croce CM (1987) Tp44 molecules involved in antigen-independent T cell activation are expressed on human plasma cells. J Immunol 138:4128-4132.
8. Linsley PS, Brady W, Grosmaire L, Aruffo A, Damle NK, Ledbetter JA (1991) Binding of the B cell activation antigen B7 to CD28 costimulates T cell proliferation and interleukin 2 mRNA accumulation. J Exp Med 173:721-730.
9. Freeman GJ et al (1993) Uncovering of functional alternative CTLA-4 counter-receptor in B7-deficient mice. Science 262:907-909.
10. Boussiotis VA, Freeman GJ, Gribben JG, Daley J, Gray G, Nadler LM (1993) Activated human B lymphocytes express three CTLA-4 counterreceptors that costimulate T cell activation. Proc Natl Acad Sci USA 90:11059-11063.
11. Linsley PS, Brady W, Urnes M, Grosmaire LS, Damle NK, Ledbetter JA (1991) CTLA-4 is a second receptor for the B cell activation antigen B7. J Exp Med 174:561-569.

12. Azuma M, Yssel H, Phillips JH, Spits H, Lanier LL (1993) Functional expression of B7/BB1 on activated T lymphocytes. J Exp Med 177:845-850.
13. Yokochi T, Holly RD, Clark EA (1982) B lymphoblastoid antigen (BB-1) expressed on Epstein-Barr virus activated B cell blasts, B lymphoid lines, and Burkitt's lymphomas. J Immunol 128:823-827.
14. Hata H, Xiao HQ, Petrucci MT, Woodliff, Epstein J (1993) Interleukin-6 gene expression in multiple myeloma: a characteristic of immature tumor cells. Blood 81:3357-3364.
15. Kornbluth J, Spear B, Raab SS, Wilson DB (1985) Evidence for the role of class I and class II HLA antigens in the lytic function of a cloned line of human natural killer cells. J Immunol 134:728-735.

B-1 (CD 5+) B-Cells as a Marker of Immune Dysregulation in Multiple Myeloma

M. R. MacKenzie and T. G. Paglieroni
Division of Hematology and Oncology, University of California, Davis School of Medicine and Sacramento Medical Foundation Center for Blood Research

Our group has a long standing interest in the immune abnormalities present in patients with multiple myeloma. In particular the poor primary immune response to new antigen challenges exemplified by recurrent upper respiratory infections. This aberrant physiology has an in vitro correlate of diminished levels of serum immunoglobulins other than the monoclonal protein which marks the expression of the plasma cell tumor. Attempts to define the mechanisms of this immunodeficiency traditionally use the manipulation of production of immunoglobulins by peripheral blood mononuclear cells stimulated by pokeweed mitogen (PWM). In this system the inhibition of Ig production by added cell populations or culture supernatants is determined. Recently the use of Staphylococcal Aureus protein (SAC) as a stimulant proved a satisfactory system as well.

Approximately 20 years ago we defined that one of the cells circulating in the peripheral blood of myeloma patients and capable of inhibiting PWM driven Ig production by normal PBL had an unusual phenotype by the methods then available. The cells were non plastic adherent, SRBC, EAC3, SIg negative cells which had receptors for anti Rh Human IgG bound to human RBC.[1,2,3] We have subsequently identified these cells as belonging to the CD5+ (now called B-1) B cells [4]. The cells active in immunosuppression were found only in patients with diagnosed multiple myeloma. There was no evidence for immunosuppressive cells in normal individuals, individuals with benign monoclonalgammopathy and very importantly not in patients with macroglobulinemia and in only 1 to 2% of patients with CLL [5].

We have determined the percentage of B cells that are CD5+ in patients with myeloma and their distribution as compared to normal individuals. Table 1 provides a summary of the % of CD5+ CD19 + (B-1) cells in peripheral blood

Table 1. % CD19 Cells Coexpressing CD5*

Tissue	Myeloma	Normal
Peripheral Blood	46.9 ± 21.4% (15)	13.2 ±8.1% (20)
Spleen	50.0 ± 22.1% (15)	24.1 ± 6.8% (20)
Lymph Node	10.9 ± 3.0% (7)	12.5 ± 7.2% (8)
Tonsil	13.6 ±1.8% (6)	10.8 ± 3.1% (5)
Bone Marrow	<1%	<1%

* expressed as a % of CD19 cells

in normal individuals and myeloma patients as well as the % of B cells that are B-1 in various lymphoid tissues. An important caveat when examining this data is that all of the myeloma patient tissue data is derived from late disease, i.e. these are from immediate autopsy. Note the increased % of B-1 cells in peripheral blood and spleen and the normal distribution in lymph node, tonsil and the absence in bone marrow. The normal spleens are from trauma patients undergoing splenectomy.

The data from myeloma patients expressed as % of lymphocytes is presented in Figure 1. Immunosuppressive B-1 cells in myeloma patients were only found in the peripheral blood or spleen populations not in tonsil or lymph node. This data suggests that immunosuppressive B-1 cells are only a subpopulation of the B-1 cells as a whole.

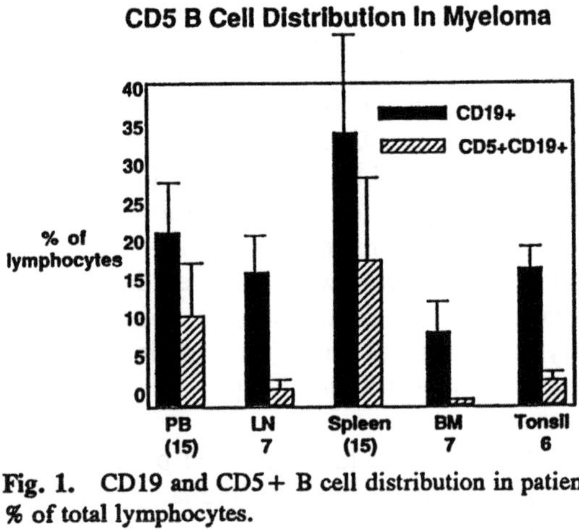

Fig. 1. CD19 and CD5+ B cell distribution in patients with advanced myeloma expressed as a % of total lymphocytes.

Table 2. Patients with MGUS 26

Patients with normal level of CD5+ cells	17	Patients with increased CD5+ cells	9
Immunosuppressive CD5+ cells	0	Immunosuppressive CD5+ cells	6
Progressed to myeloma	1	Progressed to myeloma	7

Are these cells part of the neoplastic clone? Or could they be part of an immune regulatory system which serves to control B cell physiology and thus arise in response to either high levels of immunoglobulin or increased numbers of plasma cells? Finally do they precede or follow the neoplasm?.

We examined our data on a series of patients with monoclonalgammopathy of uncertain significance (MGUS). We studied 26 individuals with this label at the time of diagnosis and with serial studies for up to 10 years. Our initial interest was in the increased % of B cells that were CD5+. However when we determined whether or not these cells were immunosuppressive a distinct pattern emerged.

Nine of the 26 individuals had increased % of B-1 cells and 6 of these patients had immunosuppressive B-1 cells. All 6 progressed to myeloma by 5 years. One additional patient in the increased % of B-1 cells category progressed. Thus 7 of 9 individuals with increased % of their B cells were B-1 cells and in particular if these were immunosuppressive progressed to overt myeloma.. Where as only 1 of the other 17 MGUS patients with normal % of B-1 and no immunosuppressive cells progressed (Table 2). The importance of immunosuppressive B-1 cells is strengthened by studies of a group of 23 individuals with isolated plasma cytoma These patients were followed for at least 5 years after their initial therapy of local radiation. Eight had demonstrable immunosuppressive B-1 cells at time of initial presentation and 5 progressed to disseminated disease. None of the individuals without such cells have progressed at the current time. We have reported an additional instructive case. A blood donor was shown to have immunosuppressive CD 5+ B cells present 10 years prior to the development of overt myeloma.[6] Thus the presence of persistent immunosuppressive CD5+B (B-1) cells in patients with plasma cell dyscrasia is a marker for progressive disease. Further the cells are present up to several years prior to the demonstration of overt dissemination.

If immunosuppressive B cells are part of a normal B cell physiologic system we should be able to demonstrate their presence in some normal immune response. We chose to examine the changes in IgM as well as the presence and % of CD5+ B cells in patients receiving blood transfusions as compared to normal age and sex matched adults. Two groups were studied.

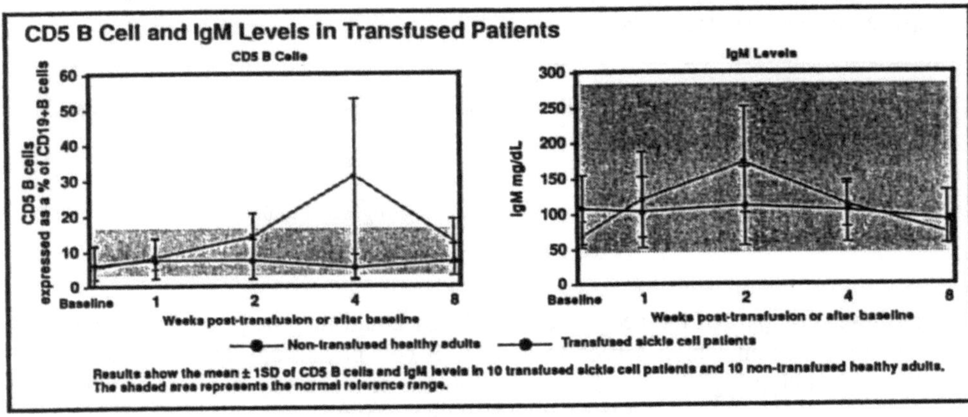

Fig. 2. Time course of CD5+ B cells and IgM levels in transfused patients as compared to non transfused adults over an 8 week period.

Eight transfused sickle cell patients not in crisis followed by a second group of 10. Blood was sampled before transfusion and at weekly intervals for 8 weeks. The results are identical in both and are illustrated for the second group in Figure 2.

The data show that post transfusion IgM levels rose and returned to normal by 8 weeks in both sets of patients. Peak CD5+ B cells occurred at 4 weeks and returned to normal by 8 weeks. There were no such changes in the matched not transfused controls. This data is consistent with the hypothesis that an increased level of immunoglobulin triggers a response in CD5+ B (B-1) cells which then in turn release cytokines which inhibit further Ig production as part of normal immunoregulatory feed back. This hypothesis was tested with the second group of sickle cell patients. This time in addition to the determination of the % of CD5+ cells the samples were examined for the presence of immunosuppressive cells. No such cells were demonstrable before transfusion. In this group of patients the kinetics of IgM and CD5+ (B-1) cells over time was the same as the previous study. However, immunosuppressive activities were present at 4 weeks corresponding to the peak in % of CD5+ B cells (Figure 3). The inhibition disappeared by 8 weeks. CD5+ B cells were isolated by use of magnetic beads as follows. First anti CD3 magnetic bead labeled cells were removed. Then anti CD5 magnetic beads added and the cells were positively selected by removal with a magnet. These cells were found to mediate the immunosuppression. Thus immunosuppressive CD5+ B cells can be demonstrated to occur in non neoplastic disorders in response to an immune stimulus. The data suggests that immunosuppressive CD 5+ (B-1) cells are not part of the neoplastic clone. They appear before overt neoplasm. They are present in non neoplastic conditions.

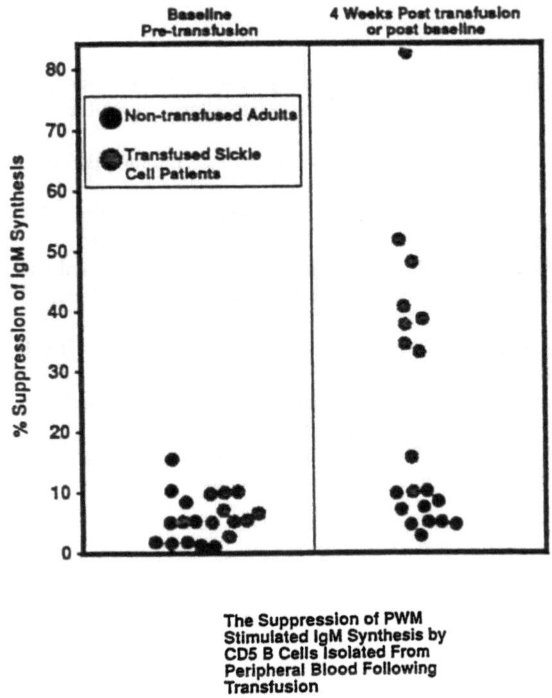

Fig. 3. CD5+ B cells isolated from transfused patients acquire the capacity to suppress IgM synthesis by normal stimulated PBL. This capacity was lost by 8 weeks post transfusion.

We propose the following hypothesis The presence of immunosuppressive B-1 cells in myeloma are an exaggeration of a normal response. Its persistence reflects the dysregulation of the neoplastic plasma cell. The consequence is suppression of normal primary B cell responses and immunodeficiency. The demonstration of the persistence of these cells may be used as a marker for eventual disseminated disease in MGUS and plasmacytoma patients. The absence of these immunoregulatory cells in macroglobulinemia and CLL may reflect that these diseases do not involve the terminally differentiated plasma cell. We have attempted to identify if cytokines with known immunosuppressive capacity were the effector molecules. These studies were conducted with the supernatant from a cell line which displays the CD5 protein on its surface and secretes an immunosuppressive protein [7]. A combination of direct assay, biologic assay, antisera inhibition and probing for specific mRNA was used. The following molecules have been eliminated, Interleukin 1, IL-4, IL-10, IL-11, interferon alpha, beta, and gamma, TNF alpha and beta, and TGF beta. The elucidation of stimulus for immunosuppressive CD5+ (B-1) cells as well as the mediator of the suppression will add to our understanding of B cell physiology.

References

1. Froland SS, Wisloff F, Michaelsen TE (1974) Human lymphocytes with receptors for IgG. A population of cells distinct from T and B lymphocytes. Int Arch Allergy Appl Immuno 47:124.
2. Paglieroni T, MacKenzie MR. (1977) Studies on the pathogenesis of an immune defect in multiple myeloma. J Clin Invest 59:1120.
3. Paglieroni T, MacKenzie MR (1980) Multiple myeloma: an immunologic profile. III. cytotoxic and suppressive effects of the EA rosette-forming cell. J Immunol 124:2563.
4. MacKenzie MR, Paglieroni TF, Warner NL (1987) Multiple myeloma: an immunologic profile. IV. The EA rosette-forming cell is a Leu-1 positive immunoregulatory B cell. J Immunol 139(1):24-28.
5. Paglieroni T, Caggiano V, MacKenzie MR (1988) CD5 positive immunoregulatory B cell subsets. Am J Hematol 28:276-278.
6. Paglieroni T, Caggiano V, MacKenzie MR (1992) Abnormalities in immune regulation precede the development of multiple myeloma. Am J Hematol 40:51-55.
7. Scibienski RJ, Paglieroni TF, MacKenzie MR (1990) Establishment and characterization of a CD5 positive immunosuppressive human myeloma cell line. Leukemia 4:775-780.

Genes Expressed Selectively in Murine and Human Plasma Cell Neoplasms

P.L. Bergsagel, L.A. Brents, J. B. Trepel* and W.M. Kuehl
NCI-Navy Medical Oncology Branch and *Medicine Branch.
National Cancer Institute, Bethesda, MD 20889-5105

Genes Isolated from Murine Plasmacytomas

For the past several years, we have been using a novel subtractive cDNA cloning method to identify genes that are differentially expressed in murine plasmacytoma cell lines and tumors but not in murine mature B cell lymphomas [1,2,3,4,5,6]. The murine cDNAs have been screened against a panel of plasmacytoma cell lines, B and pre-B lymphoma cell lines, other hematopoietic and non-hematopoietic cell lines, normal tissues, and two models of normal plasma cells (spleen cells from mice reconstituted with hematopoietic stem cells infected with an IL-6 retrovirus and spleen cells cultured for 5 days in vitro with LPS).

Most of the differentially expressed murine cDNAs that we have identified are expressed in both plasmacytomas and normal plasma cells, although expression levels are generally higher in plasmacytomas than in the two normal plasma cell models that we have used. Examples of these genes include syndecan (an adhesion molecule), EGP314 (a pan-epithelial glycoprotein that has been given the official name M1s1 by the Mouse Genome Informatics Group), placental alkaline phosphatase, XLR3 (defining a new acidic subfamily of the XLR gene family) [6], and pig 'hep' (a gene that is markedly upregulated - perhaps by IL-6 - in liver cells during cardiogenic shock). Several of these genes are expressed in pre-B and plasma cells but not at intermediate stages of B cell maturation. Examples of this pattern of expression include syndecan, placental alkaline phosphatase, and a human gene (isolated as described below) with possible homology to mb-1.

Other genes are expressed in plasmacytomas but not normal plasma cells. Our best studied example of such a gene is PC326, a new member of the ß-transducin repeat "mosaic" protein family that is X-linked, with normal expression detected only in testis [5]. Similar to the situation for all XLR gene family members, we have been unable to identify a human homologue of PC326. In contrast to the other genes we have characterized, the expression of PC326 is dysregulated in plasmacytomas, with some indirect evidence suggesting that its expression is related to the dysregulation of intracisternal A particle sequences in murine plasmacytomas.

One other murine gene, which has an enigmatic pattern of expression, is a new X-linked member of the hematopoietic growth factor receptor gene family, although demonstrating only minimal homology to any member of this gene family in available data bases [4]. This gene (designated PC251 at present) is most closely related to the IL3, IL5, and GMCSF α receptor subunits, and is expressed in all plasmacytomas, most non-hematopoietic cell lines, and all normal tissues, but in no other hematopoietic cell line. It is expressed in normal, resting B cells, with preliminary results suggesting that expression is down-regulated in B cells stimulated to proliferate during 2 days of incubation with LPS. The expression pattern and functional significance of this gene are presently unclear.

Genes Isolated from Human Multiple Myeloma Cell Lines

Recently we have extended the subtractive cDNA approach to human myeloma to identify genes that are expressed in two human myeloma cell lines but not two B lymphoblastoid cell lines (LCL). First strand cDNA was prepared from the 8226 myeloma cell line by priming with an adaptor-dT17-(G,A,C) oligonucleotide. The cDNA was subtracted by hybridization with a tenfold excess of mRNA derived from two B lymphoblastoid cell lines, and then positively selected by hybridization to a two fold excess of mRNA from the unrelated H929 myeloma cell line; in each case double-and single-stranded polynucleotides were separated by hydroxyapatite chromatography. A 3' homopolymeric stretch of dC residues was added with terminal transferase, and the positively selected subtractive cDNA was PCR amplified to provide a renewable source of subtractive probe that appears to be enriched approximately 200 fold (compared to a first strand cDNA) for sequences that are differentially expressed in myeloma cells. This probe was used to screen 84,000 clones in a conventional unidirectional H929 cDNA library. We identified 101 clones that hybridized to the subtractive probe but not a first strand cDNA probe synthesized from LCL mRNA. The 3' sequence of these clones enabled us to eliminate related clones, resulting in the identification of 45 apparently different genes. By screening Northern blots containing 1 µg of mRNA each from the two plasmacytoma cell lines and the two B lymphoblastoid cell lines, it was determined that 40 genes are selectively expressed in the two plasmacytoma cell lines (qualitatively subtractive), 2 are expressed at a significantly higher level in the plasmacytoma cell lines (quantitatively subtractive), and 3 are expressed at similar levels in all four cell lines (non subtractive).

Four of the subtractive genes are known markers of myeloma cells, i.e. Ig lambda light chain and J chain, each of which is expressed at a higher level in the myeloma cell lines than the LCL lines, as well as CD28 and CD56 (N-CAM). Five other qualitatively subtractive genes were also known entities, i.e. ornithine aminotransferase, lipoprotein lipase, cystathionine beta-synthase, N-cadherin, and the apparent human homologue of avian c-maf. Although 5' sequence was obtained from the remaining 35 qualitatively subtractive cDNAs, we found only limited homologies to known sequences for 8 of these genes, including one gene with a possibly significant homology to mb-1 (see above). At this time we have only a minimum of additional information on the expression patterns of c-maf and N-cadherin, as indicated below.

The human c-maf cDNA probe detects a major 3.5 kb mRNA species in the myeloma cell lines. Sequence of approximately one third of the coding region indicates that the apparent human c-maf homologue is highly homologous to avian c-maf (>90% amino acid sequence identity). Preliminary expression studies indicate that c-maf mRNA is expressed at a relatively high level in 9 of 12 human myeloma cell lines listed in Table 1, but at a low (or undetectable) level in 7 B lymphoblastoid cell lines, 4 Burkitt's lymphoma cell lines, one T cell line, and two pre-B cell lines. It is worth noting that c-maf is a member of the bZIP transcription factor family, which includes jun, fos, and NF-IL6. The heterodimerization of many of these proteins, e.g., jun/fos, jun/NF-IL6, jun/c-maf, raises the possibility that c-maf may perturb the IL-6 pathway that is implicated in human multiple myeloma [7,8,9].

Table 1. Phenotypic Marker Expression in Myeloma and B Lymphoblastoid Cell Lines

		CD19	DR	CD28	CD38	CD56	N-cadherin	Syndecan
A.	8226	0	0	3	3	1	+	+
	H929	0	0	0	2	1	+	+
	JJN3	0	+		+	0	+	+
	EJM	0	1	1	1	0	+	
	KMM1	0	0	0	3	0	0	
	OPM2	0	1	1	1	1	+	+
	U266	0	1	3	3	1	0	
	SK-MM2	0	0	0	0	0	0	
	ark	1	1	2	3	0	+	+
	H1112	0	0	2	2	0	0	
	SK-MM1	0	0	3	1	0	+	
	GM2132	2	3	0	3	1	0	
B.	ram	2	2	0	1	0	0	
	ARH77	+	+		+	+	0	
	CESS	2	2	0	0	2	0	
	CB32	3	2	0	1	0	0	
	CB33	3	2	0	1	0	0	

The twelve human myeloma cell lines (section A) include: 8226 [10], H929 [11], JJN3 [12], EJM [13], KMM1 [14], OPM2 [15], U266 [16], SK-MM1 and SK-MM2 [17], ark [18], H1112 (unpublished, H. Oie), GM2132 [10,19]. The five B LCL (section B) include two lines (ram [18], ARH77 [20]) that are sometimes listed as human myeloma cell lines, but are included here as LCL lines because they contain EBV DNA and have markers like other LCL: CESS [21], CB32 and CB33 [22]. By Southern blotting, all LCL lines (section B) contain EBV sequences whereas all myeloma cell lines (section A) do not contain detectable EBV sequences (Brents and Bergsagel, unpublished). The surface expression of CD19, DR, CD28, CD38, and CD56 (N-CAM) was determined by flow cytometric analysis, and scored as 0-4 based on fluorescent intensity. The boxed results for JJN3 and ARH77 were taken from the literature. The expression of N-cadherin was determined on three occasions with two pairs of oligonucleotides using a semi-quantitative RT-PCR assay. The expression of syndecan was determined on two occasions for the five indicated cell lines using an RT-PCR assay.

N-cadherin is a calcium-dependent homotypic adhesion molecule that is expressed in a limited number of tissues, including neural tissues and bone marrow stromal cells [23]. Table 1 summarizes the expression pattern of a number of markers in 12 human myeloma cell lines and 5 B lymphoblastoid cell lines. Using a semi-quantitative RT-PCR assay, we detected similar levels of expression of N-cadherin in 7 of 12 myeloma cell lines but none of the B lymphoblastoid cell lines. In a more preliminary survey of 5 human myeloma cell lines, we confirmed that syndecan is often expressed in myeloma cells. The data in Table 1 suggest that there is not a clear correlation between the expression of CD56 (N-CAM) and either N-cadherin or syndecan, whereas there appears to be a correlation between the expression of syndecan and N-cadherin.

It is noteworthy that 4 adhesion molecules (syndecan, M1S1 [EGP314], N-cadherin, and N-CAM [CD56]) are differentially expressed in plasmacytomas, with at least the first two also being expressed in normal plasma cells (based on the apparent correlation of expression of syndecan and N-cadherin in myeloma cell lines, it seems likely that N-cadherin will prove to be a marker of normal plasma cells although this remains to be determined)[3,18,23,24]. This suggests that one of the key distinguishing features of plasma cells compared to mature B cells is their interaction with the extracellular matrix, and stromal or epithelial cells. The expression of CD56 in malignant but not normal human plasma cells remains unexplained [24].

References

1. Timblin C, Battey J, Kuehl WM (1990) Application for PCR technology to subtractive cDNA cloning: identification of genes expressed specifically in murine plasmacytoma cells. Nucleic Acids Res 18:1587-93
2. Timblin C, Bergsagel PL, Kuehl WM (1990) Identification of consensus genes expressed in plasmacytomas but not B lymphomas. Curr Top Microbiol Immunol 166:141-7
3. Bergsagel PL, Victor-Kobrin C, Timblin CR, Trepel J, Kuehl WM (1992) A murine cDNA encodes a pan-epithelial glycoprotein that is also expressed on plasma cells. Journal of Immunology 148:590-596
4. Bergsagel PL, Victor-Kobrin C, Brents L, Kuehl WM (1992) Genes expressed selectively in plasmacytomas: markers of differentiation and transformation. Curr Top Microbiol Immunol 182:223-228
5. Bergsagel PL, Timblin CR, Eckhardt L, Laskov R, Kuehl WM (1992) Sequence and expression of a murine cDNA encoding PC326, a novel gene expressed in plasmacytomas but not normal plasma cells. Oncogene 7:2059-64
6. Bergsagel PL, Timblin CR, Kozak CA, Kuehl WM (1994) Sequence and expression of murine cDNAs encoding Xlr3, a new member of the xlr gene family. Gene (in press)
7. Hsu W, Kerppola TK, Chen PL, Curran T, Chen-Kiang S (1994) Fos and Jun repress transcription activation by NF-IL6 through association at the basic zipper region. Mol Cell Biol 14:268-76
8. Kataoka K, Noda M, Nishizawa M (1994) Maf nuclear oncoprotein recognizes sequences related to an AP-1 site and forms heterodimers with both Fos and Jun. Mol Cell Biol 14:700-12
9. Kerppola TK, Curran T (1994) Maf and Nrl can bind to AP-1 sites and form heterodimers with Fos and Jun. Oncogene 9:675-84
10. Matsuoka Y, Moore GE, Yagi Y, Pressman D (1967) Production of free light chains of immunoglobulin by a hematopoietic cell line derived from a patient with multiple myeloma. Proc Soc Exp Biol Med 125:1246-50
11. Gazdar AF, Oie HK, Kirsch IR, Hollis GF (1986) Establishment and characterization of a human plasma cell myeloma culture having a rearranged cellular myc proto-oncogene. Blood 67:1542-9
12. Jackson N, Lowe J, Ball J, Bromidge E, Ling NR, Larkins S, Griffith MJ, Franklin IM (1989) Two new IgA1-kappa plasma cell leukaemia cell lines (JJN-1 & JJN-2) which proliferate in response to B cell stimulatory factor 2. Clin Exp Immunol 75:93-9

13. Hamilton MS, Ball J, Bromidge E, Lowe J, Franklin IM (1990) Characterization of new IgG lambda myeloma plasma cell line (EJM): a further tool in the investigation of the biology of multiple myeloma. Br J Haematol 75:378-84
14. Togawa A, Inoue N, Miyamoto K, Hyodo H, Namba M (1982) Establishment and characterization of a human myeloma cell line (KMM-1). Int J Cancer 29:495-500
15. Katagiri S, Yonezawa T, Kuyama J, Kanayama Y, Nishida K, Abe T, Tamaki T, Ohnishi M, Tarui S (1985) Two distinct human myeloma cell lines originating from one patient with myeloma. Int J Cancer 36:241-6
16. Nilsson K, Bennich H, Johansson SG, Ponten J (1970) Established immunoglobulin producing myeloma (IgE) and lymphoblastoid (IgG) cell lines from an IgE myeloma patient. Clin Exp Immunol 7:477-89
17. Eton O, Scheinberg DA, Houghton AN (1989) Establishment and characterization of two human myeloma cell lines secreting kappa light chains. Leukemia 3:729-35
18. Ridley RC, Xiao H, Hata H, Woodliff J, Epstein J, Sanderson RD (1993) Expression of syndecan regulates human myeloma plasma cell adhesion to type I collagen. Blood 81:767-74
19. Goldstein M, Hoxie J, Zembryki D, Matthews D, Levinson AI (1985) Phenotypic and functional analysis of B cell lines from patients with multiple myeloma. Blood 66:444-6
20. Burk KH, Drewinko B, Turjillo JM, Ahearn MJ (1978) Establishment of a human plasma cell line in vitro. Cancer Res 38:2508-13
21. Bradley TR, Pilkington G, Garson M, Hodgson GS, Kraft N (1982) Cell lines derived from a human myelomonocytic leukaemia. Br J Haematol 51:595-604
22. Lombardi L, Newcomb EW, Dalla-Favera R (1987) Pathogenesis of Burkitt lymphoma: expression of an activated c-myc oncogene causes the tumorigenic conversion of EBV-infected human B lymphoblasts. Cell 49:161-70
23. Takeichi M (1991) Cadherin cell adhesion receptors as a morphogenetic regulator. Science 251:1451-5
24. Van Camp B, Durie BG, Spier C, De Waele M, Van Riet I, Vela E, Frutiger Y, Richter L, Grogan TM (1990) Plasma cells in multiple myeloma express a natural killer cell-associated antigen: CD56 (NKH-1; Leu-19). Blood 76:377-82

Molecular and Biological Role of CD40 in Multiple Myeloma

J.J. Westendorf[1], G.J. Ahmann[2], J.A. Lust[2], R.C. Tschumper[1], P.R. Greipp[2], J.A. Katzmann[3], and D.F. Jelinek[1]

Departments of [1]Immunology, [2]Internal Medicine, and [3]Laboratory Medicine and Pathology, Mayo Clinic and Foundation, Rochester, Minnesota 55905

Introduction

Multiple myeloma is a progressive and fatal disease characterized by the accumulation of malignant plasma cells in the bone marrow. Although the cause of myeloma remains unknown, interleukin 6 (IL-6) has been demonstrated to be a primary growth factor for malignant plasma cells *in vitro* [1, 2]. Elevated levels of IL-6 in myelomatous bone marrow and a direct correlation of IL-6 levels with disease severity suggest that IL-6 may be a crucial factor for tumor expansion *in vivo* as well [2, 3]. At the present time, it remains uncertain, however, as to whether paracrine or autocrine sources are responsible for the significant increases in IL-6 levels that occur during disease progression (reviewed in [4]).

The biological significance of elevated levels of IL-6 and tumor cell proliferation in response to IL-6 is best understood by comparing the effects of IL-6 on normal B cells and plasma cells with the effects of IL-6 on myeloma cells. Table 1 summarizes these important differences. First, because normal plasma cells are terminally differentiated cells, they do not proliferate in response to any stimulus, including IL-6. By contrast, malignant plasma cells or myeloma cells may proliferate in an inducible and/or unregulated fashion. It is also important to note that normal activated B cells do not proliferate in response to IL-6; IL-6 instead functions as a differentiation factor for B cells [5-9]. Of equal interest, although the IL-6 gene can be induced during normal B cell activation, it is no longer expressed in plasma cells [10]. In myeloma cells, by contrast, IL-6 gene expression may be inducible and/or constitutive [11]. Deregulation of the IL-6 gene in a malignant cell that has gained the capacity to proliferate aberrantly in response to this cytokine therefore suggests that an autocrine mechanism could play a role in tumor expansion in myeloma patients.

Our studies have focused on understanding possible mechanism(s) of autocrine IL-6 secretion by myeloma tumor cells. Preliminary findings [12, 13] indicated that myeloma cells express the important B cell signaling molecule, CD40 (reviewed in [14]). Because an association between CD40 and IL-6 signaling was previously demonstrated in normal B cells [15], we have specifically studied CD40 and its role in autocrine IL-6 secretion by malignant plasma cells.

TABLE 1. Summary of Selected Differences Between Normal B Cells, Plasma Cells, and Myeloma Cells

Phenotype	B Cells	Plasma Cells	Myeloma Cells
Proliferation	Inducible and regulated	None	Inducible and/or unregulated
Proliferation in response to IL-6	None	None	Responsive
Expression of IL-6	Inducible and regulated	None	Inducible and/or constitutive

Materials and Methods

Proliferation assays

The various myeloma cell lines were maintained in RPMI 1640 (Sigma) supplemented with 10% heat-inactivated FCS, antibiotics, and 1 ng/ml IL-6. Before all assays, the cells were washed three times and resuspended in media. For proliferation assays, cells were cultured at 2-5x10^4 cells/well in round-or flat-bottom microtiter plates in a final volume of 200 µl. Cultures were incubated for 3 days at 37°C and in the presence of 5% CO_2. One microcurie of [^{3}H]thymidine was added to the cultures 18 hours before harvesting and [^{3}H] thymidine incorporation was quantitated by liquid scintillation spectroscopy. Values represent the mean of triplicate samples ± the standard error of the mean (SEM).

IL-6 ELISA

IL-6 levels in each supernatant were determined using a Cytokine™ Human IL-6 ELISA (Biosource) according to the manufacturer's instructions.

Flow cytometry

For three-color flow cytometric analysis, cells were labeled at 4°C for 30 minutes with three fluorescently tagged monoclonal antibodies (mAb) recognizing the cell surface molecules CD40, CD38, and CD45 (FL1 = CD40-FITC, Biosource; FL2 = CD38-PE, Becton Dickinson; FL3 = CD45-PerCP, Becton Dickinson). Analysis was performed on a FACScan flow cytometer

(Becton Dickinson). For experiments utilizing sorted cells, cells were labeled with CD38-PE and CD45-FITC for 30 minutes at 4°C, prior to selecting the $CD38^{hi}CD45^{lo}$ population using a FACSStar IV flow cytometer (Becton Dickinson).

Assays with the CD40L

Baculovirus engineered to express recombinant membrane-bound murine CD40 ligand (CD40L) was the generous gift of Dr. Kathy Meek (UT Southwestern). Wild type AcMNPV virus was also provided by Dr. Meek and was used as a negative control. Sf9 insect cells were maintained at 27°C in Grace's insect medium (Gibco/BRL) and 7.5% FCS. Sf9 cells were infected with wild type or CD40L virus in serum-free EX-CELL 400 medium (JRH Biosciences) using standard procedures [16]. Sf9 cells were harvested 72 hours post-infection and cultured with myeloma cells as described in the text. For some experiments, membranes from infected Sf9 cells were isolated as previously described [17] and were added to myeloma cell cultures instead of intact Sf9 cells.

Results

Previous studies by our group demonstrated that a recently established myeloma cell line, ANBL-6 [12], expressed the important signaling molecule CD40. This result was initially surprising because it had been previously suggested that myeloma cells as well as normal plasma cells do not express CD40 [18, 19]. Additional studies by our group demonstrated that the ANBL-6 cell line is dependent on exogenous IL-6 for growth [12]. Of great interest, the addition of an anti-CD40 mAb (G28-5; kindly provided by Dr. Ed Clark, U. Washington) also stimulated the proliferation of these cells, and did so in a manner that involved the induction of an autocrine IL-6 growth loop [13].

The focus of the studies described herein was to extend our characterization of the effects of CD40 stimulation to include normal plasma cells, fresh patient myeloma cells, and newly established myeloma cell lines.

Characterization of new myeloma cell lines

The characterization of the ANBL-6 cell line has been reported previously [12]. In these studies, a subclone of the ANBL-6 cell line, designated ANBM-6 (established from the same patient; described in [12]), was used. We have recently established 3 new human myeloma cell lines that we have designated DP-6, KAS-6, and KP-6. Before extending our CD40 studies to include these new cell lines, it was important to characterize the cell surface phenotype of these cell lines and determine patterns of IL-6 responsiveness. Phenotypic characterization of these cell lines is summarized in Table 2.

It may be seen that all 4 of the cell lines express CD40 in addition to CD38 and CD54. The cell lines were uniformly negative for CD19, CD20, and the Epstein Barr virus nuclear antigen (EBNA). The phenotype of each cell line is consistent with phenotypes displayed by plasma cells [20]. To continue characterization of these cell lines, the responsiveness of the cells to exogenous IL-6 was also determined (Table 3).

TABLE 2. Phenotypic analysis of human myeloma cell lines

	Myeloma Cell Lines			
Antigen	ANBM-6	DP-6	KAS-6	KP-6
CD19	-	-	-	-
CD20	-	-	-	-
CD38	+	+	+	+
CD40	+	+	+	+
CD45	-	+/-	-	+
CD54	+	+	+	+
CD56	-	+	-	+
EBNA	-	-	-	-

TABLE 3. IL-6 responsiveness of myeloma cell lines

				Anti-IL-6 + Anti-IL-6 R	
Expt. No.	Cell Line	Nil	IL-6	Nil	IL-6
		$[^3H]$ Thymidine Incorporation (cpm x 10^{-3}/SEM)			
1	ANBM-6	0.7 ± 0.1	52.7 ± 1.7	0.3 ± 0.1	4.1 ± 0.5
	DP-6	4.1 ± 0.2	27.5 ± 0.6	1.4 ± 0.1	3.9 ± 0.1
	KP-6	38.9 ± 0.1	58.8 ± 2.1	26.5 ± 0.4	28.9 ± 0.7
2	KAS-6	6.6 ± 0.5	30.2 ± 1.4	5.7 ± 0.1	8.2 ± 0.4

Myeloma cells were cultured at a density of 5.0×10^4 cells per round bottom microtiter well in RPMI + 10% FCS in Expt. 1 and at 2.5×10^4 cells per well and in RPMI + 2% FCS in Expt. 2. Cells were stimulated with 2.5 ng/ml (Expt. 1) and 1 ng/ml (Expt. 2) IL-6 in the presence or absence of neutralizing mAb to IL-6 and the gp80 component of the IL-6 receptor. DNA synthesis was assayed on day 3.

Each of the 3 new cell lines, as well as the ANBL-6 cell line, displayed an increase in DNA synthesis when cultured in the presence of exogenous IL-6. Importantly, the augmentation in proliferation of all the cell lines could be blocked by the simultaneous addition of neutralizing anti-IL-6 (Biosource) and anti-IL-6 receptor (gp80 subunit; Biosource) mAb. Of interest, differences were observed in the amount of DNA synthesis by each of the cell lines cultured in media alone. For example, the KP-6 cell line displayed an elevated proliferative rate in the absence of exogenous IL-6. It is important to note, however, that some of this background proliferation could be blocked by the presence of the neutralizing mAb. This result supports the conclusion that although the KP-6 cells are responsive to exogenous IL-6, they also constitutively produce a basal level of IL-6.

CD40L stimulation

We were next interested in determining whether CD40 stimulation of IL-6 secretion by the ANBL-6 cells [13] was unique to this cell line or whether an autocrine loop could be stimulated by cross-linking CD40 in the newly established cell lines as well. Table 4 displays the results obtained when each of the cell lines were stimulated with Sf9 cells expressing membrane bound CD40L.

TABLE 4. CD40L-Induced Proliferation of Myeloma Cell Lines

Cell Line	Anti-IL-6	Nil	1 ng/ml IL-6	Control Sf9	CD40L Sf9
		72 hr [^{3}H] Thymidine Incorporation (cpm x 10^{-3} ± SEM)			
ANBM-6	-	0.4 ± 0.1	8.1 ± 0.1	0.4 ± 0.1	3.5 ± 0.3
	+	0.4 ± 0.1	0.4 ± 0.1	0.4 ± 0.1	0.6 ± 0.3
DP-6	-	0.4 ± 0.1	10.2 ± 0.5	0.5 ± 0.1	3.6 ± 0.8
	+	0.2 ± 0.1	0.8 ± 0.1	0.2 ± 0.1	0.7 ± 0.1
KAS-6	-	1.9 ± 0.1	39.8 ± 1.0	4.5 ± 0.5	49.0 ± 0.6
	+	1.4 ± 0.1	4.7 ± 0.3	1.4 ± 0.1	17.8 ± 0.7
KP-6	-	50.6 ± 1.0	68.6 ± 1.1	50.0 ± 2.3	60.4 ± 0.5
	+	32.2 ± 2.1	37.8 ± 1.5	28.4 ± 1.9	38.3 ± 1.4

ANBM-6 and DP-6 cells were cultured at $5x10^4$/flat microtiter well under serum-free conditions in the presence or absence of $2.5x10^4$ control virus or CD40L infected Sf9 cells. KAS-6 cells were stimulated with isolated Sf9 membranes instead of intact Sf9 cells in the presence of 2% FCS. KP-6 cells were stimulated with intact Sf9 cells at a 1:2 ratio and in the presence of 2% FCS.

As may be seen in Table 4, each of the cell lines proliferated in response to exogenous IL-6 in a manner that was inhibited by the addition of an IL-6 neutralizing mAb. Importantly, Sf9 cells infected with control virus were without effect on myeloma cell proliferation. Of interest however, when Sf9 cells or membranes from Sf9 cells expressing membrane-bound CD40L were added to the cultures, there was significant augmentation in proliferation, and this again occurred in a manner that was neutralized by the anti-IL-6 mAb. It is of some interest that the degree of augmentation induced by membrane-bound CD40L was least noticeable in the KP-6 cells. However, it is important to note that this cell line expressed a relatively high level of IL-6 in an apparently constitutive fashion. Thus, it may be difficult to substantially alter this baseline rate of IL-6 expression in these cells.

CD40 expression and function in normal plasma cells and freshly isolated patient tumor cells

Because of the potential importance of CD40 as a signaling molecule on myeloma cells, we were interested in determining: 1) whether normal plasma cells express CD40; and 2) if normal plasma cells express CD40, could it stimulate IL-6 production. To answer the first question, we employed 3-color immunofluorescence analysis and took advantage of the finding that plasma cells express high-density CD38, and express little to no CD45 [20]. Cells expressing high density CD38 were observed to express a range of CD45 (Figure 1). When analysis gates were placed around the high density CD38 expressing populations and then analyzed for FITC-CD40 binding, essentially 100% of the CD38^{hi}CD45^{neg} (R2), CD38^{hi}CD45^{lo} (R3) and CD38^{hi}CD45^{hi} (R4) cells were found to also express CD40. We have analyzed at least 4 other normal bone marrows using this method, and in all cases, plasma cells expressed CD40 (data not shown). These results indicate that plasma cells obtained from normal individuals also express CD40, suggesting that CD40 expression by myeloma cells may not be aberrant.

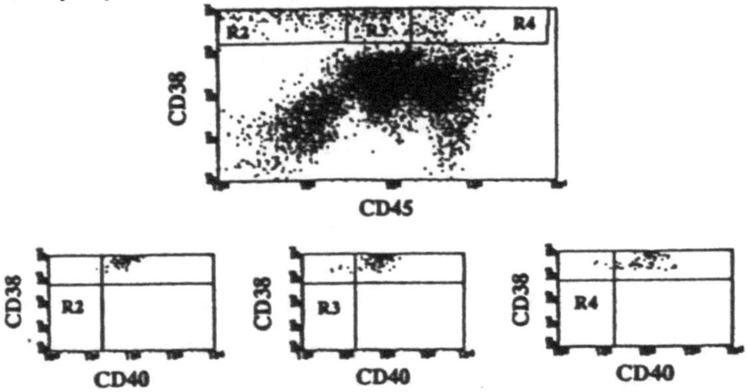

Fig. 1. CD40 Expression on Normal Plasma Cells. Plasma cells in a normal bone marrow mononuclear cell preparation were identified by high density CD38 and variable CD45 expression (top panel). CD40 expression on CD38^{hi} cells in the gated regions, R2, R3, R4, was then determined (bottom panel).

To test the possibility that CD40 signaling of IL-6 secretion was unique to myeloma cells, we carried out the experiment shown in Table 5. In this experiment, we obtained bone marrow from one normal individual and from 3 myeloma patients, isolated the mononuclear cells, and identified the plasma cells by staining with PE-CD38 and FITC-CD45. A flow cytometer was used to isolate the $CD38^{hi}CD45^{lo}$ normal and malignant plasma cells. The purity of the sorted populations was verified by preparing cytospins of the sorted cells and staining with Wright's Giemsa stain (results not shown). The ability of the anti-CD40 mAb (G28-5) to induce IL-6 secretion was then assayed.

TABLE 5. CD40 Responsiveness of Normal Plasma Cells and Patient Myeloma Cells

Patient	Nil	Anti-CD40
	(IL-6, pg/ml)	
Normal	<2	<2
MM-1	<2	140
MM-2	<2	76
MM-3	<2	32

Purified plasma cells were cultured for 3 days in the presence or absence of anti-CD40 (G28-5) prior to harvesting the supernatants and assaying for IL-6 secretion by ELISA.

Importantly, although normal plasma cells did not secrete IL-6 in response to anti-CD40, all patient myeloma cells did secrete significant levels of IL-6 in response to CD40 stimulation. It is important to note that the tumor cells obtained from these three patients also displayed a high proliferative index (data not shown). These data support the conclusion that CD40 stimulation of myeloma cells may be an important mechanism by which an autocrine growth loop may be induced. In continuation of these studies, we have begun to examine the expression of CD40L in bone marrow. Preliminary experiments indicate that CD40L expressing cells are present in myelomatous bone marrow (data not shown), thus providing a potential biological source of stimulation of the CD40-autocrine IL-6 growth loop.

Discussion

The results described herein provide evidence that CD40 expression by myeloma cells may be biologically significant because of the ability of this molecule to induce autocrine IL-6 secretion. Studies by several other groups [4, 21] have alternatively suggested that non-tumor cells, such as bone marrow stromal cells, are sources of IL-6 for the tumor population. Our data do not discount the role of paracrine sources of IL-6 in disease progression; however, we believe that the consequences of the induction of an autocrine IL-6 growth loop in the tumor cell itself are profound. For example, autocrine IL-6 may be more potent than paracrine IL-6 simply because of the proximity of the growth factor with the tumor cell after it is produced. It is also important to emphasize that considerable variability between patient tumor cells is observed *in vivo* as well as *in vitro*, albeit for reasons not yet understood. Because of this variability, CD40 stimulation of an autocrine IL-6 loop may not be uniformly observed, but instead, may be dependent on the phenotype of the tumor cells. Indeed, the KP-6 cell line described in this report was only minimally responsive to CD40 ligand stimulation. However, in marked contrast with the other cell lines, this myeloma cell line constitutively expresses the IL-6 gene at a high level; thus supporting the conclusion that further upregulation of IL-6 expression through CD40 may not be possible, or would be minimal at best, in these cells. In addition, when other myeloma cell lines that are no longer responsive to exogenous IL-6 were tested for stimulation in response to CD40L, no augmentation in proliferation was observed (data not shown).

We propose a model that is consistent with the biological variability observed between patient tumor cells obtained at various stages of disease (Figure 2). This model suggests that there are distinct biological differences between the clonal plasma cells existing in patients with monoclonal gammopathy of undetermined significance (MGUS) [22] and the malignant plasma cells from myeloma patients. Our data support the conclusion that CD40 will be expressed on all plasma cells, including normal and the clonal plasma cells from MGUS patients, but that responsiveness to CD40 will be characteristic only of myeloma cells from patients with advanced disease. Our model further speculates that responsiveness to CD40 is acquired because of an accompanying deregulation of IL-6 expression during the progression of the malignant plasma cell into one characterized by a high proliferative index. By extension, the model predicts that myeloma cells exhibiting a low growth rate may not be CD40 responsive because regulation of the IL-6 gene may still be intact and because the cells may not proliferate in response to IL-6. Lastly, it is important to point out that loss of a regulatory element controlling IL-6 gene transcription may not necessarily result in constitutive IL-6 secretion. Alternatively, these cells could be poised to respond to a stimulus such as CD40. Because both the wild type p53 and retinoblastoma (RB) tumor suppressor gene products have been demonstrated to negatively regulate IL-6 transcription [23], and because both of these genes reside on chromosomes frequently altered in multiple myeloma [24], it is tempting to speculate that loss or mutation of p53 or RB may be important genetic changes that occur during tumor progression in this disease. Studies are currently in progress to address this possibility.

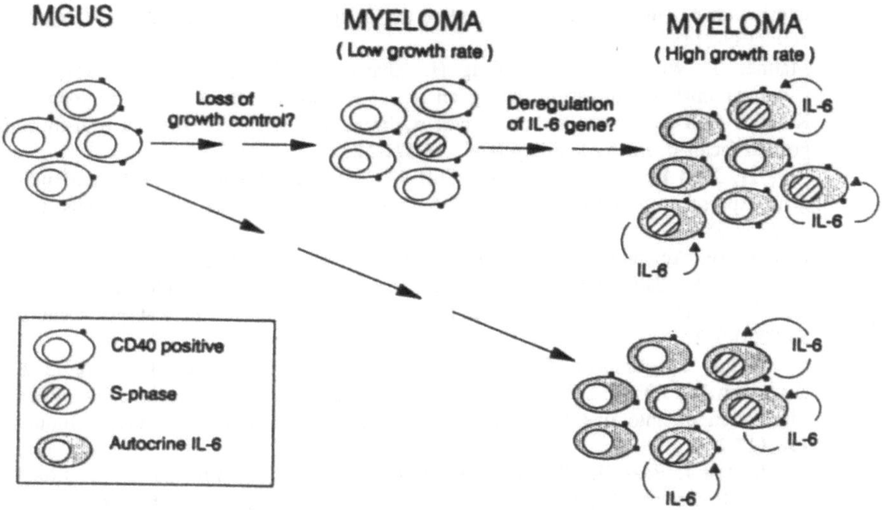

Fig. 2. Speculative model of tumor progression in the monoclonal gammopathies.

This work was supported by the Mayo Foundation.

References

1. Kawano M, Hirano T, Matsuda T, Taga T, Horii Y, Iwato K, Asaoku H, Tang B, Tanabe O, Tanaka H, Kuramoto A, Kishimoto T (1988) Autocrine generation and requirement of BSF-2/IL-6 for human multiple myelomas. Nature 332:83-85
2. Zhang XG, Klein B, Bataille R (1989) Interleukin 6 is a potent myeloma-cell growth factor in patients with aggressive multiple myeloma. Blood 74:11-13
3. Bataille R, Jourdan M, Zhang XG, Klein B (1989) Serum levels of interleukin 6, a potent myeloma cell growth factor, as a reflection of disease severity in plasma cell dyscrasias. Blood 84:2008-2011
4. Mandelli F, Avvisati G, Tribalto M (1992) Biology and treatment of multiple myeloma. Curr Opin Oncology 4:73-86
5. Muraguchi A, Hirano T, Tang B, Matsuda T, Horii Y, Nakajima K, Kishimoto T (1988) The essential role of B cell stimulating factor 2 (BSF-2/IL-6) for the terminal differentiation of B cells. J Exp Med 167:332-344
6. Roldán E, Brieva JA (1991) Terminal differentiation of human bone marrow cells capable of spontaneous and high-rate immunoglobulin secretion: Role of bone marrow stromal cells and interleukin 6. Eur J Immunol 21:2671-2677
7. Splawski JB, McAnally LM, Lipsky PE (1990) IL-2 dependence of the promotion of human B cell differentiation by IL-6 (BSF-2). J Immunol 144:562-569

8. Alderson MR, Pike BL (1989) Recombinant human interleukin 6 (B cell stimulatory factor 2) enhances immunoglobulin secretion by single murine haplen-specific B cells in the absence of cell division. Intl Immunol 1:20-28
9. Beagley KW, Eldridge JH, Lee F, Kiyono H, Everson MP, Koopman WJ, Hirano T, Kishimoto T, McGhee JR (1989) Interleukins and IgA synthesis. Human and murine interleukin 6 induce high rate IgA secretion in IgA-committed B cells. J Exp Med 169:2133-2148
10. Matthes T, Werner-Favre C, Tang H, Zhang X, Kindler V, Zubler RH (1993) Cytokine mRNA expression during an *in vitro* response of human B lymphocytes: Kinetics of B cell tumor necrosis factor α, interleukin (IL)6, IL-10, and transforming growth factor β, mRNAs. J Exp Med 178:521-528
11. Hata H, Xiao H, Petrucci MT, Woodliff J, Chang R, Epstein J (1993) Interleukin-6 gene expression in multiple myeloma: A characteristic of immature tumor cells. Blood 81:3357-3364
12. Jelinek DF, Ahmann GJ, Greipp PR, Jalal SM, Westendorf JJ, Katzmann JA, Kyle RA, Lust JA (1993) Coexistence of aneuploid subclones within a myeloma cell line that exhibits clonal immunoglobulin gene rearrangement: Clinical implications. Canc Res 53:5320-5327
13. Westendorf JJ, Ahmann GJ, Armitage RJ, Spriggs MK, Lust JA, Greipp PR, Katzmann JA, Jelinek DF (1994) CD40 expression in malignant plasma cells: Role in stimulation of autocrine IL-6 secretion by a human myeloma cell line. J Immunol 152:117-128
14. Banchereau J, Baza F, Blanchard D, Briere F, Galizzi JP, van Kooten C, Liu YJ, Rousset F, Saeland S (1994) The CD40 antigen and its ligand. Annu Rev Immunol 12:881-922
15. Clark EA, Shu G (1990) Association between IL-6 and CD40 signaling. J Immunol 145:1400-1406
16. O'Reilly DR, Miller LK, Luckow VA (1994) Baculovirus expression vectors, a laboratory manual. Oxford University Press, New York
17. Brian AA (1988) Membrane isolated procedure stimulation of B-cell proliferation by membrane-associated molecules from activated T cells. Proc Natl Acad Sci USA 85:564-568
18. Paulie S, Ehlin-Henriksson B, Mellstedt H, Kobo H, Ben-Aissa H, Perlmann P (1985) A p50 surface antigen restricted to human urinary bladder carcinomas and B lymphocytes. Canc Immunol Immunother 20:23-28
19. Ling NR, MacLennan ICM, Mason D (1987) B-cell and plasma cell antigens: New previously defined clusters. In: McMichael AJ (ed) Leucocyte typing III. Oxford University Press, Oxford, pp 302-335
20. Terstappen LWMM, Johnsen S, Segers-Nolton IMJ, Loken MR (1990) Identification and characterization of plasma cells in normal bone marrow by high-resolution flow cytometry. Blood 76:1739-1747
21. Klein B, Zhang XG, Jourdan M, Content J, Moussiau F, Aarden L, Bataille R (1989) Paracrine but not autocrine regulation of myeloma cell growth and differentiation by interleukin 6. Blood 73:517-526
22. Kyle RA, Lust JA (1989) Monoclonal gammopathies of undetermined significance. Sem Hematol 26:176-200
23. Santhanum U, Ray A, Sehgal PB (1991) Repression of the interleukin 6 gene promoter by p53 and the retinoblastoma susceptibility gene product. Proc Natl Acad Sci USA 88:7605-7609
24. Durie BGM (1992) Cellular and molecular genetic features of myeloma and related disorders. Hem/Onc Clin N Amer 6:463-477

The Proliferation-Associated Cytosolic Protein Lap18 (stathmin) is Expressed at Atypically Low Levels in BALB/c Plasmacytoma Cells

D. C. ROWLANDS[1], N. A. JONES[1], G. BROWN[1], M. POTTER[2], B. MUSCHINSKI[2] and I. C. M. MACLENNAN[1]

[1]Department of Immunology, Birmingham Medical School, Birmingham, B15 2TT, UK. and
[2]Laboratory of Genetics, National Cancer Institute, National Institutes for Health, Bethesda, MD 20892, USA.

Introduction

Lap18 is a 19kDa cytosolic protein that has been described by several groups. A plethora of synonyms have been used to describe this protein: prosolin [1] stathmin [2], 19K [3], p19 or metablastin [4], p18 or Op18 [5,6] and pp21/pp23 [7] in addition to Lap18 [8]. The cDNA and the protein sequence show a very high degree of conservation amongst mammals [9-13]. The protein can constitute upto 0.5% of total cytosolic protein [1]. Lap18 contains three known phosphorylation sites at serine 16, 25 and 38. Several forms of the protein have been described including a non-phosphorylated and mono- and bis-phosphorylated forms [1,2]. The protein has been shown to be a substrate for cyclic-AMP dependent protein kinase, p34*cdc2* and MAP kinase [14,15]. Although protein kinase C isoforms do not appear to phosphorylate the protein directly, both mono- and bis-phosphorylated forms can be induced by treatment of proliferating cells with phorbol 12-myristate 13-acetate [16,17]. In many instances such treatment is associated with growth arrest and subsequent down-regulation of Lap18 expression [16,17]. More direct evidence linking Lap18 with cell growth is provided by a study that showed inhibition of Lap18 mRNA in peripheral blood lymphocytes by antisense oligonucleotides caused delayed entry into S-phase of the cell cycle [18].

Monospecific antibodies against Lap18 have been produced [19,20] and these can be used to identify Lap18 expression in tissue sections [20]. Extensive immunohistochemical studies show that Lap18 is expressed in all proliferating cells in tissues of both mice and humans [20]. This is consistent with the finding *in vitro* that levels of Lap18 decrease following growth arrest [13,16,21] and are up-regulated in lymphoid cells that are induced to proliferate [3,13,16,17]. No obvious difference in the level of p19 expression was noted between three strains of mice (BALB/c, B10Br and CBA) [20].

Increased levels of Lap18 have been described in some leukaemia and lymphoma cells [5,22]. In addition, impaired phosphorylation of the protein has been described in leukaemic T-lymphoblasts [23]. This has prompted the suggestion that the protein may be important in the neoplastic transformation of cells.

Serial back-cross breeding experiments indicate that a gene or genes confering susceptibility of BALB/c mice to plasmacytoma is encoded on a region of chromosome 4 [24]. It is intriguing that the gene encoding Lap18 in mice is located in the same part of chromosome 4 [8].

Materials and Methods

Plasmacytoma induction in BALB/c mice

Plasmacytomas from BALB/c and BALB/c.D2 congenic mice that were given intraperitoneal silicone, intraperitoneal pristane alone, or pristane plus a transforming virus, were studied. Two were advanced plasmacytomas after induction by pristane 30 days previously followed by ABL/myc virus 9 days later. A further tumour induced by pristane in a (BALB/c x DBA/2) F2 was studied. One plasmacytoma was from a BALB/c mouse given pristane 62 days previously and J3 virus, which carries *v-myc* and *v-raf-1*, 54 days previously. One tumour was induced in a BALB/c.D2-IdhPep3 mouse by silicone, with a latency of 195 days.

Immunohistochemistry

Sections of paraffin-embedded tissue were prepared for immunostaining by a microwave-heating antigen retrieval system as described previously [25]. A polyclonal antiserum against an amino-terminal synthetic peptide was raised in rabbits and was shown to recognise Lap18 on Western blots and sections [20]. This antiserum was applied to sections at a dilution of 1:500. Control sections were incubated with the same antiserum that had been inhibited with the immunising peptide at a dilution of 4μg/ml. Sections were also stained with PC10, a mouse monoclonal antibody against proliferating cell nuclear antigen (PCNA) [26]. Staining was detected by incubation with a biotinylated secondary antibody followed by alkaline-phosphatase conjugated streptavidin-biotin complex. Alkaline phosphatase activity was identified by an azo-dye capture method using Naphthol-AS-MX-phosphate and Fast Red.

Cell culture and measurement of Lap18 levels

The mouse plasmacytoma lines studied were P3-NS1/1-Ag4-1 (NS1) [27], P3/X63-Ag8 (Ag8) [28], MOPC 460, MOPC 47A and XRPC 24. These were maintained in exponential growth in RPMI-1640 medium (Gibco) supplemented with 10% heat inactivated fetal calf serum, 100U/ml penicillin and 100μg/ml streptomycin. The medium for MOPC 460 cells was supplemented with recombinant murine IL6 (1ng/ml, British Biotechnology Products Ltd, Abingdon, UK). Cell extracts were prepared for 1-dimensional SDS-PAGE in 0.125M Tris-HCL pH6.8, 20% (v/v) glycerol, 4% (w/v) SDS, 10% (v/v) 2-mercaptoethanol and bromophenyl blue. For 2-dimensional gel electrophoresis cell extracts were prepared in buffer containing 9M urea, 2% CHAPS, 80mM dithiothreitol, 5% ampholines pH 3.5-10 and 4% Nonidet P-40. Multiple isoelectric focussing was used in the first dimension and performed for 15 hours at 700W in gels containing 4.5% ampholines (3% pH4-6; 1.5% pH3.5-10). The second dimension was SDS-PAGE (12% separating gel). All gels were loaded with equivalent amounts of protein. Proteins were transferred electrophoretically onto nitrocellulose membranes and stained for the presence of Lap18 using the antisera described above. Bound antibody was detected using 0.1μC/ml ^{125}I-labelled protein A followed by autoradiography using pre-flashed films. Autoradiographs were scanned using a laser densitometer and analysed with Gelscan XL software.

Results

The abdominal and thoracic organs from five BALB/c mice in which five different plasmacytomas were growing were studied. In each case there were multiple small tumour deposits on the serosal surface of the peritoneal cavity (Figures 1 and 2). These all showed a high growth fraction, as demonstrated by the presence of numerous mitoses. Sections stained for proliferating cell nuclear antigen indicated that the neoplastic cells had a labelling index in excess of 50%. Immunohistochemical staining with the anti-Lap18 sera showed that plasmacytoma cells were either unstained or weakly stained (Figures 1 and 2). By contrast, strong positive staining for Lap18 was seen in all normal cells that expressed proliferating cell nuclear antigen. These included the epithelial cells at the base of the intestinal villi (Figures 1 and 2); haemopoietic tissue in the splenic red pulp; germinal centre cells in lymph nodes, spleen and Peyer's patches; thymic cortical lymphocytes and ovarian granulosa cells in secondary follicles.

These *in vivo* studies were extended by assessing the expression of Lap18 in 5 mouse plasmacytoma cell lines. Cultures that were in exponential growth were taken and protein extracts prepared for 1-dimensional electrophoresis. Lap18 was detected on Western blots using the antisera described in the methods. The amount of Lap18 was measured by comparison with protein extracts of the human promyelocytic leukaemia cell line HL60 and K562 eythroleukaemia cells (Table 1). The levels of Lap18 in all five mouse plasmacytoma cell lines studied were 12% or less of that found in HL60 cells.

Phosphorylation of Lap18 in plasmacytoma cells was studied using protein extracts from the same cell lines. These were analysed by 2-dimensional electrophoresis followed by Western blotting and autoradiography. Cells in log phase of growth in standard tissue culture medium contained Lap18 mainly in the non-phosphorylated form (Table). This conforms with the situation in a wide variety of other cell lines [1,2,29,30]. When cultures were treated with the 10nM phorbol 12-myristate 13-acetate for 4 hours, Lap18 showed markedly increased phosphorylation with production of detectable levels of both mono- and bis- phosphorylated forms of the protein (Table). The proportion of the mono- and bis- phosphorylated forms was also similar to that found in cells of other lineages treated in this way [1,2,29,30]. Thus, although Lap18 levels are low in plasmacytoma cells, compared with the levels found in normal proliferating cells, no defect in Lap18 phosphorylation was identified as assessed by Western blots of 2D gels after treatment with 10nM phorbol 12-myristate 13-acetate.

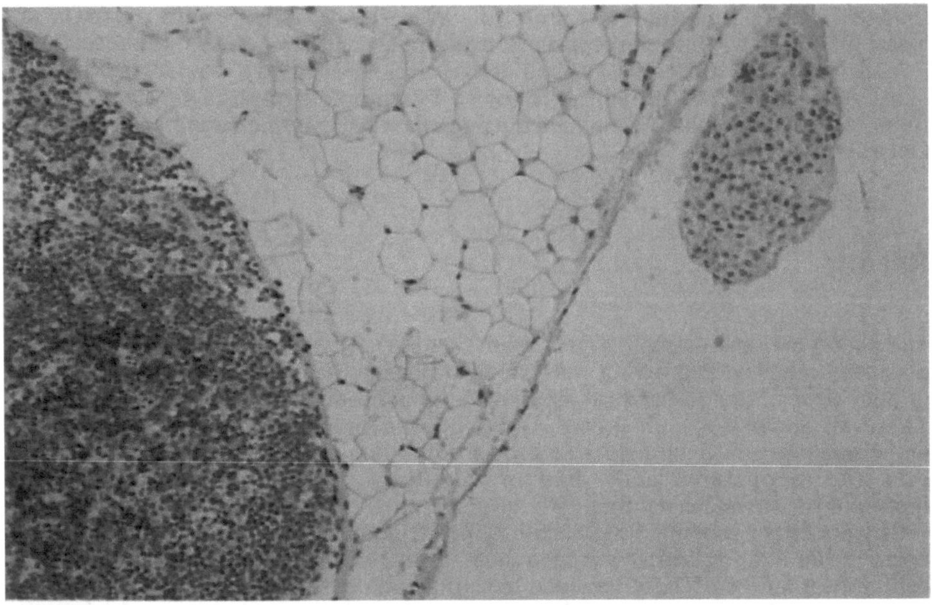

Figure 1.
Plasmacytoma on the peritoneal surface, stained for Lap18. There is strong positivity in germinal centre cells of a mesenteric lymph node. In contrast, the plasmacytoma cells are negative.

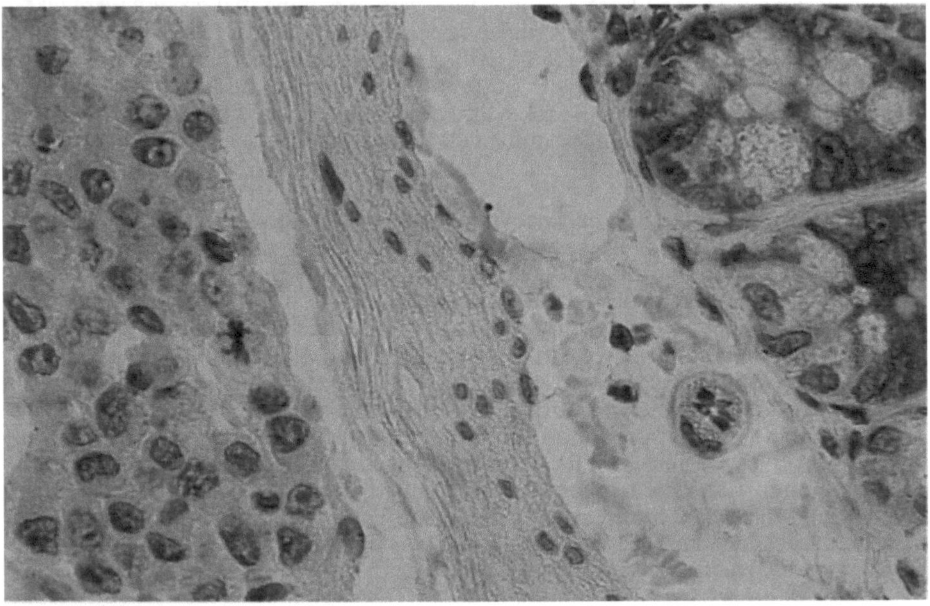

Figure 2.
Higher power view of a plasmacytoma in the serosa of the small intestine. There is only very weak positivity in the plasmacytoma cells, including one cell in mitosis. In contrast, basal crypt epithelial cells are strongly positive.

Table: Level of Lap18 expression and phosphorylation in mouse plasmacytoma cell lines compared with human myeloid cell lines.

Cell line	Relative level of Lap18 protein (HL60 = 100)	Ratio of unphosphorylated : phosphorylated Lap18	
		- PMA	+ PMA
Plasmacytoma			
P3-NS1/1-Ag4-1	4	86 : 24	53 : 47
P3/X63-Ag8	6	99 : 1	44 : 56
MOPC 460	12	98 : 2	76 : 24
MOPC 47A	5	99 : 1	97 : 3
XRPC 24	6	99 : 1	78 : 22
Myeloid			
HL60	100	96 : 4	33 : 59
K562	122	98 : 2	35 : 65

Discussion

The level of Lap 18 expression in BALB/c plasmacytoma cells growing *in vitro* and *in vivo* is abnormally low compared with that in all other proliferating mouse cells studied. Similar analysis of a wide range of normal human tissues also shows close association between Lap18 expression and that of the proliferation associated Ki-67 antigen [20]. Studies of the bone marrow from six patients with multiple myeloma were stained with the anti-Lap18 sera. Although very few or no tumour cells were stained this is consistent with the low growth fraction in these tumours (unpublished data). Six cell lines derived from patients with advanced multiple myeloma with extra-medullary disease have also been studied. When these cells are in exponential growth their levels of Lap18 has been found to be within normal limits for proliferating cells [31]. Like BALB/c plasmacytoma cells these myeloma cell lines phosphorylate Lap18 when grown in 10nM phorbol 12-myristate 13-acetate, but unlike HL60, K562 [18] and mitogen-treated normal peripheral blood T cells [16] they do not down regulate their Lap18 or growth arrest in response to this stimulus.

Plasmacytomas can only be induced regularly in a limited number of mouse strains; BALB/c mice are the classical susceptible strain [32]. The plasmacytomas are invariably associated with a translocation between the *c-myc* and Ig heavy-chain genes [33]. Although this translocation is necessary it is not sufficient for neoplastic transformation [34], indicating that additional genetic change(s) must occur. These are dependent upon chronic granuloma formation in the peritoneal cavity that has classically been induced by mineral oil injection. The genetic basis of the susceptibility of BALB/c mice to develop plasmacytomas has been studied using classical serial backcross-breeding approach where resistance is introduced to a background of a susceptible strain of mice [24,35]. This has now mapped the susceptibility gene(s) to a part of chromosome 4 in mice equivalent to a section of human chromosome 1 between bands p35 and p65 [8]. Lap18 is encoded by a gene that lies between bands p35 and 36.1 [36]. While this association is intriguing the co-mapping of plasmacytoma susceptibility and the Lap 18 gene to a very large segment of chromosome 4 may well be coincidental. Recent studies of granuloma formation in strains of mice susceptible or resistent to plasmacytoma development indicates that large granulomas are significantly more frequent in susceptible strains [24,35].

The genetic changes leading to plasmacytoma development result in defective completion of terminal differentiation of plasmablasts to plasma cells. This differentiation stage is associated with the plasmablasts coming out of cell cycle and probably losing the ability to proliferate in response to growth factors. Little is known about origin of physiological plasma cells in the peritoneal cavity. It seems likely that these arise, like plasma cells in other sites, from B cell activation in secondary lymphoid organs [37,38]. Defective signalling through Lap18 might be causally associated with this failure of plasmacytoma cells to come out of cell cycle despite their arrival at a site where they would normally differentiate to plasma cells and despite their partial differentiation to cells that secrete high levels of immunoglobulin. Futher work to test this possibility is required.

References

1. Braverman R, Battacharya B, Feuerstein N, Cooper HL (1986) Identification and characterization of the nonphosphorylated precursor of pp17, a phosphoprotein associated with phorbol ester induction of growth arrest and monocytic differentiation in HL-60 promyelocytic leukemia cells. J Biol Chem 261:14342-14348
2. Sobel A (1991) Stathmin: a relay phosphoprotein for multiple signal transduction? TIBS 16:301-305
3. Gullberg M, Noreus K, Brattsand G, Friedrich B, Shingler V (1990) Purification and characterization of a 19-kilodalton intracellular protein. An activation-regulated putative protein kinase C substrate of T lymphocytes. J Biol Chem 265:17499-17505
4. Schubart UK, Xu J, Fan W, Cheng G, Goldstein H, Alpini G, Shafritz DA, Amat JA, Farooq M, Norton WT, Owen TA, Lian JB, Stein GS (1992) Widespread differentiation stage-specific expression of the gene encoding phosphoprotein p19 (metablastin) in mammalian cells. Differentiation 51:21-32
5 Hanash SM, Strahler JR, Kuick R, Chu EHY, Nichols D (1988) Identification of a polypeptide associated with the malignant phenotype in acute leukemia. J Biol Chem 263:12813-12815
6. Hailat N, Strahler J, Melham R, Zhu XX, Brodeur G, Seeger RC, Reynolds CP, Hanash SM (1990) N-myc gene amplification in neuroblastoma is associated with altered phosphorylation of a proliferation related polypeptide (Op18). Oncogene 5:1615-1618
7. Peyron JF, Aussel C, Ferrua B, Haring H, Fehlmann M (1989) Phosphorylation of two cytosolic proteins. An early event of T-cell activation. Biochem J 258:505-510
8. Mock BA, Krall MM, Padlan C, Dosik JK, Schubart UK (1993) The gene for Lap18, leukemia-associated phosphoprotein-p18 (metablastin), maps to distal mouse chromosome 4. Mammalian Genome 4:461-462
9. Zhu XX, Kozarsky K, Strahler JR, Eckerskorn C, Lottspeich F, Melham R, Lowe J, Fox DA, Hanash SM, Atweh GF (1989) Molecular cloning of a novel human leukemia-associated gene. Evidence of conservation in animal species. J Biol Chem 264:14556-14560

10. Schubart UK, Banergee MD, Eng J (1989) Homology between the cDNAs encoding phosphoprotein p19 and SCG10 reveals a novel mammalain gene family preferentially expressed in developing brain. DNA 8:389-398
11. Doye V, Soubrier F, Bauw G, Boutterin MC, Beretta L, Koppel J, Vandekerckhove J, Sobel A (1989) A single cDNA encodes two isoforms of stathmin, a developmentally regulated neuron-enriched phosphoprotein. J Biol Chem 264:12134-12137
12. Maucuer A, Doye V, Sobel A (1990) A single amino acid difference distinguishes the human and the rat sequences of stathmin, a ubiquitous intracellular phosphoprotein associated with cell regulations. FEBS Lett 264:275-278
13. Melhem RF, Zhu XX, Hailat N, Strahler JR, Hanash SM (1991) Characterization of the gene for a proliferation-related phosphoprotein (oncoprotein 18) expressed in high amounts in acute leukemia. J Biol Chem 266:17747-17753
14. Beretta L, Dobransky T, Sobel A (1993) Multiple phosphorylation of stathmin: identification of 4 sites phopshorylated in intact cells and *in vitro* by cyclic AMP dependent protein kinase and $p34^{cdc2}$. J Biol Chem 268:20076-20084
15. Markland U, Brattsand G, Shingler V, Gullberg M (1993) Serine 25 of oncoprotein 18 is a major cytosolic target for the mitogen-activated protein kinase. J Biol Chem 268:15039-15049
16. Cooper HL, McDuffie E, Braverman R (1989) Human peripheral lymphocyte growth regulation and response to phorbol esters is linked to synthesis and phosphorylation of the cytosolic protein, prosolin. J Immunol 143:956-963
17. Strahler JR, Lamb BJ, Ungar DR, Fox DA, Hanash SM (1992) Cell cycle progression is associated with distinct patterns of phosphorylation of Op18. Biochem. Biophys. Res Commun 185:197-203
18. Melhem RF, Strahler JR, Hailat N, Zhu XX, Hanash SM (1991) Involvement of Op18 in cell proliferation. Biochem Biophys Res Commun 179:1649-1655
19. Koppel J, Boutterin MC, Doye V, Peyro-Saint-Paul H, Sobel A. (1990) Developmental tissue expression and phylogenetic conservation of stathmin, a phosphoprotein associated with cell regulations. J Biol Chem 265:3703-3707
20. Rowlands DC, Williams A, Jones NA, Guest SS, Reynolds GM, Barber PC, Brown G (1994) Lap18 / stathmin expression is a feature of proliferating cells of most, if not all, cell lineages. (submitted)
21. Brown G, Bunce CM, Rowlands DC, Williams GR (1994) All-trans retinoic acid and 1α,25-dihydroxyvitamin D_3 co-operate *in vitro* to promote differentiation of the human promyeloid leukemia cell line HL60 to monocytes. Leukemia (in press)
22. Brattsand G, Roos G, Marklund U, Ueda H, Landberg G, Nanberg E, Sideras P, Gullberg M (1993) Quantitative analysis of the expression and regulation of an activation-regulated phosphoprotein (oncoprotein 18) in normal and neoplastic cells. Leukemia 7:569-579
23. Cooper HL, Fuldner R, McDuffie E, Braverman R (1990) A specific defect of prosolin phosphorylation in T cell leukemic lymphoblasts is associated with impaired down-regulation of DNA synthesis. J Immunol 145:1205-1213
24. Potter M, Mushinski EB, Wax JS, Hartley J, Mock BA. (1994) Identification of two genes on chromosome 4 that determine resistance to plasmacytoma induction in mice. Cancer Res 54:969-975
25. Harrison RF, Reynolds GM, Rowlands DC (1993) Immunohistochemical evidence for the expression of proliferating cell nuclear antigen (PCNA) by non-proliferating hepatocytes adjacent to metastatic tumours and in inflammatory conditions. J Pathol 171:115-122
26. Waseem NH, Lane DP (1990) Monoclonal antibody analysis of the proliferating cell nuclear antigen (PCNA). Structural conservation and the detection of a nucleolar form. J Cell Sci 96:121-129
27. Kohler G, Howe SC, Milstein C (1976) Fusion between immunoglobulin-secreting and non-secreting myeloma cell lines. Eur J Immunol 6:292-295
28. Katz F, Povey S, Parkar M, Schneider C, Sutherland R, Stanley K, Solomon E, Greaves MF (1983) Chromosome assignment of monoclonal antibody defined determinants on human leukemia cells Eur J Immunol 13:1008-1013
29. Jones NA, Lord JM, Brown G (1992) Changes in the phosphorylation status of a 19KDa cytosolic protein are linked to the growth arrest of HL60 cells. Leukemia Res 16:353-361

30. Sahai A, Feuerstein N, Cooper HL, Salomon DS (1986) Effect of epidermal growth factor and 12-O-tetradecanoylphorbol-13-acetate on the phosphorylation of soluble acidic proteins in A431 epidermoid carcinoma cells. Cancer Res 46:4143-4150
31. Jones NA, Rowlands DC, Johnson WEB, MacLennan ICM, Brown G (1994) Persistent growth of BALB/c mouse plasmacytoma and human myeloma cell lines in phorbol myristate acetate is associated with continued expression of Lap18 (stathmin). (submitted)
32. Potter M (1984) Genetics of susceptibility to plasmacytoma development in BALB/c mice. Cancer Surv 3:247-264
33. Potter M, Weiner F (1992) Plasmacytogenesis in mice: model of neoplastic development dependent upon chromosomal translocations. Carcinogenesis 13:1681-1697
34. Clynes R, Stanton LW, Wax J, Smith-Gill S, Potter M, Marcu KB (1988) Synergy of an IgH promoter-enhancer-driven c-*myc*/v-Ha-*ras* retrovirus and pristane in the induction of murine plasmacytomas. Curr Top Microbiol Immunol 141:115-124
35. Janz S, Muller J, Shaughnessy J, Potter M (1993) Detection of recombinations betwen c-*myc* and immunoglobulin switch α in murine plasma cell tumors and preneoplastic lesions by polymerase chain reaction. Proc Natl Acad Sci USA 90:7361-7365
36. Ferrari AC, Seuanez HN, Hanash SM, Atweh GF (1990) A gene that encodes for a leukemia-associated phosphoprotein (p18) maps to chromosome bands 1p35-36.1. Genes Chromosome Cancer 2:125-129
37. Benner R, Hijmans W, Haajiman JJ (1981) The bone marrow: the major source of serum immunoglobulins, but still a neglected site of antibody formation. Clin Exp Immunol 46:1-8
38. MacLennan ICM, Chan EYT (1991) In Epidemiology and Biology of Multiple Myeloma. Eds. GI Obrams and M Potter, Springer-Verlag, Berlin, Heidelberg, p. 129-135.

Plasmacytoma Induction in Mice

Induction of Plasmacytomas in Genetically Susceptible Mice with Silicon Gels

M. Potter, S. Morrison and F. Miller[1]

Laboratory of Genetics, DCDBC, National Cancer Institute, National Institutes of Health, Bethesda, MD, USA; [1]CBER, Food and Drug Administration, Bethesda, MD, USA

Abstract

Silicone gels injected intraperitoneally into strains of mice related to BALB/c develop plasmacytomas in approximately the same numbers and with similar phenotypes as previously obtained with pristane. Silicone gels produce few side effects and are well tolerated for long periods. Silicone gels contain several components that are potentially biologically active: residual vinyl groups and platinum. Microscopic and histological evidence suggests the silicone gel is degraded over a long period of time. Preliminary studies with long chain liquid dimethylpolysiloxanes with viscosities of 1000 cSt and 12,500 cSt have not produced plasmacytomas as yet. The plasmacytomagenic action of the gel appears to be due to the release of liquids from the gel matrix.

Background

In January 1992 David Kessler, Commissioner of the Food and Drug Administration, called on the manufacturers of silicone-filled implants to declare a voluntary moritorium on their further use until more was known about the biological effects of the components of the implants. The silicone get implant consists of a relatively impermeable silicone elastomeric capsule that encases gelatinous contents. Silicone gel liquids that leak out of the capsules either from ruptures or seepage can induce the formation of a silicone granuloma [1,2]. Further, silicone gel-protein mixtures are potent immunological adjuvants [3]. Silicone gels have not been previously reported as plasmacytomagenic agents in BALB/c mice, a strain known to be susceptible to peritoneal plasmacytoma formation. Several different kinds of materials introduced into the peritoneal cavities of BALB/c mice, however, have been previously shown to induce plasmacytomas (PCTs). Merwin and

Algire in 1959 [4] first found that solid plastic objects, such as Millipore Diffusion Chambers, induced PCTs in BALB/c mice. Later, in 1962 Merwin and Redmon [5] found that discs made of Lucite, the Acryloid adhesive or Millipore membranes implanted intraperitoneally induced these tumors, while Lucite rings were ineffective. There was some suggestive evidence that these agents induced mechanical trauma to the peritoneal soft tissues. Paraffin oils were soon found to be as effective as the solid materials in inducing plasmacytomas [6,7,8]. Paraffin oils induced the formation of a copious peritoneal cellular exudate composed of macrophages, monocytes, neutrophils and lymphocytes [9,10]. Paraffin oil droplets engaged and complexed with macrophages, neutrophils and other cells adhered to peritoneal surfaces and caused the formation of oil granulomatous tissue [11,12]. At sites where the complexes attached to mesenteric surfaces a local angiogenesis took place, thus permitting this tissue to become permanently integrated into the submesothelial connective tissues [13]. This tissue designated the oil granuloma was extensive and important for plasmacytoma development. Morphological stages of plasmacytoma development were found in histological sections taken at various intervals after the injection of pristane [11,13,14]. Pristane also induces complicating side effects, such as the formation of a caseous material that adheres to peritoneal surfaces particularly in the upper abdomen and diaphram. This causes adhesions and obstructs the lymphatics that drain the peritoneum. This material is probably the result of emulsification of oil and peritoneal exudate. Silicone gels have provided a new type of physical material that does not have the mechanical properties of the solid plastics nor the exudate-inducing properties of the paraffin oils. We chose to test the most medicinally refined available materials for this purpose, mainly the gel contents of commercially obtained Mammary Implants. Several implants with different lot numbers were purchased for this purpose [15].

Plasmacytoma Formation with Silicone Gels

The silicone gel material proved to be difficult to manipulate. Although it was crystal clear it tenaciously adhered to almost any solid surface, the gel also had elastic properties making it difficult to cut it into small chunks. However, it was possible to stuff the gel into the barrel of a 5 ml plastic syringe which could then be pushed into the barrel of a 1 ml syringe. The gel in the 1 ml syringe could now be injected i.p. through 18 guage needles [15]. The injection step required considerable pressure to force the material through the needle. Thus to control the amount of the injection, the mice were first anesthesied with i.p. Avertin. The gel was detached from the needle by pinching the end of the needle as it was withdrawn from the peritoneal space.

All of the regimens produced plasmacytomas in BALB/cAn mice [15], but the 3 dose schedule was the most effective. Recent experiences in our laboratory using the potent paraffin oil plasmacytomagen, pristane (2,6,10,14 tetramethylpentadecane), has resulted in a declining incidence of plasmacytomas in BALB/cAnPt mice from values of greater than 60% prior to 1987 to 30-40% in 1991-1994 [16,14]. At first we attributed this decrease in incidence to the physical isolation of our conventional mouse colony and the use of more rigorous cage washing methods. This notion was strengthened by the finding that Specific Pathogen Free BALB/c AnPt mice were virtually refractory to plasmacytoma induction [17]. However, while we were obtaining relatively low incidences with BALB/cAnPt, we found that the congenic BALB/c derived strain BALB/cAnPt.DBA/2-Idh1-Pep3 (C.D2-Idh1-Pep3) consistently and concurrently developed high incidences of 60-80% plasmacytomas following 3 i.p. injections of pristane [14,15]. C.D2-Idh-Pep3 carries approximately 30 centimorgans of DBA/2 chromatin between the centromeric Idh-1 and telomeric Pep3 markers [18]. Certain other BALB/c.DBA/2 congenics that carry DBA/2 genetic material from other chromosomes also gave high incidences of plasmacytomas. A working hypothesis is that sometime during the mid to late 1980's a mutation occurred in a BALB/c progenitor that modified a susceptibility gene. Mice homozygous for this mutation were partially resistant to plasmacytoma induction. Mutations with similar effects have been described in the BALB/cJax subline [19].

For these reasons we have used C.D2-Idh1-Pep3 congenic mice to test various silicone-containing materials which at this time includes the silicone gels from 2 different implant lots [15] and 2 different liquid dimethylpolysiloxane (DMPS) polymers of 1000 and 12,500 centistoke viscosities. Each mouse recieved 0.1 mL of the respective gel or liquid on days 0, 60 and 120. The multiple dose regimen proved the most effective in experiments with pristane [16]. The result of these experiments is that the silicone gels were plasmacytomagenic while the DMPS liquids were not (Fig. 1). Also, the potency of the gels from the 2 different lots differed. The gel from the second implant was only half as effective as the first. Both types of material induced the formation of peritoneal granulomatous tissue on peritoneal surfaces. DMPS induced the formation of a silicone granuloma. One explanation for the failure of DMPS to induce PCTs in these experiments may be related to the dose given. Further experiments are in progress to test this hypothesis. Alternatively, the DMPS may be handled differently by macrophages.

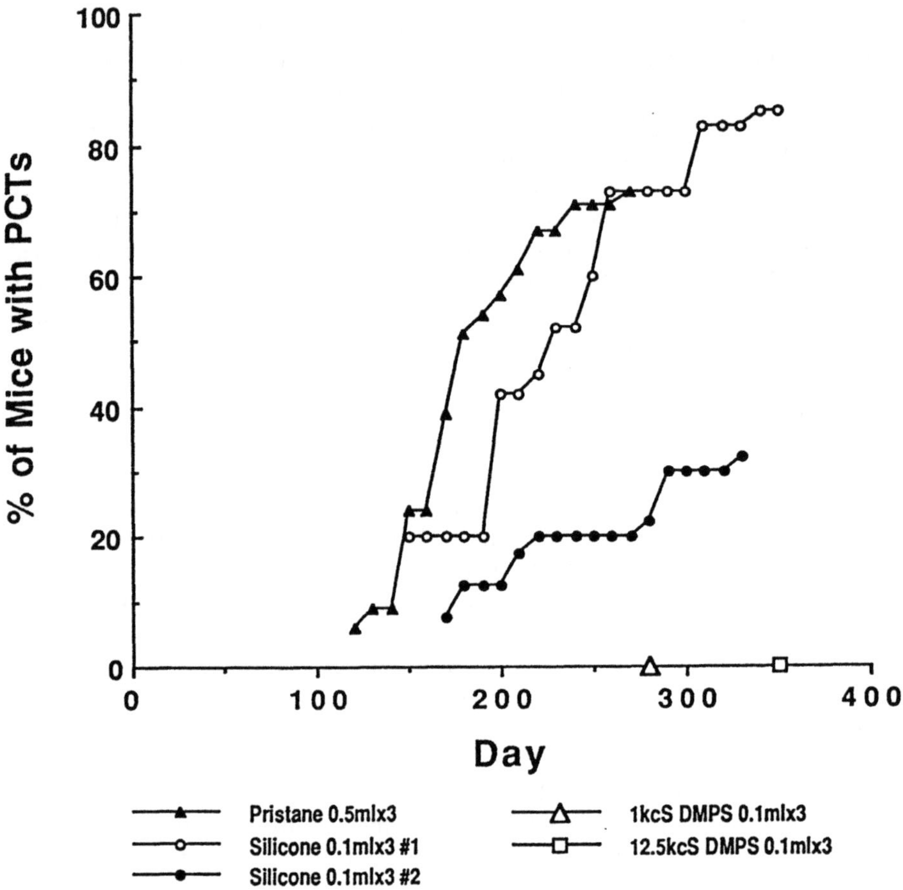

Fig. 1. Plasmacytoma induction curves in C.D2-Idh-Pep3 mice given various silicones and pristane. The pristane curve is the mean of 3 separate experiments.

When silicone gel is injected i.p. in single or multiple doses, it congeals into a single gelatious blob which becomes located in the upper ventral part of the abdominal cavity. For several hundred days this blob remains unattached by adhesions and can be easily removed as one cohesive mass. In many of the mice autopsied 400 days after injection of the gel, the blobs appear to have broken down and only stringy material could be found in the peritoneal cavity. This stringy material was distributed between the intestines and throughout the abdominal cavity.

Histological studies of mice injected with the silicone gels have shown that the mice develop a chronic granulomatous tissue on peritoneal surfaces [15]. Most of this appears as very fine deposits on the intestinal mesenteries and the omentum. The granulomatous tissue consists of small droplets of material (in the form of clear vacuoles that contain refractile material) surrounded by various kinds of reactive cells. In the early phases of silicone granuloma (SilG) formation the vacuoles are lined by reactive macrophages, neutrophils, and fibroblasts. Several layers of cells, some resembling fibroblasts, line the vacuoles, and vacuoles for the most part are separated by a rather extensive amount of interstitial inflammatory tissue containing lymphocytes, macrophages, fibroblasts, PMNLs and plasma cells. After 300-400 days the silicone granuloma changes in character. Many of the vacuoles become encased by a dense eosinophilic collageous material. Very late the large vacuoles appear to have broken down into many smaller ones that now show a 'foamy' appearence. The material inside large and small vacuoles is highly refractile. Very few, if any, interstitial inflammatory macrophages or neutrophils are present. The SilG at this stage appears to be 'burned out'.

Potential Reactive Materials in Silicone Gels

Metallic silicon is used as the starting compound in the synthesis of silicones and is reacted with various chloride compounds (hydrochloric acid, methyl chloride, vinyl chloride or mixtures) to generate various chlorosilanes [20] (Fig. 2). Following hydrolysis with H_2O the chlorine atoms are replaced by oxygen, and these polymerize to produce polysiloxanes compounds. Many of these are cyclic. When treated with alkali plus the addition of endblocking agents linear siloxane polymers are formed. Two copolymers methyl-hydrogenpolysiloxane and vinylmethylpolysiloxane are used to form crosslinks between the chains. The crosslinking reaction is catalysed by platinic chloride (Fig. 2). In a silicone gel not all of the hydrogens are linked to available vinyl groups. Further linear and cyclic compounds are trapped in the gel matrix but can diffuse out.

The potential biologically reactive components in silicone gels are: 1) platinum, which is very difficult to remove after the polymerization process; 2) residual vinyl groups; and 3) possibly low molecular weight siloxane compounds. Platinum is known to cause contact dermatitis, allergy and also to act as non-specific mitogen for lymphocytes [21]. There is very little or no information on the biological activity of silicon-vinyl compounds and these remain to be investigated. Other low molecular weight vinyl compounds such as vinyl chloride [22] or vinyl acetate [23] are known to be mutagenic or carcinogenic. An important question is whether the residual vinyl groups in the gel can be converted into low molecular weight compounds that can enter cells and induce genetic change. There is some NMR evidence that silicone gels can be physiologically degraded into low molecular weight derivatives [24].

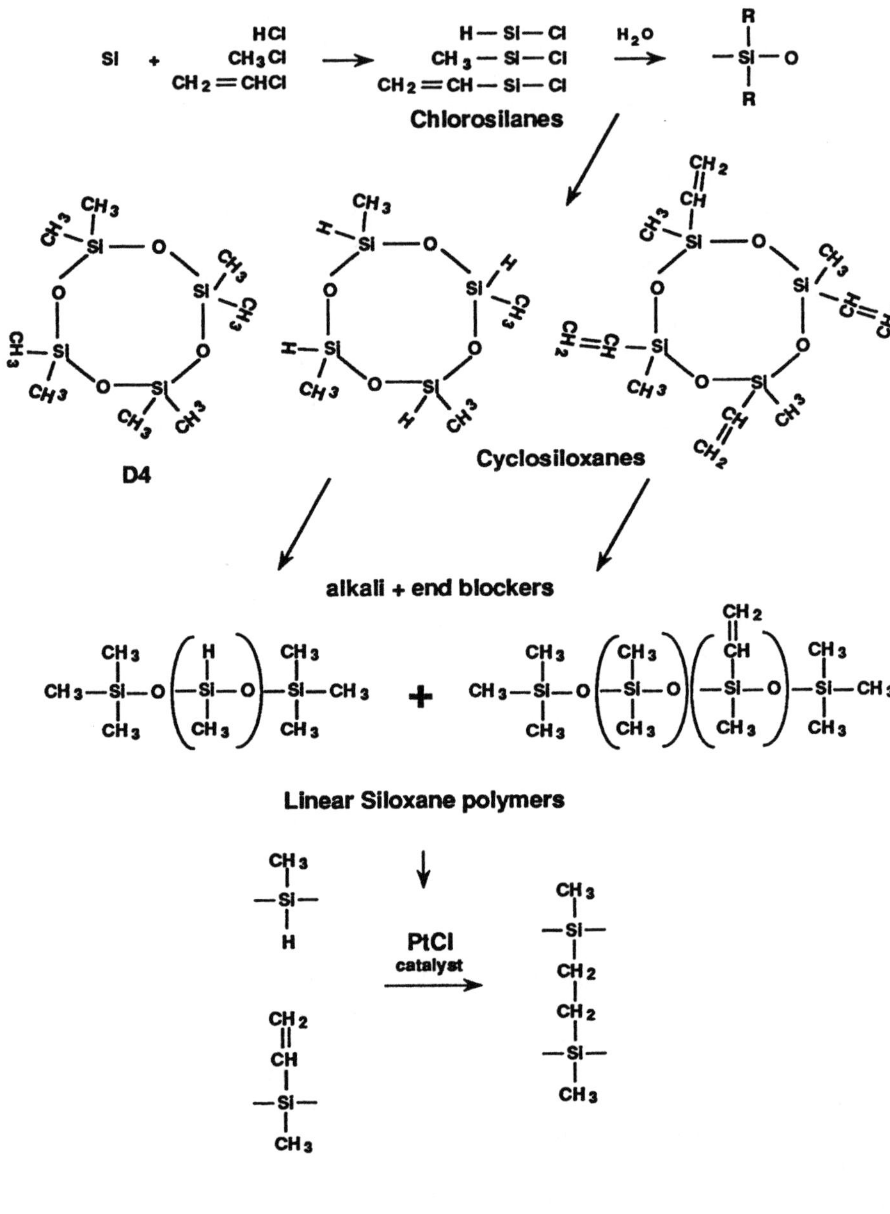

Fig. 2. Scheme showing some of the steps in silicone gel synthesis

Differences in Pristane and Silicone Gels

The injected silicone gel appears to remain intact for extensive periods, but it is clear that liquid materials seep out into the peritoneal space where they induce the formation of the silicone granuloma. While it is tempting to conclude that the liquid material derived from the silicone gel is more efficient in inducing plasmacytomas, this conclusion must await further comparative studies in which comparable volumetric doses of silicones and pristane have been administered. The chemical identity of the liquid material also presents an unresolved problem. It has been assumed that the non-polymerized linear liquid polymers are trapped in the matrix during gelation, but an alternative possibility is that it could be a degradation product of the gel. Thus far, only large (0.5 mL) doses of pristane have been compared to (0.1 mL) silicone gels and DMPS. Further compounds with similar chain length to pristane have not yet been introduced in comparable doses.

Chronic Inflammation in Peritoneal Tissues is Associated with Plasmacytopoiesis

Plasmacytopoiesis has been described in association with the intraperitoneal injection of alum-protein and alum-vaccine precipitates in an excellent morphological study by Athanassaides and Spiers [25,26]. The alum mixture becomes infiltrated with macrophages and after seveal weeks lymphocytes invade these masses and plasmacytopoiesis has been observed. Weinberg et al. have described the plasmacytogenesis in mesenteric tissues during chronic Schistosoma mansoni infestations [26]. The worms reside in the mesenteric vessels but extrude products into the peritoneal space. Eggs and other products evoke the formation of vascular polyp like structures which morphologically resemble the polyps first described in mice injected with pristane by Anderson [13]. These inflammatory responses develop around highly vascularized tissues inside these polyps. These authors noted the striking numbers of lymphocytes and proposed that these were mesenteric milky spots. Plasma cell proliferations have been found in the pristane granulomas [11,12,14]. Thus far neither the Schistosomal infections nor the alum injections have been shown to be plasmacytomagenic; however, these must be further investigated in BALB/c mice. Terminal plasma cell formation takes place in chronic peritoneal granulomatous tissues. Virgin or antigen experienced circulating B lymphocytes enter the peritoneal connective tissues through the mesenteri and omental blood vessels and then under the influence of local factors become plasma cells. It is likely that during these steps the first morphological manifestations of abnormal proliferation are evident.

The cells and tissues that react to the liquid material derived from the silicone gel provides a suitable microenvironment for this abnormal development. Our current working hypothesis on the pathogenesis of PCT development in BALB/c and susceptible mice is that B-lymphocytes which have developed c-myc-Ig switch region or Pvt-IgK illegitimate recombinations survive and proliferate in microenvironments rich in several factors, i.e., provided by oil and silicone granulomatous tissue. These persistant mutant cells develop additional new genetic or adaptative changes (as yet not defined) that establish uncontrolled growth.

References

1. Travis WD, Balogh K, Abraham JL (1985) Silicone granulomas: report of three cases and review of the literature. Hum Pathol 16:19-27
2. Dodd LG, Sneige N, Reece GP, Fornage B (1993) Fine-needle aspirationcytology of silicone granulomas in the augmented breast. Diagn Cytopathol 9:498-502
3. Naim JO, Lanzafame RJ, van Oss CJ (1993) The adjuvant effect of silicone-gel on antibody formation in rats. Immunol Invest 22(2):151-161
4. Merwin RM, Algire GH (1959) Induction of plasma cell neoplasms and fibrosarcomas in BALB/c mice carrying diffusion chambers. Proc Soc Exp Biol Med 101:437-439
5. Merwin RM, Redmon LW (1963) Induction of plasma cell tumors and sarcomas in mice by diffusion chambers placed in the peritoneal cavity. J Natl Cancer Inst 31:990-1007
6. Potter M, Robertson CL (1960) Development of plasma-cell neoplasms in BALB/c mice after intraperitoneal injection of paraffin-oil adjuvant, heat-killed staphylococcus mixtures. J Natl Cancer Inst 25:847-861
7. Potter M, Boyce C (1962) Induction of plasma cell neoplasms in strain BALB/c mice with mineral oil and mineral oil adjuvants. Nature 193:1086
8. Anderson PN, Potter M (1969) Induction of plasma cell tumours in BALB-c mice with 2,6,10,14-tetramethylpentadecane (pristane). Nature 222:994-995
9. Cancro M, Potter M (1976) The requirement of an adherent cell substratum for the growth of developing plasmacytoma cells in vivo. J Exp Med 144:1554-1567
10. Shacter E, Beecham EJ, Covey JM, Kohn KW, Potter M (1988) Activated neutrophils induce prolonged DNA damage in neighboring cells. Carcinogenesis 9:2297-2304
11. Potter M, MacCardle RC (1964) Histology of developing plasma cell neoplasia induced by mineral oil in BALB/c mice. J Natl Cancer Inst 33:497-515
12. Potter M, Wax JS, Anderson AO, Nordan RP (1985) Inhibition of plasmacytoma development in BALB/c mice by indomethacin. J Exp Med 161:996-1012
13. Anderson AO, Wax JS, Potter M (1985) Differences in the peritoneal response to pristane in BALB/cAnPt and BALB/cJ mice. Curr Top Microbiol Immunol 122:242-253
14. Potter M, Mushinski EB, Wax JS, Hartley J, Mock BA (1994) Identification of two genes on chromosome 4 that determine resistance to plasmacytoma induction in mice. Cancer Res 54:969-975
15. Potter M, Morrison S, Wiener F, Miller FW (1994) Induction of plasmacytomas with silicone gel in genetically susceptible strains of mice. J Natl Cancer Inst
16. Potter M, Wax JS (1983) Peritoneal plasmacytomagenesis in mice: comparison of different pristane dose regimens. J Natl Cancer Inst 71:391-395
17. Byrd LG, McDonald AH, Gold LG, Potter M (1991) Specific pathogen-free BALB/cAn mice are refractory to plasmacytoma induction by pristane. J Immunol 147:3632-3637

18. Mock BA, Holiday DL, Cerretti DP, Darnell SC, O'Brien AD, Potter M (1994) Construction of a series of congenic mice with recombinant chromosome 1 regions surrounding the genetic loci for resistance to intracellular parasites (Ity, Lsh, and Bcg), DNA repair responses (Rep-1), and the cytoskeletal protein villin (Vil). Infect Immun 62:325-328
19. Potter M, Wax JS (1981) Genetics of susceptibility to pristane-induced plasmacytomas in BALB/cAn: reduced susceptibility in BALB/cJ with a brief description of pristane-induced arthritis. J Immunol 127:1591-1595
20. LeVier RR, Chandler ML, Wendel SR (1978) The pharmacology of silanes and siloxanes. In: Bendz G, Lindqvist I (eds) Biochemistry of silicone and related problems. Plenum Publishing Corp, New York, pp 473-514
21. Schuppe H-C, Haas-Raida D, Kulig J, Bomer U, Gleichmann E, Kind P (1992) T-cell-dependent popliteal lymph node reactions to platinum compounds in mice. Int Arch Allergy Immunol 97:308-314
22. Green T (1990) Chloroethylenes: a mechanistic approach to human risk evaluation. Annu Rev Pharmacol Toxicol 30:73-89
23. Norppa H, Tursi F, Pfaffli P, Maki-Paakkanen J, Jarventaus H (1985) Chromosome damage induced by vinyl acetate through in vitro formation of acetaldehyde in human lymphocytes and Chinese hamster ovary cells. Cancer Res 45:4816-4821
24. Garrido L, Pfleiderer B, Papisov M, Ackerman JL (1993) In vivo degradation of silicones. Magn Reson Med 29:839-843
25. Athanassiades TJ, Speirs RS (1968) Formation of antigen-induced granulomas containing plasma cells: a light and electron microscopic study. J Reticuloendothelial Soc 5:485-497
26. Athanassiades TJ, Speirs RS (1972) Granuloma induction in the peritoneal cavity. A model for the study of inflammation and plasmacytopoiesis in nonlymphatic organs. J Reticuloendothelial Soc 11:60-76

Strain-Related Cellular Mechanisms as a Determinant for Susceptibility and Resistance to PC Induction

Hiroyuki Sugiyama[1], Santiago Silva, Yisong Wang and George Klein

Department of Tumor Biology, Karolinska Institutet, S-171 77 Stockholm, Sweden
[1]Present address: Division of Internal Medicine, Saiseikai Noe Hospital, Imafuku-Higashi 2-2-33, Joto-ku, Osaka 536, Japan

Introduction

Plasmacytomas (PCs) can be induced in the susceptible BALB/c strain by intraperitoneal (i.p.) injection of pristane oil. They regularly carry an immunoglobulin (Ig)/*myc* translocation. The post-pristane latency period is usually more than 120 days. The minimal latency period can be shortened to one third by an injection of Abelson murine leukemia virus (A-MuLV), a potent pre-B lymphoma-inducing agent. The notion of co-operative interaction between deregulated *myc* and v-*abl* has been supported by our previous result that A-MuLV can rapidly induce translocation-free PCs in Eμ-c- or N-*myc* transgenic mice at nearly 100% with or without pristane (Sugiyama et al., 1990; Wang et al., 1992). We recently found that rapid PC induction by A-MuLV occurs only in conjunction with deregulated *myc*, and by comparison of v-*abl* and a c-*abl* mutant, ΔXB, that the activity of the virus to induce pre-B lymphomas is related to its high titer presumably achieved by *gag* sequences (Sugiyama et al., in press). We further report here that the lymphoma-inducing activity of A-MuLV in N-*myc* mice is related to the upregulation of endogenous c-*myc* in contrast to the downregulation in rapidly induced PCs. We also show that strain-related susceptibility and resistance to PC induction is at least partly determined by cellular mechanisms other than the deregulation of *myc*.

Results and Discussion

As in our previous work (Wang et al., 1992), we have used a BALB/c-backcrossed (N3) line of N-*myc* transgenic mice that have very low spontaneous lymphoma incidence. *In vitro* immortalized cells were obtained from bone marrow or spleen cells of these transgenic mice and their normal littermates infected with A-MuLV (v-*abl*) or retroviruses carrying c-*abl* derivatives, ΔXB and *bcr-abl* (Sugiyama et al., in press). The structure of the *abl* proteins are illustrated in Figure 1.

As reported elsewhere (Sugiyama et al., in press), ΔXB accelerates the development of PCs in N-*myc* transgenic mice similarly to A-MuLV, although it has a deletion of only the SH3 domain of c-*abl*. ΔXB has also the lymphoid cell-immortalizing activity *in vitro* on bone marrow cells from the N-*myc* transgene carriers and non-carriers. In contrast, *bcr-abl* lacks the PC-accelerating activity, although it has lymphoid cell-immortalizing activity irrespective of the presence of the N-*myc* transgene. Immortalized cells have a pre-B phenotype and may later become tumorigenic in BALB/c mice. We have previously shown that PCs induced in the N-*myc* transgenic mice by A-MuLV express no detectable endogenous c-*myc* (Wang et al., 1992). To see whether v-*abl* or ΔXB co-operates with the N-*myc* transgene in lymphoid transformants as in PCs, we studied the expression of the endogenous c-*myc* and the transgene in these cells. Figure 2 shows that

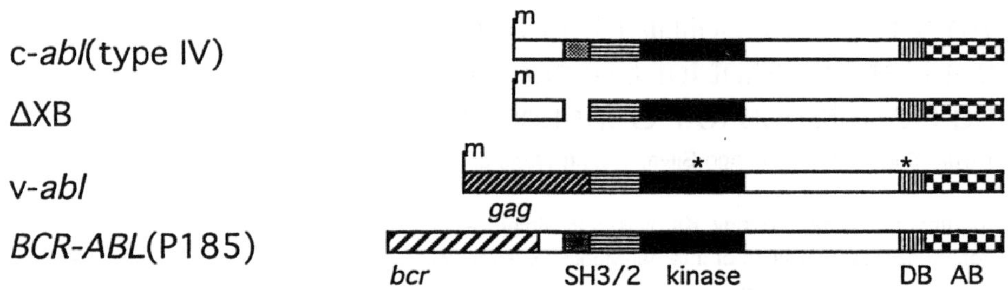

Figure 1. Protein structures of abl-derivatives. *point mutations. DB=DNA-binding domain. AB=actin-binding domain. m=myristylation site.

lymphoid transformants, both *in vitro* immortalized cells and transplanted pre-B lymphomas, uniformly express endogenous c-*myc*, albeit at low levels, irrespectively of the presence or absence of the high levels of the N-*myc* transgene expression. This confirms the difference between lymphoid transformation and PC-acceleration; the former being a direct function of activated *abl* that upregulates endogenous c-*myc*, and the latter being a co-operative interaction of activated *abl* and deregulated *myc*.

We then investigated the strain-related difference in susceptibility and resistance to PC-induction. N-*myc* transgenic BALB/c (N3) mice were crossed with PC-resistant DBA/2 mice. The offspring CDF1 mice and further backcrossed progeny, CDBC1 mice, were monitorred for PC development after pristane treatment. As shown in Figure 3(a), only BALB/c (N3), but not CDF1 nor CDBC1 transgenic mice were susceptible for PC induction, suggesting that the deregulated *myc* alone does not by itself lead to plasmacytomagenesis. This finding argues against the speculation that susceptibility may be determined by the proneness of the *myc*/Ig translocation.

Pristane plus A-MuLV accelerated PC development in 21 of 25 BALB/c (N3) transgenic mice, as shown in Figure 3(b). Two PCs that developed early were associated with a lymphoma component. Only one pre-B lymphoma developed in 6 non-transgenic BALB/c (N3) littermates after the same treatment, indicating that the presence of the N-*myc* transgene was a crucial factor for PC development. All 12 CDF1 transgenic mice developed tumors. Seven were pre-B lymphomas, and one was a mixture of lymphoma and PC. Four PCs developed later than these lymphomas. Only one pre-B lymphoma developed in 18 non-transgenic CDF1 littermates. All 9 CDBC1 transgenic mice developed exclusively pre-B lymphomas, while 5 of 9 normal littermates developed the same type of tumors, after longer latency. These findings suggest that the transgenic mice contained more target cells for A-MuLV transformation than their normal littermates, even though the N-*myc* transgene does not contribute the process of pre-B lymphoid transformation, as we discussed above, and also that susceptibility to lymphoid transformation as well as to PC induction is determined by strain-related cellular factor.

One can argue that PC development may have masked by earlier onset pre-B lymphomas in A-MuLV-infected mice. As we reported previously, ΔXB induces exclusively PCs in N-*myc* transgenic mice (Sugiyama et al., in press). Therefore, we infected pristane-treated N-*myc* mice with ΔXB virus instead of A-MuLV. As shown in Figure 3(c), ΔXB virus induced almost exclusively PCs irrespectively of the genetic background. The latency period, however, is shortest in BALB/c (N3)

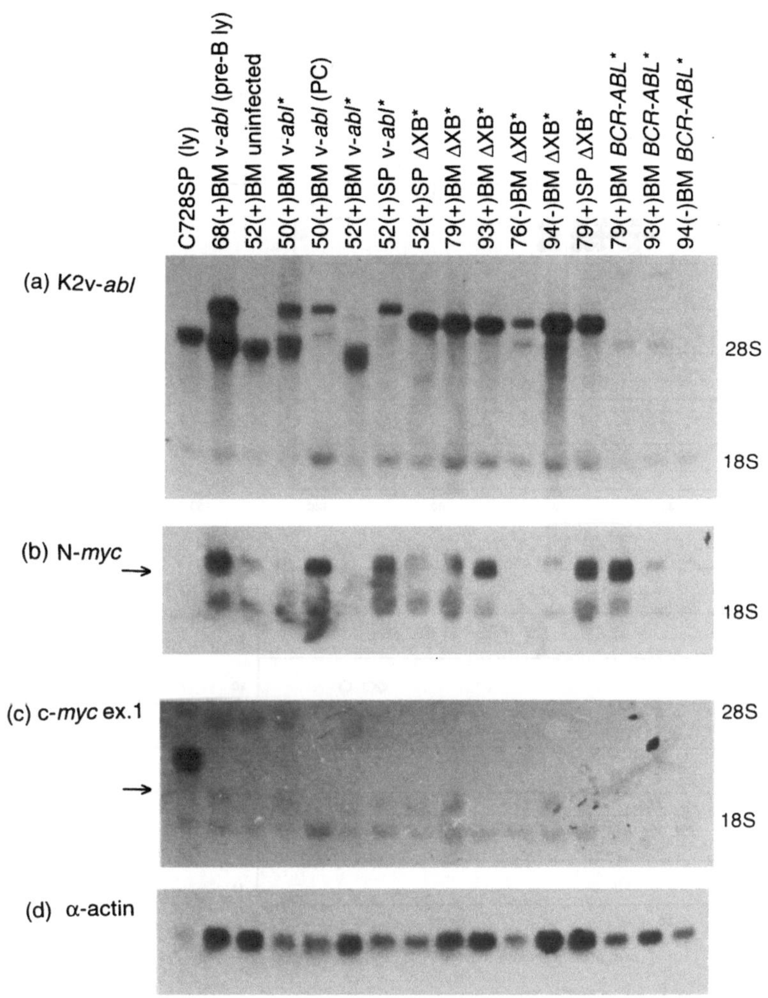

Figure 2. Northern blot analysis of total RNA from *in vitro* immortalized cells and tumors derived from virus-infected bone marrow or spleen cells of N-myc transgenic mice (+) and their normal littermates (-). *=in vitro immortalized cells. Transplanted primary tumors, a lymphoma and a PC were included. C728SP (ly) is a lymphoma developed in a c-*myc* transgenic mouse after A-MuLV infection.

(mean=49±7 days, n=7), longer in CDF1 (mean=68±11 days, n=13, t=4.128, P=0.001), and even longer in CDBC1 mice, suggesting that resistance of CDF1 and CDBC1 is a relative one, and cellular mechanisms other than the presence of the N-*myc* transgene may have a decisive role in the difference of susceptibility vs. resistance. F1 transgenic hybrids (CBF1) of BALB/c and C57BL/6, another PC-resistant parental strain, showed post-virus latency periods similar to CDF1.

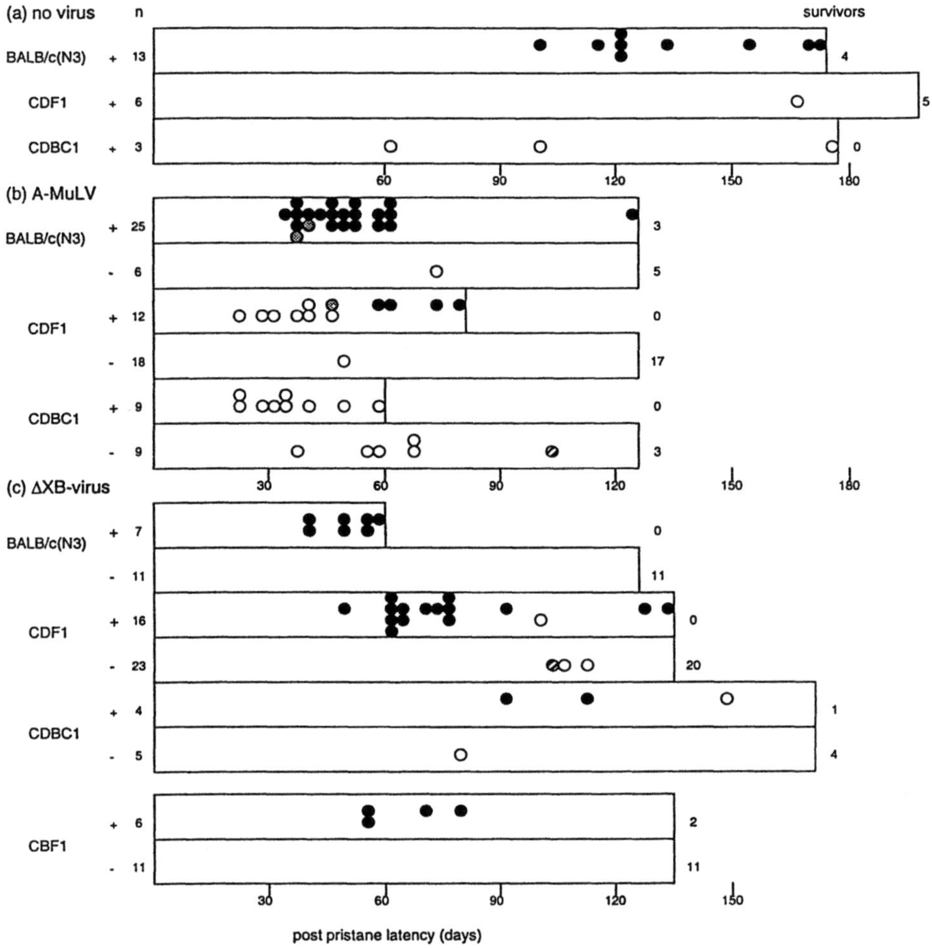

Figure 3. Tumor onset in N-myc transgenic BALB/c (N3) and hybrid mice. CDF1=(BALB/cxDBA/2)F1, CDBC1=(CDF1+DBA/2)F1, and CBF1=(BALB/cxC57BL/6)F1. Transgene-positive (+) and negative (-) littermates were treated with 0.5 ml pristane alone (a) or followed 7-15 days by A-MuLV (b) or ΔXB virus (c) infection. n=number of mice treated. ● =PCs; ○ =lymphomas; ● =mixtures of a PC and a lymphoma; ⊘ =myeloid tumor.

Conclusion

Certain contrasts between PC-acceleration and lymphoid cell transformation by the *abl*-derivatives have become clear. The PC-accelerating activity depends on co-operative interaction with deregulated *myc*, while the lymphoid cell-transforming activity is a direct effect of the *abl* derivatives. These two activities are differencially achieved apparently by the help of cellular mechanisms, which are different from strain to strain. Thus, the BALB/c background may support the co-operative interaction of *myc* and *abl* to induce PCs.

Acknowledgements

We thank Margareta Hagelin, Maj-Lis Solberg and Sofie Nilsson for technical assistance. This investigation was supported by PHS grant 5 RO1 CA 14054-15 from the National Cancer Institute, NIH, USA, and by a grant from the Swedish Cancer Society. H.S. was supported by fellowships from UICC (International Union Against Cancer) and the Swedish Institute.

References

1. Sugiyama H, Silva S, Wang Y, Weber G, Babonits M, Rosén A, Wiener F and Klein G (1990) Abelson murine leukemia virus transforms preneoplastic Eμ-*myc* transgene-carrying cells of the B-lymphocyte lineage into plasmablastic tumors. Int J Cancer 46: 845-852

2. Sugiyama H, Wang Y, Jackson P, Sawyers CL and Klein G, Molecular requirements for rapid plasmacytoma and pre-B lymphoma induction by Abelson murine leukemia virus in *myc*-transgenic mice. Int J Cancer (in press)

3. Wang Y, Sugiyama H, Axelson H, Panda CK, Babonits M, Ma A, Steinberg JM, Alt FW and Klein G (1992) Functional homology between N-*myc* and c-*myc* in murine plasmacytomagenesis: plasmacytoma development in N-*myc* transgenic mice. Oncogene 7:1241-1247.

Human and Mouse B-Cell Lymphomas

Alterations of the BCL-6 Gene in Diffuse Large-Cell Lymphoma

B.H. Ye[1], F. Lo Coco[1], C.-C. Chang[1], J. Zhang[1], A. Migliazza[1], K. Cechova[1], D.M. Knowles[2], K. Offit[3], R.S.K. Chaganti[3] and R. Dalla-Favera[1],

[1]Department of Pathology, College of Physicians and Surgeons, Columbia University, New York, NY 10032; [2]Department of Pathology, Cornell University, New York, NY 10021; [3]Cell Biology and Genetics Program and the Department of Pathology, Memorial Sloan-Kettering Cancer Center, New York, NY 10021.

Introduction

The molecular analysis of specific chromosomal translocations has improved our understanding of the pathogenesis of various non-Hodgkin lymphoma (NHL) subtypes including follicular lymphoma (FL), Burkitt's lymphoma (BL) and mantle-cell lymphoma, which are characterized by chromosomal translocations causing the deregulated expression of the BCL-2, C-MYC, and BCL-1, respectively (for review see Korsmeyer 1992, Dalla-Favera 1993, and Gaidano and Dalla-Favera 1992). However, relatively little is known about the molecular pathogenesis of diffuse large cell lymphoma (DLCL), the most frequent and most lethal human lymphoma (Magrath 1990). DLCL accounts for ~40% of initial NHL diagnoses and is often the final stage of progression of FL (Magrath 1990). Several molecular alterations have been detected at variable frequency in these tumors, but none has been specifically or consistently associated with the disease (Ladanyi et al. 1991, Offit et al. 1989a).

Recently substantial progress has been made in this area by cloning the chromosomal junctions of chromosomal translocations involving band 3q27 (Ye et al. 1993a, Kerckaert et al. 1993 and Miki et al. 1994), which are common in DLCL. A candidate proto-oncogene, BCL-6, has been identified at 3q27 which is frequently and specifically altered in DLCL (Ye et al. 1993b). This report reviews recent findings on the structure and expression of the BCL-6 gene, on the consequence of BCL-6 rearrangements in DLCL, and on the potential significance of BCL-6 lesions as markers in NHL clinical diagnosis.

Chromosomal translocations affecting 3q27 in DLCL: cloning of chromosomal breakpoints

Cytogenetic analysis of large panels of DLCL cases have revealed relative frequent (10-12%) chromosomal alterations affecting band 3q27 in this NHL subtype (Offit et al. 1989b, Bastard et al. 1992). These alterations involve reciprocal translocations between the 3q27 region and various

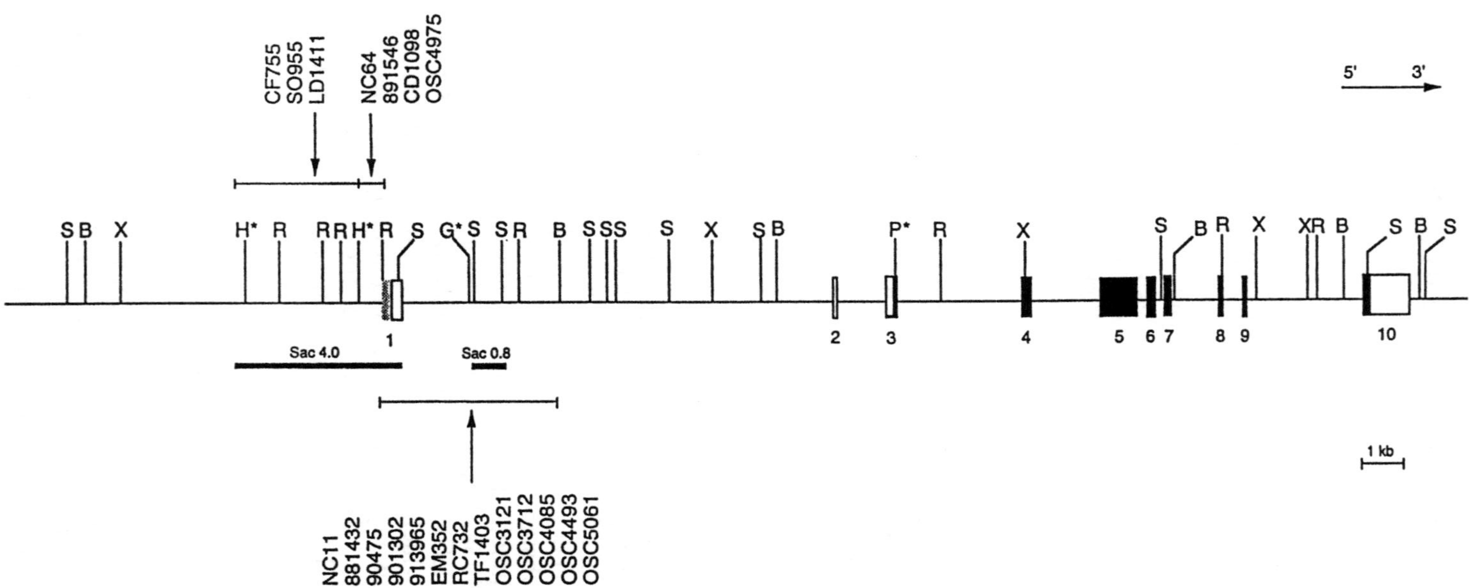

Fig. 1. Exon-intron organization of the BCL-6 gene and mapping of breakpoints detected in DLCL. Coding and non-coding exons are represented by filled and empty boxes, respectively. The putative first exon has been entirely sequenced in its portion overlapping the cloned cDNA sequences. The initiation of transcription has not been precisely mapped (shaded box on 5' side of first exon). Patient codes (such as NC11 and 891546) are grouped according to the rearranged patterns displayed by tumor samples. Restriction sites marked by asterisks have been only partially mapped within the BCL-6 locus. Restriction enzyme symbols are: S: Sac I; B: BamH I; X: Xba I; H: Hind III; R: EcoR I; G: Bgl II; P: Pst I.

alternative chromosomal partners including, but not limited to, those carrying the Ig heavy-(14q32) or light- (2p12, 22q11) chain loci. These observations suggested that 3q27 may be the site of a proto-oncogene whose structural lesion may be critical for DLCL pathogenesis.

As a first step toward the molecular characterization of the putative 3q27 proto-oncogene, we have cloned the chromosomal breakpoints of several cases of (3;14)(q27;q32) translocations in which the involvement of the immunoglobulin locus on 14q32 provided a probe for the cloning of the translocation junctions. The same genomic region was cloned from all the analyzed (3;14)(q27;q32) cases (Ye et al. 1993b, Kerckaert et al. 1993, and Miki et al. 1994). Furthermore, molecular probes from this region identified rearrangements in this locus in 13 of 17 cases carrying 3q27 alterations, irrespective of the partner chromosomes involved in the translocations (Ye et al. 1993b). These results indicated that the chromosomal breakpoints clustered within a restricted genomic region in various DLCL cases (Fig. 1) further suggesting that they may affect a genetic locus important for lymphomagenesis.

The BCL-6 gene and its protein product

A search for transcribed sequences led to the identification of a gene, called BCL-6, which was expressed in B cells and whose sequences span the translocation breakpoints (Ye et al. 1993b). The corresponding cDNA was cloned, sequenced and demonstrated to code for a novel protein containing six C_2H_2 zinc-finger motifs and a conserved stretch of seven amino acids (the H/C link) connecting the successive zinc-finger repeats as typical of various members of the *Krüppel*-like subfamily of zinc-finger proteins (Rosenberg et al. 1986). The NH_2-terminal region of BCL-6 has homologies with some zinc-finger transcription factors including the human ZFPJS protein, a putative human transcription factor that regulates the major histocompatibility complex II promoter (Sugawar et al. unpublished sequence in GeneBank), the *Tramtrack* (*ttk*) (Harrison and Travers 1990, Read and Manley 1992) and Broad-complex (*Br-c*) proteins in *Drosophila* that regulate developmental transcription (DiBello et al. 1991), the human KUP protein (Chardin et al. 1991), and the human PLZF protein, which is occasionally involved in chromosomal translocations in human promyelocytic leukemia (Chen et al. 1993). These structural homologies suggest that BCL-6 may function as a DNA-binding transcription factor that regulates organ development and tissue differentiation. Zinc-finger encoding genes are plausible candidate oncogenes as they have been shown to participate in the control of cell proliferation, differentiation, and organ pattern formation (for review see El-Baradi and Pieler 1991). In addition, alterations of zinc-finger genes have been detected in a variety of tumor types including *PLZF* and *PML* in acute promyelocytic leukemia (de The'H et al. 1991, Kakizuka et al. 1991, and Pandolfi et al. 1991), *EVI-1* in mouse and human myeloid leukemia (Morishita et al. 1988, Fichelson et al. 1992), *TTG-1* in T-cell ALL (McGuire et al. 1989), *HTRX* in acute mixed-

Table 1. Expression of the BCL-6 gene in cell lines and normal tissues

Cell line or tissue	Phenotype	BCL-6 RNA
B lineage		
697	pre-B (acute lymphoblastic leukemia)	±
Daudi	B cell (EBV+ Burkitt Lymphoma)	++
Ramos	B cell (EBV- Burkitt Lymphoma)	++
Bjab	B cell (B cell Lymphoma)	++
LCL (6 tested)	B cell (EBV-immortalized)	-
U266	plasmacell (multiple myeloma)	-
T lineage		
CEM	pre-T (acute lymphoblastic leukemia)	-
Hut-102	T cell (adult T cell leukemia)	-
Other		
K562	erythroid (chronic myelogenous leukemia)	-
U937	monocyte (acute monocytic leukemia)	-
HL-60	myeloid (acute promyelocytic leukemia)	-

lineage leukemia (Djabali et al. 1992, Tkachuk et al. 1992, and Gu et al. 1992), and *WT-1* in Wilm's tumor (Haber et al. 1990).

The BCL-6 gene is expressed in mature B cells

The cDNA clone was used as a probe to investigate BCL-6 RNA expression in a variety of human cell lines by Northern blot analysis. A single 3.8 kb RNA species was detectable in all cell lines derived from mature B-cells except for EBV-immortalized lymphoblastoid cell lines (LCL). No BCL-6 RNA was found in pro-B-cells, plasma cells, T cells, other hematopoietic cell lineages (see Table 1) or other normal tissues, except for skeletal muscle in which low level expression was seen (Ye et al., unpublished data). The selective expression of BCL-6 in differentiated B-cells suggests that BCL-6 may play a role in the control of normal B-cell differentiation and lymphoid organ development.

Rearrangements of the BCL-6 gene in DLCL

In order to define the incidence of BCL-6 rearrangements in various lymphoproliferative diseases, we cloned the entire BCL-6 genomic locus. Fig. 1 shows that this gene contains 10 exons spanning approximately 26 kb of genomic DNA. Sequence analysis indicated that the first exon is non-coding and that the transcription initiation codon is located within the third exon (Ye et al. 1993b). Using various probes from this locus, we analyzed a panel of cases not previously selected on the basis of 3q27 breakpoints but representative of the major subtypes of NHL as well as of other lymphoproliferative diseases including NHL (125 cases), ALL (45), CLL (51) and MM (23) (Lo Coco et al. 1994). The NHL series was representative of low- (41), intermediate- (45) and high-grade (24) subtypes according to the Working Formulation. Fifteen cases of cutaneous T-cell NHL were also included.

The results of this analysis are summarized in Table 2. All cases of ALL, CLL and MM showed a normal BCL-6 gene. Eighteen of the 125 NHL cases displayed BCL-6 rearrangements. Among distinct NHL histologic subtypes, rearrangements were detected in 16/45 (35%) DLCL, but significantly less frequently in FL (2/31; 6%). All of the DLCL cases displaying BCL-6 rearrangements lacked BCL-2 rearrangements which were found in only two 2 DLCL cases. Although cytogenetic data were not available for the panel of tumors studied, the frequency of BCL-6 rearrangements far exceeds that expected for 3q27 aberrations (10-12% in DLCL), suggesting that BCL-6 rearrangements can occur as a consequence of submicroscopic chromosomal aberrations.

Table 2. Rearrangements of the BCL-6 gene in lymphoid tumors.

Tumor[a]		Rearranged/Tested	%
NHL			
	FL	2/31	6
	DLCL	16/45	35
	SNCL	0/22	0
ALL		0/45	0
CLL		0/51	0
MM		0/23	0

[a] NHL, non-Hodgkin's lymphoma; ALL, acute lymphoblastic leukemia; CLL, chronic lymphocytic leukemia; MM, multiple myeloma; FL, follicular lymphoma; DLCL, diffuse large cell lymphoma; SNCL, small non-cleaved cell lymphoma.

Rearrangements affect the 5' non-coding region of BCL-6

The positions of the breakpoints within the BCL-6 locus have been mapped allowing for a preliminary understanding of their effect on the structure/function of the BCL-6 gene (Ye et al. 1993b). All the observed breakpoints could be mapped within the putative 5' flanking region, within the first exon or within the first intron of BCL-6 (Fig.1). In all the observed rearrangements, the coding domain of BCL-6 is left intact whereas the 5' regulatory region, presumably containing the promoter sequences, is either completely removed (in case of truncation within the first exon or intron) or truncated. As a result, all of the coding exons of the BCL-6 gene are linked downstream to heterologous sequences which, based on cytogenetic analysis, can originate from different chromosomes in different cases. It is predicted that the functional consequence of these truncations is the expression of a normal BCL-6 protein under the control of heterologous regulatory sequences leading to the loss of its normal pattern of regulation.

BCL-6 rearrangements are associated with a distinct clinico-pathologic subset of DLCL

DLCL are highly heterogeneous in terms of pathologic manifestations and prognosis. In order to determine whether BCL-6 rearrangements could identify any significant subset of tumors, 102 DLCL cases were analyzed for rearrangements and the results correlated with histology, BCL-2 rearrangement status, age, stage, clinical status and treatment outcome (Offit et al. 1994). BCL-6 rearrangements were noted in a statistically significant subset of cases characterized by involvement of extranodal tissues, lack of bone marrow involvement and favorable prognosis (Table 3). Overall, BCL-6 status was found an independent prognostic marker of survival and freedom from disease progression in the multivariate model, and added predictive value to established clinical prognostic models (Offit et al. 1994). These findings suggest that BCL-6 rearrangements may identify a biologically distinct subset of DLCL and may serve as a diagnostic and prognostic marker in the management of patients with this disease.

Table 3. Clinico-pathologic features of DLCL cases displaying BCL-6 rearrangements

BCL-6	No of cases	BCL-2[a]	Extranodal[b]	Prognosis[c]
Rearranged	23	0 (0%)	16 (70%)	19 (83%)
Germ-line	79	21 (26%)	27 (34%)	40 (50%)

[a] Number of cases displaying BCL-2 rearrangements.

[b] Number of cases in which the tumor appeared outside lymphnodes as dominant extranodal disease.

[c] Number of cases free of disease progression 36 months after diagnosis.

References

Bastard C, Tilly H, Lenormand B, et al. (1992) Translocations involving band 3q37 and Ig gene regions in non-Hodgkin's lymphoma. Blood 79: 2527-2531

Chardin P, Courtois G, Mattei MG, and Gisselbrecht S (1991) The KUP gene located on human chromosome 14 encodes a protein with two distinct zinc fingers. Nucleic Acid Res 19:1431-1436

Chen Z, Brand NJ, Chen A et al. (1993) Fusion between a novel Krüppel-like zinc finger gene and the retinoic acid receptor-alpha locus due to a variant t(11:17) translocation associated with acute promyelocytic leukaemia. EMBL J 12:1161-1167

Dalla-Favera R (1993) Chromosomal translocations involving the c-myc oncogene in lymphoid neoplasia. In: Kirsch IR (ed) The causes and consequences of chromosomal aberrations. Boca Raton: CRC Press, pp313-332 de The' H, Lavau C, Marchio A, et al. (1991) The PML-RARα fusion mRNA generated by the t(15;17) translocation in acute promyelocytic leukemia encodes a functionally altered RAR. Cell 66:675-684.

DiBello PR, Withers DA, Bayer CA, Fristrom JW and Guild GM (1991) The Drosophila Broad-complex encodes a family of related proteins containing zinc fingers. 129:385-397

Djabali M, Selleri L, Parry P, et al. (1992) A trithorax-like gene is interrupted by chromosome 11q23 translocations in acute leukemias. Nature Genet 2:113-118

El-Baradi T and Pieler T (1991) What we know and what we would like to know. Mech Dev 35:155-169

Fichelson S, Dreyfus F, Berger R, et al. (1992) Evi-1 expression in leukemic patients with rearrangements of the 3q25-q28 chromosomal region. Leukemia 6:93-99

Gaidano G and Dalla-Favera R (1992) Oncogenes and tumor suppressor genes. In: Knowles DM (ed) Neoplastic hematopathology. Baltimore: Wilkins & Wilkins, pp245-261

Gu Y, Nakamura T, Alder H, et al. (1992) The t(4;11) chromosome translocation of human acute leukemias fuses the ALL-1 gene, related to Drosophila trithorax, to the AF-4 gene. Cell 71:701-708

Haber DA, Buckler AJ, Glaser T, et al. (1990) An internal deletion within an 11p13 zinc finger gene contributes to the development of Wilms' tumor. Cell 61:1257-1269

Harrison SD and Travers AA (1990) The tramtrack gene encodes a Drosophila finger protein that interacts with the ftz transcriptional regulatory region and shows a novel embryonic expression pattern. EMBO J 9:207-216

Kakizuka A, Miller, Jr, WH, Umesono K, et al. (1991) Chromosomal translocation t(15;17) in human acute promyelocytic leukemia fuses RARα with a novel putative transcription factor, PML. Cell 663-674

Kerckaert JP, Deweindt C, Tilly H et al. (1993) LAZ-3, a novel zinc-finger encoding gene, is disrupted by recurring chromosome 3q27 translocations in human lymphomas. Nature Genet 5:66-69

Korsmeyer SJ (1992) Bcl-2 initiates a new category of oncogenes: Regulators of cell death. Blood 80:879-886

Ladanyi M, Offit K, Jhanwar SJ, et al. (1991) Myc rearrangements and translocations involving band 8q24 in diffuse large cell lymphomas. Blood 77:1057-1063

Lo Coco F, Ye BH, Lista F et al. (1994) Rearrangements of the BCL6 gene in diffuse large cell non-Hodgkin's lymphoma. Blood 83:1757-1759

Magrath IT (1990) The non-Hodgkin's lymphomas: An introduction. In: Magrath IT (ed) The Non-Hodgkin's lymphomas. Lodon: Edward Arnold, pp1-14

McGuire E, Hockett RD, Pollock KM, et al. (1989) The t(11;14)(p15;q11) in a T-cell acute lymphoblastic leukemia cell line activates multiple transcripts, including Ttg-1, a gene encoding a potential zinc finger protein. Mol Cell Biol 9:2124-2132

Miki T, Kawamata N, Hirosawa S, and Aoki N (1994) Gene involved in the 3q27 translocation associated with B-cell lymphoma, BCL5, encodes a Krüppel-like zinc-finger protein. Blood 83:26-32

Morishita K, Parker DS, Mucenski ML, et al. (1988) Retroviral activation of a novel gene encoding a zinc finger protein in IL3-dependent myeloid leukemia cell lines. Cell 54:831-840

Offit K, Koduru PRK, Hollis R, et al. (1989a) 18q21 rearrangements in diffuse large cell lymphoma: Incidence and clinical significance. Brit J Haematol 72: 178-182

Offit K, Jhanwar S, Ebrahim S A D, et al. (1989b) t(3;22)(q27;q11): A novel translocation associated with diffuse non-Hodgkin's lymphoma. Blood 74: 1876-1879

Offit K, Lo Coco F, Louie DC et al. (1994) Rearrangement of the bcl-6 gene as a prognostic marker in diffuse large-cell lymphoma. N Engl J Med (in press)

Pandolfi PP, Grignani F, Alcalay M, et al. (1991) Structure and origin of the acute promyelocytic leukemia myl/RARα cDNA and characterization of its retinoid-binding and transactivation properties. Oncogene 6:1285-1292

Read D and Manley J (1992) Alternative transcripts of the Drosophila tramtrack gene encode zinc finger proteins with distinct DNA binding specificities. EMBO J 11:1035-1044

Rosenberg UB, Schroder C, Preiss A, et al. (1986) Structural homology of the product of the Drosophila Krüppel gene with Xenopus transcription factor IIIA. Nature 319:336-339

Tkachuk DC, Kohler S, Cleary ML. (1992) Involvement of a homolog of Drosophila Trithorax by 11q23 chromosomal translocations in acute leukemias. Cell 71:691-700

Ye BH, Rao PH, Chaganti RSK, and Dalla-Favera R (1993a) Cloning of bcl-6, the locus involved in chromosomal translocations affecting band 3q27 in B-cell lymphoma. Cancer Res. 53:2732-2735

Ye BH, Lista F, Lo Coco F et al. (1993b) Alterations of a zinc finger-encoding gene, BCL-6, in diffuse large-cell lymphoma. Science 262:747-750

Classification of Mouse Lymphomas

T. N. Fredrickson,[1] J. W. Hartley,[2] H. C. Morse III,[2] S. K. Chattopadhyay,[2] and K. Lennert[3]

[1]Registry of Experimental Cancers, National Cancer Institute and [2]Laboratory of Immunopathology, National Institute of Allergy and Infectious Diseases, National Institutes of Health, Bethesda, Maryland, 20892 USA, and [3]Zentrum für Pathologie und angewandte Krebsforschung, Kiel, FRG

We have applied the Kiel Classification of human lymphomas (Lennert and Feller 1992) to a series of mouse lymphomas (Table 1). In our survey, some came from other sources but most were from mice congenic for ecotropic murine leukemia virus induction loci and expressing high levels of endogenous leukemia viruses. Although such expression is associated with lymphoma development (Fredrickson et al. 1984), the proposed classification is considered generally applicable irrespective of etiology, since thymic lymphomas induced chemically with virus (Joshi and Frei 1970) or occurring in transgenic mice (Pattengale 1994) are morphologically similar. The classification attempts to provide a diagnosis based on morphology supported by molecular and immunologic analysis.

The morphologic features, combined with immunophenotypic and molecular data, used as diagnostic criteria are outlined in Table 2. Lymphocytic lymphomas are characterized by fairly uniform sheets of small lymphocytes, usually most prominently in the spleen, with minimal numbers of mitotic figures (Fig. 1). Other lymphomas are clearly derived from lymphoid cells that can be distinguished in normal germinal centers: immunoblastic, centroblastic, and centrocytic. Immunoblastic lymphomas are characterized by large cells with a distinctive, vesicular nucleus containing a single large, central nucleolus (Fig. 2). In addition,

Table 1. Diagnoses in two surveys of spontaneous lymphoma

Type	1984	1993
Lymphocytic	5	5
Immunocytoma	ND*	7
Centrocytic	3	1
Centrocytic-Centroblastic	5	2
Centroblastic	22	13
Immunoblastic	3	7
Lymphoblastic	4	50
Marginal Zone	ND*	13
Total	42	98

* ND = not diagnosed because these types of lymphomas were not covered by the Lukes and Collins classification used in 1984 (Fredrickson et al. 1985).

there are variable numbers of lymphoid cells of other types scattered throughout, but immunoblasts clearly predominate. Since immunoblastic lymphomas can be either of B or, rarely, of T cell type, and these are morphologically similar, a genetic or phenotypic analysis is necessary to distinguish the cell of origin. Centroblastic lymphomas are composed of uniform sheets of medium-sized cells containing a vesicular nucleus (Fig. 3) and prominent, often paired, nucleoli frequently adhered to the nuclear membrane. Infiltration of nonlymphoid organs is common, particularly in periportal areas of the liver. Centrocytic lymphomas, found only occasionally in our series, are composed fairly uniformly of small cells with irregularly shaped nuclei and small nucleoli (Fig. 4). We have not seen a sufficient number of centrocytic lymphomas to characterize them more fully, but malignant B cell immunoblasts and centroblasts have all had a mature B cell phenotype ($B220^+$, M^+, K^+) and J_H rearrangements only. T immu-noblastic and lymphoblastic lymphomas have a similarly mature T phenotype and TCRβ rearrangements (Table 2). As had been shown for lymphoblastic T cell lymphomas (Herr et al. 1983; Fredrickson et al. 1993), about 60 percent of T lymphomas also have rearranged J_H genes.

Table 2. Diagnostic criteria for common mouse lymphomas

Morphology in H & E-stained tissue sections		Immunoblastic lymphoma*	Centroblastic lymphoma	Centrocytic lymphoma	Immunocytoma
Growth Pattern		Variably mixed with lymphocytes and plasmacytoid cells	Solid sheets of similarly sized cells	Solid sheets	Diffuse, multiple cell types often with plasmacytoid differentiation
Cell Size		Large	Medium	Small	Variable
Cytoplasm		Basophilic, abundant	Basophilic, sparse to abundant	Not distinguishable	Variable
Nucleus		Vesicular, ovoid, prominent membrane	Vesicular, round	Fine chromatin, irregular shape	Variable
Nucleolus		Single, large, central basophilic	Two medium-size adhered to nuclear membrane, basophilic	Multiple, small, central	Variable
Mitoses		+++	+++	+ / −	++
SURFACE MARKERS	B220	+ }(B)	+ or −		+
	Kappa	+ }(B)	+		+
	CD4	+ or − }(T)	−		−
	CD8	+ or − }(T)	−		−
GENE REARRANGE-MENTS	J_H	+ (B) + or − (T)	+		+
	Tcrβ	+ (T)	−		−

* These lymphomas can be derived from either T or B cells and are morphologically indistinguishable; the other types of

Immunocytomas are much more heterogeneous B cell lymphomas comprising a mixed population of immunoblasts, small lymphocytes, variable numbers of centroblasts and centrocytes, and often a considerable component of plasma cells. The case shown to represent this type of lymphoma is of an aged SJL/J mouse (Fig. 5) and is typical of this strain, although it also occurs in mice of other strains.

Plasmacytomas growing within granulomatous lesions in the mesentery and omentum after intraperitoneal inoculation with mineral oil, pristane, plastics or other agents (Potter and Weiner 1992), are mainly composed of plasma cells; however, early foci (Fig. 5) growing on peritoneal surfaces vary in morphology from those with a high proportion of plasmablasts and more immature plasma cells to those composed of mature plasma cells.

The most common lymphoma of our second series was lymphoblastic in morphology, an unexplained change from the study conducted ten years earlier (Table 1). Three-quarters of them were composed of B cells, as distinguished by J_H rearrangement, the others being clearly T cell by both phenotype and the

Table 2, Continued

Morphology in H & E-stained tissue sections		Lymphocytic lymphoma*	Plasmacytic	Lymphoblastic lymphoma	Marginal zone lymphoma
Growth Pattern		Solid sheets with occasional immunoblasts	Solid sheets, plasmablasts to plasma cells	Cohesive, streaming outside splenic or nodal capsule; starry sky; CNS involvement	Invasion of splenic red pulp from enlarged marginal zone
Cell Size		Small	Medium	Medium	Medium
Cytoplasm		Narrow rim, palely basophilic	Basophilic, abundant	Basophilic, scanty to moderate	Abundant, grey
Nucleus		Coarse chromatin	Coarse chromatin	Ovoid to round, sometimes cleaved	Pleomorphic, vesicular
Nucleolus		Small, single	Not prominent	One to several, variable size	Small, single
Mitoses		+ / −	++	++++	+ / ++
SURFACE MARKERS	B220	− } (B)	−	Dull (B)	+ or −
	Kappa	+	−	−	+
	CD4	+ or − } (T)	−	+ or − } (T)	−
	CD8	+ or −	−	+ or −	−
GENE REARRANGE-MENTS	J_H	+ (B) + or − (T)	+	+ (B) + or − (T)	+
	Tcrβ	+ (T)	−	+ (T)	−

lymphomas are of B cell origin.

presence of TCRβ rearrangements. T lymphoblastic lymphomas have several possible immunologic phenotypes (Table 2), whereas the B cell lymphoblasts, by immunochemistry, usually stained B220 dull but were negative for immunoglobulin markers. A recent report of B lymphoblastic lymphomas occurring in SL mice suggests that such tumors are of bone marrow origin because of large numbers of marrow cells positive for B220 and 6C3, the latter a marker found only on pre-B and early B cells (Okamoto et al. 1994). It was clear in our series that infiltration of lymph nodes or thymus, in the case of T lymphoblastic lymphomas, occurred before splenic involvement; however, we have not established the bone marrow as the site of primary growth for the B lineage cells. Lymphoblastic lymphomas of either T or B cells are morphologically indistinguishable, and both grow in a characteristically contiguous fashion, forming solid sheets interspersed by tingible body macrophages (Fig. 7).

A mouse lymphoma newly identified in this study is one of marginal zone origin in the spleen. The small- to medium-sized cells, with plentiful grey cytoplasm by Giemsa staining (Fig. 8), were uniformly of a mature B cell phenotype and B genotype and showed advanced invasion of the red pulp. These lymphomas appear to be of splenic origin and morphologically resemble those described in the human.

While this classification does not include some seen less frequently, it does deal with the great majority of mouse lymphomas liable to be encountered. Our survey has shown that morphology alone can serve as the primary diagnostic criterion, but further analyses are sometimes necessary to establish lineage and are a necessity for lymphoblastic lymphomas. More detailed analyses may be required for other reasons, such as to differentiate a monoclonal immunocytoma from a polyclonal, reactive proliferation of B cells. This is particularly true in dealing with mouse lymphomas of B cell origin, which often are composed of mixtures of different kinds of lymphoid cells, as well as macrophages, which makes the malignant component in lymph nodes or spleen difficult to discern. In such cases, a clearer picture can often be obtained from studies of infiltrates of nonlymphoid organs, since these tend to be more monomorphic. Finally, and in contrast to human lymphomas, there is not a clear microscopic distinction in mouse lymphomas between those that are follicular compared to a diffuse pattern of growth. Most mouse lymphomas are diffuse in their growth patterns, although from the standpoint of their gross appearance, immunoblastic, centroblastic, marginal zone lymphomas, and immunocytomas tend to be composed of solid, white, fleshy, demarcated tumors. This is in contrast to the lymphoblastic type, in which the nodes and spleen have uniformly dispersed growth with extensive involvement of other organs. Thus, while there are distinct similarities between human lymphomas and those of mice, there are also some differences. Nevertheless, we feel that the criteria of the Kiel Classification, as compared to others (Fredrickson et al. 1985; Lukes and Collins 1988; Pattengale 1994) are more adaptable to lymphomas of mice and provide a basis for a clear classification which is both biologically sound and useful for diagnosis.

Acknowledgments

The skillful work of Brenda Rae Marshall in preparing this manuscript is gratefully acknowledged.

References

Fredrickson TN, Morse H, Rowe W (1984) Spontaneous tumors of NFS mice congenic for ecotropic murine leukemia virus induction loci. J Natl Cancer Inst 73:521-524.

Fredrickson TN, Morse HC III, Yetter RA, Rowe WP, Hartley JW, Pattengale PK (1985) Multiparameter analyses of spontaneous non-thymic lymphomas occurring in NFS/N mice congenic for ecotropic murine leukemia viruses. Am J Pathol 121:349-360.

Fredrickson TN, Hartley JW, Morse HC III (1993) Early divergence of erythroid lineage suggested by gene rearrangements in mouse hematopoietic neoplasms. Exp Hematol 21:354-357.

Herr W, Perlmutter AP, Gilbert W (1983) Monoclonal AKR/J thymic leukemias contain multiple J_H immunoglobulin gene rearrangements. Proc Natl Acad Sci USA 80:7433-7436.

Lennert K, Feller AC (1992) Histopathology of Non-Hodgkin's Lymphomas (Based on the Updated Kiel Classification). Springer-Verlag, Berlin.

Joshi VV, Frei JV (1970) Gross and microscopic changes in the lymphoreticular system during genesis of malignant lymphoma induced by a single injection of methylnitrosourea in adult mice. J Natl Cancer Inst 44:379-394.

Lukes RJ, Collins RD (1988) Tumors of the Hematopoietic System. Atlas of Tumor Pathology, Second Series, Fascicle 28 Armed Forces Institute of Pathology, Washington, DC.

Okamoto K, Yamada Y, Ogawa MS, Toyokuni S, Nakakuki Y, Ikeda H, Yoshida O, Hiai H (1994) Abnormal bone marrow B-cell differentiation in pre-B lymphoma-prone SL/Kh mice. Cancer Res 54:399-402.

Pattengale PK (1994) Tumours of the lymphohaematopoietic system. In: Turusov V, Mohr U (eds) Pathology of Tumours in Laboratory Animals, Tumors of the Mouse, 2nd edition. International Agency for Research on Cancer (WHO) and International Life Sciences Institute, Lyon, pp 651-670.

Potter M, Wiener F (1992) Plasmacytomagenesis in mice: model of neoplastic development dependent upon chromosomal translocations. Carcinogenesis 13:1681-1697.

Legends to Figures

Fig. 1 Lymphocytic lymphoma in a 202-day-old Akv-2 congenic NFS/N mouse that had a 0.91 g spleen and slightly enlarged lymph nodes. There were three J_H rearrangements and none for TCRβ. The lymphoma was of a mature B cell phenotype and was $CD5^+$. Hematoxylin and eosin stain, × 1,000.

Fig. 2. Immunoblastic lymphomas from an Akv-1 congenic NFS/N mouse, 349 days of age, with a spleen weight of 1.91 g. Histology and histocytochemistry showed that the IgM^+, K^+ immunoblasts constituted the malignant component indicated by the presence in a Southern blot of a single J_H rearranged band (shown much more clearly in the DNA sample from the kidney) in addition to the germline band. This splenic section contains immunoblasts (arrows), numerous macrophages (arrowheads), and lymphocytes. Giemsa stain, × 1,000.

Fig. 3. Centroblastic lymphoma in the 3.94 g spleen of a 373-day-old C58v-1 congenic NFS/N mouse showing the fairly uniform population of medium-sized cells with poorly defined cytoplasm and a round, vesicular nuclei containing, most typically, two peripheral nucleoli (arrows). There were two rearranged J_H bands in addition to the germline band in splenic DNA. Hematoxylin and eosin stain, × 1,000.

Fig. 4. Centrocytic lymphoma in the periportal area of the liver of a 431-day-old Akv-2 congenic NFS/N mouse that had a 2.54 g spleen. The tumor uniformly comprises centrocytes with irregularly shaped nuclei and indistinct nucleoli and cytoplasm. There were three rearranged J_H bands, plus germline, in the DNA from this spleen. Hematoxylin and eosin stain, × 1,000.

Fig. 5. Immunocytoma in a 532-day-old SJL/J mouse with immunoblasts (arrows), plasmacytoid cells (arrowheads), and small lymphocytes. The spleen weighed 0.41 g, but a cervical lymph node was 1.59 g. These organs and the thymus, which was enlarged, had similar lesions. Both the thymus and the spleen had two identical rearranged J_H bands plus germline. Hematoxylin and eosin stain, × 1,000.

Fig. 6. Plasmacytoma growing as a very early focus on the mesentery of a BALB/c mouse typical for this strain given an intraperitoneal injection of material such as pristane or silicone inducing a granulomatous response. A fairly high proportion of plasma cells and cells intermediate in differentiation to mature plasma cells (arrowheads) are seen. Hematoxylin and eosin stain, × 1,000.

Fig. 7. Lymphoblastic lymphoma from a 322-day-old Akv-2 NFS/N mouse with a spleen of 0.37 g but a cervical lymph node of 2.08 g, which had a single J_H rearrangement with none for TCRβ. Lymphoblastic lymphomas of T cell origin have similar morphology. Hematoxylin and eosin stain, × 1,000.

Fig. 8. Marginal zone lymphoma from a 326-day-old C58v-1 congenic NFS/N mouse with a spleen of 0.70 g. The spleen had two J_H but no TCRβ rearrangements, and the marginal zone cells were $B220^+$, IgM^+, and K^+. Their characteristic morphology is a small to medium size, a round to ovoid nucleus containing indistinct nucleoli, and pale, grayish cytoplasm. Giemsa × 1,000.

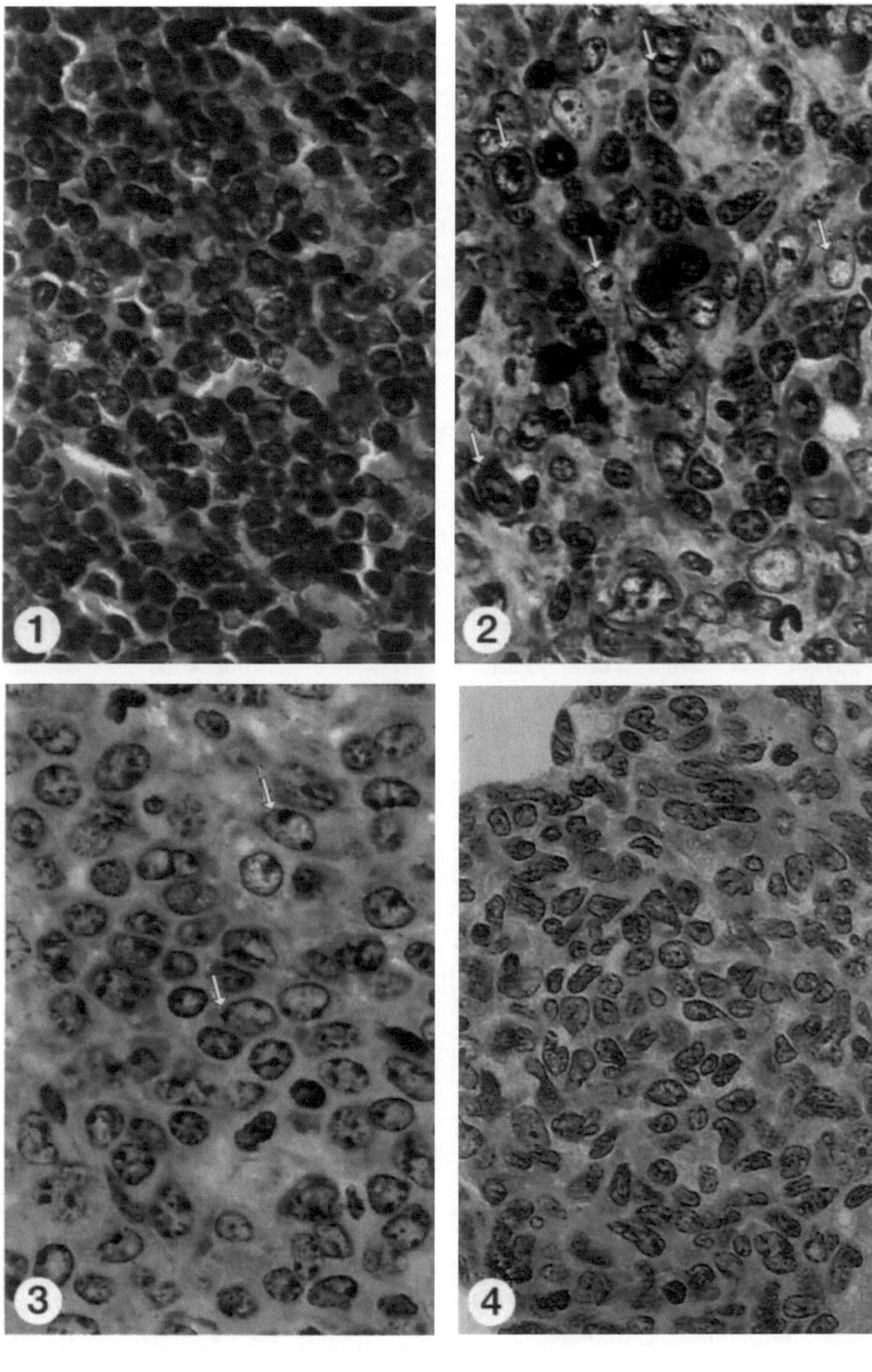
1
2
3
4

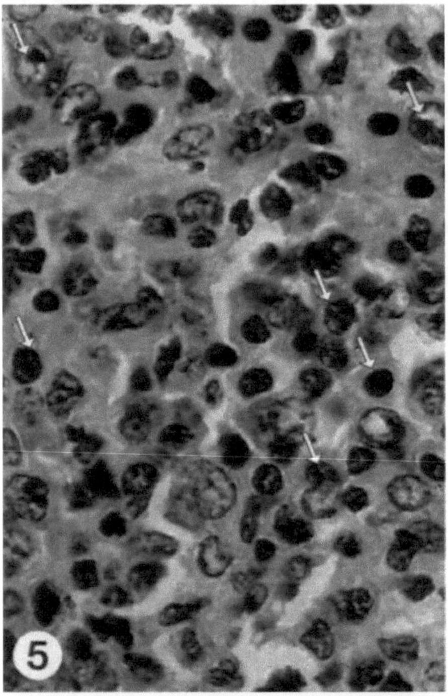
5

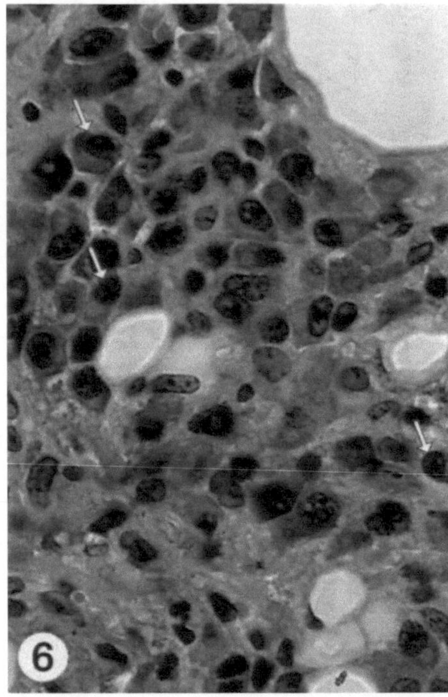
6

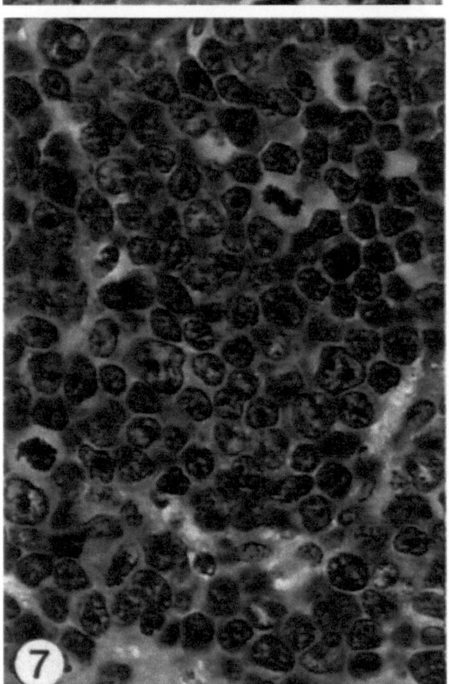
7

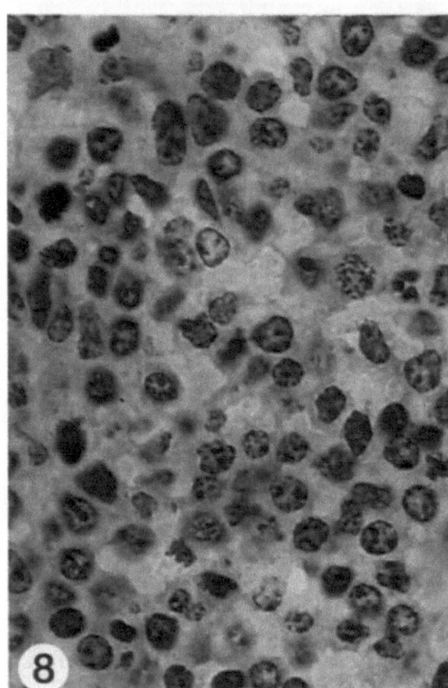
8

The Natural History of a Lymphoproliferative Disorder in Aged NZB Mice

G E. Marti[1], R.A. Metcalf[1], and E. Raveche[2]

[1] Section of Flow and Image Cytometry, Laboratory of Medical and Molecular Genetics, Division of Cellular and Gene Therapies, Center for Biologics Evaluation and Research, Food and Drug Administration, NIH, Bldg 29, Room 502, 8800 Rockville Pike, Bethesda, MD 20892; [2] Laboratory Medicine and Pathology, University of Medicine and Dentistry of New Jersey, New Jersey Medical School, 185 South Orange Avenue, University Heights, Newark, NJ 07103 -2714

Abstract

The molecular lesions of human familial and common B-CLL remain unknown. As an approach to this problem, aged NZB mice with a B cell lymphoproliferative disorder were chosen as a murine model. Three groups of NZB mice (2 months, 6 months and >18 months) for a total of nineteen were studied. A complete autopsy including a CBC was performed on each mouse. Spleen cells were immunophenotyped and cell cycle analysis was performed. Spleen weight, peritoneal cell counts and absolute lymphocytes counts were all elevated in the oldest group. All mice showed evidence of extramedulary hematopoiesis and the older group showed lymphocytic infiltrates in the lacrymal glands, kidneys, liver and lungs. Two of the seven aged mice had a malignant lymphoma. One was a marginal zone lymphoma and the other a lymphocytic lymphoma. Splenic immunophenotyping showed a loss of T cells with an increase in B cells as the mice age. Cell cycle analysis revealed hyperdiploidy in all of the aged mice with a decrease in the percentage GoG1 cells. This disease appears to involve an absolute lymphocytosis of the peritoneum and the peripheral blood compartment. This is associated with splenic aneuploidy. The infiltration of the spleen by malignant cells of varying morphology is a late event. The aged NZB mouse continues to be a model for human B-CLL.

Introduction

We have studied common and familial B cell chronic lymphocytic leukemia (B-CLL) for the last several years (1-4). Raveche and co-workers have studied autoimmunity in the NZB mouse and more recently tumor formation. They have suggested that the aged NZB mouse may be a model for human B-CLL (5-10). Some of the characteristic features of this mouse are: hypergammaglobulinemia, autoantibodies, polyclonal B cell activation, CD5 B cell lymphocytosis in peritoneal cavity and or spleen, hyperdiploidy (aneuploidy), and a transplantable, splenic lymphoma. A peripheral blood lymphocytosis variably appears and represents a late event. Both *in vitro* tumor cell lines and T cell lines have been established from these mice. The striking splenomegaly and long clinical course suggest that this is a model for splenomegalic B-CLL. The purpose of this study was to confirm and extend the previous observations of Raveche and co-workers concerning the aged NZB mouse as a model for human B-CLL (10). The overall goal is to develop a set of assays that will not only trace the earliest development of this disease but will permit or define a convenient way to detect and phenotype the presence or absence of neoplastic disease in F2 intercrosses and Fl backcrosses. This phenotyping is necessary for a planned genetic linkage analysis of NZB crossed with DBA/2 mice.

Materials and Methods

Necropsy Protocol

Two groups of six NZB mice each (2 months, 6 months) and seven NZB mice (18-24 months) were studied. The two month and six month old mice were obtained from Jackson Laboratories. The seven aged NZB mice were obtained from Dennis Klinman (CBER, FDA). The mice were anesthetized with methoxyflurane (Metofane®, inhalation anesthetic, Pitman-Moore) and were exsanguinated via the right brachial artery. A pediatric device (Microtainer with EDTA or Microtainer serum separator tube) was used to collect the blood. Blood was collected for a CBC, platelet count and white cell differential. Blood films were prepared to estimate reticulocytes (Anlytics, Inc., Gaithersburg, MD).

The abdominal cavity was lavaged with 10 ml phosphate buffered saline (PBS). To insure death, cervical dislocation was performed. The spleens were excised, weighted and sections were taken for histology. The remaining spleen was processed into a cell suspension and lyzed once with ammonium chloride. After a cell count, the spleen cell suspension was used for cell surface immunophenotyping and cell cycle analysis.

A complete autopsy consisted of the examination of all body cavities. The following organs were removed and fixed in 10% neutral buffered formalin: brain, spinal cord (cervical, thoracic and lumbar), pituitary, peripheral nerve (siatic), salivary glands, all cervical lymph nodes, thyroid, parathyroids, adrenals, eyes, lacrymal glands, skeletal muscle (biceps femorus), tongue, mammary gland, lungs (with major bronchi), trachea, esophagus, heart, aorta, thymus, mediastinal lymph node, axillary and inguinal lymph nodes, spleen, pancreas, liver, gall bladder, kidneys, stomach, mesenteric lymph node, peritoneum, duodenun, jeujunum, ileum, cecum, colon, rectum, skin, urinary bladder, prostate, testes, epidimis, bone (sternum), bone (femur), bone marrow smears, ovaries, uterus and all gross lesions. For histological preparation, the above listed tissues were dehydrated in graded ethanol, cleared in xylene, embedded in paraplast, microtomed at 5 microns, and mounted on glass slides. The slides were stained with hematoxylin and eosin and the tissues protected with coverslips and reviewed by a pathologist [Experimental Pathology Laboratories (EPL) Inc., Herndon, VA 22170]. An additional five to seven one year old NZB mice were sacrified for the further evaluation of three color flow cytometry and to evaluate the effect of red cell lysis upon the spleen cell preparation. This protocol has been approved by the CBER Animal Care and Use Committee. The spleens of all seventeen mice were photomicrographed and indidivual composite photos were prepared.

Flow Cytometry Surface Markers and Cell Cycle

The reagents used to stain spleen cells were directly conjugated and consisted of the following: anti-CD3, anti-T cell receptor (alpha, beta), anti-Thy1.2, anti-CD4, anti-CD8, anti-IgG, anti-IgM and anti-Class II. Cells were stained and fixed in 0.5% paraformaldehye. These reagents were obtained from either Caltag or Pharmagen. In some experiments, a combination of FITC conjugated goat anti-IgM (F(ab)'2, anti-CD5 conjugated with phycoerythrin (PE) and anti- B220 conjugated with PE-Cy5 were used for three color analysis. A FACScan flow cytometer using Lysis II software was used to acquire 10-20,000 events. For cell cycle analysis, cell suspensions (1×10^6 cells per ml final concentration) were fixed in cold 70% ETOH overnight. In some of the cell cycle analysis experiments, cells were stained with the FITC anti-IgM reagent, fixed in 70% ETOH and then stained with propidium iodine (PI). At the time of analysis, the ethanol fixed cells are incubated with propidium iodide (PI, 50 ug/ml) and ribonuclease (1 mg/ml) for 30 min at 37°C and analyzed immediately on a FACScan flow cytometer using CELLFIT software for both the acquisition and analysis of 20,000 events. FDAplot was used to prepare composite displays.

Results

General Necropsy

Table 1 contains a selected summary of autopsy and hematological values. Spleen weights are a convenient way of measuring splenomegly while the peritoneal WBC can be used as an indicator of a lymphocytosis. All of the aged mice demonstrated splenomegly and a peritoneal lymphocytosis. There is a progressive decline with age in circulating red cells with a slight increase in the MCV. The blood absolute lymphocyte count also increases as a function of age. Review of the blood films confirmed the lymphocytosis and a slight reticulocytosis. The red cells contained a significant number of inclusions.

Table 1. Selected NZB necropsy and hematologic values

Characteristic	2 Months	6 Months	18 Months
spleen weight	67 ± 23 mg	107 ± 18 mg	405 ± 177 mg
Peritoneal WBC	3.2 x 10^6 ± 0.7 cells/ml	5.6 x 10^6 ± 2.3 cells/ml	24 x 10^6 ± 25cells/ml
RBC count 10^6 per ul	8.5 ± 0.6	7.8 ± 0.5	6.1 ± 1.4
MCV (femtoliter)	48.6 ± 0.7	47 ± 0.5	57 ± 9
ALC x 10^3 cells per ul	2.5 ± 1.2	3.3 ± 0.6	15.4 ± 4.4

Histology

The major autopsy finding in the 2 month old mice was extramedullary hematopoiesis (EMH) in the spleen. At six month of age, splenic EMH plus mononuclear infiltrates in the lacrymal glands were noted. At 18-24 months of age, splenic EMH, and mononuclear infiltrates in the lacrymal glands, kidneys, liver and lungs were noted. In addition, a splenic lymphoproliferative disorder with or without lymph node hyperplasia was seen in two aged mice. Aside from enlargement, the spleens were grossly normal at dissection; isolated and/or confluent nodules were not seen. Of interest, salivary glands showed no signs of mononuclear infiltration.

EMH appeared to increase as a function of age. The red pulp was uniformly expanded and at low power sheets of megakaryocytes and red cell precursors could be seen. There was white pulp hyperplasia with one or more of the white pulp areas being confluent at all ages. In some of the mice the marginal zone appeared to be slightly enlarged. Many of the lymphoid aggregates contained germinal centers. A definite lymphoproliferation characterized by an infiltration or sheets of lymphoid cells were noted in two of seven aged NZB spleens. These lymphoid cells appeared to infiltrate or replace the red pulp areas with compression of the remaining normal remnant lymphoid aggregates. Two different morphological patterns were seen. In one of the aged NZB mice (Fig. 1), the red pulp was replaced by a marginal cell lymphoma. The cells varied in size from small to medium and showed plasmacytoid differentiation with abundant cytoplasm. In the second aged NZB mouse (Fig. 2), the red pulp was replaced by a lymphocytic lymphoma. The cells were medium sized with slightly irregular nuclei, smooth chromatin and an occasional nucleolus. In both cases there appeared to be an increase in white pulp. In a third aged mice, lymphocytic lymphoma was noted in a lymph node without splenic involvment.

Flow Cytometry

In the initial experiments, spleen cell preparations were not lyzed. Ammonium chloride lysis seemed to permit better resolution of subpopulations. Initial difficulty in detecting CD5+ B cells clones resolved itself with experience. Surface immunophenotyping showed a generalized decrease in T cells associated with B cell hyperplasia and activation (data not shown). Cell cycle analysis using single parameter PI content showed the presence of aneuploidy and increased S phase cells in all of the aged mice. Using two parameter data, this contrast is even sharper between the age groups. In addition to being able to detect aneuploidy, the other significant finding is the differences between percent GoG1 cells as a function of age. At 2 months, 6 months and 18-24 months, the GoG1 fraction (mean percentage plus or minus 1 standard deviation) is 86 ± 5, 81 ± 6, 65 ± 10, respectively. This progressive decrease is associated primarily with a corresponding increase in S phase cells or B cell activation.

Discussion

In the present study nineteen NZB mice were examined for disease progression as a function of age. The only abnormality found in the youngest mice was EMH. In the next age group of six month old mice, EMH and mononuclear infiltrates in the lacrymal glands were noted. The origin of this mononuclear infiltrate is unknown. It may be related to the underlying autoimmune dysfunction of the NZB or to a coincidental viral infection.

In the aged NZB, EMH that was more marked in the older individuals. Mononuclear infiltrates were seen in the lacrymal glands and the lungs, kidneys, and liver. In addition two of the aged NZB spleens showed involvement by a malignant lymphoma. One was a marginal cell lymphoma and the other a lymphocytic lymphoma. Immunophenotyping showed a decrease in T cells and an increase in B cells. Cell cycle analysis showed the presence of hyperdiploidy in all of the aged NZB. Futhermore, the progressive decrease in GoG1 cells with increasing age correlates with an increase in S phase cells. Although as noted above, decreased GoG1 may be associated with either an increase in S phase cells or G2M cells or both.

Autoimmunity in the NZB mouse has been studied for many years and is used by many investigators as a model for systemic lupus erythematosus. Autoimmune hemolytic anemia and complex membranous glomerulonephritis have been extensively studied. The lymphoproliferative disorder in NZB mice has not been studied as extensively. East *et al.* originally described this NZB splenic tumor as a primitive reticular neoplasia using spleen touch imprints (11-13). Mellors found four malignant neoplasms arising in lymphatic tissues out of twenty selected complete autopsies (14). The animals were 9 to 11 months old. Two histological patterns were noted: reticular sacrcoma and pleomorphic malignant lymphoma. Sugai *et al.* re-ported that 30% of NZB/NZW F1 mice have monoclonal serum IgM (15). Greenspan *et al.* have shown that transplantable lymphomas from these mice express surface immunoglobulin (16). Yumoto *et al.* in a study of 255 animals, 23 had non thymic lymphomas and five histological forms were noted (17). Sadahira *et al.* detected malignant lymphoma in eight of thirty-four mice over the age of 13 months (18). Much later Phillips et al. described it as CLL-like using hematoxylin and eosin stained, paraffin embedded sections and electron microscopy (10). Stall et al. has demonstrated the early appearance of homogenous clonal populations in the NZB (19). In an attempt to define immunologic oncogenesis, Fialkow *et al.* detected abnormal karyotypes in 14 of 20 adult NZB spleens but not in their bone marrow (20). Raveche *et al.* have further characterized these chromosomal abnormalities by cytogenetics and flow cytometry. Although the relationship between autoimmunity and neoplastic transformation is not well understood, the emergence of hyperdiploid, activated, CD5+ B cell clones is genetically controlled (21). The studies in this report confirm this latter observation. Hyperdiploidy was seen in all of the aged NZB mice. However, we were surprised to find evidence of malignant lymphoma in only two of seven animals. Unfortunately, peritoneal cavity cells and blood were not routinely evaluated flow cytometrically. However, all of the aged NZB mice had elevated peritoneal WBC and ALC (see Table I). Repeat step cuts of spleen parafin blocks did not reveal any further evidence of lymphoproliferative disease. In addition to finding only two out of seven spleens to contain clear histological evidence of malignancy, the two lymphomas were distinctly different.

In conclusion, we have confirmed the scattered reports concerning the number of animals developing a lymphoid malignancy and variation in the histological pattern. There is undoubtedly some degree of peritoneal and blood involvement. The initiation of disease appears to involve an absolute lymphocytosis of the peritoneum with progression to the spleen and the peripheral blood compartment. This is associated with splenic aneuploidy. The infiltration of the spleen by malignant cells is a late event which may then undergo a second or continuous neoplastic transformation. With or without the presence of histological evidence of a lymphoid tumor, these abnormal cells form tumors in the appropriate host. Thus the NZB remains an attractive murine model to study human B-CLL. A larger study of aged NZB mice including immunoperoxidase assays and molecular techniques should contribute to our understanding of the molecular lesions involving human B-CLL.

Acknowledgments

We are grateful to E. Jaffe for reviewing the photomicrographs and to T. N. Fredrickson for discussions and reviewing the manuscript. We acknowledge the technical assistance of P. Carter and G. Washington for the flow cytometric analysis. We also acknowledge the work of two summer students, J. Schafer and L. Lin.

References

1. Marti GE, Zenger V, Caproaso NE, Brown M, Washington GC, Carter P, Schechter G, Noguchi P (1989) Antigenic expression of B-cell chronic lymphocytic leukemia lymphocytes. Anal Quan Cytol Histol 11:315-323
2. Caparoso NE, Whitehouse J, Bertin P, Amos C, Papadopoulos N, Miller J, Whang-Peng J, Tucker MA, Fleisher TA, Marti GE (1991) A 20 year clinical and laboratory study of familial B-chronic lymphocytic leukemia in a single kindred. Leuk Lymphoma 3:331-342.
3. Marti GE, Faguet G, Bertin P, Agee J, Washington G, Ruiz S, Carter P, Zenger V, Vogt R, Noguchi P (1992) CD20 and CD5 expression in B-chronic lymphocytic leukemia (B-CLL). Proc NY Acad Sci USA 651:464-466
4. Marti GE, Faguet GB, Stewart C, Branham P, Carter PH, Washington GC, Bertin P, Muller J, Capraosa N, Whitehouse J, Amos CI, Fleisher TA, Vogt R (1992) Evolution of leukemic heterogeneity of human B-CLL lymphocytes between and within patients. Curr Top Microbiol Immunol 182:303-311
5. Raveche ES, Tjio JH, Steinberg AD (1979) Genetic studies in NZB hyperdiploidy in the spleen of NZB mice. Cytogenet Cell Genet 23:182-193

6. Raveche ES, Alabaster O, Taurog J, Tjio JH, Steinberg AD (1981) Analysis of NZB hyperdiploid spleen cells. J Immunol 126:154-160
7. Raveche ES, Novotny EA, Hansen CT, Tjio JH, Steinberg AD (1981) Genetic studies in NZB mice. V. Recombinant inbred lines demonstrate that genes control autoimmune phenotype. J Exp Med 153:1187-1197
8. Seldin MF, Conroy J, Steinberg A, D'Hoosteleare L, Raveche ES (1987) Clonal expansion of abnormal B cells in old NZB mice. J Exp Med 166:1585-1590
9. Raveche E, Lalor P, Stall A, Conroy J (1988) In vivo effects of hyperdiploid Lyl+ B cells of NZB origin. J Immunol 141:4133-4139
10. Phillips J, Mehta K, Fernandez C, Raveche E (1992) The NZB mouse as a model for chronic lymphocytic leukemia. Cancer Res 52:437-433
11. East J, de Sousa MAB, Parrott DMV (1965) Immunopathology of New Zealand Black (NZB) mice. Transplantation 3:711-729
12. East J, de Sousa MAB, Prosser PR, Jaquet H (1967). Malignant changes in New Zealand Black mice. Clin exp. Immunol 2:427-443
13. East J (1970) Immunopathology and neoplasms in New Zealand Black (NZB) and SJL/J mice. Prog Exp Tumor Res 13:84-134
14. Mellors RC (1966) Autoimmune disease in NZB/Bl mice. II. Autoimmunity and malignant lymphoma. Blood 27: 435-448
15. Sugai S, Pillarisetty R, Talal N (1973) Monoclonal macroglobulinemia in NZB/NZW F1 mice. J Exp Med 138:989-1001
16. Greenspan JS, Gutman GA, Talal N, Weissman IL, Gugai S (1974) Thymus-antigen and immunoglobulin-positive cells in lymph nodes, thymus and malignant lymphomas of NZB/NZW mice. Clin Immunol Immunopath 3:32-51
17. Yumoto T, Yoshida Y, Yoshida H, Ando K, Matsui K (1980) Prelymphomatous and lymphomatous changes in splenomegaly of new zealand black mice. Acta Pathol Jpn 30:171-86
18. Sadahira Y, Mori M, Ozaki M, Awai M (1987) Characteristics of histiocytic lesions in the reticuloendothelial system of NZB mice. Acta Pathol Jpn 37:1719-1729
19. Stall AM, Farinas MC, Tarlinton DM, Lalor PA, Herzenberg LA, Strober S, Herzenberg LA (1988) Ly-1 B-cell clones similar to human chronic lymphocytic leukemias routinely develop in older normal mice and young autoimmune (New Zealand Black-related) animals. Proc Natl Acad Sci USA 85:7312-7316
20. Fialkow PJ, Paton GR, East J (1973) Chromosomal abnormalities in spleens of New Zealand Black mice, a strain characterized by autoimmunity and malignancy. Proc Nat Acad Sci USA 70:1094-1098
21. Okada T, Takiura F, Tokushige K, Nozawa S, Kiyosawa, T, Nakauchi H, Hirose S, Shirai T (1991) Major histocompatability complex controls clonal proliferation of CD5+ B cells in H-2-congenic New Zealand mice: a model for B cell chronic lymphocytic leukemia and autoimmune disease. Eur J Immunol 21:2743-2748

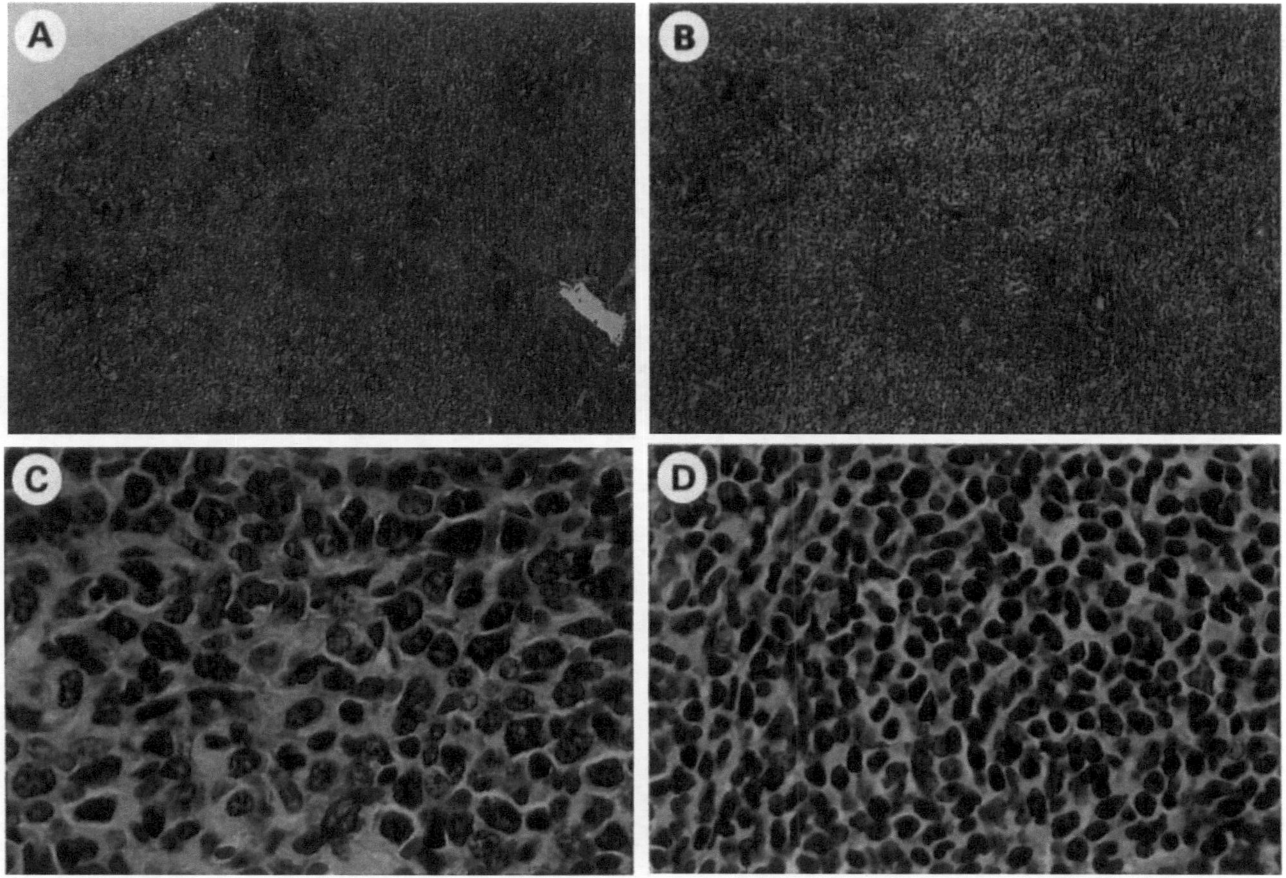

Fig. 1. Marginal cell lymphoma in an aged NZB. A, B, C and D are 5x, 20x, 100x oil and 100x oil respectively. In A the red pulp is replaced, and the interface between the red and white pulp is shown in B. C shows the lymphoid infiltration, while D shows the remaining remnant white pulp.

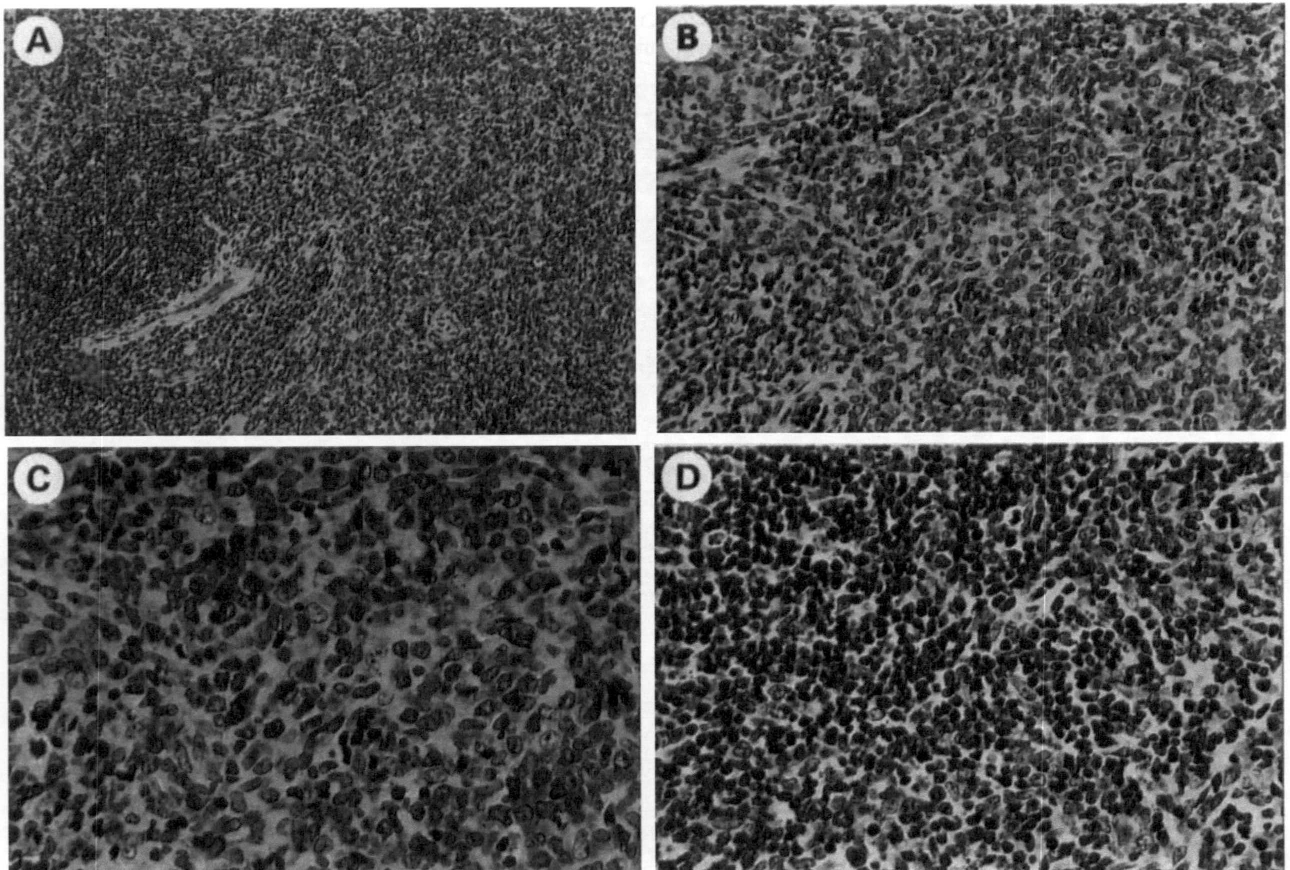

Fig. 2. Lymphocytic lymphoma in an aged NZB. A, B, C and D are 20x, 40x, 63x oil and 63x oil, respectively. In A the red pulp is replaced, and in B there is some invasion into the white pulp. C and D are higher magnifications of the lympoid infiltrate and the remaining normal white pulp, respectively.

EBV Infection of B-CLL Cells in Vitro Potentiates Their Allostimulatory Capacity if Accompanied by Acquisition of the Activated Phenotype

E. Klein and J. Avila-Cariño
Microbiology and Tumorbiology Center (MTC), Karolinska Institutet, S-171 77 Stockholm, Sweden.

Introduction

The immune response is mainly responsible for the harmless EBV carrier state in humans, as indicated by the elevated risk for virus genome carrying B cell malignancies in immunosuppressed individuals (Nalesnik et al. 1988). The role of cellular immunity in the control mechanism is substantiated by *in vitro* results showing that T cells can inhibit the virus induced transformation of B cell cultures and that cytotoxic T cells (CTL) specific for autologous immortalized B cells (lymphoblastoid cell lines, LCLs) are detected in experiments performed with cells of seropositive individuals (Thorley-Lawson et al. 1977; Moss et al. 1979). In search of the CTL targets, all the known EBV encoded proteins expressed in LCLs, except for one, EBNA-1, provided peptides which were recognized by T cells when presented by the appropriate HLA molecules on the cell surface (Masucci and Ernberg 1994). These CTLs represent an EBV specific cellular memory because they can be generated *in vitro* only in experiments performed with lymphocytes of seropositive individuals (Moss et al. 1979). On the other hand, autologous LCL can stimulate T lymphocytes *in vitro* even in experiments with cells of seronegative individuals (Weksler 1976). It is likely that the efficient interaction of EBV genome carrying B cells with T lymphocytes is important in controlling their proliferative potential prior to development of the specific immunity.

The influence of the B-cell phenotype and its interaction with T lymphocytes has been shown earlier in several types of experiments, using also normal B cells (Crow and Kunkel 1985). B-type chronic lymphocytic leukemia (B-CLL) cells exhibit the resting phenotype and do not stimulate T lymphocytes *in vitro*. However when CLL cells were treated with phorbol ester, 12-O-tetradecanoylphorbol-13-acetate (TPA), their phenotype changed and they could trigger allogeneic and even autologous T cells for DNA synthesis (Kabelitz et al. 1984).

Though they proliferate, EBV negative Burkitt lymphoma (BL) lines resemble resting B cells and their allostimulatory capacity is very weak. They can be infected with the virus *in vitro*. In experiments with 5 EBV negative BL lines we found that acquisition of the viral genome (B95-8 or P3HR-1 virus) was accompanied by phenotypic changes. The cells expressed activation markers and they became more efficient allogeneic stimulators. The degree of phenotypic changes induced by EBV varied between the various BL lines. Usually they did not reach the fully lymphoblastoid phenotype of the LCLs. Their allostimulatory capacity was also weaker when compared to the LCL established from the same individual. There was no significant difference between the response of the lymphocytes collected from 9 EBV-seropositive and 3 seronegative individuals (Avila-Cariño et al. 1987). These experiments did not allow to decide whether the presence of the EBV genome, or the phenotypic change induced by the virus, is responsible for the acquisition of T-cell-stimulatory capacity of the cells.

We performed similar experiments with sublines of one EBV-carrying BL selected for phenotypic differences with regard to the interaction of cells, growing as solitary or aggregated cells. Surface marker analysis showed that the former resembled resting, the latter resembled activated B lymphocytes. Expression of the EBV-encoded proteins was shown to differ in these sublines; the cells with the resting phenotype expressed only EBNA-1, while those with the activated phenotype expressed EBNA-1-6 and LMP-1 (Gregory et al. 1990). Only the latter stimulated allogeneic T lymphocytes. Analysis of the surface markers suggested that the stimulatory capacity is correlated with expression of adhesion molecules LFA-1 and LFA-3. Since both samples carried the viral genome, these results established the importance of the cell phenotype for the T cell stimulatory capacity (Avila-Cariño et al. 1991).

In both series of experiments the stimulators were lymphoma cells adapted to grow *in vitro*. To approach the question whether the strong T-cell-stimulatory potential of the LCLs is coupled to their immortalized state we used CLL cells. The majority of B-CLL clones, represented by different patients, carry EBV receptors and can be infected with the virus but they rarely give rise to immortalized lines (Walls et al. 1989). In addition, B-CLL cells can be activated by B cell mitogens (Carlsson et al. 1989). Thus, when exposed *in vitro* to EBV, and/or to mitogens, B-CLL cells can provide activated B cell populations without and with EBV infection. We compared the T cell stimulatory capacity of such cells derived from 12 patients. We show here the results obtained with 2 patients, GHW and WT. Their cells responded differently when exposed to EBV. The results with GHW are representative for the majority of experiments.

Experimental Results

Blood lymphocytes of CLL patients were isolated on Ficoll-Isopaque gradient and B cell-enriched populations were obtained after passage through nylon wool. Contamination with T cells and macrophages was less than 1 %. The cells were cultured in RPMI-1640 medium containing 2mM L-glutamine, 100 IU/ml penicillin, 100 μg/ml streptomycin and 10% heat-inactivated FCS. Four cultures were initiated, each containing 3-5 x 10^6 cells/ml. One contained untreated cells, the others were treated as follows: (1) Exposure to EBV containing supernatants from the virus producer B95-8 cell line for 1 hr followed by washing. (2) Exposure to the "activation mixture, AcMx" which contained 0.1% heat-inactivated, formalin-treated *S. aureus* Cowan I (SAC), 10 units of recombinant IL-2 (a gift from Ajinomoto, Kawasaki, Japan) and 25% (v/v) of the supernatant from the T cell hybridoma MP6 (Carlsson et al. 1989), which contains an isoform of thioredoxin. (3) Combination of (1) and (2) i.e. EBV infection and AcMx. The cells were cultured at 37°C in 5% CO2 and after 5 days, the viable cells were separated on Ficoll-Isopaque. The populations were analysed for immunophenotype, expression of EBNA and the capacity to induce proliferation of allogeneic blood lymphocytes. (Detailed results are in press, Avila-Cariño et al. 1994)

Results with the Cells of Patient GHW (Table 1)
The cultures contained mainly CLL cells as indicated by the expression of the CD5 marker. The recovery of cells 5 days after explantation was highest in the AcMx treated samples. The morphology of the EBV infected culture was not altered, while the AcMx-treated cells without and with EBV infection were enlarged and they aggregated. In the EBV infected cultures about 30% EBNA-positive cells were detected. Both EBNA 1 and 2 bands were visualised in western blots.

Blood lymphocytes from 6 healthy individuals were tested for proliferative response in mixed cultures with the CLL cells. The EBV infected CLL cells induced no or only very low level of DNA synthesis, while the AcMx-treated samples stimulated the allogeneic lymphocytes, though weakly. The AcMx-treated and EBV-infected cells had stronger stimulatory potential, the DNA synthesis was 5 fold higher than that induced by the activated but non virus infected samples. An LCL established from normal B cells of the patient with the B95-8 virus was used as stimulator for one responder population. It induced much higher level of DNA synthesis than the CLL sample treated with EBV-AcMx; the R.I. was 400.

We searched for phenotypic traits, such as the expression of activation markers and adhesion molecules for possible association with the stimulatory capacity of the different CLL populations. Here we show the percentages of positive cells reactive with the various reagents, estimated from the FACS analysis. These numbers provide a good information about the differences between the samples. The reactivity profiles indicated that the EBV infected cultures contained a subpopulation with activated phenotype as shown by the increased expression of CD23 and CD39 on a fraction of cells. The profiles of the AcMx treated populations showed a shift to the right indicating a general increase of CD23 and CD39 markers. After the combined treatment (EBV and AcMx) the profiles reflected both the change induced by EBV and by AcMx, in that the culture contained also a fraction of cells with higher fluorescence intensity. The adhesion molecules CD11a, and CD58 were expressed in all 3 cultures. Expression of CD54 was high on the cells which were exposed to the activation mixture. Importantly, the AcMx cells without and with EBV infection had similar reactivity profiles with all reagents. The amounts of BB-1 and MHC Class II molecules were increased on the cells of both populations.

Table 1. Experiments performed with the cells of patient GHW[1]

	EBV	AcMx	EBV-AcMx
Survival of cells[2]	17, 17	45, 76	33, 44
% EBNA positive cells[3]	37, 35	0	32, 25
Stimulation of lymphocytes[4]	1 (1-2)	4 (2-6)	17 (6-36)
Immunophenotype[5]			
CD3	7	6	6
CD5	85	80	78
CD21	52	62	47
CD23	19	38	52
CD39	19	39	58
CD11a	64	70	79
CD54	32	88	87
CD58	75	83	83
BB-1	6	26	39
sIgM	5	11	7

1. 10^7 cells were infected with EBV and/or activated with AcMx (SAC-IL2-MP6). Two independent experiments.
2. Recovery of viable cells after 5 days. % of the input.
3. Proportion of EBNA-positive cells scored by anticomplement immunofluorescence (Reedman and Klein, 1973).
4. Reactivity Index. Mean of 6 experiments performed with responder lymphocytes collected from 6 different healthy donors, in parenthesis the range.
5. Percentage of positive cells.

Results with the Cells of Patient WT (Table 2)

The untreated CLL cells of WT expressed neither activation markers nor cell adhesion molecules. They were CD23, CD39, CD11a, CD18, CD54 and CD58 negative. They were highly susceptible for infection with EBV as shown by the presence of EBNA in almost all cells. The recovery of the cells in the 5 days old cultures was higher than in the cultures of patient GWH. It is important to note that the culture did not represent immortalized cells. The cells expressed activation markers and cell adhesion molecules. They stimulated the allogeneic lymphocytes. Activation of the cells was slightly less efficient in the culture exposed only to AcMx, as indicated by the levels of the various markers. Stimulation of the allogeneic lymphocytes was stronger by the EBV carrying cells.

Table 2. Experiments performed with the cells of patient WT[1]

	EBV	AcMx	EBV-AcMx
Survival of cells	75, 45	22, 65	52, 56
% EBNA positive cells	100, 77	0	98, 63
Stimulation of lymphocytes	41 (21-52)	13 (3-24)	39 (20-53)
Immunophenotype			
CD3	22	25	28
CD5	92	87	86
CD21	92	76	81
CD23	88	37	62
CD39	90	39	86
CD11a	75	40	70
CD54	90	82	94
CD58	91	70	90

1. See details in table 1.

The results obtained with these 2 and with 10 additional CLL patients showed that activated CLL cells can stimulate allogeneic T cells. Provided they were activated, the EBV infected samples were more efficient allo-stimulators. Comparison of the phenotypes did not reveal any particular trait imposed by EBV infection which was associated with the potentiation of allostimulatory capacity.

Thus, EBV contributed to the allostimulatory capacity of B cells but only when the cells exhibited the activated phenotype. The effect of EBV was superimposed on the phenotype dependent interaction of B cells with the T lymphocytes. The mechanism which is responsible for the superior B cell stimulatory activity of EBV carrying B (CLL) cells compared to those with similar phenotype but devoid of the virus is not known yet.

Discussion

Our results showed the contribution of EBV in the interaction of B (CLL) and T cells in allogeneic combinations which is based on the recognition of alloantigens. In order to allow conclusions which may be relevant for *in vivo* events we attempted to perform similar experiments with autologous combinations using thus the T cells of the CLL patients as responders (Tomita et al. to be published).

Based on earlier reports we expected an impaired T cell function of the CLL patients (Wolos et al. 1980). Technical factors hampered also these tests. In order to minimize the contribution of normal B cells in the EBV infected CLL samples the patients who have entered our study had high leukocyte counts. Thus the T cell populations were not comparable to those derived from healthy individuals which were used in the allogeneic combinations. Indeed, T cells separated from the patients had very low or no responses to allogeneic EBV-AcMx treated CLL samples.

Experiments with T cells of 8 patients were performed, including GHW. The autologous EBV-AcMx CLL cells stimulated the lymphocytes in only 2 cases and in the same experiments the AcMx treated cells did not stimulate. Thus the difference between the 2 samples, without and with EBV, was seen also in the induction of the autologous T cell response.

The lymphocytosis in the acute infectious mononucleosis (IM) syndrome is at least in part, the consequence of the T cell stimulation by activated B lymphocytes. The first event in the EBV infection of normal B cells is their activation (Thorley-Lawson and Mann 1985). Results with experimental infection showed that only a fraction of infected B cells enters the immortalized state (Tosato 1987). Based on the establishment of proliferating B cell cultures from the blood of IM patients under conditions which detect B cells with indegenous proliferative capacity, such cells are rare (Lewin et al. 1990). It can be envisaged therefore that upon the primary infection a proportion of B cells acquire the virus, and they become activated. These elicit a T cell response, involving their activation and proliferation. The recognition of activated B cells represent a regulatory mechanism within the immune response. It is likely that the harmless maintenance of EBV is possible because the main target of the virus, the B lymphocyte is under a strict control. Extrapolated to normal cells, our present results suggest that the presence of the EBV in the B cells potentiates their T cell stimulatory capacity when they acquire the activated phenotype. Taken together, it seems that the EBV-induced immortalization of B cells does not have to be implicated for generation of the early T cell response. This event is probably of great importance because the activated T cells can eliminate or suppress EBV genome carrying potentially malignant and/or virus producing B cells before the development of an EBV specific immune response.

References

Avila-Cariño J, Lewin N, Yamamoto K, Tomita Y, Mellstedt H, Brodin B, Rosén A, Klein E (1994) EBV infection of B-CLL cells in vitro potentiates their allostimulatory capacity if accompanied by acquisition of the activated phenotype. Int J Cancer in press

Avila-Cariño J, Torsteinsdottir S, Ehlin-Henriksson B, Lenoir G, Klein G, Klein E, Masucci MG (1987) Paired Epstein-Barr virus (EBV)-negative and EBV-converted Burkitt lymphoma lines: stimulatory capacity in allogeneic mixed lymphocyte cultures. Int J Cancer 40:691-697

Avila-Cariño J, Torsteinsdottir S, Ehlin-Henriksson B, Masucci MG, Klein E (1991) Search for the critical characteristics of phenotypically different B cell lines, Burkitt lymphoma cells and lymphoblastoid cell lines, which determine differences in their functional interaction with allogeneic lymphocytes. Cancer Immunol Immunother 34:128-132

Carlsson M, Tötterman TH, Rosén A, Nilsson K (1989) IL-2 and a T cell hybridoma (MP6) derived B cell-stimulatory factor act synergistically to induce proliferation and differentiation of human B-chronic lymphocytic leukemia cells. Leukemia 3:593-601

Crow MK, Kunkel HG (1985) Activated B lymphocytes: stimulators of an augmented autologous mixed leukocyte reaction. Cell Immunol 90:555-568

Gregory CD, Rowe M, Rickinson AB (1990) Different Epstein-Barr virus-B cell interactions in phenotypically distinct clones of a Burkitt's lymphoma cell line. J Gen Virol 71:1481-1495

Kabelitz D, Tötterman TH, Nilsson K, Gidlund M (1984) Phorbol ester treated chronic B lymphocytic leukaemia cells induce autologous T cell proliferation without generation of cytotoxic T cells. Clin Exp Immunol 57:461-466

Lewin N, Åman P, Åkerlund B, Gustavsson E, Carenfelt C, Lejdeborn L, Klein G, Klein E (1990) Epstein-Barr virus-carrying B cells in the blood during acute infectious mononucleosis give rise to lymphoblastoid lines in vitro by release of transforming virus and proliferation. Immunol Lett 26:59-66

Masucci MG, Ernberg I (1994) Epstein-Barr virus: adaptation to a life within the immune system. Trends in Microbiology 125:125-130

Moss DJ, Rickinson AB, Pope JH (1979) Long-term T-cell-mediated immunity to Epstein-Barr virus in man. III. Activation of cytotoxic T cells in virus-infected leukocyte cultures. Int J Cancer 23:618-625

Nalesnik MA, Jaffe R, Starzl TE, Demetris AJ, Porter K, Burnham JA, Makowa L, Ho M, Locker J (1988) The pathology of posttransplant lymphoproliferative disorders occurring in the setting of cyclosporin A - prednisone immunosuppression. Am J Pathol 133:173-192

Reedman B, Klein G (1973) Cellular localization of an Epstein-Barr virus (EBV)-associated complement-fixing antigen in producer and non-producer lymphoblastoid cell lines. Int J Cancer 11:499-520

Smith JB, Knowlton RP, Koons LS (1977) Immunologic studies in chronic lymphocytic leukemia: defective stimulation of T-cell proliferation in autologous mixed lymphocyte culture. J Natl Cancer Inst 58:579-585.

Thorley-Lawson DA, Chess L, Strominger JL (1977) Suppression of in vitro Epstein-Barr virus infection. A new role for adult human T lymphocytes. J Exp Med 145:495-508

Tosato G (1987) The Epstein-Barr virus and immune system. In: Van de Woude GF, Klein G (ed) Academic Press, New York, pp 75-124 (Advances in Cancer Research, vol 49

Walls EV, Doyle MG, Patel KK, Allday MJ, Catovsky D, Crawford DH (1989) Activation and immortalization of leukaemic B cells by Epstein-Barr virus. Int J Cancer 44:846-853

Weksler ME (1976) Lymphocyte transformation induced by autologous cells. III. Lymphoblast-induced lymphocyte stimulation does not correlate with EB viral antigen expression or immunity. J Immunol 116:310-314

Wolos JA, Davey FR (1980) Function of lymphocyte subpopulations in chronic lymphocytic leukemia: activity in the allogeneic and autologous mixed lymphocyte reaction. Cancer 45:893-898

Acknowledgements

This work was suported by the Swedish Cancer Society and the Cancer Research Institute/Concern Foundation, USA.

A New Form of Epstein-Barr Virus Latency in vivo

E. Miyashita and D.A. Thorley-Lawson
Dept. of Pathology, Tufts University School of Medicine, 136 Harrison Ave., Boston MA 02111 USA.

Introduction

Epstein-Barr virus is a human herpesvirus that is associated with a number of tumors including Burkitts lymphoma, immunoblastic lymphoma, Hodgkins lymphoma, rare T cell lymphomas and nasopharyngeal carcinoma, suggesting a relatively broad tissue tropism for the virus in vivo [1,2]. Immunosuppression through allograft transplantation or HIV infection is known to promote the development of EBV positive tumors, particularly immunoblastic lymphomas. In vitro the virus has a strong tropism for B cells which it infects and causes to become latently infected, immortalized lymphoblasts. This is the only model system for EBV latency in a normal cell. These latently infected cells express 9 known latent proteins and high levels of cell surface markers, such as CD23, that are characteristically expressed on activated B cells [3,4].

Infected cells are present for life in the circulation of healthy seropositive individuals. Surprisingly little is know about the nature of these cells although some are known to be B cells. Table I summarizes a list of pertinent questions about the nature of these cells and what has been definitively described in the literature.

Table I. What we know about EBV bearing cells in the circulation?

Normal Individuals.	
Absolute numbers of infected cells in blood?	Not known.
Phenotype of infected cells in blood?	Not known.
What viral genes are expressed?	Not known.
What changes with immunosuppression?	
The absolute numbers of infected cells in blood?	Not known.
The phenotype of infected cells in blood?	Not known.
The nature, latent or lytic, of the infection?	Not known.

The lack of definitive information arises due to the nature of the assays used to look for the virus. Most early work exploited the fact that EBV immortalized B cell lines grow directly from peripheral blood [5,6]. It has been shown that this occurs by an indirect method involving release of virus and reinfection in culture. Therefore, this technique can only detect cells capable of releasing virus. It will not detect other cell types that either cannot grow directly in culture or will not release infectious virus. Although the frequency of cells detected by this method increases upon immunosuppression, it has not been established if this reflects a true increase in the number of virus infected cells or simply a change in the fraction of infected cells able to release infectious virus. More recently this question has been addressed by PCR based techniques. However, to date the assays used have not been sensitive enough to detect single virus infected cells in large enough numbers of uninfected cells. The most sensitive so far reported is 1 in 1.5x 10^5 [7]. This has required that analysis be performed on bulk populations of cells - essentially estimates of viral burden. However, it is not known what the genome copy number per cell is in vivo. From in vitro studies we know latently infected cells carry variable numbers (usually 5-100) of the viral genome as episomes [8] whereas rare cell lines with integrated genomes often only carry 1 or 2 copies per cell [9] and lytically infected cells carry thousands of copies per cell. Therefore, it is not possible to convert estimates of viral burden into reliable estimates of infected cell number. The PCR signal for EBV DNA increases upon immunosuppression, but it is not known if this increase in viral burden reflects an increase in latently infected cells or a change in the fraction of cells that are lytically infected. More surprising, no attempt has been made to test if the PCR signals are from bona fide infected cells or simply represent virions or viral fragments attached to viral receptors on the outside of B cells.
We initiated our studies to precisely quantitate the number of infected cells in the circulation of healthy adults and then use this information to analyze the phenotype of the cells and the nature, latent or lytic, of the infection.

Results

Development of a reproducible and quantitative DNA PCR assay.

To make definitive statements about the number and kind of EBV infected cells in, for example, peripheral blood, it was essential to develop a PCR based assay that could routinely detect a single copy of the viral genome in a defined number of cells. Conditions for the assay were defined by analyzing limiting dilutions of Namalwa, which has only 1 or 2 copies of the genome, in the presence of 10^6 EBV negative cells. Through

Poisson analysis of multiple samples it was possible to demonstrate that we could readily and reproducibly detect a single Namalwa cell, i.e. 1-2 genomes, in the presence of 10^6 uninfected cells. By processing up to 50 aliquots of 10^6 cells per sample it is therefore possible to detect as low as 1 genome in $50x10^6$ cells. With this sensitivity and using $\leq 10^6$ cells per sample, we can identify every infected cell in a given sample and, therefore, when we compare observations between subpopulations of cells and individuals, we know we are comparing absolute numbers not numbers compromised by a limit in the sensitivity.

In what lymphoid compartment does the EBV reside?
Table II demonstrates an example of an experiment on subfractions of peripheral blood mononuclear cells from a single individual.

Table II. Frequency of EBV bearing cells in subfractions of peripheral blood mononuclear cells from a healthy seropositive donor.

Population	#Cells/Sample	#positive/#tested	Est. Frequency
E-	$1x10^6$	14/15	>1 in $1x10^6$
B	$2x10^5$	7/7	
	$2x10^4$	5/7	1 in $1.5x10^4$
M	$1x10^6$	1/10	~1 in $1x10^7$
T	$1x10^6$	1/40	~1 in $4x10^7$
BJAB	$1x10^6$	0/36	

Cells were aliquoted in multiple samples of 10^6 down. Frequencies were estimated by Poisson statistics on the limiting dilution analysis. All the cells were subjected to PCR using primers from the Bam W region that were chosen to have high annealing temperatures. PCR conditions were optimized for temperature and Mg++ concentration. Reactions were usually performed in sets of 36, of which 10 were BJAB controls. The PCR reactions were resolved on an agarose gel and the specific product identified with a probe internal to the sequence that was amplified. The probe was labelled by random priming. The type of cells tested, the number of cells per sample and the fraction positive are indicated for each set. An example of a Southern blot is shown in Fig. 1. E-, B cells + monocytes; B, B cells (CD19+); M, Monocytes (E-, CD19-); T, T cells (E rosette+)

Peripheral blood mononuclear cells were fractionated by rosetting with sheep red blood cells - the positively selected population are T cells (E+) and the negative population is predominantly a mixture of B cells and monocytes (E-). Whereas 14 of the 15 E- samples were positive only 1 of 40 T cell samples were positive. The E- population was further fractionated into CD19+ (B cells) and CD19- (predominantly monocytes) using Dynal magnetic beads coated with a monoclonal antibody to the pan-B cell marker CD19. Virtually all of the EBV infected cells reside in the B cell compartment. Only 1 of 10 samples were positive for the E-,CD19- population. All samples of B cells were positive until cell number

per sample was reduced to $2x10^4$ when 5 of 7 were positive. In parallel, 36 samples of the EBV negative BJAB cell line were tested in this experiment with 36 of 36 being negative.
The in vitro tropism of the virus would predict that the virus would reside in the B cell compartment. However, the demonstration of EBV in T cell lymphomas [10] and the presence of the receptor on thymocytes [11] formally raises the possibility that EBV could also be present in other cell types such as T cells. This experiment demonstrates that the EBV infected cells reside predominantly in the B cell compartment. The purity of all the populations was confirmed in parallel by immunofluorescence and FACS analysis with the three monoclonal antibodies T3 (anti-CD3, T cell specific), B1 (anti-CD20, B cell specific) and Mo-2 (anti-CD14, monocyte specific). For this particular donor, the peripheral blood white cells were about 70% T cells, 10% B cells and 10% monocytes and 10 % undefined. Therefore, it can be estimated that greater than 95% of all the EBV infected cells reside in the B cell compartment. We have repeated the same analysis on a second donor with the same result. Although this is the expected conclusion, this is the first time it has been formally shown to be true for peripheral blood of healthy donors - it remains to be tested if it is also true for immunosuppressed donors or for other lymphoid organs.

Are the signals detected in the T and monocyte populations significant?
FACS analysis with the B1 antibody revealed that B cell contamination of both of these population was less than 1%. However, it can be estimated that contamination $\leq$ 0.2% of B cells could account for the signals seen. Therefore, we cannot formally distinguish between contamination and the presence of very rare EBV infected T cells and monocytes.

Are the signals real?
At the limiting cell dilution we are probably detecting 1-10 molecules of EBV DNA (assuming multiple genomes in a single infected cell) in the presence of DNA from as many as 10^6 cells. How do we know the signals are real?
a. For every set of PCR reactions ~10 samples of BJAB (EBV negative cell line) were included. We have never obtained a positive result.
b. Negative samples could be amplified by PCR for a cellular gene, bcl-2 which suggests that the negative samples are truly negative.
c. The signal is enriched in specifically purified populations and the fraction of negative samples increases when the number of cells per sample decreases, for a given population, suggesting that both positive and negative signals are real.
d. We have cloned and sequenced the PCR products from three individuals. The sequences demonstrate significant diversity from the laboratory strain, B95-8, indicating that they are not due to laboratory

contamination and from each other indicating that we are not detecting contamination from saliva aerosols of virus from the experimenter.

Is the viral DNA from latently infected cells?

Since we are detecting only a few molecules of EBV DNA we cannot be sure of the form of the DNA.

The DNA could be present in virions bound to the EBV receptor on the surface of the cell and the cells not really infected. If this were true, we would also expect to see enrichment of the signal in the B cell compartment, since this is the cell type that bears the receptor at high density. To eliminate this possibility we have compared the frequency of EBV positive samples in cells and nuclei prepared from E- cells from the same donor at the same time. The results indicate that there is no difference in the frequency, consistent with the viral DNA being in the nucleus not attached to the outside of the cell.

Another possibility is that the estimates are low because the virus infected cells are clumped together. However this is also ruled out by the nuclei experiment since clumps of cells should be dissociated by the detergent.

The question also arises as to whether the cells are lytically or latently infected. Based on cell lines, latently infected cells have genome copy numbers in the range of 5-100 whereas lytically infected cells have copy numbers in the thousands. The PCR signals from the peripheral B cells are comparable to those obtained with controls containing 2 copies of the viral genome, much lower than with controls containing 10^4 copies (see Fig. 1), as expected for a latent rather than a lytic infection.

Can EBV be detected in the blood of all seropositive normal adults?

Since we can detect 1 infected cell in as few as $5x10^7$ uninfected cells we have reassessed the issue of whether all seropositives have virus infected cells in their blood. We have analyzed cells from a panel of 9 seropositive individuals and find that infected cells are detectable in all those tested. The estimated frequencies for whole peripheral blood range from around 1 in $2x10^5$ to 1 in $4x10^6$. The common occurrence of low frequencies, 6 of the nine donors have frequencies lower than 1 in 10^6, may explain why previous studies have failed to detect the virus in all normal donors.

Does the frequency vary with time for a given donor?

It appears that the frequency remains quite constant, at least over the course of 1-2 years. The high frequency donors remain high and the low frequency donors remain low. For example, the frequency of EBV containing B cells in donor 5 (high frequency) was estimated at 63 per 10^6 (11/9/92), 29 per 10^6 (6/4/93) and 43 (5/31/94), a mean of 45 with a variation of less than 2 over the course of two years. Whereas the frequency in donor 9 (low frequency) was estimated as 5 per 10^6 (11/30/92) and 9 per 10^6 10 months later (10/15/93). This suggests that the frequency may be intrinsic to the particular donor and supports the conclusion that our measurements are real and reproducible.

The virus carrying B cells are not typical lymphoblasts.
EBV immortalized B cells in vitro express a characteristic lymphoblastoid phenotype. We have used our assay to characterize the phenotype of the EBV infected B cells in vivo to see if they are the same. We have fractionated peripheral B cells on the basis of CD23 expression because CD23 is expressed at a high level on EBV immortalized B cells in vitro and should positively select for such cells. Dynal magnetic beads coupled with sheep anti-mouse antibody were precoated with an anti-CD23 monoclonal antibody (optimal coating with the antibody was determined by the ability of the beads to bind to the EBV immortalized cell line ER).

We wished to manipulate the cells as little as possible and we knew that >95% of the EBV carrying cells were in the B cell population, therefore the analysis was performed on whole E- cells. FACS analysis was performed for CD23, to confirm the efficiency of separation, and CD20, to determine the percentage of B cells in each fraction. The analysis demonstrated that the CD23+ B cells were > 99% depleted, but the CD23- B cells were not depleted. The results of the PCR analysis on the CD23 sorted populations from two donor are summarized in Table III and the actual Southern blot data for one donor is shown in Fig. 1.

Table III. The EBV carrying B cells in peripheral blood are phenotypically distinct from in vitro immortalized lymphoblasts - analysis by CD23 expression.

	Frequency/10^6 Cells		
Donor #	Whole B cells	CD23+ B cells*	CD23- B cells
7	46	<2	74
9	9	<1	47

Donor 7 and 9 were selected specifically because they had, respectively, high and low frequencies of EBV carrying cells in their blood.

* Since FACS analysis cannot be done on this population the calculation was performed on the assumption that the CD23+ population are all B cells since this marker is highly specific for B cells. Nevertheless this population is likely contaminated with significant numbers of monocytes.

Using the information from the FACS and PCR analysis it was possible to estimate frequencies of EBV containing cells in CD23 positive and negative B cell populations as shown in Table III. The result is unequivocal, the EBV infected B cells in peripheral blood are CD23-.

Therefore the cells in peripheral blood are not typical immortalized lymphoblasts.

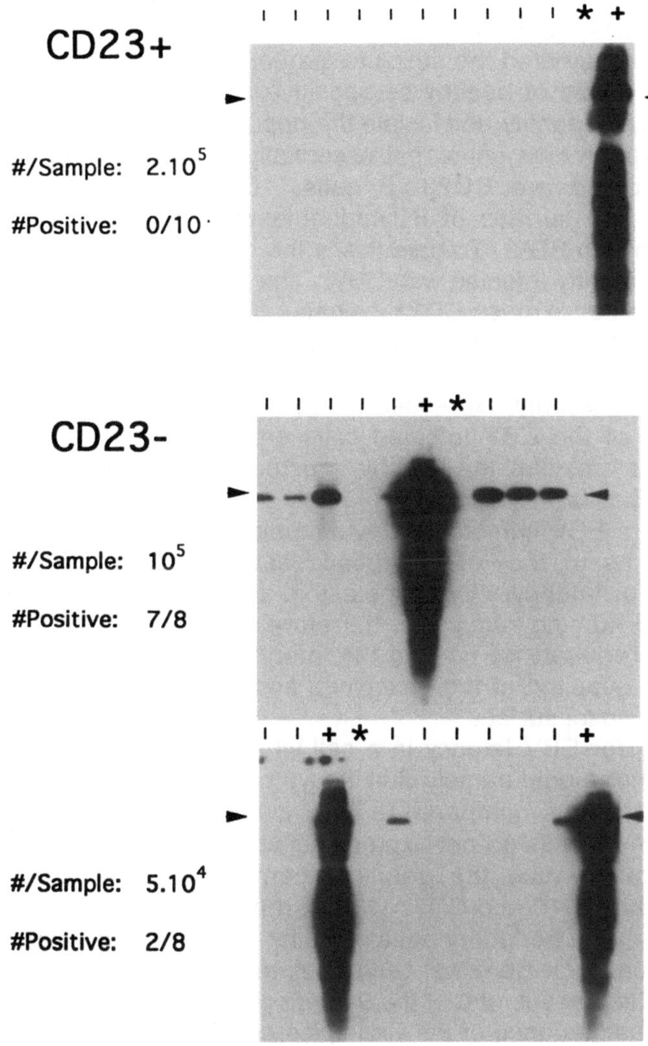

Fig 1. PCR analysis of E- Peripheral lymphocytes from a single donor fractionated on the basis of CD23 expression. All the cells were subject to PCR analysis as described in the legend to Table I. The type of cells tested, the number of cells per sample and the fraction positive are indicated on the left for each set.

► -Denotes Expected Size of the PCR product.

I -Location of lanes containing experimental samples.

+ -Positive control Namalwa cell DNA in PCR reaction. Equivalent to 10^4 genomes.

- -No DNA in the PCR reaction.

* -No PCR sample in the lane. Usually unlabelled mol. wt. markers.

Conclusions.

We have developed a DNA PCR assay that is sensitive enough to detect a single EBV genome in the presence of 10^6 uninfected cells. With this assay we have estimated the absolute frequency of virus infected cells in the peripheral blood of healthy seropositive individuals. We have also used the assay to identify and isolate the population of cells that carry the virus. Our experiments show that essentially all of the latently infected cells in the blood are CD23- B cells. CD23 is characteristically expressed on the surface of B lymphoblasts immortalized in vitro by latent infection with EBV. To date this is the only known phenotype for a normal B cell latently infected with EBV. The demonstration that infected cells in vivo do not express CD23 indicates that they are of a phenotype never seen before with a normal B cell that maintains latent EBV. Recently, we have repeated this observation with a second marker, CD80 (B7), that is only expressed on activated B cells [12] and again found that all of the EBV infected cells are CD80 (B7)- B cells. We conclude therefore, that the latently infected cells in vivo are probably small resting B cells.

The absence of EBV immortalized lymphoblasts in the circulation raises the question as to the role of these cells in EBV biology and the maintenance of latency. Lymphoblastoid cells express high levels of adhesion and homing receptors, therefore, one possibility is that this phenotype is required for homing the latently infected cell to the basal epithelium for infection of the epithelium and subsequent release of the virus at the mucosal surface.

A precedence for EBV latency in a cell which lacks activation markers exists in a tumor model namely Burkitts lymphoma (BL). The tumor cells are small and simple compared to EBV lymphoblasts, which are large and complex. BL cells do not express activation markers like CD23 and are phenotypically like the germinal center centrocyte/centroblasts expressing both CD10 and CD77 [13]. It remains to be seen if the virus carrying cells in the periphery have a similar phenotype although it would be very unusual for a germinal center cell to be in the periphery. In BL the virus only expresses one of the 9 latent proteins (EBNA-1) [14] that is essential for maintenance of the viral episome. This raises the possibility that a minimal form of latency could also exist in normal latently infected B cells in vivo. This could take the form of EBNA-1 only, like the BL, or even no viral gene transcription at all if the cells are not proliferating.

Since the latently infected cells we have detected in vivo are not activated and not replicating the virus they pose no threat to the host and do not need to be eliminated by the immune response. Indeed if they are expressing no viral proteins they may not even be detected. One prediction of this hypothesis is that the frequency of such cells will not increase upon immunosuppression, since they are predicted to not be under immunosurveillance. We have not detected significant numbers of EBV infected B cell that have the lymphoblastoid phenotype. This is probably because these cells are efficiently eliminated by cytotoxic T cell surveillance, as has been previously suggested [15]. It follows that, upon immunosuppression, the increase in infected cells may represent the emergence of lymphoblastoid cells into the circulation. These predictions can be tested by monitoring the phenotype and number of infected cells before and after immunosuppression, for example with allograft patients.

References.

1. Kieff E and Liebowitz D (1990) Epstein-Barr virus and its replication. In: Virology, Second Edition. Fields, BN, et al., eds. Raven Press, p. 1889-1920
2. Miller G (1990) Epstein-Barr virus: Biology, pathogenesis and medical aspects. In: Virology, Second edition, Fields BN, et al., eds. Raven Press, p. 1921-1958
3. Thorley-Lawson DA and Mann KP (1985) Early events in Epstein-Barr virus infection provide a model for B-cell activation. J Exp Med 162:45-59
4. Thorley-Lawson DA Nadler LM Bhan AK and Schooley RT (1985) Blast 2 (EBVCS), an early cell surface marker of human B-cell activation, is superinduced by Epstein-Barr virus. J Immunol 134:3007-3012
5. Lewin N Aman P Masucci M Klein E Klein G Oberg B Strander H Henle W Henle G 1987 Characterization of EBV-carrying B-cell populations in healthy seropositive individuals with regard to density, release of transforming virus and spontaneous outgrowth. Int J Cancer 39:472-476
6. Yao Q Czarnecka H Rickinson A (1991) Spontaneous outgrowth of Epstein-Barr virus-positive B-cell lines from circulating human B cells of different buoyant densities. Int J Cancer 48:253-257
7. Wagner HJ Bein G Bitsch A Kirchner H (1992) Detection and quantitation of latently infected B lymphocytes in Epstein-Barr virus-seropositive, healthy individuals by polymerase chain reaction J Clin Micro 30:2826-2829
8. Adams A (1979) The state of the virus genome in transformed cells and its relationship to host cell DNA. In: Epstein MA Achong B (ed) The Epstein-Barr virus Springer-Verlag Berlin pp155-183
9. Hurley EA Agger S McNeil JA Lawrence JB Calendar A Lenoir G Thorley-Lawson DA (1991) When Epstein-Barr virus persistantly infects B-cell lines, it frequently integrates J Virol 65:1245-1254
10. Anagnostopoulos I Hummel M Finn T Tiemann M Korbjuhn P Dimmler C Gatter K Dallenbach F Parwaresch MR Stein H (1992) Heterogeneous Epstein-Barr virus infection patterns in peripheral T-cell lymphoma of angioimmunoblastic lymphadenopathy type. Blood 80:1804-1812
11. Tsoukas CD Lambris JD (1988) Expression of CR2/EBV receptors on human thymocytes detected by monoclonal antibodies. Euro. J. Immunol.18:1299-1302
12. Freedman AS Freeman G Horowitz JC Daley J Nadler LM (1987) B7, a B-cell-restricted antigen that identifies preactivated B cells. J Immunol139:3260-3267.

13. MacLennan IC Liu YL Ling NR (1988) B cell proliferation in follicles, germinal centre formation and the site of neoplastic transformation in Burkitt's lymphoma. Curr Top Microbiol Immunol 141:138-148.
14. Kerr BM Lear AL Rowe M Croom-Carter D Young LS Rookes SM Gallimore PH Rickinson AB (1992) Three transcriptionally distinct forms of Epstein-Barr virus latency in somatic cell hybrids: cell phenotype dependence of virus promoter usage. Virol 187:189-201.
15. Rickinson AB Yao QY Wallace LE (1985) The Epstein-Barr virus as a model of virus-host interactions. Brit Med Bull 41:75-79.

Pharmacologic Activation of Expression of Immunodominant Viral Antigens: A New Strategy for the Treatment of Epstein-Barr-Virus-Associated Malignancies

K. D. Robertson[1,2], J. Barletta[3], D. Samid[4], and R. F. Ambinder[1,2,3]

From the Departments of Oncology[1], Pharmacology and Molecular Sciences[2], Pathology[3], Johns Hopkins School of Medicine, Baltimore, Maryland and the Pharmacology Branch[4] of the National Cancer Institute, Bethesda, Maryland.

Introduction

Although many B-cell malignancies are sensitive to chemotherapy and radiotherapy and some may be cured, the majority relapse and prove fatal. Even at the extremes of dose escalation requiring bone marrow transplantation or peripheral blood stem cell rescue, there are resistant cells in many tumors that will ultimately lead to relapse. Immunotherapeutic approaches to the destruction of tumors have attracted interest because resistance to chemotherapy and radiotherapy probably does not confer resistance to immune-mediated killing. In addition, a highly selective attack on tumor cells might avoid the morbidity associated with more conventional approaches. Immunotherapy is not yet a standard part of the approach to malignancy because of the difficulties identifying antigens that are truly tumor specific and inducing a host response to these antigens. Epstein-Barr virus (EBV)-associated lymphomas present some interesting opportunities in this regard.

Epstein-Barr virus (EBV) is a ubiquitous virus that infects the vast majority of adults worldwide [1]. A variety of B-cell tumors carry the EBV genome including African Burkitt's lymphoma, immunoblastic lymphomas in AIDS patients, and B-cell lymphoproliferative disease arising in the setting of organ transplantation. In addition, a variety of other lymphomas and carcinomas also carry the EBV genome. These include nasopharyngeal carcinoma, Hodgkin's disease - particularly mixed cellularity tumors, and undifferentiated gastric carcinomas [1,2,3]. Although in EBV seropositive donors, there is a chronic low level EBV infection and EBV may be detected in the saliva and B cells, for practical purposes the viral genome is a tumor specific target. Only $1/10^6$ non-malignant B cells are infected with the virus and there is no strong evidence to support the existence of other loci of latent infection.

EBV immortalizes and transforms the peripheral blood B cells such that they will grow indefinitely in culture and will produce tumors in mice with severe combined immunodeficiency (SCID). In vitro, EBV-immortalization of B cells is demonstrated by the addition of culture supernatant containing infectious virus to peripheral blood mononuclear cells from an EBV-seronegative donor. Over a period of several weeks EBV-transformed B cell lines appear.

A strong cytotoxic T cell response against EBV latency antigens is an important element in protecting individuals from uncontrolled proliferation of EBV-immortalized B lymphocytes [4,5]. The T cell response was first recognized in the regression assay. The addition of virus to the peripheral blood mononuclear cells from an EBV-seropositive patient will lead to transient B cell proliferation--followed by T-cell mediated regression of B cells. If T cells are physically removed by T cell rosetting or are rendered incapable of activation by a pharmacologic manipulation such as the addition of cyclosporine A, then regression will not occur and EBV-immortalized B cell lines will result. The frequency of T cells in healthy seropositive donors that are capable of mediating regression has been estimated at approximately one in 1000 [4,5]. A major component of this T cell activity is attributable to CD8(+) cytotoxic T cells directed against EBV latency antigens expressed in these immortalized B cells.

Uncontrolled EBV-driven B cell proliferation is seen in lymphoproliferative disease arising following organ transplantation. This complication is associated with immunosuppression, particularly with agents such as OKT3 (a monoclonal antibody directed against the pan-T cell antigen CD3) and cyclosporine A. Withdrawal of immunosuppression is in many instances associated with regression of the lymphoproliferative disease and presumably reflects reconstitution of an immune response. Recently the transfusion of donor lymphocytes in bone marrow transplant recipients with lymphoproliferative disease has resulted in the regression of such tumors [6].

In contrast to immortalized B cells, most EBV-associated malignancies are associated with restricted patterns of viral antigen expression. The most extreme case is African Burkitt's lymphoma in which only the EB nuclear antigen 1 (EBNA-1) is expressed but viral antigen expression in Hodgkin's disease and nasopharyngeal carcinoma is also restricted. The restricted pattern of viral latency antigen expression may render these tumors invisible to cytotoxic T cell immune surveillance. Global immunocompromise such as is seen in the post-transplant setting, is not necessarily characteristic of patients with other EBV-associated malignancies. Thus cytotoxic T cell responses against EBNA-1--the only viral antigen expressed in all EBV-associated malignancies--have not been detected. In contrast, cytotoxic T cell responses to EBNA-2, -3A, -3B, or -3C are detected in the majority of healthy EBV seropositives [7,8]. These latter antigens, hereafter referred to as the immunodominant EBNAs, are not expressed in African Burkitt's lymphoma, most EBV-associated AIDS lymphomas, nasopharyngeal carcinoma, or EBV-associated Hodgkin's disease. Patients with these tumors may have an intact cytotoxic T cell response against EBV antigens not expressed by their tumors.

Patterns of viral latentcy gene expression reflect alternative patterns of viral promoter usage [1]. Shortly after EBV infection of B cells, transcription is initiated from the W_P promoter yielding EBNA-2 transcripts. EBNA-2 transactivates an alternative latency promoter, C_P. Thereafter in lymphoblastoid cell lines, C_P is the dominant or exclusive promoter driving transcription of all of the EBNA genes. In a variety of tumor cell lines, transcription of EBNA-1 is driven from a third promoter, tentatively identified as F_P. In these cells, W_P and C_P are silent and the immunodominant EBNA's are not expressed.

Failure to express the immunodominant EBNA's may be a consequence of methylation of transcriptional regulatory elements. In lymphoblastoid cell lines in which all six EBNA's are expressed from a common 5' promoter, there is no detectable methylation of the viral genome as assessed by Southern blot hybridization following digestion with methylation sensitive enzymes such as HpaII [10]. However, in Burkitt's cell lines and primary tumor tissues, the viral genome (including transcriptional regulatory elements around C_P) is heavily methylated. DNA methylation is associated with repressed transcription in a variety of systems [9]. Repression may reflect closed chromatin structure or altered affinity of DNA binding proteins that regulate transcription. In particular, several proteins that bind to methyl-CpG sequences are known and evidence has been presented suggesting that transcriptional repressors may specifically bind to methylated DNA sequences. The potential role of methylation as a determinant of viral gene expression is highlighted by the demonstration that 5-azacytidine, an inhibitor of DNA methyltransferase, leads to demethylation and activation of EBV transcriptional regulatory elements in a Burkitt's lymphoma cell line [10].

Table 1. Viral Gene Expression in Tumors

Tumor	EBNA-1	Immunodominant EBNA's	LMP-1
Burkitt's	yes	no	no
EBV(+) immunoblastic and diffuse large cell lymphomas in the immunocompromised	yes	variable	variable
Hodgkin's disease	yes	no	yes
Nasopharyngeal carcinoma	yes	no	variable
Post-transplant lymphoma	yes	variable	variable

If patterns of gene expression in tumors in vivo could be similarly manipulated with pharmacologic agents, then tumors might be destroyed by a preexisting cellular immune response to the viral antigens. In preliminary studies, we demonstrate that 5-azacytidine and a related inhibitor of methylation, 5-aza-2'deoxycytidine, lead to demethylation and activation of latency viral promoters in cell lines, while a congener that does not inhibit DNA methyltransferase does not lead to latency gene activation. A clinical trial is planned to determine the efficacy of these agents in bringing about viral genome demethylation, latency antigen expression in tumors in patients, and clinical tumor responses. In anticipation of this trial we have developed PCR-based approaches to detecting changes in methylation and gene expression that are applicable to small needle biopsy specimens.

Materials and Methods

The Rael and B95-8 cell lines were grown in RPMI 1640 medium supplemented with 10% fetal calf serum and 2mM L-glutamine. Cultures were passaged every 3-4 days. One day prior to drug administration, cells were diluted to $5x10^5$ cells /ml. A single dose of drug was added and cells were incubated for 5 days. 5-azacytidine, 5-aza-2'deoxycytidine, and 6-azacytidine were used at 5 μM. Retinoic acid was used at 1 μM.

For genomic Southern blotting, 10 μg of total cellular DNA was digested with 100 U each of BamHI, BamHI and HpaII, or BamHI and MspI overnight. The DNA was resolved on a 1.5% agarose gel, transferred to nylon filter and hybridized by standard methods to cloned EBV BamHI-C or BamHI-W fragments labeled by nick translation.

For the methylation PCR assay, 1 μg of cell DNA was digested with 20 U of either HpaII or MspI. After heat inactivation at 65°C for 20 minutes, 50 ng of DNA from each digest, along with the same amount of uncut DNA, were used for PCR with primers flanking 4 HpaII sites immediately upstream of the C promoter (Fig.1). The sequences of the PCR primers used were: 5'-GTGGGCGGGAAGGGGCACAAG-3' and 5'-GTGGAATATGTGAGTGGACA-3'. After amplification, the samples were separated on an agarose gel, transfered to nylon membrane, and hybridized with an end-labeled synthetic oligonucleotide (5'-ATGCCGGCGAAACTGGACCAC-3').

The reverse transcriptase (RT)-PCR assay used primers described by Qu and Rowe [11] and the thermostable polymerase rTth (Perkin-Elmer, Norwalk, CT). 300ng of total RNA was used for each reaction in addition to 5' and 3' primers specific for the desired EBV transcript. The reverse transcription reaction was carried out in the presence of RNA, dNTP's, rTth, and a manganese containing buffer at 60°C for 15 minutes. Following addition of a 5' primer and a magnesium containing buffer with manganese chelator, PCR was carried out for 35 cycles. Manganese and magnesium buffers were as suggested by Perkin-Elmer. The product was electrophoresed on an agarose gel and visualized with ethidium. The identity of the amplified product was confirmed by transfer to a nylon filter and hybridization with an oligonucleotide probe spanning a splice site unique to the transcript being studied.

Results

Southern blot hybridization of DNA extracted from the Rael cell line demonstrated very different restriction patterns with methylation sensitive (HpaII) and insensitive (MspI) isoschizomers (Fig.2). This confirmed that the Rael cell line, an EBV-Burkitt's line with a highly restricted pattern of viral latency antigen expression, is heavily methylated. The probes used were recombinant plasmids containing the promoters driving transcription of the immunogenic EBNA's. In contrast to the results with the Burkitt's lymphoma cell line, Southern blot hybridization of DNA from the B95-8 cell line with unrestricted expression of viral latency antigens showed no evidence of methylation. DNA extracted from Rael cells following treatment with inhibitors of DNA methyltransferase (5-azacytidine and 5-aza-2'deoxycytidine), showed no evidence of methylation. Demethylation appeared to be a direct consequence of inhibition of the methyltransferase insofar as 6-azacytidine, a congener of these cytidine analogues, known not to inhibit the methyltransferase, did not lead to demethylation. Similarly, retinoic acid, a differentiating agent, did not alter patterns of methylation.

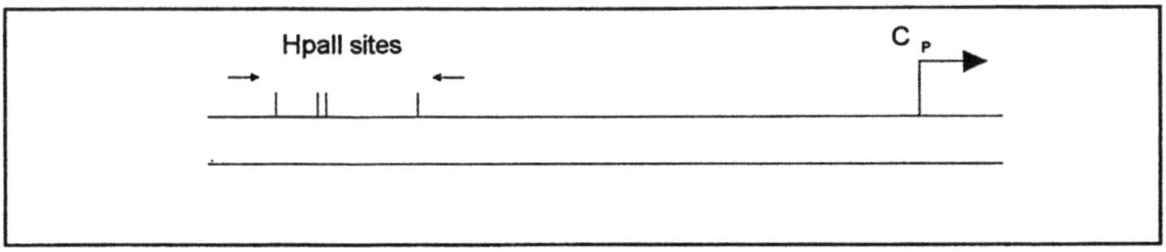

Fig. 1. PCR primer positions bracketing HpaII sites immediately upstream of the C_p are shown.

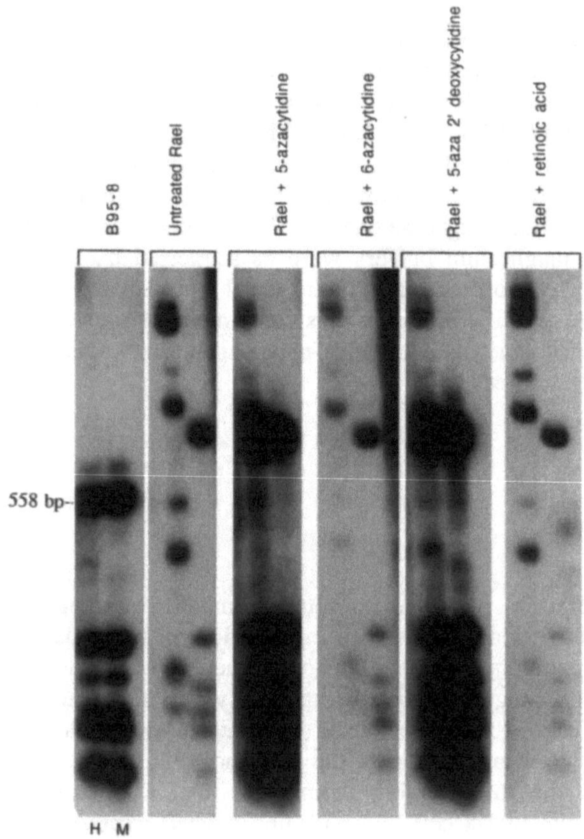

Fig. 2. Genomic Southern Blot: Total cellular DNA from untreated cultures or cultures treated as shown above were isolated and digested with BamHI (B), BamHI and HpaII (H), or with BamHI and MspI (M). After separation on an agarose gel and transfer to nylon filter, the blot was probed with BamHI-W region from EBV. 5-azacytidine and 5-aza-2'deoxycytidine induced demethylation as indicated by an increased susceptibility to HpaII cleavage. Neither 6-azacytidine nor retinoic acid induced demethylation.

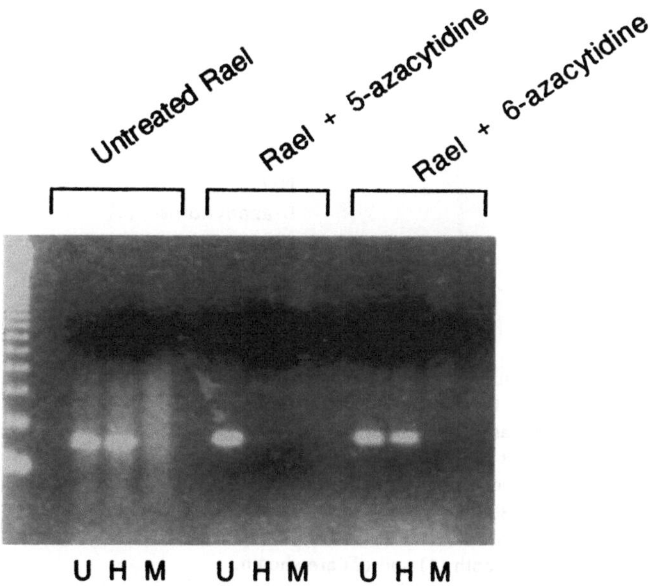

Fig. 3. Methylation PCR. Cellular DNA was extracted from untreated or treated Rael cells. PCR with primers bracketing the HpaII sites was used to amplify uncut DNA (U), HpaII digested DNA (H), or MspI DNA. Demethylation of HpaII sites by 5-azacytidine is indicated by the lack of amplification in the HpaII lane. Uncut and MspI cut DNA act as positive controls for amplification of the DNA and digestibility of the DNA respectively.

that activates EBV lytic cycle genes, did not alter patterns of methylation.

This approach to the analysis of patterns of methylation of the viral genome in tumor tissue requires approximately 10 μg of DNA per digest and is not readily applicable to very small biopsy specimens available in the context of a clinical trial. Therefore we developed a PCR-based assay to characterize patterns of methylation. The PCR assay is conceptually similar to conventional Southern blot hybridization except that it targets one or a few particular methylation sensitive restriction enzyme sites. After digestion with methylation sensitive and insensitive restriction enzymes, the presence of DNA that cannot be cleaved by methylation-sensitive enzymes is detected by polymerase chain reaction amplification using primers that bracket the restriction sites recognized by these enzymes (Fig.1). The results parallel those of conventional Southern blot hybridization in demonstrating the absence of methylation in the region of the viral genome immediately upstream of C_P in the B95-8 lymphoblastoid cell line, the presence of methylation in this region of the genome in the untreated Rael cell line, and the reversal of methylation in the viral genomes of Rael cells treated with inhibitors of methyltransferase (data not shown).

In order to assess gene expression, RT-PCR was applied to RNA extracted from cell lines. Transcription of the EBNA's from the C_P and transcription of LMP-1 were demonstrated in the control B95-8 cell line, but not in untreated Rael cells. Following treatment with 5-azacytidine or 5-aza-2'deoxycytidine, however, each of these families of transcripts were detected.

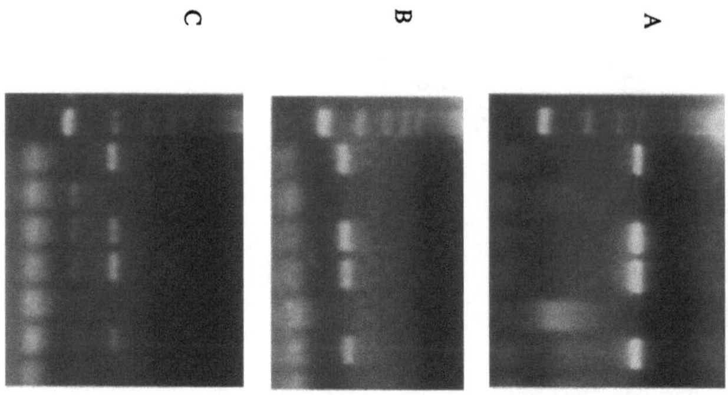

Fig. 4. Reverse Transcriptase PCR. Ethidium bromide stained agarose gels show amplification of RNA for LMP-1 (A), EBNA-3C (B), and Cp usage (C). The lane labeled "B95-8" shows the amplification products from this lymphoblastoid cell line that expresses all of the latency viral transcripts of interest in this experiment. It is included as a positive control. The lane labeled "ND" (no drug) shows the characteristic absence of amplification products in untreated Rael cells. The lanes labeled with the names of drugs show the amplification products of Rael cells treated with these drugs. 5-azacytidine treatment of Rael cells was carried out in two separate experiments. The amplification products of both (1) and (2) are shown.

Discussion

5-azacytidine and 5-aza-2'deoxycytidine are analogues of cytidine with nitrogen at the 5-position of the pyrimidine ring, that by virtue of the substitution cannot be methylated. The analogues non-competitively inhibit DNA methyltransferase, causing a block in cytosine methylation in newly replicated DNA [12]. First synthesized in 1964, 5-azacytidine has been in clinical trials since 1967. Both drugs have been used in combination with other agents to treat acute leukemia, and as single agents to treat myelodysplastic syndrome. In addition, methyltransferase inhibitors have been used as treatment for hemoglobinopathy including sickle cell anemia. Hemoglobinopathies manifest in the perinatal period after β-globin synthesis replaces γ-globin synthesis and hemoglobin production changes from HbF $(\alpha_2\gamma_2)$ to HbA($\alpha_2\beta_2$) [13]. In the fetus, cytosine residues in transcriptional regulatory elements of the γ-gene are hypomethylated but become fully methylated in adult bone marrow where the γ-genes are transcriptionally repressed. Treatment of patients with continuous infusion 5-azacytidine at 2mg/kg/day x 7 days or similar regimens resulted in a 4 to 6-fold increase in γ-messenger RNA in bone marrow of 8/9 patients without an increase in reticulocyte count [13]. This was accompanied by a decrease in methylation of regulatory regions upstream of the γ gene in erythroblasts, a rise in the production of fetal hemoglobin and clinical improvement. Similar effects were achieved with 5-azacytidine administered orally with tetrahydrouridine, an inhibitor of cytidine deaminase [14]. This apparently promising approach to treatment of hemoglobinopathy by promoter "switching" has been pursued and is a major focus of current therapeutic research. However, 5-azacytidine was abandoned as a "switching agent" for the treatment of hemoglobinopathy because of concerns with potential mutagenicity and carcinogenicity.

The ability of 5-azacytidine and 5-aza-2'deoxycytidine to induce a switch in EBV latency promoters and bring about transcription and expression of the immunodominant genes suggests a new therapeutic strategy: pharmacologic modulation of viral antigen expression so as to increase the susceptibility of these tumors to already extant EBV-specific cytotoxic T cells. In initial studies, it will be necessary to determine whether the changes in methylation and patterns of gene expression observed with treatment in vitro can also be achieved in vivo. With this goal in mind, we are about to embark on a trial in patients with EBV-associated malignancies characterized by restricted patterns of gene expression who have failed conventional therapies. Only patients with tumor masses accessible to biopsy will be eligible. It is anticipated that tumors will be biopsied before and after treatment, and tumor tissues studied by the techniques described here, as well as antigen detection, to determine whether 5-azacytidine has the desired effects on methylation and gene expression. If so, then a trial will be undertaken to determine the therapeutic efficacy of this approach.

The therapeutic strategy motivating these investigations is not limited to inhibitors of methyltransferase. Other agents may also activate expression from C_P, W_P, and other viral promoters driving immunodominant genes. Furthermore, the strategy does not necessarily require a fully intact host response to the upregulated viral antigens. Transfer of HLA-matched allogeneic lymphocytes, cytotoxic T cell lines, or cytotoxic T cell clones from healthy donors with intact responses may be adequate to induce tumor regression if the tumors are expressing the appropriate target antigens. A similar strategy is also being explored for developmental tumor antigens such as MAGE-1 associated with melanoma [15].

Supported by R03 CA62696-01 (RFA)

References

1. Liebowitz D, Kieff E (1993) Epstein-Barr virus. In: The Human Herpesviruses (Roizman B, Whitley RJ, Lopez C eds), New York: Raven Press. pp.107-172.
2. MacMahon EME, Glass JD, Hayward SD, Mann RB, Becker PS, Charache P, McArthur JC, Ambinder RF (1991) Epstein-Barr virus in AIDS-related primary central nervous system lymphoma. Lancet 338:969-973.
3. Ambinder RF, Browning PJ, Lorenzana I, Leventhal BG, Cosenza H, Mann RB, MacMahon EME, Medina R, Cardona V, Grufferman S, Olshan A, Levin A, Petersen EA, Blattner W, Levine PH (1993) Epstein-Barr virus and childhood Hodgkin's disease in Honduras and the United States. Blood 81:462-467.
4. Carmichael A, Jin X, Sissons P, Borysiewicz L (1993) Quantitative analysis of the human immunodeficiency virus type 1 (HIV-1)-specific cytotoxic T lymphocyte (CTL) response at different stages of HIV-1 infection: Differential CTL responses to HIV-1 and Epstein-Barr virus in late disease. J Exp Med 177:249-256.
5. Bourgault I, Gomez A, Gomard E, Levy JP (1991) Limiting-dilution analysis of the HLA restriction of anti-Epstein-Barr virus-specific cytolytic T lymphocytes. Clin Exp Immunol 84:501-507.
6. Papadopoulos EB, Ladanyi M, Emanuel D, Mackinnon S, Boulad F, Carabasi MH, Castro-Malaspina H, Childs BH, Gillio AP, Small TN, Young JW, Kernan NA, O'Reilly RJ (1994) Infusions of donor leukocytes to treat Epstein-Barr virus-associated lymphoproliferative disorders after allogeneic bone marrow transplantation. N Engl J Med 33:1185-1191.
7. Khanna R, Burrows SR, Kurilla MG, Jacob CA, Misko IS, Sculley TB, Kieff E, Moss DJ (1992) Localization of Epstein-Barr virus cytotoxic T cell epitopes using recombinant vaccinia: Implications for vaccine development. J Exp Med 176:169-176.
8. Murray RJ, Kurilla MG, Brooks JM, Thomas WA, Rowe M, Kieff E, Rickinson AB (1992) Identification of target antigens for the human cytotoxic T cell response to Epstein-Barr virus (EBV): Implications for the immune control of EBV-positive malignancies. J Exp Med 176:157-168.
9. Bird A (1992) The essentials of DNA methylation. Cell 70:5-8.
10. Masucci MG, Contreras-Salazar B, Ragnar E, Falk K, Minarovits J, Ernberg I, Klein G (1989) 5-Azacytidine up regulates the expression of Epstein-Barr virus nuclear antigen 2 (EBNA-2) through EBNA-6 and latent membrane protein in the Burkitt's lymphoma line Rael. J Virol 63:3135-3141.
11. Qu L, Rowe DT (1992) Epstein-Barr virus latent gene expression in uncultured peripheral blood lymphocytes. J Virol 66:3715-3724.
12. Chabner BA (1990) Cytidine analogues. In: Cancer chemotherapy: principles and practice (Chabner BA, Collins JM eds), Philadelphia: Lippincott. pp.154-179.
13. Humphries RK, Dover G, Young NS, Moore JG, Charache S, Ley T, Nienhuis AW (1985) 5-Azacytidine acts directly on both erythroid precursors and progenitors to increase production of fetal hemoglobin. J Clin Invest
14. Dover GJ, Charache S, Boyer SH, Vogelsang G, Moyer M (1985) 5-Azacytidine increases HbF production and reduces anemia in sickle cell disease: dose-response analysis of subcutaneous and oral dosage regimens. Blood 66:527-32.
15. Weber J, Salgaller D, Samid S, Johnson B, Herlyn M, Lassam N, Treisman J, Rosenberg SA (1994) Expression of the MAGE-1 tumor antigen is up-regulated by the demethylating agent 5-aza-2'deoxycytidine. Can Res 54:1766-71.

Rapid Identification of Novel Human Lymphoid-Restricted Genes by Automated DNA Sequencing of Subtracted cDNA Libraries

L. M. Staudt[1], A. Dent,[1] C. Ma[1], D. Allman[1], J. Powell[2], R. Maile[1], P. Scherle[1] and T. Behrens[1]

[1] Metabolism Branch, National Cancer Institute, National Institutes of Health, Bethesda, MD

[2] Division of Computer Research and Technology, National Institutes of Health, Bethesda, MD

The long term interest of our laboratory has been in identifying nuclear regulatory factors that control the differentiation and function of B lymphocytes. This interest began with the identification of a lymphoid-restricted transcription factor, Oct-2, which binds to the highly conserved octamer motif in the promoter regions of immunoglobulin genes (1,2). Through the work of many laboratories, this factor was cloned and found to contain a DNA binding domain belonging to the homeodomain family (3-7). In view of the profound importance of homeodomain proteins in many developing systems, it was initially hoped that Oct-2 would be required for the development of the B lymphoid lineage. Subsequent analysis has suggested, however, that Oct-2 must act in concert with other transcription factors to fully direct B cell development. Firstly, Oct-2 is not confined in expression to B lymphocytes: it is readily detected in the central nervous system (8) and in T lymphocytes (9) and thus may play a role in these lineages. Most telling was the inactivation of Oct-2 by homologous recombination in the mouse (10). These Oct-2 "knock-out" mice died shortly after birth for unknown reasons but at the time of death, B cell development appeared relatively normal. The B lymphocytes did, however, have a functional deficit in their response to bacterial lipopolysaccharide. Thus, Oct-2 appears to be dispensable for much of B cell development but it does control an aspect of mature B cell function.

These findings clearly suggested that B cell development is controlled by a constellation of regulatory factors acting in concert (11). The daunting challenge therefore is to rapidly clone and characterize the remaining lymphoid-restricted transcription factors in order to provide a deeper understanding of the regulation of B cell development and function. Two traditional approaches can be taken to this problem. First, cis-acting sequences can be identified in the transcriptional regulatory elements of lymphoid-restricted genes and DNA binding proteins can then be identified and cloned which interact with these motifs. A benefit of this approach is that it may lead to the identification of novel families of sequence-specific DNA binding proteins but this advantage is counterbalanced by the relatively slow rate at which these novel factors can be cloned. Furthermore, this approach will only identify, at least initially, the regulatory proteins which interact directly with DNA. Thus, an important class of transcription factors will be missed, namely the co-activators and co-repressors which do not themselves bind to DNA but which cooperate with other DNA binding proteins to activate or repress transcription. A second method involves the use of degenerate oligonucleotide primers in the polymerase chain reaction (PCR) to amplify novel members of known transcription factor families. While this approach is potentially fast, there is no guarantee that the factors cloned by this approach will be lymphoid-restricted.

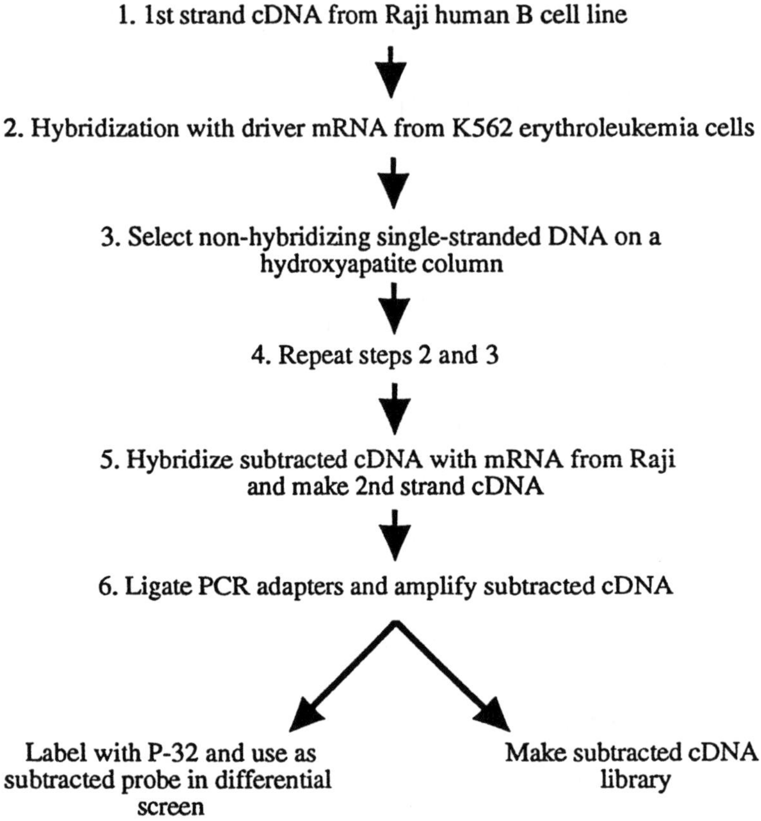

Fig. 1 Flow chart for the Raji subtracted cDNA library.

We decided, therefore, to adopt a third strategy, namely large-scale automated DNA sequencing of subtracted cDNA libraries enriched in lymphoid-restricted genes. The DNA sequences resulting from this approach are then analyzed automatically for homology with previously cloned genes. Among the many advantages of this approach is its ability to rapidly identify multiple new potential transcriptional regulatory factors which are lymphoid-restricted by virtue of the subtractive hybridization. Furthermore, this approach potentially allows for the identification of co-activators and co-repressors of transcription in addition to sequence-specific DNA binding proteins. Finally, it is possible to identify nuclear regulatory factors which might regulate aspects of B cell function other than gene transcription such as mRNA splicing and DNA replication.

The subtracted library that we analyzed most extensively started with cDNA from the Burkitt lymphoma cell line, Raji, from which we subtracted genes which were also expressed in the erythroleukemia cell line, K562. This choice of subtraction partners was based on the observation that few transcription factors are exquisitely restricted to only one stage in a cell lineage. Instead, most regulatory factors are expressed in multiple cell lineages and the specificity of gene regulation at a given developmental stage is achieved by the cooperative action of the constellation of transcription factors that are present at that particular stage. Several lymphoid-restricted factors, like Oct-2, are expressed throughout the B cell lineage and in T lymphocytes. Other factors like PU.1 are expressed in both B cells and in the myeloid lineage. Based on these considerations, we chose K562 as the "driver" in the subtractive hybridization in order to retain genes that are expressed in lymphoid and myeloid lineages but still remove genes that are expressed in all hematopoietic cells.

The subtractive method that was employed was based on the procedure of Timblin et al (12) and is illustrated in Figure 1. Following hybridization of the Raji first strand cDNA with K562 mRNA, the non-hybridizing cDNA, enriched for lymphoid-restricted genes, was selected on hydroxyapatite columns. This selection was repeated and second strand cDNA was synthesized resulting in a very small amount of subtracted cDNA (2 ng). The cDNA was ligated to PCR adapter oligonucleotides and amplified by PCR to obtain several micrograms of cDNA which was used in one of two ways. First, the subtracted cDNA was directly radiolabeled and used to differentially screen a cDNA library from the Burkitt' lymphoma cell line BJAB. For this differential screen, a second subtracted probe was prepared in analogous fashion to the Raji probe except that cDNA from the mature human B cell line EW was used as the starting material. The BJAB cDNA library was differentially screened for clones which hybridized with the both the Raji and EW subtracted cDNA probes but not with a radiolabelled first strand cDNA probe from K562. 618 clones that passed this test were arrayed in a grid and negatively screened for abundant lymphoid-restricted genes such as immunoglobulin, class II MHC and invariant chain. The remaining clones were sib-selected and sequenced to identify the unique and novel genes. Northern blot analysis of the expression pattern of the novel genes confirmed that 11 of the novel genes were lymphoid-restricted. We have characterized two of these genes extensively: Ly-GDI is a negative regulator of nucleotide exchange for the rho family of ras-like GTP binding proteins (13) and JAW1 is a integral membrane protein of the endoplasmic reticulum (14). Both of these genes are abundantly expressed in lymphocytes which reinforces the view that differential screening is only capable of identifying highly expressed genes. The remaining 9 genes showed no homology to known genes. Furthermore, as a quick assay for DNA binding proteins, these genes were translated in vitro and the resultant protein products were tested for their ability to bind non-specifically to double stranded DNA cellulose columns. None of the novel genes bound to DNA cellulose appreciably.

In order to broaden our analysis of the Raji subtracted cDNA, we cloned the cDNA into the pBluescript SK- plasmid vector to create a subtracted cDNA library. This library was screened using blue-white screening on IPTG/Xgal plates and recombinant clones were grown up for high purity plasmid DNA minipreps (Qiagen). The recombinant clones were further screened for those with the largest insert sizes by agarose gel electrophoresis. The quality of the subtraction was good as evidenced by the fact that approximately one third of randomly chosen clones from the subtracted library were found by Northern blot analysis to be derived from genes preferentially expressed in lymphoid cells. Using Northern blots, we have confirmed, to date, that 33 of the novel genes from the subtracted library were derived from lymphoid-restricted genes. Based on the high percentage of lymphoid-restricted clones in the library, we decided to choose random clones for automated DNA sequencing. Between 24 and 36 clones were sequenced daily using a Catalyst 800 Molecular Biology LabStation (Applied Biosystems) to prepare the sequencing reactions and a 373A automated DNA sequencer (Applied Biosystems). All clones were sequenced using the M13 forward sequencing primer and some clones were also sequenced using the M13

Table 1. Summary of homology searches of the Raji subtracted cDNA library sequence database

Total clones sequenced	Exact Matches (%)	Novel Sequences (%)	Strong Homology (%)
1290	673 (52%)	617 (48%)	148 (11%)

reverse sequencing primer. The resultant sequences were analyzed automatically for homology with previously cloned sequences using the BLASTN and BLASTX algorithms (15) using previously described software (16). The sequence information and results of the homology searches were stored in a relational SYBASE database modified slightly in structure from the ESTDB database (16).

A summary of the sequencing results of 1290 cDNA clones is presented in Table 1. Roughly half of the sequences showed an exact match to a previously cloned gene. The diversity of these exact matches was great, thus precluding any efficient use of hybridization screening to remove known genes prior to sequencing. Many of the clones appeared to be chimeric, containing a fragment of a known gene fused to an unknown sequence. This chimerism most likely resulted from ligation of short cDNA fragments to each other during the process of ligation of PCR adapters to the subtracted cDNA. The inadvertent benefit of this chimerism is that the sequencing effort was effectively sampling a larger number of total genes. The obvious disadvantage of this chimerism is that sequences from the subtracted clones must be confirmed to be derived from only one gene by using PCR primers to amplify the sequence from mRNA.

The exact matches allowed us to assess to what extent the library contained coding sequence versus 3' untranslated sequence. The cDNA apparently became fragmented during the subtractive hybridization process resulting in cDNA sequences ranging in size from 100 base pairs to more than 700 base pairs. Thus, although the initial Raji cDNA synthesis was oligo-dT primed, the sequences of some of the cDNA clones matched known genes at positions more than 2 kilobases 5' of the poly-A tail. Therefore, the library effectively samples the coding regions of genes that have less than a 2 kilobase 3' untranslated region. A final insight into the subtracted library is provided by an analysis of the subcellular distribution and/or function of the exact matches. Most of the exact matches were to membrane proteins (including a sizable fraction of the CD antigens expressed in B cells) which reflects the intensive cloning efforts directed towards these proteins in the immune system. Importantly, the subtracted library contained clones from several nuclear proteins that have been shown to be relatively restricted to lymphoid cells including Oct-2, BSAP/Pax5 and several members of the ets family of transcription factors. Thus, it seemed reasonable to expect that the novel genes in the library would include nuclear factors as well, an expectation that we have subsequently confirmed.

A significant number of the novel sequences showed homology to previously cloned sequences, thus providing clues to their biological function. Overall, 24% of the novel sequences (11% of all sequences) showed amino acid homologies with known genes which gave scores of 75 or more with the BLASTX algorithm or gave lower scores which nevertheless appeared to be biologically meaningful. In these latter cases, the regions of homology included amino acid identity at positions that are highly conserved in a family of proteins.

Table 2. Putative function or subcellular location of novel genes in the Raji subtracted cDNA library

	Number of clones (%)
Plasma membrane	29 (20%)
Nucleus	22 (15%)
Intracellular signaling	21 (14%)
Enzyme	16 (11%)
Secreted	13 (9%)
Cytoskeleton	10 (7%)
Protease or protease inhibitor	9 (6%)
Cytoplasmic vesicles	6 (4%)
Transporter	5 (3%)
Other	17 (12%)

Zinc finger domains fall into this category since 6 residues within each finger are virtually invariant.

The amino acid homologies in many cases gave insight into the probable subcellular distribution and/or function of the novel genes (Table 2). Many genes showed homology to proteins that have been recently implicated in intracellular signaling. This includes a variety of kinases, phosphatases, GTP binding proteins and their regulators. Furthermore, motifs that are commonly found in signaling molecules such as SH2 domains were also represented. Many of these presumptive signaling molecules have been confirmed to be lymphoid-restricted in their expression by Northern blot analysis. These results reinforce the view that although many components of signaling pathways are shared by diverse cell types, there is considerable regulation of signaling that is cell-type specific.

A major goal of this approach was achieved with the identification of a number nuclear proteins among the novel genes. The largest subset of these potential nuclear proteins was the zinc finger class which presumably reflects the exceedingly large size of this transcription factor family. In addition, we have identified novel members of the homeobox and LIM domain families of transcription factors. Interestingly, we have found homologues of nuclear proteins which have uncertain function at present, including components of the nuclear matrix. We are currently investigating three of the novel lymphoid-restricted nuclear factors in depth to determine their role in B cell biology.

One zinc finger gene from our library was found to be expressed at high levels in human mature B tumor cell lines and was almost undetectable in non-lymphoid cell lines. This gene showed the most clearly differential expression pattern of all the zinc finger genes that we tested from our library and for this reason we focused our attention on its role in B cell physiology. Subsequently, this gene was independently cloned by several groups as a gene at chromosomal position 3q27 that is rearranged in at least 30% of cases of diffuse large cell lymphoma and has been dubbed BCL-6 (also LAZ-3 and BCL-5) (17-19). In the translocations analyzed thus far, the BCL-6 locus is rearranged to immunoglobulin loci resulting in no obvious changes in the BCL-6 coding region but presumably juxtaposing the BCL-6 gene next to the strong immunoglobulin transcriptional enhancers. Diffuse large cell lymphoma constitutes the largest subgroup of non-Hodgkins lymphoma and thus the mechanism of action of BCL-6 as an oncogene is a pressing clinical problem. Interestingly, we have found that virtually all human B cell tumor lines express high levels of BCL-6, even those that do not have cytogenetic evidence of chromosome 3 translocations. Several possibilities could explain this observation. First, these cell lines could have non-cytogenetically detectable rearrangements of the BCL-6 gene, which would extend the contribution of BCL-6 rearrangements to other lymphoid malignancies. Second, the BCL-6 gene could be transcriptionally up-regulated in these cells due to the actions of other oncoproteins. Finally, BCL-6 may be up and down-regulated at discrete stages of B cell development and the elevated BCL-6 expression in mature B cell tumor cell lines may reflect this normal regulation.

Another approach that we have taken to identify potential nuclear proteins has been to screen our subtracted cDNA library sequence database for nuclear localization signals (NLS's). NLS's are short peptide motifs that are present in all nuclear proteins which direct the efficient accumulation of those proteins in the nucleus. Two general types of NLS's have been described (20). One form resembles the NLS found in SV40 virus large T antigen and consists of 5 basic residues in a row, usually preceded by a proline or glycine residue. Another form, the bipartite NLS's, has 2 basic residues followed by a random spacer of 10 amino acids and ending with 5 residues, 3 of which are basic. In order to increase the specificity of this screen, we also used the GRAIL computer program (21) to predict whether a putative NLS was in a potential coding region of the sequence. The GRAIL program is a neural network program that has been trained to recognize coding regions by extensive analysis of the different sequence characteristics of exonic coding regions and intronic regions in genomic DNA. We identified a small subset of the Raji subtracted cDNAs which encoded NLS's in reading frames predicted by GRAIL to have a high probability of being coding regions. We expressed one such coding region in bacteria and used this bacterially expressed protein to raise rabbit polyclonal antisera. By confocal immunofluorescence microscopy, we found that this protein was indeed localized to the nucleus of human B lymphoid cells and was not detectable in other cell types. Furthermore, we have been able to show that this particular protein also contains potent transcriptional activation domains and thus this approach was apparently successful in identifying a novel lymphoid-restricted nuclear protein that is a potential transcription factor.

The approach that we have taken has several advantages over other methods aimed at identifying the differences in gene expression between two cell types. The most critical advantage of the present approach is the ability to analyze the coding regions of many differentially expressed genes early on in the effort. Another powerful recent method, differential display (22), is based on the PCR amplification of the 3' ends of a subsets of mRNAs. These PCR products are then separated by electrophoresis through a denaturing polyacrylamide gel. By comparing the pattern of PCR fragments obtained from different cell types, it is possible to rapidly identify the genes which are differentially expressed between the cell types. One advantage that differential display has over subtracted cDNA libraries is that differential display offers the opportunity to compare multiple cell types simultaneously whereas subtracted libraries are inherently binary. However, one particularly cumbersome aspect of the differential display technique is that the initial cDNA clone obtained by this procedure is almost invariably derived from the 3' untranslated region of a mRNA and thus its sequence is not informative as to the potential biological function of the protein encoded by the mRNA. The coding region sequence can only be obtained by the arduous route of isolating long cDNA clones from the same gene. In contrast, we found that automated DNA sequencing of a subtracted cDNA library was a highly efficient means of identifying coding regions of differentially expressed genes and thereby predicting the functions of their gene products.

Acknowledgments

The authors wish to acknowledge the fine technical assistance of Jaya Jagadeesh. This work was supported in part by a Cancer Research Institute Investigator Award (L.M.S.).

References

1. Staudt,LM, Singh,H, Sen,S, Wirth,T, Sharp,PA and Baltimore,D (1986) Nature, 323:640-643.
2. Landolfi,NF, Capra,DC and Tucker,PW (1986) Nature, 323:548-551.
3. Staudt,LM, Clerc,RG, Singh,H, LeBowitz,JH, Sharp,PA and Baltimore,D (1988) Science, 241:577-580.
4. Ko,H-S, Fast,P, McBride,W and Staudt,LM (1988) Cell, 55:135-144.
5. Clerc,RG, Corcoran,LM, LeBowitz,JH, Baltimore,D and Sharp,PA (1988) Genes Dev, 2:1570-1581.
6. Muller,MM, Ruppert,S, Schaffner,W and Matthias,P (1988) Nature, 336:544-551.
7. Scheidereit,C, Cromlish,JA, Gerster,T, Kawakami,K, Balmaceda,C, Currie,RA and Roeder,RG (1988) Nature, 336:551-557.
8. He,X, Treacy,MN, Simmons,DM, Ingraham,HA, Swanson,LW and Rosenfeld,MG (1989) Nature, 340:35-42.
9. Kang,SM, Tsang,W, Doll,S, Scherle,P, Ko,H-S, Tran,AC, Lenardo,MJ and Staudt,LM (1992) Mol Cell Biol, 12:3149-3154.
10. Corcoran,LM, Karvelas,M, Nossal,GJV, Ye,Z-S, Jacks,T and Baltimore,D (1993) Genes and Development, 7:570-582.
11. Staudt,LM and Lenardo,MJ (1991) Annu Rev Immunol, 9:373-398.
12. Timblin,C, Battey,J and Kuehl,WM (1990) Nucleic Acids Res, 18:1587-1593.
13. Scherle,P, Behrens,T and Staudt,LM (1993) Proc Natl Acad Sci USA, 90:7568-7572.
14. Behrens,TW, Jagadeesh,J, Scherle,P, Kearns,G, Yewdell,J and Staudt,LM (1994) J Immunol, in press.
15. Altschul,S, Gish,W, Miller,W, Meyers,E and Lipman,D (1990) J Mol Biol, 215:403-410.
16. Kerlavage,AR, Adams,MD, Kelley,JC, Dubnick,M, Powell,J, Shanmugan,P, Venter,JC and Fields,C (1993) Analysis and management of data from high-throughput expressed sequence tag projects. In: Proceedings of the Twenty-Sixth Annual Hawaii International Conference on System Sciences. IEEE Computer Society Press, Los Alamitos, CA, pp 585-594.
17. Ye,BH, Lista,F, Lo Coco,F, Knowles,DM, Offit,K, Chaganti,RSK and Dalla-Favera,R (1993) Science, 262:747-750.
18. Kerckaert,JP, Deweindt,C, Tilly,H, Quief,S, Lecocq,G and Bastard,C (1993) Nat Genet, 5:66-70.
19. Miki,T, Kawamata,N, Hirosawa,S and Aoki,N (1994) Blood, 83:26-32.
20. Dingwall,C and Laskey,RA (1991) TIBS, 16:478-481.
21. Uberbacher,E and Mural,R (1991) Proc Natl Acad Sci USA, 88:11261-11265.
22. Liang,P and Pardee,AB (1992) Science, 257:967-971.

Tumorigenesis in Mice with an SV40 T Antigen Transgene Driven by the Immunoglobulin γ1 Heavy Chain Germline Promoter

Wesley Dunnick[1], Laura Elenich[1], Kirk Cunningham[1], Clarence Chrisp[2], and Latham Claflin[1].
[1]Department of Microbiology and Immunology and [2]Unit for Laboratory Animal Medicine, The University of Michigan Medical School, Ann Arbor, MI 48109-0620

Introduction

The immunoglobulin heavy chain switch is a genetic recombination whereby a V(D)J region is rearranged from association with the Cμ gene to association with a downstream Cγ, Cε, or Cα gene [1,2]. The recombination event occurs in switch (S) regions found upstream of each CH gene except Cδ. Switch recombination is preceded by a change in the "accessibility" of the switch region DNA, as indicated by the induction of transcription of the CH locus prior to switch recombination [1-4]. Both induction of germline transcripts and switch recombination of any CH locus are specifically regulated by cytokines.

There are two general approaches to generate either tumors or immortalized B cells at specific stages of differentiation - transgene oncogenesis and retroviral infection [5,6]. These approaches have yielded a spectrum of tumors and cell lines characteristic of either early or late B lineage cells, *i.e.*, pro-B, pre-B and plasma cells. Few mature B cell tumors and lines have been produced. We have developed a strategy to target a relatively rare cell population, those undergoing antigen-driven differentiation, by constructing a fusion gene composed of the germline γ1 promoter region and the SV40 T antigen oncogene.

During antigen-driven differentiation of B cells, transcription of the γ1 gene in its germline arrangement precedes switch recombination to the γ1 heavy chain gene. These germline transcripts are particularly abundant when the B cells are also exposed to IL-4 [7-9]. Transcription begins 5' of the Sγ1 region and continues through the constant region exons (Cγ1). A 5' exon, termed "Iγ1", is spliced in nuclear RNA to the Cγ1 exons. Our construct included 2100 bp of 5' γ1 germline promoter sequences [10] and 240 bp of the Iγ1 exon fused to the SV40 T antigen coding sequences. One AUG in the Iγ1 exon was mutated so that if transcripts were initiated at the same places as the germline transcripts of the endogenous γ1 gene, the AUG initiator codon at the beginning of the T antigen gene would be the first encountered by the translation machinery. Four lines of transgenic mice were derived using this construct. All four lines have many (about 30) copies of the transgene.

T Antigen Expression in Transgenic Mice

Expression of the T antigen transgene was studied by S1 nuclease analysis of RNA derived from several tissues (Table 1). If expression of T antigen was regulated in exactly the same way as expression of the endogenous γ1 gene, then T antigen RNA would not be expressed in B cells treated with the mitogen LPS, but it would be expressed in B cells treated with LPS and IL-4. Younger mice (2-3 months of age) of three lines (founders #237, 273, and 274) expressed T antigen RNA in thymus, fresh splenic B cells, B cells stimulated with LPS, and B cells stimulated with LPS and IL-4. The level of expression in RNA from B cells treated with LPS and IL-4 is at least 20-fold less than that from the endogenous γ1 gene. (Expression of T antigen transcripts was detected in founder #220, but in none of its offspring.) Expression was not observed in several non-lymphoid tissues, for example liver, kidney, and brain. We have observed the same pattern of tissue expression in nine other lines of transgenic mice that use similar Iγ1 promoter regions, but different reporter genes (usually luciferase). Taken together, these results indicate that the DNA sequences flanking the 5' end of Iγ1 allow some degree of regulation, in that the transgenes are expressed only in lymphoid tissue. On the other hand, the transgenes are not expressed exactly like the endogenous gene, which is silent in T cells and in unstimulated or LPS-stimulated B cells [7-9]. These transgenes seem to be lacking some negative element that prevents expression in T cells and in B cells that are not exposed to IL-4.

Table 1:RNA Expression in γ1 Promoter:T Antigen Transgenic Mice

Tissue	T Antigen RNA in Young Mice	T Antigen RNA in Old Mice	Endogenous γ1 Gene RNA
Fresh B cells	+	++	-
B cells, LPS	+	++	-
B cells, LPS + IL4	+	++	+++
Thymus	+	++	-
Brain	-	+++	-
Heart	-	++	-
Liver	-	+++	-
Kidney	-	++	-
Muscle	-	-	-
Harderian gland	+	+++	-

Relative expression levels: -, none detectable; +, 1X; ++, 5X; +++, 20X.

We have observed T antigen RNA expression in the Harderian gland, the gland that wets the eyeball. This expression is significant in young mice and can be very high in old mice. We have also observed two Harderian gland adenomas in one-year-old transgenic mice; both of these tumors expressed very high levels of T antigen RNA. Tumors of the Harderian gland can be common in older mice of some strains [11]. Examination of histologic sections confirmed that the tumors originated in the Harderian gland and were not derived from lymphoid tissue.

We have preliminary evidence that the transcription start sites used in the transgene are different than those used in the endogenous gene [10]. Because the abundance of transcripts from the transgenes is relatively low, we have not yet localized the exact start sites, but we suspect that the transcripts initiate at least 1 kb 5' of the normal start sites. Transcripts that include such a large leader region are unlikely to direct T antigen protein expression. Many AUG codons will be encountered in the first 1 kb of the transcript. Since the mammalian translation machinery usually initiates protein synthesis at the first AUG encountered [12], in-frame translation of T antigen would not be expected. It is possible that some stimuli (for example, IL4) initiate a small amount of transcription at the normal start sites, and that these transcripts allow the translation of active T antigen protein.

In old mice (8-15 months) with γ1 promoter:T antigen transgenes, RNA from the transgene is expressed in almost all tissues tested, including liver, kidney, and brain (Table 1). The basis of this age-related T antigen RNA expression is unknown.

Germline γ1 Expression in Transgenic Mice

We reported that a few T antigen transgenic mice had splenic B cells that expressed their endogenous γ1 gene at high levels, even without IL-4 stimulation [13]. We hypothesized that endogenous germline γ1 transcripts were expressed by long-lived B cells, which arose by the following mechanism. The γ1 promoter:T antigen transgenes may have many of the <u>positive</u> elements needed for expression of germline γ1 transcripts. If B cells in the γ1 promoter:T antigen transgenic mice are induced by IL-4 or other stimuli *in vivo* to transcribe the endogenous γ1 germline gene, they would also begin to transcribe the T antigen transgene. This T antigen expression would in turn immortalize the B cells, and the cells would constitutively produce the positive factors that activate the γ1 promoter, since they need these factors to express T antigen. In the presence of these factors, the B cells should also transcribe their endogenous γ1 gene constitutively. Unfortunately, after a 3 month hiatus in breeding to rid our colony of several murine viruses, this phenotype has not been observed in γ1 promoter:T antigen B cells. We hypothesize that the agent that stimulated the

B cells to produce γ1 germline transcripts and T antigen is no longer present in our colony.

Tumors in γ1 promoter:T Antigen Transgenic Mice

We have checked about 200 γ1 promoter:T antigen transgenic mice between 1 and 2 years of age for tumors. The vast majority of these mice had been backcrossed from the founders (FVB) to BALB/c for one to three generations. In spite of the T antigen RNA expression pattern (above), no tumors of the thymus or of non-lymphoid tissues, excepting the two Harderian gland tumors discussed above, were found. We have observed a total of 13 lymphoid tumors in the spleen or mesenteric lymph node. Histologic sections demonstrated that many of these tumors were of the follicular center cell type, and that tumor cells comprised the majority of the cells in the spleen or lymph node. Four of these lymphoid tumors were from a small group of transgenic mice in which both SJL and C57Bl/6 genes had been introduced. Since both of these strains are susceptible to lymphomas [14,15], many of these tumors could be spontaneous and not caused by T antigen expression. In fact, we have found two similar tumors in nontransgenic littermates.

Less than one-half of these tumors express T antigen RNA, consistent with the notion that many of them are spontaneous tumors (Table 2). Those tumors that express T antigen RNA usually express large amounts, approaching the amount of RNA from the fully induced endogenous γ1 promoter of normal B cells.

In an attempt to increase the rate of tumorigenesis, we crossed our γ1 promoter:T antigen mice with mice with a *ras* transgene driven by a composite SV40:Eμ enhancer, a kind gift of Alan Harris and colleagues [16]. Virtually all of the γ1 promoter:T antigen/Eμ:*ras* doubly transgenic mice have lymphoid tumors by eight months of age. The kinetics of tumorigenesis is almost exactly that reported by Harris *et al.* [16]. The doubly transgenic mice developed tumors about one week faster, a minor difference. Furthermore, the types of tumors observed were, in the majority of cases, the same as described for Eμ:*ras* transgenic mice [16].

Table 2: Summary of Tumor Characteristics

Transgenes	# of Tumors	Age of Onset	T Antigen RNA Expr.		Iγ1 Expr.	Rearrangements	
			+	++		JH	Sγ1
T Ag only	13	11-24 mo.	1/11	4/11	0/13	5/9	3/12
T Ag & *ras*	27	2-6 mo.	3/7	0/7	0/18	2/8	1/13
none	2	24 mo.	not appl.		0/2	0/2	0/2

"Age of onset" is for 90% of the tumors observed. RNA expression levels and DNA rearrangements are scored as number of tumors that were positive/number of tumors tested. + and ++ are defined as in Table 1.

As discussed above, if these tumors were transformed while beginning the pathway toward γ1 switch recombination, they might constitutively express germline transcripts of their endogenous γ1 gene. We have tested this prediction by analysis of RNA derived from total spleen or lymph node of tumor bearing mice. Since these organs are grossly enlarged, we assume that many of the cells are tumor cells, and that the RNA derived from them is representative of the tumor. Of the 31 tested, none express endogenous germline transcripts.

One of the hallmarks of the tumor we seek is evidence of switch recombination to γ1. DNA from two of our tumors has a single restriction fragment that hybridizes to sequences from both the μ and γ1 switch regions--an SμSγ1 rearrangement. This is evidence for switch recombination from μ to γ1. The SμSγ1 rearrangement is accompanied by a JH rearrangement. Two other tumors have Sγ1 rearrangements that we have not further characterized. Several other tumors have at least one JH rearrangement. These JH rearrangements are not present in all the cells in the tissue sample, as they are subgenomic in quantity.

We attempted long-term culture of splenic and/or lymph node tumors from nine mice. In some cultures LPS or LPS and IL-4 were added. All tumors survived *in vitro* for approximately four weeks during which time a significant but variable proportion of cells were viable. In four cases, cultures could be expanded weekly or biweekly, All cell lines, however, died between four to eight weeks of culture.

Conclusions

The γ1 promoter:T antigen construct we are presently using is expressed at a very low level using inappropriate transcription start sites. Since it is expressed in all lymphoid cells, we have not targeted expression to the exact cells of choice. Even though we have obtained a few interesting tumors, including those with an SμSγ1 rearrangement, the overall rate of tumorigenesis is very low. Some of the tumors we have obtained are probably spontaneous. In another line of investigation, we have recently found that important γ1 regulatory elements lie 3' to the γ1 switch region, in sequences not included in our γ1 promoter:T antigen constructs. We plan to redesign our T antigen construct, using more of the γ1 gene, in hopes of better regulated and higher level T antigen expression. Such expression may in turn lead to tumors of B cells undergoing antigen-driven differentiation.

Acknowledgements

We thank Jan Berry for construction of these transgenes and Drs. Sally Camper and Thom Saunders of the University of Michigan Transgenic Facility for timely advice and assistance. This work was supported by a grant from the National Cancer Institute (CA39068).

References

1. Lutzker SG, Alt FW (1989) Immunoglobulin heavy chain class switching. In: Berg D E, Howe M (ed) Mobile DNA. American Society for Microbiology, Washington,D.C., pp 693-714
2. Esser C, Radbruch A (1990) Immunoglobulin class switching: molecular and cellular analysis. Ann Rev Immunol 8:717-735
3. Stavnezer-Nordgren J, Sirlin S (1986) Specificity of immunoglobulin heavy chain switch correlates with activity of germline heavy chain genes prior to switching. EMBO J 5:95-102
4. Yancoupoulos GD, DePinho RA, Zimmerman KA, Lutzker SG, Rosenberg N, Alt FW (1986) Secondary rearrangement events in pre-B cells: VHDJH replacement by a LINE-1 sequence and directed class switching. EMBO J 5:3259-3266
5. Adams JM, Cory S (1991) Transgenic models of tumor development. Science 254:1661-1167
6. van Lohuizen M, Berns A (1990) Tumorigenesis by slow-transforming retroviruses--an update. Biochim Biophys Acta 1032:213-235
7. Stavnezer J, Radcliffe G, Lin Y-C, Nietupski J, Berggren L, Sitia R, Severinson E (1988) Immunoglobulin heavy-chain switching may be directed by prior induction of transcripts from constant-region genes. Proc Natl Acad Sci USA 85:7704-7708
8. Esser C, Radbruch A (1989) Rapid induction of transcription of unrearranged Sγ1 switch regions in activated murine B cells by interleukin 4. EMBO J 8:483-488
9. Berton MT, Uhr JW, Vitetta ES (1989) Synthesis of germ-line γ1 immunoglobulin heavy-chain transcripts in resting B cells: induction by interleukin 4 and inhibition by interferon γ. Proc Natl Acad Sci USA 86:2829-2833
10. Xu M, Stavnezer J (1990) Structure of germline immunoglobulin heavy-chain γ1 transcripts in interleukin 4 treated mouse spleen cells. Dev Immunol 1:11-17
11. Huff JE, Haseman JK, Eustis S, Maronpot RR, Peters AC, Persing RL, Chrisp CE, Jacobs AC (1989) Multiple-site carcinogenicity of benzene in Fischer 344 rats and B6C3F1 mice. Environ Health Perspect 82:125-163
12. Kozak M (1983) Translation of insulin-related polypeptides from mRNAs with tandemly reiterated copies of the ribosome binding site. Cell 34:971-978
13. Dunnick W, Elenich L, Berry J, Albrecht D, Stavnezer J, Claflin JL (1992) Regulation of switch recombination to the murine γ1 gene. In: Potter M, Melchers F (ed) Mechanisms in B Cell Neoplasia 1992. Springer Verlag, Berlin, pp 143-147

14. Seigler R, Rich MA (1968) Pathogenesis of reticulum cell sarcoma in mice. JNCI 41:125-143
15. Heston WE (1964) Induction of mammary gland tumors in strain C57BL/He mice by isografts of hypophyses. JNCI 32:947-955
16. Haupt Y, Harris AW, Adams JM (1992) Retroviral infection accelerates T lymphomagenesis in Eμ-N-*ras* transgenic mice by activating c-*myc* or N-*myc*. Oncogene 7:981-986

Circulating B-Cells in Follicular Non-Hodgkin' s Lymphoma Show Variant Expression of L-Selectin Epitopes

G. S. JENSEN [1], J. L. PO [1], P. HUERTA [1], AND C. SHUSTIK [2].

Departments of [1] Surgery and [2] Hematology, Royal Victoria Hospital, McGill University, Montreal, Canada.

Introduction.

The L-selectin receptor on normal lymphocytes is involved in homing to lymph nodes via interaction with high endothelial venules [1,2]. Lymphoma cell expression of L-selectin and recognition of corresponding ligand on high endothelial venule cells may influence the dissemination of lymphomatous B cells into nodal and extranodal sites [3-9]. Pals et al [7] found significant differences in L-selectin expression by malignant cells between nodal and gastrointestinal non-Hodgkin's lymphomas, with frequent (84/116 cases, or 72%) expression in nodal biopsies compared to 24% of cases with gastrointestinal tumors. Moller et al [8,9] found low or absent expression of L-selectin on lymphoma regardless of site of involvement. Discordance in these results may be partly explained by the use of an immunohistochemical assay to evaluate L-selectin expression on freshly frozen tissue rather than flow cytometric analysis of fresh tumor cells. The presence of circulating malignant cells with a t(14;18) translocation detected by PCR has been demonstrated in 15 out of 22 (68%) of patients [10], including patients in apparent clinical remission [11]. We have analyzed peripheral blood B cells from patients with follicular NHL and normal donors, and demonstrated that the L-selectin on the surface of circulating B cells in follicular non-Hodgkin's lymphoma is different from L-selectin expressed on normal blood B cells. Different epitopes of the L-selectin molecule were studied by a series of nine monoclonal antibodies (MoAbs), of which 6 are directed against the ligand-binding site. We detected circulating B cells with altered L-selectin expression in the blood of 9 of 10 patients with follicular lymphoma (FL). A subset of blood B cells from patients with FL were negative for the Leu 8 epitope, but highly positive for TQ1, and negative or only dimly positive for Lam1.1 and Lam1.6. In 4 of these patients analyzed, these B cells expressed the bcl-2 protein and included 15-25% of cells in S+G2/M phase. If the B cells are confirmed as part of the malignant clone, the altered constellation of L-selectin epitopes may determine particular patterns of dissemination.

Certain epitopes of L-selectin are altered on a subset of peripheral blood B cells from follicular lymphoma patients.

Using multiparameter flow cytometry, we analyzed the expression of the L-selectin homing molecule on the circulating B cells from 10 patients with FL, using nine different MoAbs directed towards various epitopes on the L-selectin molecule.

TABLE 1. L-SELECTIN MONOCLONAL ANTIBODIES USED IN THIS STUDY.

MoAb:	Source:	MoAb affects on PBL function:
LEU8	Becton-Dickinson	Inhibits B cell differentiation [12]
TQ1	Coulter Electronics	Blocks binding to PPME [2]
LAM1.1	Dr. T. Tedder	Blocks HEV binding, enhances PPME binding [2]
LAM1.2	Dr. T. Tedder	Blocks binding to HEV, PPME [2]
LAM1.3	Dr. T. Tedder	Blocks binding to HEV, PPME [2,13]
LAM1.4	Dr. T. Tedder	Blocks binding to HEV, PPME [2]
LAM1.5	Dr. T. Tedder	Enhances binding to PPME [2]
LAM1.6	Dr. T. Tedder	Blocks binding to HEV [2]
FMC46	Dr. H. Zola	Expressed on dividing cells [14]

L-selectin on normal peripheral blood lymphocytes (PBL), as well as circulating malignant B cells from B-CLL showed an equivalent pattern of reaction with all MoAbs. We demonstrated that the circulating CD19+ B cells in patients with FL express medium/high density of L-selectin recognized by the MoAbs TQ1, LAM1.3, LAM1.4, and FMC46 (Fig. 1 and 2). However, most of the L-selectin molecules expressed on the circulating lymphoma B cells are poorly recognized by five other MoAbs, with Leu8 and Lam1.6 showing the poorest binding. While B cells expressing normal L-selectin reactive with all MoAbs are present, there is a distinctive subset on which the L-selectin has an abnormal constellation of epitopes. Peripheral blood T cells from these patients express L-selectin which is recognized equally well by all monoclonal antibodies. At least six of the MoAbs against L-selectin used in this study are directed towards the ligand-binding portion of the molecule (see Table 2) [2, 12-14]. We assume that the changes in L-selectin on FL blood B cells must include the ligand-binding area, possibly leading to altered affinity or specificity of the aberrant L-selectin molecule. Abnormal L-selectin epitope expression on FL blood B cells is not limited to patients with active disease, since patient #3 in clinical remission had a very high percentage of TQ1+ B cells, and almost no Leu8+ Lam1.6+ circulating B cells (Table 2).

TABLE 2. FOLLICULAR LYMPHOMA PATIENTS INCLUDED IN THIS STUDY:

Patient #:	Pathology:	Stage:	Disease status:	Marrow involvement:
1	FSCC	IV	Remission	Yes
2	FSCC	IV	Active	Yes
3	FSCC	IV	Active	Yes
4	FSCC	IV	Active	No
5	FSCC	IV	Active	Yes
6	FSCC	III	Active	No
7	FSCC	IV	Active	Yes
8	FSCC	IV	Active	Yes
9	FSCC	IV	Active	Yes
10	FLC	III	Active	No

FSCC: Follicular small cleaved cell lymphoma, FLC: Follicular large cell lymphoma.

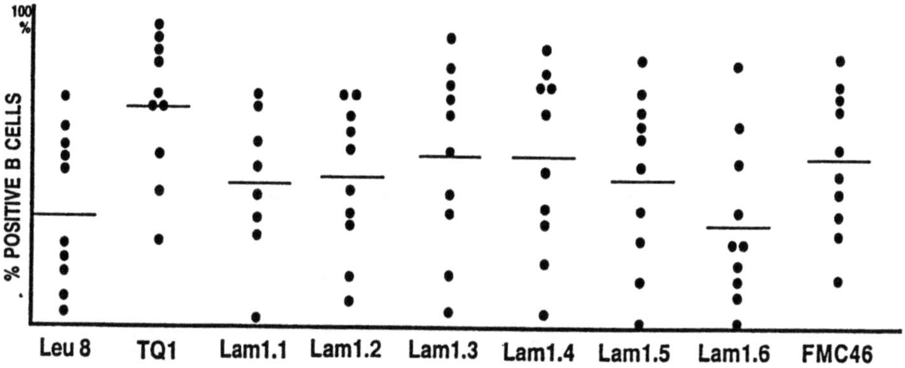

Fig. 1. Circulating CD19+ B cells from patients with follicular non-Hodgkin's lymphoma express varying amounts of an abnormal form of L-selectin. The B cells from most patients are 50-100% brightly positive for TQ1, indicating the presence of L-selectin on most B cells. Note that only a fraction of the L-selectin recognized by TQ1 is recognized by the Leu8, Lam1.1, and Lam1.6 antibodies directed towards other epitopes on the L-selectin molecule.

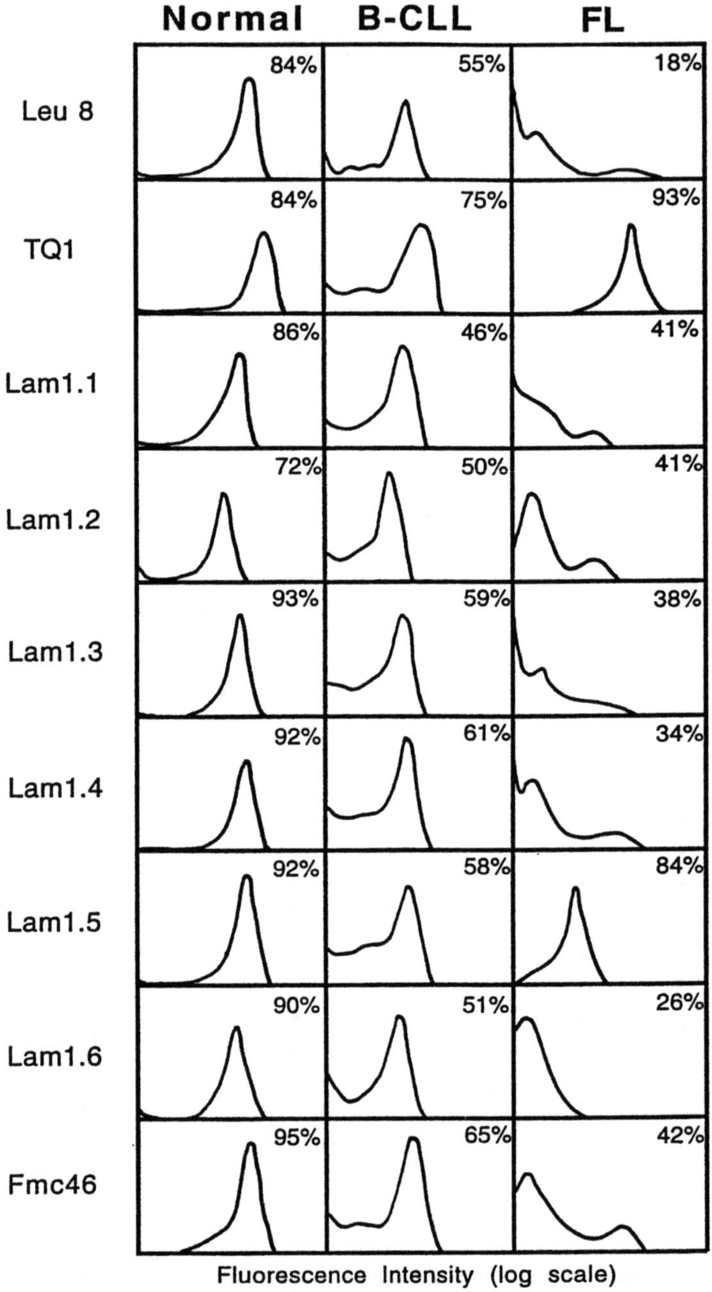

Fig. 2. Expression and percentage of L-selectin on peripheral blood CD19+ small B cells from 1) a normal donor, 2) a patient with B-CLL, and 3) a patient with FL (patient # 3). Histograms show L-selectin epitope expression as analyzed by 9 different monoclonal antibodies and multiparameter flow cytometry. The percentages of positive cells indicated in each histogram are based on comparison to the appropriate isotype control for each MoAb.

Molecular heterogeneity of the L-selectin molecule expressed on circulating B cells from patients with follicular lymphoma.

Using a gentle protein extraction and a non-reducing acrylamide gel system, we have been able to demonstrate that the L-selectin on PBL from FL patients, but not from normal donors, comprise cells with an aberrant L-selectin molecule (Fig. 3). Since a non-reducing gel system is being used, the variation may not correlate with molecular weight, but could also be the result of conformational changes. Under these conditions, normal PBL express a form of L-selectin that yields only a single band, while the patient PBL results in a smear, indicating the possibility of a heterogeneity of L-selectin forms. The altered epitopes on the L-selectin expressed by a subset of FL blood B cells may be a result of altered glycosylation, or a conformational change which could mask certain epitopes, or from an altered protein core similar to that reported for the CD44 molecule [15-17].

Fig. 3. Heterogeneity of the L-selectin expressed by peripheral blood lymphocytes from a patient with follicular lymphoma, compared to L-selectin on normal PBL. Protein was extracted by 0.5% Chaps and run on a non-reducing acrylamide gel. After blotting onto PVC membranes, L-selectin was visualized by FMC46 monoclonal antibody and anti-Mouse- biotin/ Alkaline Peroxidase.

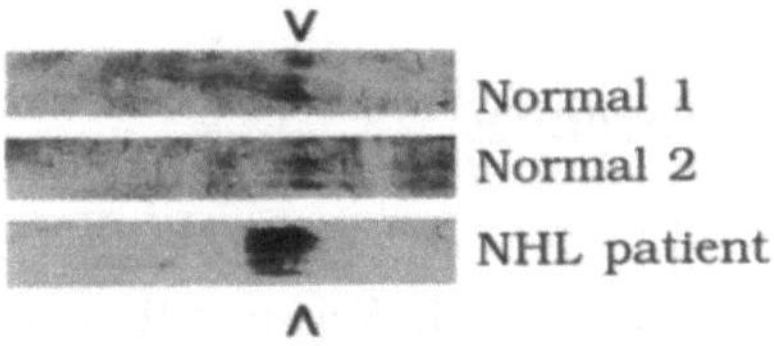

Circulating B cells in follicular lymphoma comprise a large subset of cells in S+G2/M phase expressing bcl-2.

The cell cycle profile was evaluated on peripheral blood samples from 4 FL patients using Propidium Iodide (PI) labeling and flow cytometric analysis. We found increased numbers of circulating B cells in the S and G2+M phases, as defined by double labeling with a CD20 monoclonal antibody and PI. The proliferating cells were almost entirely limited to the CD20 positive B cell population. In the same population, we found increased expression of the bcl-2 protein, which supports the association of these B cells with the malignant clone [18-21].

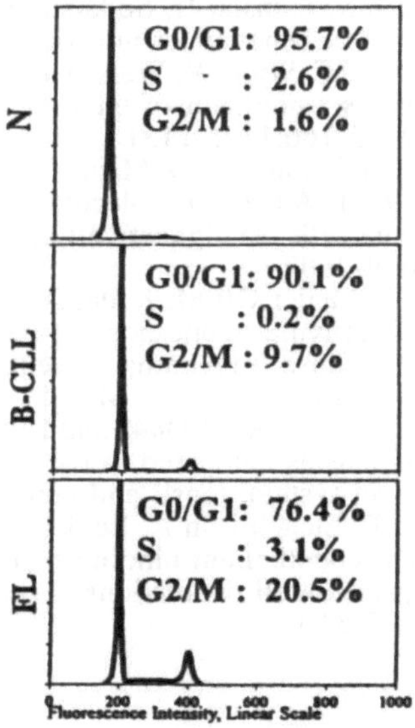

Conclusion.

The presence of circulating malignant cells in non-Hodgkin's lymphoma has been established by several groups, using PCR [18-21]. However, the percentage of malignant B cells relative to the normal polyclonal blood B cells is unknown. We have shown that a population of circulating B cells in follicular lymphoma exhibit (an) abnormal form(s) of the L-selectin homing molecule. Since several of the altered epitopes appear to include the ligand-binding portion of the L-selectin molecule, we suspect that the abnormal form of L-selectin may have the ability to bind to ligand(s) not normally recognized by L-selectin. Further studies of tissue specific binding are in progress to elucidate the mechanism by which an altered L-selectin receptor may affect migratory patterns of lymphoma B cells.

Acknowledgements. *We thank Dr. T. F. Tedder for permission to use the LAM1.1-6 antibodies and Dr. H. Zola for the FMC46 antibody. This study was supported by a grant from the Cancer Research Society Inc. Quebec.*

References.

1. Gallatin WM, Weissman IL, Butcher EC (1983) A cell-surface molecule involved in organ-specific homing of lymphocytes. Nature 304:30-34
2. Spertini O, Kansas GS, Reimann KA, Mackay CR, Tedder TF (1991) Function and evolutionary conservation of distinct epitopes on the leukocyte adhesion molecule-1 (TQ-1,Leu8) that regulate leukocyte migration. J Immunol 147:942-949
3. Stauder R, Hamader S, Fasching B, Kemmler G, Thaler J, Huber H (1993) Adhesion to high endothelial venules: A model for dissemination mechanisms in non-Hodgkin's lymphoma. Blood 82:262-267
4. Jalkanen S, Joensuu H, Soderstrom K-O, Klemi P (1991) Lymphocyte homing and clinical behavior of non-Hodgkin's lymphoma. J Clin Invest 87:1835-1840
5. Bargatze RF, Wu NW, Weissman IL, Butcher EC (1987) High endothelial venule binding as a predictor of the dissemination of passaged murine lymphomas. J Exp Med 166:1125-1131
6. De Rossi G, Zarcone D, Mauro F, Cerruti G, Tenca C, Puccetti A, Mandelli F, Grossi CE (1993) Adhesion molecule expression on B-cell chronic lymphocytic leukemia cells: Malignant cell phenotypes define distinct disease subsets.Blood 81:2679-2687
7. Pals ST, Meijer CJLM, Radaszkiewicz T (1991) Expression of the human peripheral lymph node homing receptor (LECAM-1) in nodal and gastrointestinal non-Hodgkin's lymphoma. Leukemia 5:628-631
8. Moller P, Eichelmann A, Mechtersheimer G, Koretz K (1991) Expression of b1 integrins, H-CAM (CD44) and LECAM-1 in primary gastro-intestinal B cell lymphomas as compared to the adhesion receptor profile of the gut-associated lymphoid system, tonsil and peripheral lymph node. Int J Cancer 49:846-855
9. Moller P, Eichelmann A, Leithauser F, Mechtersheimer G, Otto HF (1992) Venular endothelium binding molecules CD44 and LECAM-1 in normal and malignant B-cell populations.A comparative study. Virchows Archiv A Pathol Anat 421:305-313

10 Lambrechts AC, de Ruiter PE, Dorssers LCJ, van 't Veer MB (1992) Detection of residual disease in translocation t(14;18) positive non-Hodgkin's lymphoma, using the polymerase chain reaction: A comparison with conventional staging methods. Leukemia 6:29-34
11 Price CGA, Meerabux J, Murtagh S, Cotter FE, Rohatiner AZS, Young BD, Lister TA (1991) The significance of circulating cells carrying t(14;18) in long remision from follicular lymphoma. J Clin Oncol 9:1527-1532
12 Murakawa Y, Strober W, James SP (1991) Monoclonal antibody against the human peripheral lymph node homing receptor homologue (Leu8) inhibits B cell differentiation but not B cell proliferation. J Immunol 146:40-46
13 Spertini O, Kansas GS, Munro M, Griffin JD, Tedder TF (1991) Regulation of leukocyte migration by activation of the leukocyte adhesion molecule-1 (LAM-1) selectin. Nature 349:691-694
14 Pilarski LM, Turley EA, Shaw ARE, Gallatin WM, Laderoute MP, Gillitzer R, Beckman IGR, Zola H (1991) FMC46, a cell protrusion-associated leukocyte adhesion molecule-1 epitope on human lymphocytes and thymocytes. J Immunol 147:136-143
15 Jackson DG, Buckley J, Bell JI (1992) Multiple variants of the human lymphocyte homing receptor CD44 generated by insertions at a single site in the extracellular domain. J Biol Chem 267:4732-4739
16 Mackay CR, Terpe H-J, Stauder R, Marston WL, Stark H, Gunthert U (1994) Expression and modulation of CD44 variant isoforms in humans. J Cell Biol 124:71-82
17 Salles G, Zain M, Jiang W, Boussiotis VA, Shipp MA (1993) Alternatively spliced CD44 transcripts in diffuse large cell lymphoma: Characterization and comparison with normal activated B cells and epithelial malignancies. Blood 82:3539-3547
18 Gribben JG, Freedman AS, Woo SD, Blake KB, Shu RS, Freeman G, Longtime JA, Pinkus GS, Nadler LM (1991) All advanced stage non-Hodgkin's lymphomas with a polymerase chain reaction amplifiable breakpoint of bcl-2 have residual cells containing the bcl-2 rearrangement at evaluation and after treatment. Blood 78:3275-3280
19 Liu J, Johnson RM, Traweek ST (1993) Rearrangement of the bcl-2 gene in f ollicular lymphoma. Diagnostic Mol pathol 2:241-247
20 Gulley ML, Dent GA, Ross DW (1992) Classification and staging of lymphoma by molecular genetics. Cancer suppl 69:1600-1606.
21 Hickish TF, Purvies H, Mansi J, Soukop M, Cunningham D (1991) Molecular Cancer 64:1161-1163

Light Chain Loss and Reexpression Leads to Idiotype Switch. Surrogate Light Chains are Probably Responsible for this Process

Y. Caspi, M. Taya, N. Hollander and J. Haimovich

Department of Human Microbiology, Sackler Faculty of Medicine, Tel-Aviv University, Tel Aviv 69978, Israel.

Summary

The murine B-lymphocyte cell line 38C-13 is characterized by several cell surface markers typical for an early stage of B-cell differentiation. Cells of this cell line posses cell surface membrane IgM molecules composed of μ and κ polypeptide chains. They also produce "surrogate" or "pseudo" light chains (ΨL) coded by the λ5 and VpreB genes. Variants of the 38C-13 cell line which do not synthesize κ chains can be isolated from the 38C-13 population by the use of anti-idiotype antibodies *in vivo* and *in vitro*. In some κ chain-deficient variant cell lines, cells which have regained surface IgM expression but have lost the original idiotype specificity, can be isolated. This idiotype switch is probably due to a secondary rearrangment of the κ gene. In the κ chain-deficient variant cells, the μ chains assemble with the surrogate light chains but the assembled IgM-like molecules are not expressed on the cell surface. It is suggested that surrogate light chains play an important role in the induction of κ gene rearrangement but that surface expression of μ–ΨL complexes is not required for this process.

Introduction

Idiotype-negative variants of the 38C-13 murine B-lymphocyte cell line [1-3] have been isolated and characterized previously by us and by others [4-8]. *In vivo,* these variants appear in mice which had been treated with antibodies specific for the 38C-13 IgM idiotype or vaccinated by the idiotypic immunoglobulin and challenged with 38C-13 tumor cells. Among the survivors of such animals, some will develop tumors, usually a long time after the first challenge, some of which are 38C-13-idiotype negative. *In vitro*, this has been achieved by the treatment of 38C-13 cells with idiotype-specific antibodies and complement [4] or immunotoxins [7] and the subsequent cloning of the surviving cells. It is also possible to obtain these variants by simply, albeit more tediously, recloning the 38C-13 cultures ([8] and our unpublished results). It has been assumed from these experiments that idiotype-negative variants appear spontaneously in 38C-13 cultures *in vitro* and *in vivo* and that idiotype-specific immunotherapy merely selects these variants. Molecular characterization of idiotype-negative variants revealed the stability of expression of the original μ heavy chain and the "switch" to the expression of alternative genes for κ light chains [4,6]. In addition, several variants have been isolated that do not produce light chains at all [4,7]. Characterization of these light chain-negative variants reveals that their production of μ chains is quantitatively similar to that of either the original 38C-13 line or the light chain-positive, idiotype-negative variant cell lines. In addition to μ chains, these variants synthesize polypeptides coded by the λ5 and VpreB genes, recently denoted "surrogate" or "pseudo" light chains (ΨL, for review see [9]). Although μ chains in these cells assemble with the ΨL chains into molecules with a molecular weight similar to that of monomeric IgM [5], the assembled IgM-like molecules do not reach the cell surface and are intracellularly degraded [5,10]. The importance of the λ5 gene for B-cell maturation and κ chain gene rearrangement has been recently proven by experiments using gene targeting [11]. It has been suggested by several investigators that this maturation is due to a signal transduction through IgM-like molecules (composed of μ and ΨL chains) on the cell surface [12-14]. Since in our system κ chain-deficient variants do switch to production of light chains and surface IgM expression but do not deposit IgM-like molecules on their cell surface, we would like to propose an alternative possibility, namely, that the initiation of κ chain gene rearrangement by the IgM-like molecules is an intracellular process.

Materials and Methods

Cell Lines.

The murine B-cell lymphoma 38C-13 is a carcinogen induced tumor, originally obtained in a T-cell depleted C3H/eB mouse [1]. This tumor has been characterized previously [1-3]. Idiotype-negative variant cell lines were obtained *in vivo* as a result of idiotype-specific immunotherapy against the tumor, and *in vitro* as a result of treatment of 38C-13 cells with anti-idiotype antibodies and complement [4].

Immunofluorescence Analysis.

The analysis was done according to a method described elsewhere [15]. Briefly, 5 x 10^5 cells were washed twice in PBS and incubated for 30 min at 4°C with rat monoclonal anti-mouse IgM or anti-38C-13 idiotype antibodies developed in our laboratory. The cells were then washed and incubated with fluorescein-isothiocyanate - labeled mouse anti-rat IgG antibodies (Jackson Immuno Research Laboratories). The cells were washed, resuspended in PBS and analyzed by a fluorescence activated cell sorter (FACS 440, Becton Dickinson).

Biosynthetic Labeling of Cells.

Cells (1-2 x 10^7 cells/ml) were labeled with [^{35}S]methionine for 2 h as described [10]. The IgM was immunoprecipitated from lysates with rabbit anti-mouse IgM (ICN) and adsorbed on protein A sepharose (Pharmacia). The labeled IgM was eluted from the adsorbent and analyzed by sodium dodecyl sulfate-polyacrylamide gel electrophoresis (SDS-PAGE).

Results and Discussion

Immunofluorescence analysis of the 38C-13 murine B-lymphocyte cell line, of a surface-IgM-negative variant derived from it and of a surface-IgM-positive but idiotype-negative "secondary" variant derived from the latter, is described in Fig.1. Immunofluorescence has been performed with the culture media of hybridomas secreting either rat antibodies specific for the idiotype of 38C-13 IgM (2B6) or rat antibodies specific for mouse μ chain (4F6). These hybridomas have been obtained by a procedure previously described by us [15]. The use of rat antibodies specific for murine κ chains has resulted in flow cytometry patterns similar to those obtained with the μ chain specific ones (data not shown). It is clear from the results described in Fig.1 that the switch from the original idiotype of the parental 38C-13 cell line to that of the secondary variant was

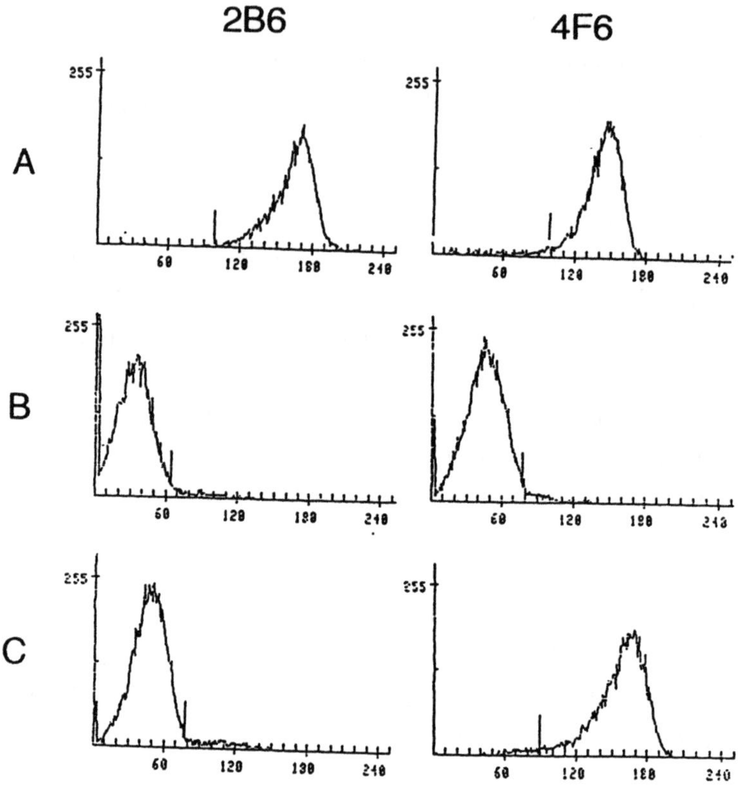

Fig.1. Fluorescent Activated Cell Sorter analysis of A, 38C-13 cell line; B, a variant cell-line derived from 38C-13 *in vivo* in a mouse treated with a monoclonal rat anti-38C-13 idiotype antibody and subsequently challenged with 38C-13 cells and C, a variant cell-line obtained *in vitro* by the limiting dilution cloning of the cell line described in B. 4F6 is a monoclonal rat antibody specific for mouse μ chain; 2B6 is a rat monoclonal antibody specific for the 38C-13 IgM idiotype.

through an intermediary stage of surface IgM-negative cells. Previous reports by others and by us have shown that the idiotype switch in the 38C-13 cell line is due to a secondary rearrangement of the gene for the κ chain while the gene coding for μ chain remains the same [4,6]. Moreover, further characterization by Levy *at all* [8] has shown that the idiotype switch is due to the rearrangment of upstream V genes to downstream J regions in these variants.

In order to follow the synthesis and assembly of immunoglobulins in 38C-13

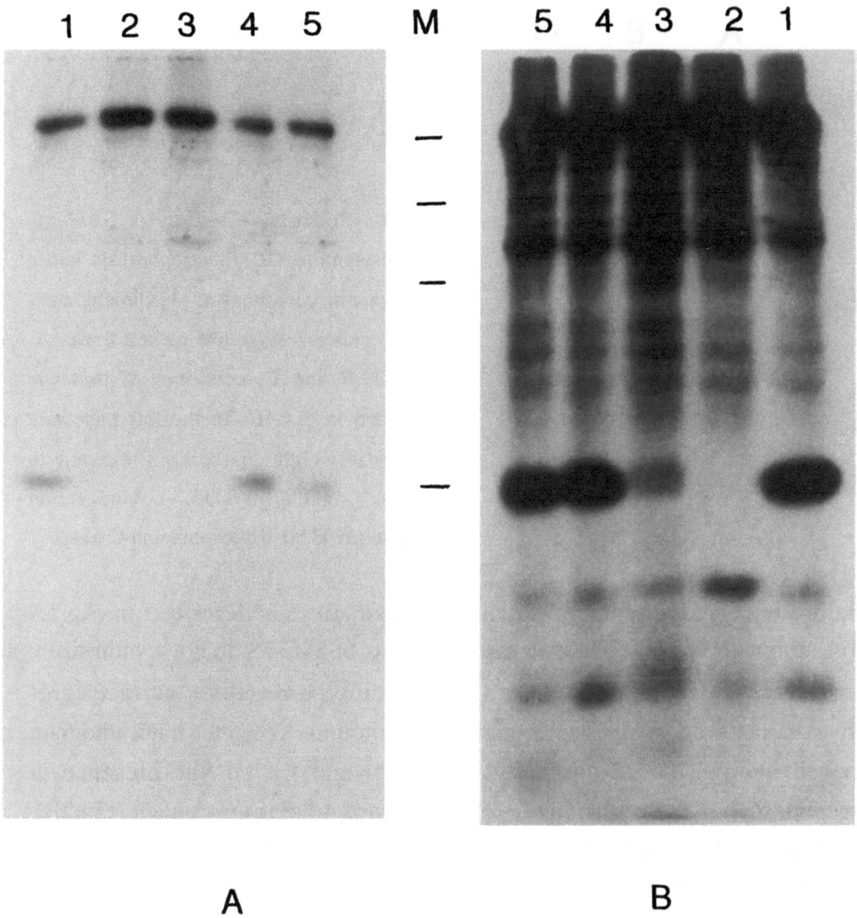

Fig.2. Sodium Dodecyl Sulfate-Polyacrylamide Gel Electrophoresis analysis under reducing conditions of [^{35}S]methionine-labeled cells of the following cell lines: 1, 38C-13; 2, a variant cell line described in Fig.1B; 3, a variant cell line derived from 38C-13 which lost production of light chains but which starts to accumulate light chain-positive cells again; 4 and 5, variant cell lines of the type described in Fig.1C. A and B are autoradiograms of the same gel exposed for different lengths of times. Markers (M) are from top to bottom: μ chain (76 kDa); bovine serum albumin (67 kDa); ovalbumin (42 kDa); and L chain (23 kDa)

and its variants, cells were labelled with [^{35}S]methionine and the labeled IgM was isolated by immunoprecipitation with anti-IgM and subjected to SDS-PAGE

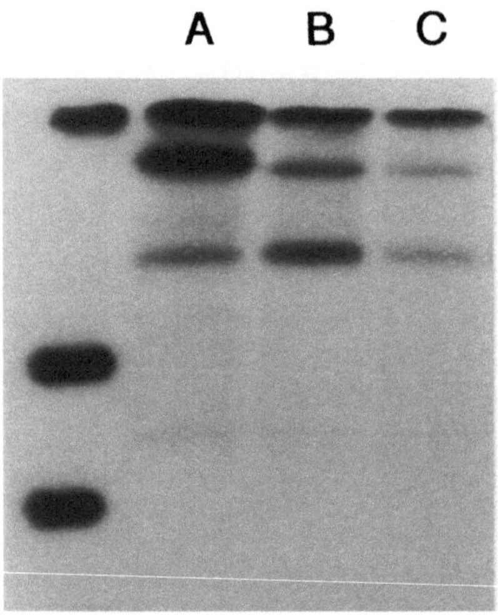

Fig.3. Sodium Dodecyl Sulfate-Polyacrylamide Gel Electrophoresis under non-reducing conditions of [^{35}S]methionine-labeled cells of the following cell lines: A, 38C-13; B and C, cell lines of the type described in Fig.1B. In the left panel are molecular weight markers. From top to bottom: IgG (150 kDa), bovine serum albumin (67 kDa) and ovalbumin (42 kDa).

in the presence and absence of a reducing agent. Results described in Fig.2A clearly show that surface-IgM-negative variants of 38C-13 do not synthesize κ chains whereas 38C-13 as well as 38C-13 idiotype-negative, surface-IgM-positive variants do synthesize them in large amounts. When the autoradiogram described in Fig.2A is exposed for a much longer period, the presence of polypeptides with molecular masses of 18 kDa and 14 kDa are evident (Fig.2B). These polypeptides have been previously described in pre-B cell lines and normal pre-B cells and were shown to be the products of the λ5 and VpreB genes [9,16].

It is of interest to note that incorporation of [^{35}S] methionine into the ΨL chains is very weak. A rough estimation of the extent of labeling can be achieved by comparing the intensity of the λ5 and VpreB bands in Fig.2B with those of the κ polypeptide chains in Fig.2A. For roughly equal intensities it is necessary to expose the gels for about 50-fold longer times. This is probably due to a much lower specific activity of the ΨL chains as compared to that of κ chains and not to a limitation in their amount since the extent of assembly of μ chains with the ΨL chains is comparable to their assembly with κ chains (Fig.3). This lower specific activity is probably the reason why ΨL chains have not been detected in previous studies in which radioactively- labeled immunoglobulins of 38C-13 cells were followed.

Contrary to previous reports by other investigators who described the presence of ΨL polypeptides on the surface of pre-B cell lines [12-14, 17], our results clearly show that complexes of μ and ΨL chains do not reach the cell surface of the κ chain-deficient cell lines. The μ chains do however assemble with the pseudo light chains into molecules with molecular weights similar to those of μ–κ and $(\mu-\kappa)_2$ complexes (Fig.3).We have recently reported that the assembled IgM-like molecules in surface IgM-negative variants of 38C-13 are degraded in the rough endoplasmic reticulum by a "quality-control" mechanism [10]. The recently reported absence of μ–ΨL complexes on the majority of pre-B cells in the bone marrow of normal mice [18,19] is in line with our observation of their absence in the κ-negative variants of our 38C-13 B-lymphocyte cell line.

As seen in Fig.2, in addition to the κ-negative variant cell lines, the κ-positive, idiotype-negative ones as well as the parental 38C-13 also produce the ΨL polypeptides. These are probably co-expressed on the cell surface together with the μ and the κ polypeptide chains [17] but their role, if any, in B-lymphocyte differentiation is not clear at the moment.

Studies on the importance of the ΨL polypeptide chains for B-lymphocyte development have clearly shown that when λ5 gene is "knocked out" by gene targeting, B-cell development is severely impaired [11]. It is assumed that μ–ΨL complexes in the κ chain-deficient variants of 38C-13 play a similar role to that of μ–ΨL complexes in pre-B cells, both leading to κ gene rearrangement. Since in the κ-negative variants of 38C-13, the μ–ΨL complexes do not reach the cell surface, it is concluded that κ gene rearrangement does not require a signal transduction through a receptor on the cell surface. Rather, it is suggested that the induction is through an intracellular "signal transduction" process.

References

1 Bergman Y, Haimovich J (1977) Characterization of a carcinogen-induced murine B-lymphocyte cell line of C3H/eB origin. Eur J Immunol 7:413-417
2. Bergman Y, Haimovich J, Melchers F (1977) An IgM-producing tumor with biochemical characteristics of a small B-lymphocyte. Eur J Immunol 7:574-579
3. Bergman Y, Haimovich J (1978) B-lymphocytes contain three species of μ chains. Eur J Immunol 8:876-880
4. Taya M, Haimovich J (1991) Immunotherapy of B-lymphoma by anti-idiotype antibodies: Characterization of variant tumor cells appearing a long time after the initial tumor inoculation. Cancer Immunol Immunother 34:43-48
5. Taya M, Rabinovich E, Haimovich J (1992) Characterization of IgM molecules in light-deficient variants of a B-cell tumor. Immunol Lett 33:173-178
6. Starnes CO, Caroll WL, Campbell MJ, Houston LL, Apell G, Levy R (1988) Heterogeneity of a murine B cell lymphoma. Isolation and characterization of idiotypic variants. J Immunol 141:333-339
7. Caroll WL, Starnes CO, Levy R, Levy S (1988) Alternative Vκ gene rearrangements in a murine B cell lymphoma. J Exp Med 268:1607-1620

8. Levy S, Campbell MJ, Levy R (1989) Functional immunoglobulin light chain gene are replaced by ongoing rearrangements of germline Vκ genes to downstream Jκ segment in a murine B cell line. J Exp Med 170:1-13
9. Melchers F, Karasuyama H, Haasner D, Bauer S, Kudo A, Sakaguchi N, Jameson B, Rolink A (1993) The surrogate light chain in B cell development. Immunol Today 14:60-68
10. Rabinovich E, Bar-Nun S, Amitay R, Shachar I, Gur B, Taya M, Haimovich J (1993) Different assembly species of IgM are directed to distinct degradation sites along the secretory pathway. J Biol Chem 268:24145-24148
11. Kitamura D, Kudo A, Schaal S, Muller W, Melchers F, Rajewsky K (1992) A critical role of λ5 protein in B cell development. Cell 69:823-831
12. Tsubata T, Reth M (1990) The products of pre-B cell-specific genes (λ5 and VpreB) and the immunoglobulin μ chain form a complex that is transported onto the cell surface. J Exp Med 172:973-976
13. Nishimoto N, Kubagawa H, Ohno T, Gartland GL, Stankovic AK, Cooper MD (1991) Normal pre-B cells express a receptor complex of μ heavy chains and surrogate light-chain proteins. Proc Natl Acad Sci USA 88:6284-6288
14. Brouns GB, de Vries E, van Noesel CJM, van Lier RAW, Borst J (1993) The structure of the μ/pseudo light chain complex of human pre-B cells is consistent with a function in signal transduction. Eur J Immunol 23:1088-1097
15. Maloney DG, Kaminski MF, Burowski D, Haimovich J, Levy R (1985) Monoclonal anti-idiotype antibodies against the murine B cell lymphoma 38C13: charaterization and use as probes for the biology of the tumor *in vivo* and *in vitro*. Hybridoma 4:191-209
16. Cherayil B, Pillai S (1991) The ω/λ5 surrogate immunoglobulin light chain is expressed on the surface of transitional B lymphocytes in murine bone marrow. J Exp Med 173: 111-116
17. Karasuyama H, Rolink A, Melchers F (1993) A complex of glycoproteins is associated with VpreB/λ5 surrogate light chain on the surface of μ heavy chain-negative early precursor B cell lines. J Exp Med 178:469-478
18. Karasuyama H, Rolink A, Shinkai Y, Young F, Alt FW, Melchers F (1994) The expression of VpreB/λ5 surrogate light chain in early bone marrow precursor B cells of normal and B cell-deficient mutant mice. Cell 77:133-143
19. Guelpa-Fonlupt V, Tonnelle C, Blaise D, Fougereau M, Fumoux F (1994) Discrete early pro-B and pre-B stages in normal human bone marrow as defined by surface pseudo-light chain expression. Eur J Immunol 24:257-264

Growth Regulation: IL-6

Regulation of Terminal differentiation of Human B-Cells by IL-6

Selina Chen-Kiang
Brookdale Center for Molecular Biology, Mount Sinai School of Medicine, New York, NY, 10029.

The Hallmarks of Terminal Differentiation of B Cells

Terminal differentiation of B cells to plasma cells is characterized by a marked change in the cellular morphology as the cells develop the capacity to synthesize and secrete large quantities of immunoglobulin (Ig). Concomitantly, the B cells decrease the expression of Ig and MHC class II on the cell surface. The terminally differentiated plasma cells cease to proliferate. Despite these well orchestrated series of events that characterize B cell terminal differentiation, the molecular mechanisms that underlie this process are not well understood.

Induction of B Cell Terminal Differentiation by IL-6

Enhancement of Ig Synthesis and Secretion

Cytokines can regulate certain aspects of B cell differentiation. IL-6 had previously been shown to induce a 2-3 fold increase in Ig secretion by activated peripheral mouse B cells and by the EBV-immortalized, clonal IgG1-bearing human B lymphoblastoid cell line, CESS [22]. This level of change, however, was not sufficient for molecular analysis. By selecting for CESS cells expressing a high density of surface IgG, we obtained a population of less differentiated cells which responded to IL-6 by markedly increasing Ig synthesis and secretion (greater than 50-fold) [36, 40]. The molecular basis for this increase lies in transcriptional activation of the Ig $\gamma1$ heavy and λ light chain genes, as well as differential accumulation of the secreted form-specific $\gamma1$ mRNA (20-40 fold) relative to the alternatively-spliced membrane form-specific $\gamma1$ mRNA (2-fold) [40].

The IL-6 responsiveness is not restricted to IgG-bearing B cells. IL-6 induced a 20-25 fold increase of the secreted form-specific μ mRNA in an IgM-positive human lymphoblastoid cell line SKW6-cl4 [32, 36]. Curiously, at the single cell level, the IL-6-induced changes were observed in only 20-40% of cells of each cloned cell line, despite recloning by limited dilution. The reason for this restricted response is currently not clear. Freshly isolated tonsillar B lymphocytes were also inducible by IL-6, leading to enhanced synthesis of IgG and IgM. This induction,, however, requires activation of tonsillar B lymphocytes into the cell cycle, such as immortalization with EBV. As in CESS cells, the enhancement was due to an increase in the secreted form-specific Ig heavy chain mRNA. The IL-6-induced cells were also plasmacytoid in appearance [36]. Since tonsillar B lymphocytes are representative of polyclonal B lymphocytes of secondary lymphoid organs, it appears that a subset of human B lymphocytes, once activated to an IL-6-responsive state, are poised to mature to Ig-secreting cells by IL-6 signaling.

Morphological Maturation

By transmission electron microscopy, the IL-6-induced cells display an enlarged cytoplasm, well developed rough endoplasmic reticulum and an eccentric nucleus [36], prominent features of plasma cells. Immunocytochemical analysis further suggests that several endoplasmic reticulum proteins that are important for protein transport and secretion, such as ribophorin I and ribophorin II, are increased in cells that are enhanced in Ig synthesis [32]. Thus, coordinated with activation of Ig gene expression, IL-6 signals lead to morphological maturation of CESS cells *in vitro* that are characteristic of terminal differentiation of B cells *in vivo*.

Reduction of Surface MHC Class II Expression

The close resemblance of IL-6-induced CESS cells with plasma cells suggests that IL-6 signals may lead to other changes that are observed during terminal differentiation of B cells *in vivo*. Plasma cells do not express MHC class II molecules on the cell surface [16]. The expression of MHC class II was found to be markedly reduced in IL-6-induced CESS cells in which Ig synthesis was enhanced. Analysis of simultaneous Ig and class II expression at the single cell level by immunocytochemistry confirmed these results [32, 36]. IL-6 simulation of CESS cells therefore results in inverse regulation of Ig and surface MHC class II expression, as observed in plasma cells.

Growth Arrest of IL-6 Differentiated B Cells

IL-6 stimulation of Ig synthesis is transient, dose- and time-dependent, reaching a maximal level at 4 days after IL-6 induction [40]. However, the increase is not sustained when IL-6 treatment is prolonged [36]. These results may be explained by two possibilities that are not mutually exclusive: feedback control of IL-6 signaling and growth arrest of IL-6-induced cells. Preliminary results support both possibilities. The cells that are rendered non-responsive by prolonged IL-6 treatment can be restimulated again after removal of IL-6 for a period of time [36]. The IL-6-responsive cells, selected by their reduced expression of surface MHC class II, undergo growth arrest [32]. Thus, IL-6 signaling is subject to negative regulation. Stimulation of CESS cells with IL-6 leads to cessation of cell growth that is also a hallmark of terminal differentiation of B cells.

The Two IL-6 Signal Transduction Pathways

To address how IL-6 coordinately induces multiple differentiated phenotypes in B cells, it is essential to understand the mechanisms of IL-6 signaling. IL-6 is a multifunctional cytokine that regulates immune response, acute phase reactions, haematopoiesis and viral infection in a cell type-dependent manner [7, 24, 45]. The high affinity IL-6 receptor complex is comprised of a ligand-binding IL-6-R-α and a signal transducing gp130 (IL-6Rβ), which is shared by leukemia inhibitory factor, oncostatin M and ciliary neurotrophic factor [14, 21] (Fig. 1).

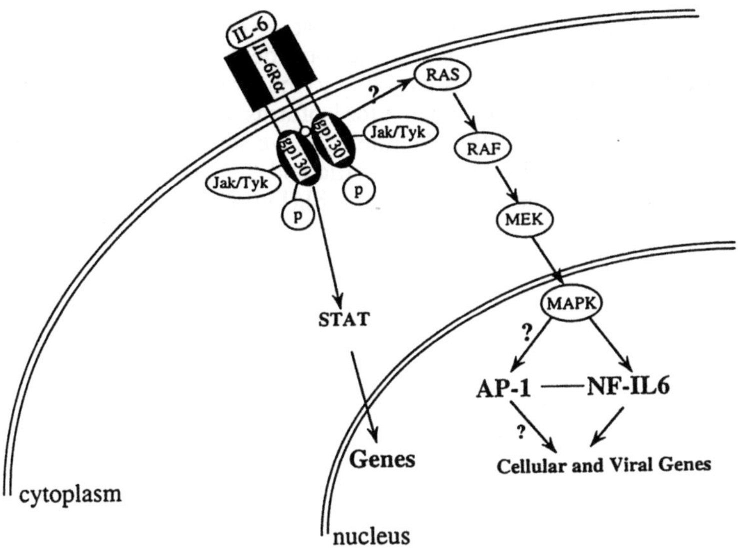

Fig. 1. The Jak-STAT and the delayed IL-6 signaling pathways

The Jak-STAT Pathway

Signaling by IL-6 begins with homodimerization of gp130 [9, 33] and phosphorylation of gp130 by Jak-Tyk tyrosine kinases (Fig. 1) [30, 46]. This is followed by rapid activation and nuclear translocation of latent transcription factors STAT-3 [2, 56], and a related STAT-1 [13, 44] initially identified in the signaling pathways of interferon α and interferon γ [2, 42, 55, 56]. The STATs (signal transducers and activators of transcription) [8] are likely to mediate rapid and transient activation of a subset of IL-6 responsive genes, such as those encoding the acute phase response, by directly interacting with STAT-binding sites in the promoters [52].

The Delayed Pathway

The Jak-STAT activity declines rapidly, however, usually within minutes after activation [55]. IL-6 regulation of B cell differentiation is therefore mediated by a delayed pathway that stably propagates the IL-6 signals (Fig. 1). This pathway appears to involve the ras-dependent, mitogen-activated protein kinases [35]. It leads to activation of transcription factors of the AP-1 family and the CCAAT/enhancer binding protein (C/EBP) family, which in human cells is comprised of NF-IL6 [1] and a minor species NF-IL6β [23] [19, 20, 29, 34, 43]. Most IL-6 responsive genes, including the Ig genes, contains NF-IL6 binding sites in their promoters and enhancers, suggesting that NF-IL6 may mediate the IL-6 signals.

The NF-IL6 gene does not contain introns but it encodes three proteins that are translated from in-frame AUGs of the same mRNA species (NF-IL6-1, NF-IL6-2, and NF-IL6-3, Fig.2), analogous to its rodent counterpart LAP (also called IL-6DBP, AGP/EBP, C/EBPβ, rNF-IL6, or CRP-2) [5, 6, 11, 31, 39, 54]. NF-IL6 regulates target genes by binding as obligate dimers through the basic-leucine zipper (b-ZIP) regions to specific sequences in the promoters [1, 11, 26]. Transactivation by NF-IL6 can be enhanced by phosphorylation at serine/threonine residues in the b-ZIP region and the activating domain that lies outside the b-ZIP region [35, 49, 53]. The most novel feature of NF-IL6 is its ability to function as an activator or inhibitor according to the ratio of activator to inhibitor isoforms, as first proposed by Schibler [12] and confirmed by us [19, 25]. Additionally, we showed that the expression of the three NF-IL6 isoforms is cell type-specific and that IL-6 regulates the ratio of NF-IL6 isoforms to favor the activator form [18].(Fig. 2).

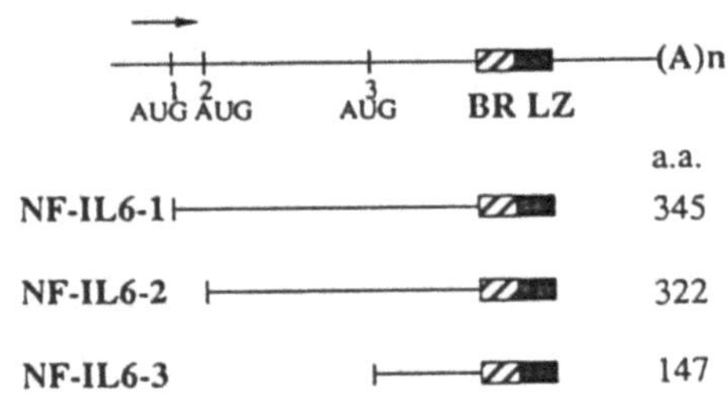

Fig. 2. The NF-IL6 mRNA and three isoforms. The arrow indicates the direction of transcription of the NF-IL6 mRNA. (A)n represents the poly (A) tail. NF-IL6-1, NF-IL6-2 and NF-IL6-3 represent the three isoforms translated from the first, second and third AUG, respectively. BR: basic region, LZ: leucine zipper region.

Less is known about the role of Fos and Jun in IL-6 signaling. In common with NF-IL6, Fos and Jun related proteins contain the b-ZIP structure and are also activated by IL-6 according to lineage and cell type. For example, *jun* related genes are activated in the B and myeloid lineages whereas *fos* is activated in the neuronal lineage [19, 20, 29, 34, 43].

Modulation of the Delayed IL-6 Signaling Pathway by Dimerization Between NF-IL6 and AP-1

The simultaneous activation of NF-IL6 and AP-1 [19] suggests that they may interact to mediate the IL-6 signals. This hypothesis is strongly supported by experimental evidence. We showed that recombinant NF-IL6 proteins directly associate with Jun and with Fos in the absence of their cognate recognition DNA elements *in vitro*. Deletion and substitution mutagenesis analysis further established that all three NF-IL6 isoforms dimerize with Jun and Fos via coiled coil interactions at the leucine zipper regions [19, 20] (Fig. 3).

Cross-family dimerization leads to two functional consequences: acquisition of new DNA binding specificities by the heterodimers and modulation of transactivating activities of NF-IL6 and Fos/Jun [19, 20]. NF-IL6 binds to all AP-1-binding sites with varying efficiency, whereas Fos and Jun do not bind to most NF-IL6-binding sites [19]. The binding of NF-IL6 to NF-IL-6 sites, for example that of the IL-6 promoter, is reduced in the presence of Fos or Jun (Fig. 3). This correlates with reduction of NF-IL6 site-mediated activation by NF-IL6 in transiently transfected cells [19] Thus, heterodimerization with Fos/Jun diminishes NF-IL6 DNA binding activity and regulatory functions.

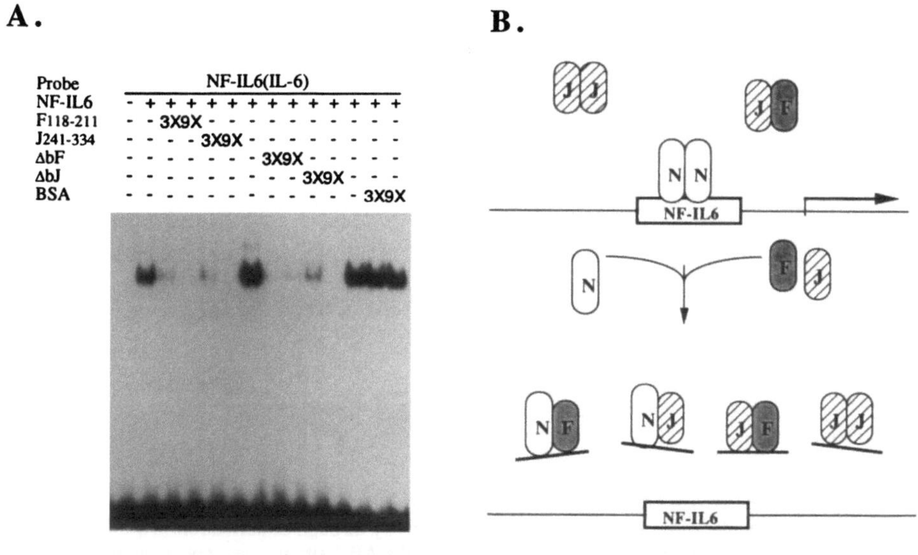

Fig. 3. Fos/Jun reduces the NF-IL6 DNA binding activity by dimerization. (A). The binding of recombinant NF-IL6-3 protein to an oligonucelotide probe containing the NF-IL6 site of the IL-6 promoter was analyzed by electrophoretic mobility assay in the presence (+) or absence (-) of recombinant Fos or Jun. F118-211 and J241-334 contain the b-ZIP regions of Fos and Jun, respectively. ΔbF lacks residues 139-144 in the basic region of F118-211 and ΔbJ lacks residues 260-266 in the basic region of J241-334. Fos and Jun proteins were added at 3 times (3x) or 9 times (9x) the molar concentration of NF-IL6 as indicated, as was the control BSA. (B). Schematic representation of the results shown in panel (A). N: NF-IL6, F: Fos, J: Jun.

Conversely, activation of AP-1 site mediated transcription by Fos and Jun is also modulated by cross-family dimerization (Fig. 4). In a sequence-dependent manner, the NF-IL6/Fos heterodimers, but not the NF-IL6/Jun heterodimers, can compete for binding to AP-1 sites [20], suggesting that cross-family dimerization modulates the Fos and Jun transcriptional properties. The functional consequences would depend on the transactivation potential of the NF-IL6 isoforms, as well as the relative DNA binding affinity of the various homo and heterodimers for a given AP-1 site. Consistent with this possibility, NF-IL6 modulates AP-1 activation of the collagenase AP-1 site according to the transcriptional properties of the NF-IL6 isoforms in transiently transfected cells [20].

Most importantly, association of NF-IL6 and Jun has been demonstrated *in vivo* in IL-6-induced macrophage differentiation, but only when both NF-IL6 and Jun were activated by IL-6 [20]. Fos, which was not regulated, did not associate with NF-IL6. These results provide the first example of regulated interactions between transcription factors of the NF-IL6 family and the AP-1 family. They suggest that through the interplay of the NF-IL6 isoforms and cross-family dimerization, NF-IL6 and Fos/Jun play important physiologic roles in determining the specificity of IL-6-responsive promoters.

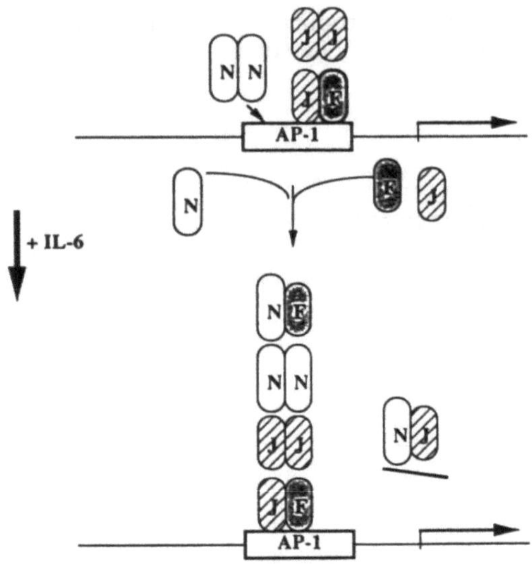

Fig.4. NF-IL6 regulates the Fos/Jun DNA binding activity through dimerization. N: NF-IL6, F: Fos, J: Jun. The distance between the protein and the AP-1 site is inversely correlated with their affinity.

The Mechanisms of IL-6 Signaling in Late Stage B Cells

Recapitulation of the hallmarks of plasma cells in IL-6-differentiated IgG and IgM-bearing human B cells suggests that the molecular mechanisms for terminal B cell differentiation may now be studied in clonal cell lines *in vitro*. Elucidation of the immediate events of IL-6 signaling has defined the molecular structure of the IL-6 receptor complex and, revealed the convergence of the immediate IL-6 signaling pathway with those of other cytokines and peptides. Studies on the delayed pathway further suggest that the cross-talk between NF-IL6 and AP-1 is dynamic and likely to be important for determining the promoter specificity, and hence the cell type specificity, of IL-6 signaling. Investigating the relationship between the two pathways should provide a link between the initiation of IL-6 signaling and the propagation of the signals.

The synthesis and secretion of Ig for humoral immune responses are destined physiologic functions of cells in the B lineage. We have shown that the marked enhancement of Ig synthesis by IL-6 is primarily due to transcriptional activation of the Ig heavy and light chain genes [40]. Since IL-6 induces newly immortalized polyclonal tonsillar B cells to differentiate *in vitro*, activation of Ig synthesis by IL-6 is most likely mediated by conserved sequences present in the Ig promoters and enhancers, such as the E boxes and the octamer motif. Consistent with this possibility, IL-6 enhances the synthesis of NF-IL6 and Oct-2, which bind to the E boxes and the octamer motif, respectively, and NF-IL6 activates Ig heavy chain enhancer-directed transcription in transiently transfected B cells [36, 55].

However, NF-IL6 may also negatively regulate the transcription of Ig gene, depending on the ratio of the activator to inhibitor NF-IL6 isoforms and their dimerization with Fos/Jun [19, 20, 25]. The functions of NF-IL6 may be further modulated by protein-protein interactions with other transcription factors that are important for Ig transcription such as NFκB. Although association between NFκB and NF-IL6 *in vivo* has not been demonstrated, NF-IL6 associates NFκB *in vitro* [28, 47]. Further investigations of the interplays of the NF-IL6 isoforms and their interactions with other transcription factors in IL-6 signaling, in particular the AP-1 family transcription factors, should provide new insight into the mechanism by which IL-6 regulates Ig gene expression in late stage B cells.

Stimulation of CESS cells by IL-6 has provided an example of simultaneous and inverse regulation of Ig synthesis and surface MHC class II expression. Extinction of MHC class II expression has been observed in LPS- or corticosteroid-treated mouse B lymphocytes [10, 15] and in cell fusions [27, 50]. It is thought to be due to transcriptional repression [27]. However, the repressor has not been identified. Since surface MHC class II expression is also controlled posttranscriptionally at multiple levels such as the assembly and transport of MHC class II dimers [3, 38, 51], the reduction of surface MHC class II expression may not simply be a reversal of activation of MHC class II genes. The mechanism that governs the reduction of surface MHC class II expression may now be elucidated in the context of terminal differentiation of human B cells by IL-6 induction.

Induction of growth arrest of CESS cells by IL-6 suggests that this clonal system may be useful for investigating the mechanism for growth control exhibited in terminal differentiation of B cells. IL-6 has been significantly linked to multiple myeloma. It promotes the growth of plasmacytoma cells *in vitro* and causes plasmacytomas in BALB/c mice, due either to accelerated terminal differentiation or deregulated cell growth [37]. Ectopic expression of IL-6 in transgenic mice of the same genetic background leads to chromosomal translocation and plasmacytomas, thus confirming a role of IL-6 in B cell differentiation and neoplasia [48]. Hypophosphorylation of the retinoblastoma gene product has been associated with growth arrest in IL-6 induced macrophage differentiation [41]. The activator form of NF-IL6 (NF-IL6-2) was shown to cause growth arrest of hepatocytes [4]. However, the mechanism by which IL-6 regulates cell proliferation is not known. Identifying the component(s) in the IL-6 signaling pathway that controls the progression of the cell cycle in late stage B cells should be instrumental for understanding the molecular basis for multiple myeloma.

Acknowledgments

I thank Ruth Abramson for a critical reading of this manuscript and members of my laboratory for helpful discussions. The work is supported by a grant (IM548) from the American Cancer Society.

References

1. Akira S, Isshiki H, Sugita T, Tanabe O, Kinoshita S, Nishio Y, Nakajima T, Hirano T, Kishimoto T (1990). A nuclear factor for IL-6 expression (NF-IL6) is a member of a C/EBP family. EMBO J. 9:1897-1906.
2. Akira S, Nishio Y, Inoue M, Wang X-J, Wei S, Matsusaka T, Yoshida K, Sudo T, Naruto M, Kishimoto T (1994). Molecular cloning of APRF, a novel IFN-stimulated gene factor 3 p91-related transcription factor involved in the gp130-mediated signaling pathway. Cell 77:63-71.
3. Anderson MS , Miller J (1992). Invariant chain can function as a chaperon protein for class II major histocompatibility complex molecules. Proc. Natl. Acad. Sci. USA 89:2282-2286.
4. Buck M, Turler H, Chojkier M (1994). LAP (NF-IL-6), a tissue-specific transcriptional activator, is an inhibitor of hepatoma cell proliferation. EMBO J. 13:851-860.
5. Cao Z, Umek RM, Mcknight S (1991). Regulated expression of three C/EBP isoforms during adipose conversion of 3T3/L1 cells. Genes & Dev. 5:1538-1552.
6. Chang CJ, Chen TT, Lei H-Y, Chen DS, Lee S-C (1991). Molecular cloning of a transcription factor, AGP/EBP, that belongs to members of the C/EBP family. Mol. Cell. Biol. 10:6642-6653.
7. Chen-Kiang S, Hsu W, Natkunam Y , Zhang X (1993). Nuclear signaling by interleukin-6. Curr. Opin. in Immunol. 5:124-128.
8. Darnell JE, Kerr IM , Stark GR (1994). Jak-STAT pathways and transcriptional activation in response to IFNs and other extracellular signaling proteins. Science 264:1415-1421.
9. Davis S, Aldrich TH, Stahl N, Pan L, Taga T, Kishimoto T, Ip NY, Yancopoulos GD (1993). LIFRβ and gp130 as heterodimerizing signal trandsducers of the tripartite CNTF receptor. Science 260:1805-1808.
10. Dennis JG, Mond JJ (1986). Corticosteroid-induced suppression of murine B cell immune response antigens. J. Immunol. 136:1600-1604.
11. Descombes P, Chojkier M, Lichsteiner S, Falvey E , Scheibler U (1990). LAP, a novel member of the C/EBP gene family, encodes a liver-enriched transcriptional activator protein. Genes & Dev. 4:1541-1551.
12. Descombes P, Scheibler U (1991). A liver-enriched transcriptional activator protein, LAP, and a transcriptional inhibitory protein, LIP, are translated from the same mRNA. Cell 67:569-579.
13. Fu X, Schindler C, Improta T, Aebersold R , Darnell JE (1992). The proteins of ISGF-3, the IFN-a induced transcription activator, define a new gene family of signal transducers. Proc. Natl. Acad. Sci. USA 89:7840-7843.
14. Gearing DP, Comeau MR, Friend DJ, Gimpel SD, Thut CJ, McGourty J, Brasher KK, King JA, Gillis S, Mosley B, Ziegler SF , Cosman D (1992). The IL-6 signal transducer, gp130: an oncostatin M receptor and affinity converter for the LIF receptor. Science 255:1434-1437.
15. Gravallese EM, Boothby MR, Smas CM, Glimcher LH (1989). A lipopolysaccharide-induced DNA-binding protein for a class II gene in B cells is distinct from NFkB. Mol. Cell. Biol. 9:3184-3192.
16. Halper J, Fu SM, Wang CY, Winchester R, Kunkel HG (1978). Patterns of expression of "Ia like" antigens during the terminal stages of B cell development. J. Immunol. 120:1480-1484.
17. Horikoshi N, Maguire K, Kralli A, Maldonado E, Reinberg D, Weimann R (1991). Direct interaction between adenovirus E1A protein and the TATA box binding transcription factor IID. Proc. Natl. Acad. Sci. USA 88:5124-5128.
18. Hsu W , Chen-Kiang S (1993). Convergent regulation of NF-IL6 and Oct-1 synthesis by interleukin-6 and retinoic acid signaling in embryonal carcinoma cells. Mol. Cell. Biol. 13:2515-2523.
19. Hsu W, Kerppola TK, Chen P-L, Curran T , Chen-Kiang S (1994). Fos and Jun repress transcription activation by NF-IL6 through association at the basic zipper region. Mol. Cell. Biol. 14:268-276.
20. Hsu W, Kerppola TK, Curran T , Chen-Kiang S (1994). IL-6 regulates the association of NF-IL6 with Jun *in vivo*: mutual modulation of transcriptional activity by cross-family dimerization. (submitted).
21. Ip NY, Nye SH, Boulton TG, Davis S, Taga T, Li Y, Birren SJ, Yasukawa K, Kishimoto T, Anderson DJ, Stahl N , Ynacopoulos GD (1992). CNTF and LIF act on neuronal cells via shared signaling pathways that involve the IL-6 signal transducing receptor component gp130. Cell 69:1121-1132.

22.Kikutani H, Taga T, Akira S, Kishi H, Miki Y, Saiki O, Yamamura Y, Kishimoto T (1985). Effect of B cell differentiation factor (BCDF) on biosynthesis and secretion of immunoglobulin molecules in human B cell lines. J. Immunol. 134:990-995.
23.Kinoshita S, Akira S, Kishimoto T (1992). A member of the C/EBP family, NF-IL6β, forms a heterodimer and transcriptionally synergizes with NF-IL6. Proc. Natl. Acad. Sci. USA. 89:1473-1476.
24.Kishimoto T, Akira S, Taga T (1992). Interleukin-6 and its receptor: a paradigm for cytokines. Science 258:593-597.
25.Klampfer L, Lee TH, Hsu W, Vilcek J, Chen-Kiang S (1994). NF-IL6 and AP-1 cooperatively modulate the activation of TSG-6 gene by IL-1 and TNF. (submitted).
26.Landschulz WH, Johnson PF, McKnight SL (1988). The leucine zipper: A hypothetical structure common to a new class of DNA binding proteins. Science 240:1759-1764.
27.Latron F, Jotterand-Bellomo M, Maffei A, Scarpellino L, Bermar M, Strominger JL, Accola RS (1988). Active suppression of major histocompatibility class II gene expression during differentiation from B cells to plasma cells. Proc. Natl. Acad. Aci. USA 5:2229-2233.
28.LeClair KP, Blanar MA, Sharp PA (1992). The p50 subunit of NF-κB associates with the NF-IL6 transcription factor. Proc. Natl. Acad. Sci. USA 89:8145-8149.
29.Lord KA, Abdollahi A, Thomas SM, DeMarco M, Brugge JS, Hoffman-Liebermann B, Liebermann DA (1991). Leukemia inhibitory factor and interleukin-6 trigger the same immediate early response, inducing tyrosine phosphorylation, upon induction of myeloid leukemia differentiation. Mol. Cell. Biol. 11:4371-4379.
30.Lütticken C, Wegenka UM, Yuan J, Buschmann J, Schindler C, Ziemiecki A, Harpur AG, Wilks AF, Yasukawa K, Taga T, Kishimoto T, Barbieri G, Pellegrini S, Sendtner M, Heinrich PC, Horn F (1994). Association of transcription factor APRF and protein kinase Jakl with the interleukin-6 signal transducer gp130. Science 263:89-92.
31.Metz R, Ziff E (1991). cAMP stimulates the C/EBP-related transcription factor rNF-IL6 to translocate to the nucleus and induce c-fos transcription. Genes Dev. 5:1754-1766.
32.Morse L, Chen-Kiang S Unpublished.
33.Murakami M, Hibi M, Nakagawa N, Nakagawa T, Yasukawa K, Yamanishi K, Taga T, Kishimoto T (1993). IL-6-induced homodimerization of gp130 and associated activation of a tyrosine kinase. Science 260:1808-1810.
34.Nakajima K, Wall R (1991). Interleukin-6 signals activating junB and TIS11 gene transcription in a B-cell hybridoma. Mol. Cell. Biol. 11:1409-1418.
35.Nakajima T, Kinoshita S, Sasagawa T, Sasaki K, Naruto M, Kishimoto T, Akira S (1993). Phosphorylation at threonine 235 by a ras-dependent mitogen-activated protein kinase cascade is essential for transcriptional activation of NF-IL6. Proc. Natl. Acad. Sci. USA 90:2207-2211.
36.Natkunam Y, Zhang X, Liu Z, Chen-Kiang S (1994). Simultaneous activation of immunoglobulin and Oct-2 synthesis and repression of surface MHC class II expression by IL-6. (submitted)
37.Nordan R, Potter M (1986). A macrophage-derived factor required by plasmacytomas for survival and proliferation in vitro. Science 233:566-569.
38.Peters PJ, Neefjes JJ, Oorshot V, Ploegh HL, Geuze HJ (1991). Segregation of MHC class II molecules from MHC class I molecules in the Golgi complex for transport to lysosomal compartments. Nature 349:669-676.
39.Poli V, Mancini FP, Cortese R (1990). IL-6DBP, a nuclear protein involved in interleukin-6 signal transduction, defines a new family of leucine zipper related to C/EBP. Cell 63:643-653.
40.Raynal M-C, Liu Z, Hirano T, Mayer L, Kishimoto T, Chen-Kiang S (1989). Interleukin-6 induces secretion of IgG1 by coordinated transcriptional activation and differential mRNA accumulation. Proc. Natl. Acad. Sci. USA 86:8024-8028.
41.Resnitzky D, Tiefenbrun N, Berissi H, Kimchi A (1992). Interferons and interleukin 6 suppress phosphorylation of the retinoblastoma protein in growth-sensitive hematopoetic cells. Proc. Natl. Acad. Sci. USA 89:402-406.
42.Sadowski HB, Shuai K, Darnell JE, Gilman MZ (1993). A common nuclear signal transduction pathway activated by growth factor and cytokine receptors. Science 261:1739-1744.
43.Satoh T, Nakamura S, Taga T, Matsuda T, Hirano T, Kishimoto T, Kaziro Y (1988). Induction of neuronal differentiation in PC12 cells by B-cell stimulatory factor/interleukin-6. Mol. Cell. Biol. 8:3546-3549.
44.Schindler C, Fu X-Y, Improta T, Aebersold R, Darnell JE (1992). Proteins of transcription factor ISGF-3: one gene encodes the 91-and 84-kDa ISGF-3 proteins that are activated by interferon alpha. Proc. Natl. Acad. Sci. USA. 89:7836-7839.

45.Spergel JM, Hsu W, Akira S, Thimmappaya B, Kishimoto T, Chen-Kiang S (1992). NF-IL6, a member of the C/EBP family, regulates E1A-responsive promoters in the absence of E1A. J. Virol. 66:1021-1030.

46.Stahl N, Boulton TG, Farruggella T, Ip NY, Dvis S, Witthuhn BA, Quelle FW, Silvennoinen O, Barbieri G, Pellegrini S, Ihle JN, Yancopoulos GD (1994). Association and activation of Jak-Tyk kinases by CNTF-LIF-OSM-IL-6 β receptor components. Science 263:92-95.

47.Stein B, Cogsweli PC, Baldwin AS (1993). Functional and physical associations between NF-kB and C/EBP family members: a Rel domain-bZIP interaction. Mol. Cell. Biol. 13:3964-3974.

48.Suematsu S, Matsusaka T, Matsuda T, Ohno S, Miyazaki J, Yamamura K, Hirano T, Kishimoto T (1992). Generation of plasmacytomas with the chromosomal translocation t(12;15) in interleukin 6 transgenic mice. Proc. Natl. Acad. Sci. USA 89:232-235.

49.Trautwein C, Caelles C, van der Geer P, Hunter T, Karin M, Chojkier M (1993). Transactivation by NF-IL6/LAP is enhanced by phosphorylation of its activation domain. Nature (London) 364:544-547.

50.Venkitaraman AR, Culbert EJ, Feldman M (1987). A phenotypically dominant regulatory mechanism suppresses major histocompatibility complex class II gene expression in a murine plasmacytoma. Eur. J. Immunol. 17:1441-1446.

51.Viville S, Neefjes J, Lotteau V, Dierich A, Lemeur M, Ploeph H, Benoist C, Mathis D (1993). Mice lacking the MHC class II-associated invariant chain. Cell 72:635-648.

52.Wegenka UA, Buschmann J, Lütticken C, Heinrich P, Horn F (1993). Acute-phase response factor, a nuclear factor binding to acute-phase response elements, is rapidly activated by interleukin-6 at the posttranslational level. Mol. Cell. Biol. 13:276-288.

53.Wegner M, Cao Z, Rosenfeld MG (1992). Calcium-regulated phosphorylation within the leucine zipper of C/EBPβ. Science 256:370-373.

54.Williams SC, Cantwell CA, Johnson PF (1991). A family of C/EBP related proteins capable of forming covalently linked leucine zipper dimers in vitro. Genes Dev. 5:1553-1568.

55.Zhang X, Zheng J-H , Chen-Kiang. unpublished.

56.Zhong Z, Zilong W, Darnell JE (1994). Stat3: a STAT family member activated by tyrosine phosphorylation in response to epidermal growth factor and interleukin-6. Science 264:95-98.

Sequence, Expression and Function of an mRNA Encoding a Soluble Form of the Human Interleukin-6 Receptor (sIL-6R)

J.A. Lust, D.F. Jelinek, K.A. Donovan, L.A. Frederick, B.K. Huntley, J.K. Braaten, and N.J. Maihle.
Department of Laboratory Medicine, Mayo Clinic, Rochester, MN.

Introduction

Hematopoietic growth factor receptors, in general, are integral membrane proteins composed of an extracellular ligand-binding domain, a transmembrane domain, and an intracellular signal-transducing domain. Soluble receptors that lack the transmembrane domain have been shown to be potent immunomodulators of their respective ligands [1-6]. Since IL-6 has been shown to be a central growth factor for myeloma cells [7], an sIL-6R may modulate myeloma cell growth. A novel IL-6R mRNA was isolated from human myeloma cells that exhibits a deletion of the entire transmembrane domain [8]. The corresponding cDNA was sequenced, expressed in fibroblasts, and the functional activity of the protein product was determined using an IL-6 dependent cell line.

Materials and Methods

Cell Lines

U266 is a human myeloma cell line, QT-6 is a quail fibrosarcoma cell line, and PA-1 is a human ovarian teratocarcinoma cell line. All three lines were obtained from the American Type Culture Collection (Rockville, MD). ANBL-6 is an IL-6 dependent human myeloma cell line developed from a patient with plasma cell leukemia [9]. All cell lines were maintained in culture medium consisting of RPMI-1640 with 10% fetal calf serum, 2 mM L-glutamine, and 200 U/mL penicillin. In addition, for QT-6 cells, the culture medium was supplemented with 1% chicken serum.

RNA and cDNA Isolation

Total cellular RNA was prepared from 20 - 100 million cells using the guanidine isothiocyanate/cesium chloride method. Ten - 20 μg of total cellular RNA was reverse transcribed with 20 units of reverse transcriptase and oligo dT using the Amersham (Arlington Heights, IL) cDNA synthesis system at 42°C for 90 minutes in a 20 μL reaction volume. Subsequently, the reaction was diluted to 100 μL with distilled water.

Polymerase Chain Reaction

The PCR technique was performed on a Perkin Elmer Cetus thermal cycler. Reactions were performed with 1 μL of the above cDNA reaction, 20 pmole of each primer, and appropriate amounts of buffer, dNTP's, and Taq polymerase in a 50 μL reaction volume. The thermal profile involved 30-40 cycles of denaturation at 94°C for 1 min, primer annealing at 50°C for 2 min, and extension at 72°C for 2 min. After ethanol precipitation, PCR products were fractionated on gels of 2% NuSieve GTG agarose and gels were stained with ethidium bromide. Bands of amplified DNA were isolated from the gel by electroelution and subsequent precipitation with ethanol.

Oligonucleotide primer sequences were as follows:

Number	Nucleotides (10)	Oligonucleotide Primer Sequence
1	946-963	5' ACA GAA TCC AGG AGT CCT 3'
2	1267-1284	5' GAT GAG AGG AAC AAG CAC 3'
3	1571-1588	5' CTA G<u>GA ATT C</u>AG CTG AAC AGC TGG AGC A 3'
4	147-164	5' ACT G<u>GA ATT C</u>GG TAG AGC CGG AAG ACA A 3'

Underlined nucleotides correspond to EcoRI sites used for subsequent cloning. Oligonucleotide primers #1 and #2 flank the transmembrane domain of the native IL-6R cDNA sequence and generate a 339-base pair fragment plus a smaller 245-base pair fragment using U266 cDNA.

Sequencing of PCR Amplified DNA

Amplified DNA was gel purified and 50 - 100 ng was annealed with an equimolar amount of one of the amplimer oligonucleotides (5'- end labeled with γ-^{32}P ATP). Annealing mixtures were heated to 90°C for 2 minutes and chilled on ice. The annealing mixture (11 μL) was apportioned into four tubes containing deoxy- and dideoxynucleotides (10:1), 0.25 μL of 0.1 M dithiothreitol and 1 U Sequenase and incubated at 37°C for 5 minutes. The reaction was terminated by addition of the formaldehyde/dye stop mix. Samples were heated to 90°C for 2 minutes, chilled, and electrophoresed on a 6% polyacrylamide/7M urea gel to 50°C for 2 hours. Gels were dried and exposed to Kodak X-AR 5 film at room temperature with an intensifying screen overnight.

Immunoprecipitation:

The cDNA corresponding to the transcript encoding the soluble human IL-6R was cloned into pBluescript II. Subsequently, a 1320-bp fragment beginning at the start codon and lacking the transmembrane domain, as detailed above, was inserted into an eukaryotic expression vector, pCDM8. QT6 (quail fibroblast line) or PA-1 (human ovarian line) cells were mock transfected or transfected with 15 μg of total plasmid DNA using the calcium phosphate technique. Forty-eight hours post transfection, cells were metabolically labelled with ^{35}S methionine and cystine, and cell lysates or supernates were immunoprecipitated with 1 μg of anti-IL-6R antibody (Biosource, Camarillo, CA). Protein-A/G Sepharose was added to each sample prior to incubation. Samples were analyzed by 0.1% SDS/10% polyacrylamide gel electrophoresis followed by autoradiography.

sIL-6R Functional Assay
For assessment of ANBL-6 responsiveness to cytokines/supernates, ANBL-6 cells were routinely washed three times with culture medium prior to adding cells to a 96-well, round bottom microtiter plate (Costar) at a density of 2.0 x 10^4 cells per well in a total volume of 200 μL. IL-6 and αIL-6 antibody were added at the indicated final concentration and cultures were carried out in triplicate at 37°C in a humidified atmosphere of 5% CO_2 and 95% air. ANBL-6 cells were cultured for three days and 1 μCi of ^{3}H-thymidine (5.0 Ci/mmole; Amersham, Arlington Heights, IL) was added for the last 18 hours. The cells were harvested onto glass fiber filter paper and ^{3}H-thymidine incorporation determined by liquid scintillation spectroscopy.

RESULTS

Sequence and Expression of an mRNA Encoding a Soluble Form of the Human IL-6R

In prior work, U266 RNA was reverse transcribed and using polymerase chain reaction, with oligonucleotide primers #1 and #2 that flank the transmembrane domain, both a 339-base pair fragment plus a smaller 245-base pair fragment were generated [8]. Sequence analysis of the smaller 245 bp fragment demonstrated the deletion of the entire transmembrane region from codons 356 (G-TG) to 387 (AG-G) illustrated in Fig. 1. Sequence analysis of the ligand binding domain for U266 (using primers #4 and #2) demonstrated identical sequences for the membrane-bound and soluble forms of the IL-6R cDNAs.

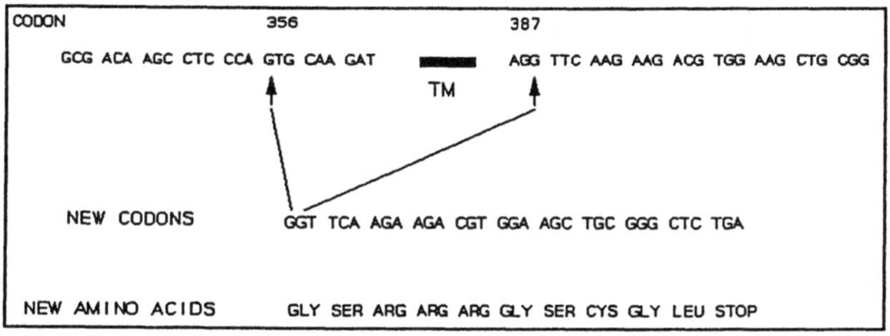

Figure 1: Schematic diagram of the sIL-6R sequence indicating the boundaries of the deletion of the transmembrane region and the resultant amino acid changes.

The deletion of the transmembrane region in the sIL-6R causes a frame shift in the translational reading frame with the insertion of ten new amino acids followed by a stop codon (Fig. 1). The soluble receptors for human G-CSF, GM-CSF, IL-5, and IL-7 are similar to soluble IL-6 receptor in that all result from an internal deletion (Table 1) [11-14]. The 5' boundary of the deletion is situated very near the transmembrane region, and therefore, does not appear to affect the structure of the extracellular (i.e., ligand-binding) domain. The deletions encompass most if not all of the transmembrane domain and cause a frame-shift in the translational reading frame such that all transmembrane and cytoplasmic C-terminal amino acids are removed. In each case, several amino acids unique to the soluble form are added. As expected from the extracellular sequences, the soluble forms of the IL-5, IL-7, and GM-CSF receptors have been shown to retain ligand-binding activity [12-14].

Table 1

Properties of Soluble Receptor Proteins

Receptor	Sequence boundaries of the transmembrane domain deletion	Extracellular amino acids missing	Number of Amino Acids unique to soluble receptor
IL-5	94-nt deletion TGG↓GAAA...AGT↓GTG	10	62
IL-6	94-nt deletion CAG↓TGC...GAG↓GTTC	3	10
IL-7	94-nt deletion TCAG↓GG...AAAG↓GATT	4	27
G-CSF	88-nt deletion CCCAG↓AG...TGCAG↓CCC	6	150
GM-CSF	97-nt deletion TTG↓GTTCT...AAG↓GTTCC	3	16

The cDNA corresponding to the transcript encoding the soluble human IL-6R was inserted into an eukaryotic expression vector, pCDM8. QT6 or PA-1 cells were mock transfected or transfected with 15 μg of total plasmid DNA using the calcium phosphate technique. QT-6 or PA-1 cells transfected with the sIL-6R cDNA yielded a single band with a molecular weight of 50-55 kDa (Fig. 2). Interestingly, this molecular weight corresponds to the size of the sIL-6R protein initially observed in normal human urine [15].

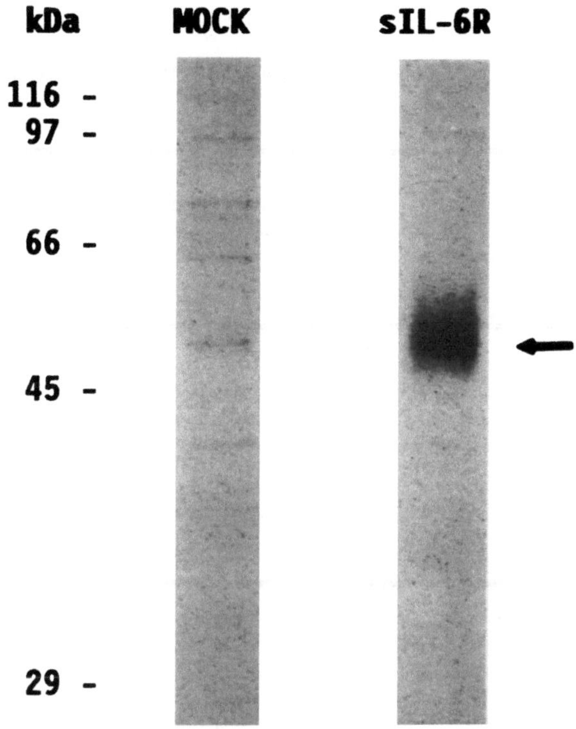

Figure 2: Immunoprecipitation analysis of sIL-6R protein products. Quail fibroblasts (QT6) were mock transfected (control) or transfected with pCDM8 containing the sIL-6R cDNA by the calcium phosphate technique. Molecular weight markers are given at the left. Similar results were obtained with the PA-1 ovarian line.

Functional Activity of Human sIL-6R

Subsequently, supernates were collected from mock or pCDM8 transfected PA-1 cells after 48 hours and assayed for their ability to stimulate or suppress the growth of an IL-6 dependent cell line, ANBL-6 [9]. As shown in Table 2, sIL-6R alone had no effect on the growth of the ANBL-6 cell line. However, the growth of ANBL-6 cells by sIL-6R was potentiated in the presence of IL-6 and could be blocked by anti-IL-6 antibody. The binding of IL-6 to its receptor triggers the association of the IL-6R subunit with a second membrane glycoprotein, gp130, which transduces the IL-6 signal (Fig. 3). Since binding of the native IL-6R with gp130 occurs extracellularly in the presence of IL-6, the above results suggest that, in the presence of IL-6, sIL-6R associates with gp130 leading to signal transduction and cell growth.

Table 2

Functional Activity of the sIL-6R Protein

Supernate Additions 50 μL/well	ANBL-6 cells @ 20,000 cells/well		
	Nil	IL-6 100 pg/mL	IL-6 (100 pg/mL) + αIL-6 antibody (300 ng/mL)
Nil	2035	5251	1686
Media	1989	6012	1825
no DNA	2608	6341	2162
pCDM8 alone	2794	6889	2727
pCDM8 + sIL-6R cDNA	2870	13168	2688

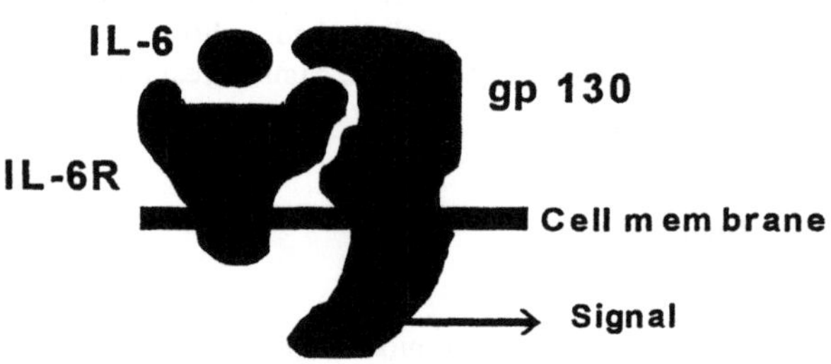

Fig. 3: Mechanism of IL-6 Signal Transduction

SUMMARY

Soluble receptors have been shown to be potent immunomodulators of their respective ligands. Since IL-6 is a central growth factor for myeloma cells, an sIL-6R may modulate IL-6 activity. We have previously reported a novel IL-6R mRNA from myeloma cells that exhibits a 94-nt deletion of the entire transmembrane domain from codons 356 (G-TG) to 387 (AG-G). The transmembrane domain deletion results in a shift in the translational reading frame with the insertion of 10 new amino acids followed by a stop codon. Sequence analysis shows the ligand-binding domain of the sIL-6R to be identical to that of the membrane-bound IL-6R up to the transmembrane domain deletion. The sIL-6R cDNA was expressed in QT-6 fibroblasts and PA-1 ovarian cells using the expression vector pCDM8. Supernates were immunoprecipitated with anti-IL-6R antibody and cells transfected with the sIL-6R cDNA produced a single band with a molecular weight of 50-55 kDa. This molecular weight corresponds to the size of the sIL-6R protein observed in normal human urine. Supernates were collected from mock or sIL-6R transfected PA-1 cells after 48 hours and assayed for their ability to stimulate or suppress the growth of an IL-6 dependent cell line, ANBL-6. Soluble IL-6R alone had no effect on the growth of the ANBL-6 cells. However, the growth of ANBL-6 cells by sIL-6R was potentiated in the presence of IL-6 and could be blocked by anti-IL-6 antibody. The above results suggest that, in the presence of IL-6, sIL-6R associates with gp130 leading to signal transduction and cell growth.

References

1. Fanslow WC, Sims JE, Sasenfeld H, Morrissey PJ, Gillis S, Dower SK, Widmer MB (1990) Regulation of alloreactivity in vivo by a soluble form of the interleukin-1 receptor. Science 248:739-742.
2. Weisman HF, Bartow T, Leppo MK, Marsh HC, Carson GR, Concino MF, Boyle MP, Roux KH, Weisfeldt ML, Fearon DT (1990) Soluble human complement receptor type 1: in vivo inhibitor of complement suppressing post-ischemic myocardial inflammation and necrosis. Science 249:146-151.
3. Ward RHR, Capon DJ, Jett CM, Murthy KK, Mordenti J, Lucas C, Frie SW, Prince AM, Green JD, Eichberg JW (1991) Prevention of HIV-1 IIIB infection in chimpanzees by CD4 immunoadhesion. Nature 352:434-436.
4. Nemerow GR, Mullen JJ, Dickson PW, Cooper NR (1990) Soluble recombinant CR2 (CD21) inhibits Epstein-Barr virus infection. J Virol 64:1348-1352.
5. Marlin SD, Staunton DE, Springer TA, Stratowa C, Sommergruber W, Merluzzi VJ (1990) A soluble form of intercellular adhesion molecule-1 inhibits rhinovirus infection. Nature 344:70-72.
6. Fanslow WC, Clifford KN, Park LS, Rubin AS, Voice RF, Beckmann MP, Widmer MB (1991) Regulation of alloreactivity in vivo by IL-4 and the soluble IL-4 receptor. J of Immunol 147:535-540.
7. Kishimoto T (1989) The biology of Interleukin-6, Blood 74:1-10.
8. Lust JA, Donovan KA, Kline MP, Greipp PR, Kyle RA, Maihle NJ (1992) Isolation of an mRNA encoding a soluble form of the human interleukin-6 receptor. Cytokine 4(2):96-100.
9. Jelinek DF, Ahmann GJ, Greipp PR, Jalal SM, Westendorf JJ, Katzmann JA, Kyle RA, Lust JA (1993) Coexistence of aneuploid subclones within a myeloma cell line that exhibits clonal immunoglobulin gene rearrangement: clinical implications. Cancer Research 53:5320-5327.
10. Yamasaki K, Taga T, Hirata Y, Yawata H, Kawanishi Y, Seed B, Taniguchi T, Hirano T, Kishimoto T (1988) Cloning and expression of the human interleukin-6 (BSF-2/IFNβ 2) Receptor. Science 241:825-828.
11. Fukunaga R, Seto Y, Mizushima S, Nagata S. (1990) Three different mRNAs encoding human granulocyte colony-stimulating factor receptor. Proc Natl Acad Sci USA 87, 8702-8706.
12. Takaki S, Tominaga A, Hitoshi Y, Mita S, Sonoda E, Yamaguchi N, Takatsu K (1990) Molecular cloning and expression of the murine interleukin-5 receptor. EMBO J 9:4367-4374.
13. Goodwin RG, Friend D, Ziegler SF, Jerzy R, Falk BA, Gimpel S, Cossman D, Dower SK, March CJ, Namen AE, Park LIS (1990) Cloning of the human and murine interleukin-7 receptors: demonstration of a soluble form and homology to a new receptor superfamily. Cell 60:941-951.
14. Raines MA, Liu L, Quan SG, Joe V, DiPersio JF, Golde DW (1991) Identification and molecular cloning of a soluble human granulocyte-macrophage colony-stimulating factor receptor. Proc Natl Acad Sci USA 88:8203-8207.
15. Novick D, Engelmann H, Wallach D, Rubinstein M: Soluble cytokine receptors are present in normal human urine. J Exp Med 170:1409-1414, 1989.

The Late B-Cell

CD5 Transgenic Mice

X. Chen, Y. Matsuura and J. F. Kearney
Division of Developmental and Clinical Immunology, Department of Microbiology, University of Alabama at Birmingham, Birmingham, Alabama

1 Introduction

CD5 is a glycoprotein expressed on all mature T cells [1], a small but distinct subset of B lymphocytes (referred to as B1a cells) [2], some mouse B cell lymphomas and most human chronic lymphocytic leukemias (CLL) [3,4]. Both human and mouse CD5 genes have been cloned. The mouse CD5 cDNA encodes a protein consisting of 471 amino acids which, within the carboxyl terminus, has one potential tyrosine and two potential threonine phosphorylation sites. The sequence surrounding this tyrosine residue is similar to the sequence surrounding the tyrosine autophosphorylation site encoded by the protooncogene c-*src* and the oncogenes *yes* and *fgr* which are conserved in both mouse and human CD5 [5]. As an accessory molecule, CD5 has been shown to be coupled to the TCR/CD3 complex on T cells [6]. Crosslinking of CD5 with anti-CD5 antibody suggests that CD5 may function as a co-stimulatory molecule to augment and sustain T cell proliferation induced by interleukins, lectins, anti-CD3 and alloantigens [8,9]. Both murine and human CD72 have been identified as a ligand for CD5 [10,11]. Recently, CD5 on B cells has also been shown to be associated with the B cell antigen receptor (BCR) complex and serves as a substrate for BCR-induced tyrosine kinase activity [13]. Despite these findings the function of CD5 on B cells is not well understood though it seems probable that CD5-CD72 interactions may play a role in B-B cell as well as in T-B cell interactions. In addition, as B1 cells express both CD5 and CD72, it is possible that interactions between these proteins contribute to the self-renewing capacity of these cells, to their expression of autoimmune specificities and to their propensity to form B cell lymphomas and human CLL. In attempts to further understand the function of the CD5 molecule on lymphocytes, we have generated CD5 transgenic mice in which all the B lineage cells express the CD5 molecule.

2 Result

2.1 Generation of the CD5 Transgene Construct and *In Vitro* Expression.

In order to express the CD5 gene in all B lineage cells, the gene was placed under the control of regulatory elements that are activated when progenitor cells differentiate into B lineage cells. To this end, we placed CD5 cDNA downstream of the immunoglobulin heavy chain (IgH) promoter and upstream of an IgH enhancer. The promoter region is included in a 3.5kb DNA fragment which was originally the 5' untranslated region of VH81X gene, the most D-proximal functional VH gene in BALB/c and the gene preferentially expressed in the fetal repertoire [14]. In addition to the promoter sequence, the 3.5kb DNA also contains other VH regulatory elements such as octamer and heptamer sequences. To ensure optimal expression, CD5 cDNA was ligated to the 3.5kb DNA in such a way that the initiation ATG codon of CD5 cDNA assumed the position of the first ATG of the leader sequence of VH81X. The downstream IgH enhancer was included in a 1.9kb BamH1-EcoR1 DNA fragment that also contained JH3-4 exons and some JH-C-μ introns. As a consequence of such an arrangement in the construct, the CD5 transgene should be transcribed and then translated just as if it were the VH81X heavy chain.

Since CD5 cDNA has its own polyadenylation signal, the mRNA transcripts should be stable. However, when we transfected the above-designed transgene into two B cell lines, S107 and GK5 of mouse and human origin respectively, we did not detect CD5 expression on long-term selected transfectants.

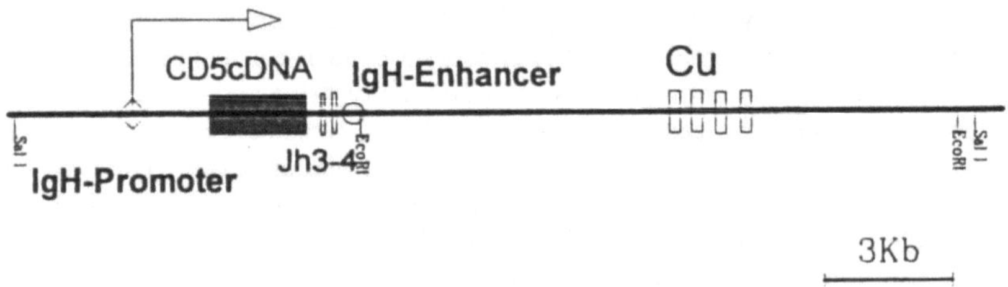

Fig. 1 Map of CD5 Transgene

In order to modify further the transgene construct, we added a 12kb EcoR1-EcoR1 DNA fragment containing the JH-Cμ intron including the switch regions and all Cμ exons downstream of the IgH enhancer to generate a second CD5 transgene construct, KC9 (Fig. 1). Upon transfecting B cell lines with this modified gene construct and selecting transfectants with mycophenolic acid, we could easily detect CD5 expression on the cell surface of transfectants (Fig. 2).

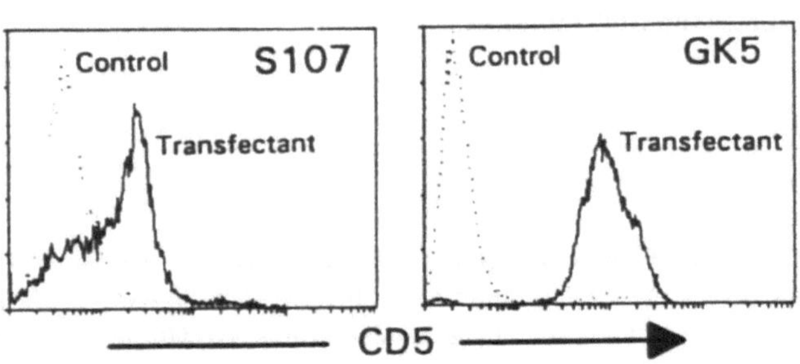

Fig. 2 FACS analysis of CDS Transfectants

2.2 Generation and Identification of Transgenic Mice.

Transgenic mice were generated by injecting the transgene (not including the plasmid portion) into fertilized mouse eggs. We identified transgenic mice by analyzing mouse tail DNA using PCR techniques. In order to distinguish endogenous from transgenic CD5, primers were used for PCR that amplified only the transgenic and not endogenous CD5. To this end, we chose a 5' primer that annealed to the VH81X promoter region and a 3' primer that annealed to a region in CD5 cDNA close to its N-terminus. By using this pair of primers, PCR will only generate a DNA 420bp fragment (Fig. 3) if the tail genomic DNA contains the transgene in which the VH81X promoter and CD5 cDNA is juxtaposed. We obtained five transgenic founders from micro injection of fertilized C571BL/6xSJL/J)F2 mice, which were then bred to BALB/c and C57BL/6 mice.

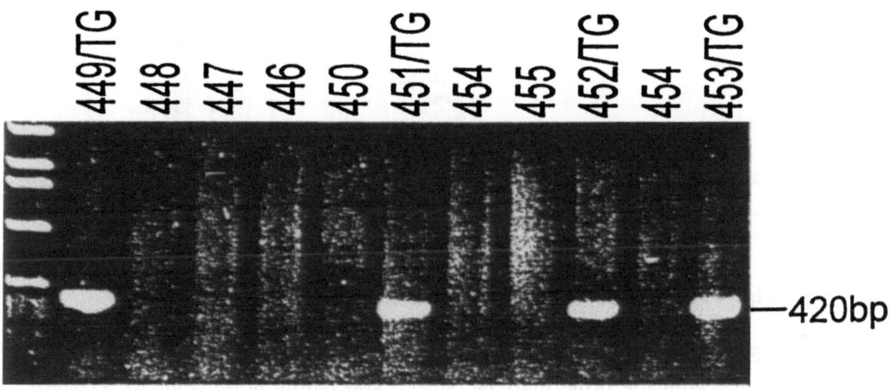

Fig. 3 Detection of CD5 Transgenic Mice

2.3 CD5 is Expressed on all B Lineage Cells in Transgenic Mice.

CD5 transgene expression was detected by staining mouse tissue with anti-mouse CD5 antibody. Analysis was done by fluorescence activated cell sorter (FACS) or fluorescence microscopy using T cells and non-transgenic B cells as positive and negative controls, respectively. In transgenic mice, all B lineage cells from pro-B to plasma cells express CD5 (it is not expressed on sections of heart, liver, kidney, or GI tract, data not shown). In bone marrow, CD5 expression occurred on almost all $B220^+$ and surface IgM^- cells including pro-B and pre-B cells (Fig. 4-B). The level of CD5 expression on pre-B and pro-B cells as well as newly formed B cells was about 5-10 times higher than that on mature T cells that recirculate to bone marrow (Fig. 4). This indicates that although it has been shown that CD5 is associated with the BCR and TCR complex on mature B and T cells [6,13], such associations are not mandatory for surface expression of CD5.

The observation that the CD5 transgene is expressed on pro-B and pre-B cells is in agreement with previous studies showing that IgH promoter and enhancer activity is already active in pro-B cells [15]. This observation is also in accordance with studies that show in *myc* transgenic mice, in which the *myc* oncogene is linked to the IgH enhancer, there is a high incidence of B cell lymphoma. A higher level of expression of CD5 on precursor B cells than on mature B cells was not expected, since it has been shown that in pre-B cells there are only small amounts of cytoplasmic μ chains correlating with the small amounts of heavy chain mRNA that is 30-50 times less than in plasma cells (15). The level of expression of transgenic CD5 on immature B cells is similar to that of pro-B and pre-B cells (Fig.4).

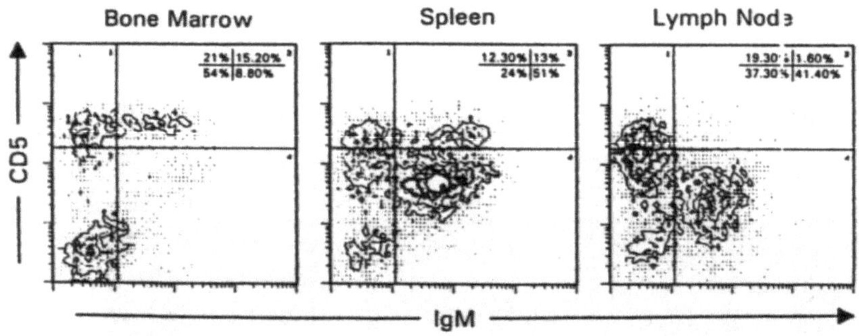

Fig. 4 Profiles of CD5 Expression in CD5 Transgenic Mice

However as B cells mature, the CD5 expression declines as shown by a comparison of B cell CD5 expression in the bone marrow, spleen and lymph node (Fig. 4). Despite this decline to lower levels on mature B cells, CD5 expression is similar to that of peritoneal B cells in three out of four of the transgenic lines. In the fourth line, the expression is much lower. However, even though the transgene CD5 is downregulated about 100-fold in the transition from immature to mature B cells, surface IgM expression remains unchanged in this mouse. When splenic B cells which expressed barely detectable levels of transgenic CD5 were put into culture with LPS, CD5 expression was induced to very high levels, while their surface IgM did not change.

3. Effects of Transgenic CD5 Expression on B Lymphocyte Phenotype and Function.

3.1 B lymphocyte phenotype

Transgenic expression of CD5 changes with the stage of B cell maturation but has little effect on most other B cell phenotypic markers, and the CD5 transgenic mouse differs very little from non-transgenic littermates with respect to standard B cell markers including IgM, IgD, class II antigen, CD38, and CD23. Likewise, serum immunoglobulin levels were similar between CD5 transgenic and normal littermates. Immunization with T-independent antigens phosphorylcholine and dextran also did not reveal any remarkable differences in total antibody levels and idiotype expression between transgenic and non-transgenic mice. Immunization of CD5 transgenic C57BL/6 mice with the T-dependent antigen NP-CG showed normal levels of anti-NP antibody though there appeared to be a marked elevation of IgG1 isotype levels.

3.2 B lymphocyte activation

A wide ranging variety of activating agents were used to determine if differential activation of CD5 transgenic versus non-transgenic littermates derived B cells could be observed. In all cases studied which included standard anti-μ (with or without IL-4), PMA, LPS, CD40L and IL-4, no differences were observed in the proliferative responses of CD5 transgenic versus normal B cells (J. Karras, T.Rothstein, unpiblished observations).

3.3 Transgenic CD5 expression has no effect on somatic hypermutation in the response to NP-CG

Since CD5 positive B cells in mice have frequently produced antibodies with lack of somatic mutations, transgenic mice and littermates were immunized with NP-CG [17]. Using appropriate primers, PCR products of the J558VH gene 186.2, which is expressed at high frequencies in anti-NP antibody and shows extensive somatic mutations, were compared between CD5 transgenic mice and normal littermates. More than 60 VH186.2 genes obtained from mice 14 days after primary immunization, were sequenced. However, there was little difference in the frequency of somatic mutations in the CDR or framework regions as shown in summary form in Table 1.

Table 1

Somatic mutation of VH in anti-NP response

		average mutation CDR (1,2)	Framework (1,2,3)	number Total	R/S ratio	W to L exchange at codon 33(*)
CD5 transgenic mouse	7	2.3	3.5	5.8±3.4	6.4:1	33.3%
	9	3.2	2.7	5.9±3.9	3.0:1	46.7%
	7,9 average	2.7	3.1	5.9±3.6	4.2:1	40.0%
Littermate mouse	6	3.1	4.1	7.2±2.2	1.5:1	0 %
	8	2.3	2.8	5.1±2.8	3.9:1	13.3%
	6,8 average	2.7	3.4	6.1±2.7	2.2:1	6.9%

The results were expressed as the mean(±SD) of 14 or 15 samples, respectively.

(*) A single replacement in CDR1,a tryptophan-to-leucine exchange at codon 33,raises the binding affinity 10-fold[17].

Still, the R/S ratio is twice as high in transgenic mice. This experiment showed that transgenic expression of CD5 does not appear, either directly or indirectly, to affect the mutation rate of the 186.2VH gene involved in the antibody response to NP. The mutation rates appeared consistent within the two groups and those already published [18,19]. However, there was a dramatic increase in the frequency of a characteristic tryptophan to leucine exchange at codon 33 in transgenic mice for reasons which are as yet unclear.

3.4 Development of splenomegaly and plasma cell hyperplasia

A striking phenotype becomes obvious in a proportion of older CD5 transgenic mice which is not evident in their non-transgenic littermates. This is marked by increased spleen weight and lymph node size, particularly in mice older than 6 months of age. Some mice develop spleens 3-5 times the weight of normal littermates. Closer examination, reveals that all transgenic mice after 6 months of age have spleens that are about 50% greater in weight.

Histological examination of these spleens reveals hyperplasia beginning in the marginal zone of the B cell follicles. This progressively increases with a striking increase in plasmablasts and, plasma cells in the red pulp, until in extremely hyperplastic spleens there is a total lack of splenic structure. Immunofluorescence staining of these spleens shows that there is a very large number of B220+ cells that are sIgM negative. Staining of frozen sections of hyperplastic spleens showed that the plasmablasts and plasma cells are frequently of one isotype.

These studies show that transgenic expression of CD5 has a dramatic effect on B cell growth and activation resulting in a high frequency of B cell hyperplasia which is not only unique to one CD5 transgenic mouse, but has been observed in the four independent sublines bred from the different founders. This observation has not been made in >300 VH81X transgenic mice using a very similar construct and has not been seen in other Ig transgenic mice. For these reasons, the phenomenon observed is most likely due to the overexpression of CD5 and is not a direct result of the site of insertion of the transgene or interference with a known oncogene.

4. Discussion

4.1. Strategy for B Cell Directed Transgenic Expression of CD5 cDNA.

It has long been realized that the μ enhancer is much more effective in driving expression of immunoglobulin rather than non-immunoglobulin genes. This is exemplified by the frequent successes in construction of IgM, IgG, IgD, or IgA transgenic mice in which expression of the transgenes is at a high level and easily detectable [20-24]. In contrast, attempts to express cDNA of human α1 anti-trypsin, coagulation factor IX, or the β-globin gene or even CD5 in mice under the control of the μ-enhancer result in only a marginal level of transgene expression [25-27]. In fact, the strategy used in constructing the CD5 transgene in the previous study was essentially similar to ours, i.e. driving the CD5 cDNA with IgH promoter and enhancer [22]. The CD5 cDNA used in both cases is the same. The only difference, which is also a common difference between the other three Ig transgene constructs described in ref. 25-27, is that Cμ genomic DNA down stream of IgH enhancer is included in our CD5 transgene. This contains a Cμ switch region in addition to exons and introns. The inclusion of a stretch of intron/exon containing DNA in a transgene has been a successful general strategy in constructing transgenes to express a cDNA structural gene. It is presumed these intron/exon structures, missing in cDNAs, will meet the need of RNA processing that occurs during RNA maturation and transportation out of the nucleus. In the case of the previous CD5 transgene construct, SV40 splice and poly A signals were employed for this purpose but the expression is very low and variable [27]. Then, what is the cause of the discrepancy between the level of expression in the two experimental systems? Our experiments suggest that it results from the proposed regulatory function of the Cμ switch region. It has been shown in IgG1 transgenic mice, that the lack of switch regions in the transgene, either of Cμ or γ1, results in extremely low production of transgenic IgG1 in mice, about 100-1000 times lower than the transgene containing a switch region [28]. Although the exact regulatory motif in the switch region and mechanism of regulation has not been established, these regions may be necessary for the full function of the IgH promoter and enhancer. This region may be neglected if the transgene product is not needed in a large amount to produce a given phenotype, similar to an oncogene whose product exists in a very small amount in normal conditions. This may be the case in *myc* and *bcl2* transgenic mice, in which the oncogene is driven by an IgH enhancer without involvement of switch regions and B cell malignancies develop [29, 30]. For a high level of expression, it may be advisable to include the switch region.

4.2 Why is Transgenic CD5 expression highest in progenitor and immature B cells?

It has been shown that pre-B cell lines produce only small amounts of μ heavy chain in the cytoplasm, corresponding to a low levels of μ chain mRNA [31]. This is in contrast to plasma cell lines which may express 30-50 fold higher levels of the Ig polypeptide chain and mRNA [31]. These well established observations stand in contrast to our findings that CD5 expression is highest in progenitor and nascent B cells and then declines to a moderate level on mature B cells. This decline in expression is not due to mutational inactivation of the CD5 transgene since *in vitro* culture of such B cells with LPS brings CD5 expression to extraordinarily high levels on all B cells. Therefore, the fluctuation of the CD5 molecule on the cell surface appears to result from the regulation of transgene expression. Since the CD5 gene is under control the of IgH enhancer and promoter, the regulatory mechanism should be similar to that of immunoglobulin genes either at a transcriptional or post-transcriptional level or both. It is unlikely that the high level of expression of CD5 on pro-B and pre-B cells is related to the integration site of the transgene because the same phenomenon is seen in all four different transgenic lines. The fact that in our VH81X transgenic mice cytoplasmic μ heavy expression is not enhanced in pre-B cells, suggests that the overall structure of the regulatory elements (including promoter and enhancer) of CD5 transgene is also unlikely to be responsible for the observed high level of expression in progenitor B cells, since the VH81X heavy chain transgene construct is very similar to the CD5 transgene except for the coding region (unpublished observations). In addition, translation of these two transgenes should be the same since both use the same initiation codon. Given these considerations, we currently assume that the CD5 level on progenitor B cells may reflect normal comparative levels of expression which then declines with maturation of B cells. It is of great interest that Hardy et. al. have recently determined that there is a high level of CD5 mRNA in bone marrow progenitor B cells in the absence of surface CD5 expression [32]. In this regard, it has been proposed from studies of CD5 knockout mice that indeed all B cells including B2 cells may express low levels of CD5 and it is only B1 cells that express sufficient amounts to be considered a subset. The various mechanisms responsible for the upregulation of CD5 on B1a cells may indeed be at the level of regulation. There may be various mechanisms involved in the downregulation perhaps by the coexpression of the CD5 ligand CD72 resulting in internalization of the ligand-receptor complex.

4.3 Effects of Transgenic Expression of CD5 on B Cell Functions

Just as in the CD5 KO mice, no striking differences have so far been detected in mice that express the CD5 transgene and their non-transgenic littermates. However, it is clear that dramatic changes occur in a subset of B cells which leads to extensive plasmablast and plasma cell formation. Further studies on the clonality of this B cell expansion, isotype expression, and regulating factors including T cell and cytokine involvement in this process remain to be elucidated.

5. Acknowledgements

We wish to acknowledge the expert assistance of Ann Brookshire in the preparation of this manuscript and Jim Karras and Tom Rothstein for their unpublished experiments on activation of CD5 transgenic mice. This work has been supported in part by USPHS grants AI14782, AI30879, and CA13148.

6. References

1. Ledbetter JA, Herzenberg LA (1979) Immunol. Rev. 47:63
2. Hardy RR, Hayakawa K, Herzenberg LA, Morse HC, III, Davidson WF (1984) Curr. Top. Microbiol. Immunol. 113:231
3. Lanier LL, Warner NL, Ledbetter JA, Herzenberg LA (1981) J. Exp. Med. 152:998
4. Boumsell L, Coppin H, Pham D, Raynal B, Lemerle J, Dausett, Bernard AJ (1980) J. Exp. Med. 152:229
5. Huang H, Herzenberg LA, et al. (1987) Proc. Natl. Acad. Sci. USA 84:204
6. Beyers AD, Spruyt LL, Williams AF (1992) Proc. Natl. Acad. Sci. USA 89:2945
7. Burgess KE, Yamamoto M, Prasad KVS, Rudd CE (1992) Proc. Natl. Acad. Sci. USA 89:9311
8. Logdberg LL, Shevach EM (1985) Eur. J. Immunol. 15:1007
9. Nishmura Y, Bierer BE, Jones WK, Jones NH., Strominger JA, Burakoff SJ (1988) Eur. J. Immunol. 18:747
10. van de Velde H, von Hoegen I, Luo W, Parnes JR, Thielemans K (1991) Nature 351:662
11. Luo W, van de Velde H, von Hoegen I, Parnes JR, Thielemans K (1992) J. Immunol. 148:1630

13. Lankester AC, van Schijindel GMW, Cordell JL, van Noesel CJM, Rene RAW (1994) Eur. J. Immunol. 24:812
14. Carlsson L, Holmberg D (1990) Inter. Immunol. 3:639
15. Gerster T, Picard D, Schaffner W (1985) Cell 45:45
16. Adams JM, Harris AW, Pinkert CA et al. (1985) Nature 318:533
17. Weiss U, Rajewsky K J. (1990) J. Exp. Med. 172:1681
18. Blier PR, Bothwell AJ J (1987) Immunol 139:3996
19. Jacob J, Kelsoe G, Rajewsky K, Weiss U (1991) Nature 358:389
20. Rusconi S, Kohler G (1985) Nature 314:330
21. Storb U, Pinkert C, Arp B, Engler P, Gollahon K, Manz J, Brady W, Brinster RL (1986) J. Exp. Med. 164:627
22. Iglesias A, Lamers M, Kohler G (1987) Nature 330:482
23. Tsang H, Pinkert C, Hagman J, Lostrum M, Brinster RL, Storb U (1988) J Immunol 141:308
24. Neuberger MS, Caskey HM, Petterson S, Williams GT Surani MA (1989) Nature 338:350
25. Pavirani A, Skern T, LeMeur M, Lutz I, Lathe R, Crystal R, Ruchs JP, Gerlinger P, Courtney M (1989) Biotechnology 7:1049
26. Gerlinger P, LeMeur M, Renard P, Wasylyk C, Wasylyk B (1986) Nucleic Acids Res 14:6565
27. Sumida T, Maeda T, Koike T, Tomioka H, Yoshida S, Taniguchi M, Rajewsky K (1990) Int. Archg Allergy Appl. Immunol 93:155
28. Gram H, Zenke G, Leuser B, Burki K (1992) Eur J Immunol 22:1185
29. Adam JM, Harris AW, Pinkert CA, Corcoran LM, Alexander WS, Cory S, Palmiter RD, Brinstger RL (1985) Nature 318:533
30. McDonnell TJ, Deane N, Platt FM, Nunez G, Jaeger U, McKearn JP, Korsmeyer SJ (1989) Cell 57:79
31. Gerster T, Picard D, Schaffner W (1986) Cell 45:45
32. Hardy, R.R., Carmack, C.E., Li, Y.S., Hayakawa, K.(1994) Immunol.Rew., 137:91-118

Activation of B-Cells by sIgM Cross-Linking Induces Accumulation of CD5 mRNA

R. S. Bandyopadhyay, M. R. Teutsch, and H. H. Wortis
Tufts University School of Medicine and Sackler School of Graduate Biomedical Sciences, 136 Harrison Ave. Boston MA 02111

ABSTRACT

The surface membrane molecule CD5 is expressed on mature T cells and on the B-1a subpopulation of B cells. These CD5 positive B cells express an antibody repertoire with a relatively high frequency of self-reactivity. There is uncertainty about the origins of CD5 B cells and the reasons for this are reviewed. Recent reports which relate to the lineage/selection debate are discussed. For instance, an increase in the frequency of CD5 B cells is a feature of several genetically determined polysystem autoimmune syndromes. In the case of *motheaten (me, me^v)* the pathogenesis of this increase in CD5 B cells is not yet understood, even though the mutation has been mapped to the *Hematopoietic cell protein-tyrosine phosphatase (Hcph)* gene. Another mutation which affects B cell development, *X-linked immunodeficiency (xid)*, encodes a point mutation in a B cell cytoplasmic tyrosine kinase. Expression of *xid* in otherwise normal mice causes a lack of CD5 B cells and a shift in the antibody repertoire. Interestingly, expression of both *xid* and *motheaten* results in an amelioration of autoantibody production. Evidence is presented that in B cells regulation of expression of CD5 can occur at the level of mRNA and that cross-linking of sIgM can induce the accumulation of CD5 mRNA. The overall concept advanced is that cells expressing natural autoantibodies are triggered via sIgM ligation to become CD5 B cells.

CD5 Protein

CD5 is a Class I surface membrane glycoprotein. The CD5 cDNA predicts a 51,500 Da polypeptide chain configured as two extracellular domains, a short transmembrane segment with a cytoplasmic domain [1]. It has 5 potential extracellular glycosylation sites that could account for the greater than predicted apparent molecular weight of 67 kD, and it's cytoplasmic domain has three potential phosphorylation sites. It's structure resembles CD6 and the macrophage scavenger molecule sufficiently to place these three molecules within an homologous family group [2].

The gene encoding CD5 maps to murine chromosome 19 and human chromosome 11. The murine CD5 locus contains 10 exons and these are distributed over 21 kb [3]. Two allelic forms of the mouse gene have been reported although no human or murine isoforms nor human alleles of CD5 have been described.

CD5 is expressed on all mature T cells and on a subset of B cells, the so-called B-1 population. Expression on developing T cells begins at about the same time that the TCR is seen as a surface protein [4]. Similarly, expression on B cells appears to coincide with the expression of sIgM [5]. Based on studies of immunoprecipitation after mild detergent solubilization [6], co-capping and fluorescent energy transfer [7] Crumpton and co-workers concluded that CD5 is part of the TCR complex. Similar studies established an association of the membrane IgM complex (B cell receptor or BCR) and CD5 [8].

In humans [9] and mice [10] CD72 (a.k.a. Lyb-2 in the mouse) has been described as a natural ligand for CD5. At the present time the functional relationships between these two molecules is unknown. Whether CD5 induces a signal through CD72 has not been established. However, experiments by Subbaro revealed that antibody binding to CD72 could induce B cell activation [11]. To date, no *in vivo* physiological function for CD5 has been established. *In vitro* experiments by several groups showed that antibodies to CD5 could induce T cell activation and proliferation [12]. In view of the more recent evidence that CD5 exists as part of the TCR complex, it remains unclear whether the observed activation was transduced by CD5 itself or whether other elements of the complex were required.

There is evidence showing that CD5 can play a role in lymphocyte activation, but in sum the data is consistent with this hypothesis rather than conclusive. Phosphorylation of tyrosine in the cytoplasmic domain of CD5 consequent to TCR-mediated stimulation of T cells was observed by Crumpton's group [13]. Inducible ser/thr kinase activity associated with T cell CD5 was recently described [14] but the catalytic target(s) of the enzyme and its role in normal cellular functions is unknown. The possibility that CD5 might serve as a negative modulator of lymphocyte activation was raised by the finding that peritoneal CD5$^+$ cells do not proliferate in response to stimulation by anti IgM antibody [15]. This may be resolved by information from Rajewsky's group as they have developed a CD5 deficient mouse [16].

CD5$^+$ and CD5$^-$ B Cells

The last few years has seen considerable debate over the origin of naturally occurring CD5 cells. This is not the place to review all the arguments and references, which were recently stated by advocates of both the lineage and selection hypotheses [17]. Although the final word is not yet revealed, we are advocates of the selection view. However, we stress that there is considerable agreement on many points and this might be useful place to restate them. It is generally agreed that there are differences between fetal/neonatal and adult B cell precursors (Figure 1). Whether the differences in expression of terminal deoxynucleotidyl transferase (TdT), with the resulting differences in repertoire, are sufficient to explain the different frequencies of resultant B cell subpopulations is a core issue of the debate. The alternative view being that fetal/neonatal precursors are precommitted to become B-1a cells.

Several groups have reported that CD5 B cells can be generated from adult precursors in cell transfer experiments. We have similar unpublished data. A crucial element in such experiments may be the use of neonatal recipients, as established by M. Elliottt [18].

Mice expressing transgenes encoding antibodies that are normally expressed by CD5 B cells have a high frequency of CD5 B cells. A detailed examination of such a mouse was published by Clarke and colleagues [19]. They demonstrated that the specificity of the antibody correlated with the phenotype of the B cells generated. It may be significant that the antibody expressed by this transgene, encoded by VH12 and Vκ4, has autoreactivity against phosphatidyl choline.

Expression of the lethal mutations *motheaten (me)* or *motheatenv (mev)* results in mice with a high frequency of CD5 B cells which produce large amounts of autoantibody. These mice accumulate plasma cells and macrophages and die from an accumulation of macrophages and granulocytes in the lungs [20]. The mutations map to the structural locus for the gene encoding *Hematopoietic cell protein-tyrosine phosphatase (Hcph)* [21]. The B cell substrate(s) of Hcph have not been identified and the pathogenesis of the overproduction of CD5 B cells is not yet understood. In contrast to the mutations affecting Hcph, the mutation *X-linked immunodeficiency (xid)* lies within the gene encoding B cell tyrosine kinase (btk) [22]. Mutation of this gene in humans results in Bruton's X-linked agammaglobulinemia [23] in which B cells fail to develop, yet affected mice have B cells. The murine *xid* syndrome involves a decrease in total B cells, an absolute deficiency of CD5 B cells, a failure to respond to thymus independent type 2 antigens (repeating

B CELL PRECURSORS

FETAL	ADULT
Arise in gut-associated tissues	Arise in bone marrow
Lack TdT Produce N-less heavy chains - High (f) of homologous recombination	Express TdT Produce N-containing heavy chains
Repertoire skewed toward 7183, Q52 V_H families	Repertoire skewed against 7183, Q52 V_H families
(f) $CD5^+$ (B-1a) precursors: HIGH	(f) $CD5^+$ (B-1a) precursors: LOW
(f) $CD5^-$ $CD11b^+$ (B-1b) precursors: MODERATE	(f) $CD5^-$ $CD11b^+$ (B-1b) precursors: MODERATE
(f) $CD5^-$ $CD11b^-$ (B-2) precursors: MODERATE	(f) $CD5^-$ $CD11b^-$ (B-2) precursors: HIGH
Do not express myosin-like light chain	Express myosin-like light chain
Late expression of Ia	Early expression of Ia

Figure 1. A summary of observed differences in the pre-B cells found in fetal/neonatal and adult mice.

unit antigens lacking intrinsic mitogenicity such as polysaccharides and their haptenated derivatives), and a decrease in IgM, IgG3, and Igλ [24]. The substrate(s) of btk have not been identified.

It is interesting that when mice express both *me* and *xid* the B cell manifestations of the former are dramatically relieved [25], raising the possibility that Hcph and btk function as part of a signal pathway that is necessary for the generation of CD5 B cells, but not essential for (most) T dependent responses. We would conjecture that this pathway would be necessary for activation induced by sIgM (TI-2 type signalling) but not CD40 (TD) ligation.

Expression of CD5 in B Cells

To better understand the basis for the differences in surface CD5 expression amongst B cells we wished to measure the level of CD5 mRNA. Since the amount of CD5 mRNA isolated from peritoneal CD5+ B cells was below the limit of sensitivity of conventional Northern blot analysis, under conditions by which mRNA in thymocytes was readily detected, (data not shown) we took a PCR based approach. The two CD5 primers for amplification were selected from exons three and five, regions in the genome that are approximately 1200 bp apart [3]. To examine the CD5 mRNA that is expressed by B cells, one ug of RNA from sorted splenic CD5+ and CD5- B cells was reverse transcribed and the cDNA was used for PCR amplification (RT-PCR). A B cell lymphoma line (CH12 [26]), that expresses high level of CD5 on its surface, was used as a positive control. PCR products were resolved on an agarose gel, blotted onto a nitran membrane and probed with a random primed product from the Ly-1 cDNA clone C5.6.1 [1]. The 365 bp band in Figure 2 corresponds to a product templated on completely processed mRNA as predicted from the primers. This was confirmed by subsequent cloning and sequencing of the 365 bp fragment (data not shown). A faint band of this size was generated from whole spleen, more abundant was seen in the CD5+ B cells whereas none was found with CD5- mRNA.

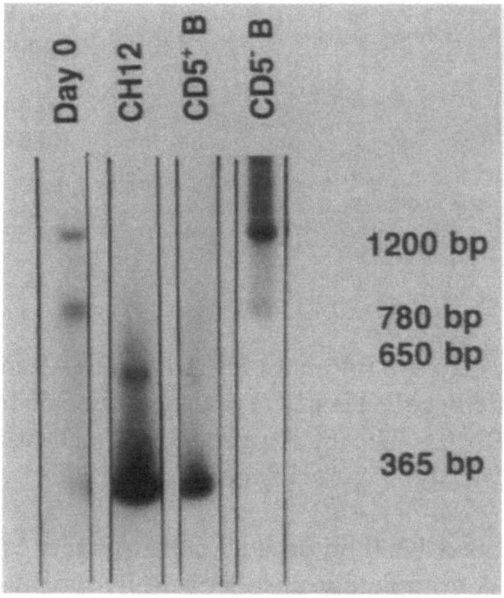

Figure 2. RT-PCR analysis of CD5 products from B-cells that were flow cytometrically sorted as CD5+ or CD5- reveals that a predicted 365 bp product is only detected from B-cell populations that express CD5 on the plasma membrane. PCR primers used for this study were selected from exons three (GGAAAACAGTGTGCAGTTCCA) and five (GTGCACCTCAGGCCTTCATG) [3]. PCR products were detected by Southern blot using a random primed Ly-1 cDNA probe [1].

We also observed a 650 bp fragment from CD5+ cells and two additional fragments from CD5- B-cells, 1200 bp and 780 bp. To examine the nature of these additional PCR products we titrated RNase into both RNA and DNA samples isolated from fresh (day 0) splenic B-cells. Genomic DNA isolated from CH12 cells was treated in the same way and used as a control. After RT-

PCR, the same two bands, 780 bp and 365 bp, were decreased by increasing concentrations of RNase while the 1200 and 650 bp bands remained (Figure 3). This result, therefore, demonstrates that the 365 and 780 bp bands are from RNA while the 1200 and 650 bp bands are, at least in part, of DNA origin. Absence of the 780 and 365 bp bands after direct amplification without reverse transcription further supports this conclusion (data not shown).

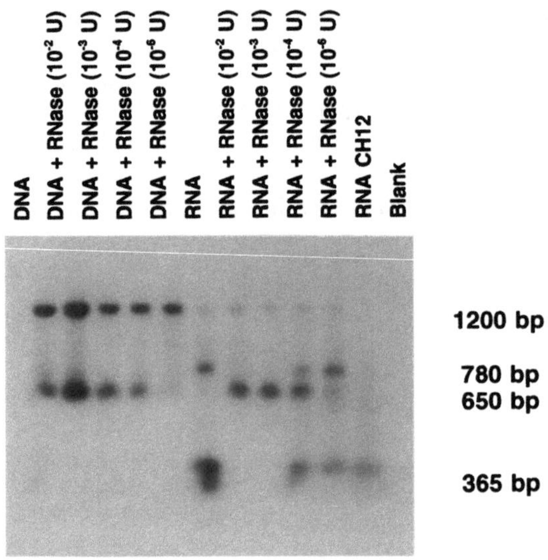

Figure 3. Titration of RNase into both RNA and DNA samples obtained from freshly isolated splenic B-cells clearly reveals that the 365 bp RT-PCR products are derived from mRNA. The PCR primers and cDNA probe used for this experiment are listed in the Figure 2 legend.

Sequencing of the cloned 1200 bp product confirmed its correspondence to the genomic fragment. A completely unspliced RNA would also be expected to generate a PCR product of that size. We have not ruled out the possibility that an unprocessed mRNA is also present but masked by the same size PCR product generated from genomic DNA. The 780 bp product was also cloned and sequenced. It was found to consist of partial and complete repeats of exons three, four and five. There was no intron sequence identified, thus ruling out the possibility that this might have been generated from a template of partially processed CD5 mRNA. Only one CD5 gene has been found, and when we examined cellular DNA by Southern blot analysis using a CD5 cDNA probe we also found no evidence for a second locus (data not shown).

Induction of CD5

Previously we had demonstrated that treatment of CD5- resting B cells with anti-IgM induced expression of surface CD5 [27]. We were interested in determining whether this induction of CD5 protein might be regulated at the level of mRNA, reflecting either induced transcription and/or message stability. T-cell (CD4+ & CD8+) and granulocyte (Gr1) depleted, high density splenic B lymphocytes from CBA mice were treated with LPS (5 ug/ml), or goat anti-Ig antibody (50 ug/ml). Surface expression of CD5 was then followed by flow cytometry for two and a half days. After 2.5 days in culture about 30% of anti-Ig treated cells became surface CD5 positive. During that period of time, however, CD5 surface expression on LPS treated cells remained at the basal level. Proliferation, as measured by thymidine incorporation was induced by both LPS and anti-Ig treatment. In the initial B cell population (time 0) 3-5% of splenic B-cells were surface CD5 positive and the mRNA was also detected. On days 0.5 to 2.5 the increased levels of CD5 mRNA in anti-Ig treated cells (Figure 4) correlated well with increased frequencies of surface expression of CD5. However, the level of 365 bp product dropped 3-5 fold in the LPS treated cells.

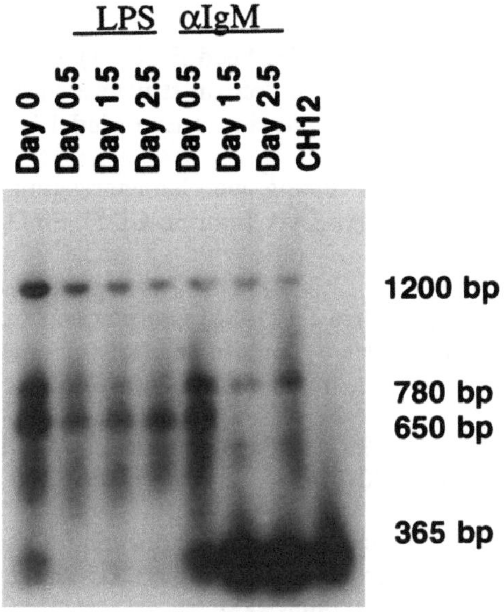

Figure 4. The level of CD5 mRNA detected by RT-PCR is increased when B-cells are activated by anti-IgM. Splenic B cells were cultured with either LPS (5μg/ml) or the $F(ab')_2$ of goat anti-mouse IgM (50 μg/ml). After the indicated time the RNA was isolated and examined by RT-PCR using the primers listed in Figure 2. The PCR products were detected by Southern blot using a random primed CD5 probe.

Assuming that the number of cycles, and the rates of elongation by the polymerase remained constant, the variations in the relative level of signals reflect the quantitative differences at the level of template CD5 mRNA in the cells. Therefore, higher levels of mRNA in the anti-Ig treated cells suggest an increased rate of transcription and/or an increase in mRNA stability.

Acknowledgments

Cassis Henry did some of the sequencing and we are grateful for her contribution to the work. We thank Diana Pierce for preparing this manuscript. Supported in part by NIH grant AI15803.

References

1 . Huang HJ, Jones NH, Strominger JL, Herzenberg LA (1987) Molecular cloning of Ly-1, a membrane glycoprotein of mouse T lymphocytes and a subset of B cells: Molecular homology to its human counterpart Leu-1/T1 (CD5). Proc. Natl. Acad. Sci. 84:204-208

2 . Aruffo A, Melnick MB, Linsley PS, Seed B (1991) The lymphocyte glycoprotein CD6 contains a repeated domain structure characteristic of a new family of cell surface and secreted proteins. J Exp Med 174:949-952

3 . Huang HJ (1987) Molecular cloning and characterization of the murine lymphocyte differentiation antigen Ly-1 (murine CD5). Ph.D. Thesis, Stanford University

4 . Lanier LL, Allison JP, Phillips JH (1986) Correlation of cell surface antigen expression on human thymocytes by multi-color flow cytometric analysis: implication for differentiation. J. Immunol 137:2501-2507

5 . Hardy RR, Hayakawa K (1991) A developmental switch in B lymphopoiesis. Proc. Natl. Acad. Sci. 88:11550-11554

6 . Osman N, Ley SC, Crumpton MJ (1992) Evidence for an association between the T cell receptor/CD3 antigen complex and the CD5 antigen in human T lymphocytes. Eur J Immunol 22:2995-3000

7 . Osman N, Lazarovits AI, Crumpton MJ (1993) Physical association of CD5 and the T cell receptor/CD3 antigen complex on the surface of human T lymphocytes. Eur J Immunol 23:1173-1176

8 . Lanester AC, van Schijndel GMW, Cordell JL, van Noesel CJM, van Lier RAW (1994) CD5 is associated with the human B cell antigen receptor complex. Eur J Immunol 24:812-816

9 . Van de Velde H, von Hoegen I, Luo W, Parnes JR, Thielemans K (1991) The B-cell surface protein CD72/Lyb-2 is the ligand for CD5. Nature 351:662-665

10 . Luo W, Van de Velde H, von Hoegen I, Parnes JR, Thielmans K (1992) Ly-1 (CD5), a membrane glycoprotein of mouse T lymphocytes and a subset of B cells, is a natural ligand of the B cell surface protein Lyb-2 (CD72). J Immunol. 148:1630-1634

11 . Subbarao B, Mosier DE (1983) Induction of B lymphocyte proliferation by monoclonal anti-Lyb2 antibody. J Immunol 130:2033-2039

12 . Alberola-Ila J, Places L, Cantrell DA, Vives J, Lozano F (1992) Intracellular events involved in CD5-induced human T cell activation and proliferation. J. Immunol. 148:1287-1293

13 . Davies AA, Ley SC, Crumpton MJ (1992) CD5 is phosphorylated on tyrosine after stimulation of the T-cell antigen receptor complex. Proc Natl Acad Sci 89:6368-6372

14 . Alberola-Ila J, Places L, Lozano F, Vives J (1993) Association of an activation inducible serine kinase activity with CD5. J Immunol 151:4423-4430

15 . Rothstein TL, Kolber DL (1988) Anti-Ig antibody inhibits the phorbol ester-induced stimulation of peritoneal B cells. J Immunol 141:4089-4093

16 . Tarakhovsky A, Müller W, Rajewsky K (1993) The CD5/Ly-1 deficient homozygous mice. J Cell Biochem Suppl 17B FZ109 (abstract)

17 . Herzenberg LA, Kantor AB, Haughton G, Arnold LW, Whitmore AC, Clarke SH, Wortis HH, Hardy R (1993) In Debate: the nature of B-cell subpopulations Immunology Today 14:79-91

18 . Elliott, M (1993) Regeneration of $CD5^+$ and $CD5^-$ B1 B cells from mouse adult bone marrow cells. J Cell Biochem Suppl 17B FZ121

19 . Arnold LW, Pennell CA, McCray SK, Clarke SH (1993) Development of B-1 cells: Segregation of phosphatidyl choline-specific B cells to the B-1 population occurs after immunoglobulin gene expression. J Exp Med 179:1585-1595

20 . Shultz LD, Sidman CL (1987) Genetically determined murine models of immunodeficiency. Ann Rev Immunol 5:367-680

21 . Shultz LD, Schweitzer PA, Rajan TV, Yi T, Ihle JN, Matthew RJ, Thomas ML, Beier DR (1993) Mutations at the murine motheaten locus are within the hematopoietic cell protein-tyrosine phosphatase (Hcph) gene. Cell 73:1445-1454

22 . Thomas JD, Sideras P, Smith CIE, Vořechovský I, Chapman V, Paul WE (1993) Colocalization of X-linked agammaglobulinemia and X-linked immunodeficiency genes. Science 261:355-358
Rawlings DJ, Saffran DC, Tsukada S, Laraespada DA, Grimaldi JC, Cohen L, Mohr RN, Bazan JF, Howard M, Copeland NG, Jenkins NA, Witte ON (1993) Mutation of unique region of Bruton's tyrosine kinase in immunodeficient XID mice. Science 261:358-362

23 . Vetrie D, Vořechovský I, Sideras P, Holland J, Davies A, Flinter F, Hammarström L, Kinnon C, Levinsky R, Bobrow M, Smith CIE, Bentley DR (1993) The gene involved in X-linked agammaglobulinemia is a member of the *src* family of protein-tyrosine kinases. Nature 361:226-233

24 . Scher, I (1982) CBA/N immune defective mice; evidence of the failure of a B cell subpopulation to be expressed. Immunol. Rev. 64:117-136

25 . Scribner CL, Hansen CT, Klinman DM, Steinberg AD (1987) The interaction of the *xid* and *me* genes. J Immunol 138:3611-3617

26 . Lanier LL, Arnold LW, Raybourne RR, Russell S, Lynes MA, Warner NL, Haughton G (1982) Transplantable B-cell lymphomas in B10.H-2^aH-4^bp/Wts mice. Immunogenetics 16:367-371

27 . Cong YZ, Rabin E, Wortis HH (1991) Treatment of $CD5^-$ B cells with anti-Ig, but not LPS, induces surface CD5: Two B cell activation pathways. Int. Immunol. 3:467-476

GP130 in Human Myeloma/Plasmacytoma

Y. Tani, N. Nishimoto, A. Ogata, Y. Shima, K. Yoshizaki and T. Kishimoto

Department of Medicine III, Osaka University Medical School,
2-2 Yamada-oka, Suita city, Osaka 565 Japan

Introduction

Interleukin 6 (IL-6) is a potent growth factor for human myeloma/plasmacytoma cells (Kawano et al. 1988, Klein et al. 1989) The evidences that 1) myeloma cells produce IL-6 and express IL-6 receptor (IL-6R), 2) IL-6 augments the *in vitro* growth of myeloma cells, 3) anti-IL-6 antibody (Ab) or anti-IL-6R Ab inhibits their growth (Klein et al. 1991, Sato et al. 1993) support the autocrine or paracrine mechanism of IL-6 for myeloma cell growth. The cytoplasmic region of IL-6R is very short (Yamasaki et al. 1988) and a signal transducing molecule, gp130 was discovered (Hibi et al. 1990). Furthermore, leukemia inhibitory factor (LIF), oncostatin M (OSM), ciliary neurotrophic factor (CNTF) and IL-11 have been shown to share the same signal transducer, gp130 in their receptor systems (for review see Kishimoto et al. 1994). On the basis of this finding, it seems likely that these cytokines as well as IL-6 may be the growth factors for human myeloma/ plasmacytoma cells (Nishimoto et al. 1994, Zhang et al. 1994). In this report, we show the stimulatory effect of LIF and OSM and the inhibitory effect of anti-gp130 monoclonal antibody (mAb) on the growth of freshly isolated human plasmacytoma cells.

Results

IL-6, LIF and OSM Augmented the *in vitro* Growth of Plasmacytoma Cells

The plasmacytoma cells were freshly isolated from biopsy specimen from a 53 year-old male patient with extramedullary plasmacytoma tumors and enriched by density gradient centrifugation. Greater than 99% of purified cells showed plasmacytoid morphology. They expressed CD38, CD54 and CD56, and low level of IL-6R and gp130, but no detectable CD3, CD5, CD10, CD20 or VLA-5 on their cell surface (Fig.1).

When these plasmacytoma cells were cultured with IL-6, LIF, OSM or IL-11 at various concentrations (1-100ng/ml), IL-6, LIF and OSM induced their growth in a dose-dependent manner (Fig. 2). The stimulatory effect of LIF and OSM was stronger than that of IL-6. IL-11, however, did not induce their proliferation.

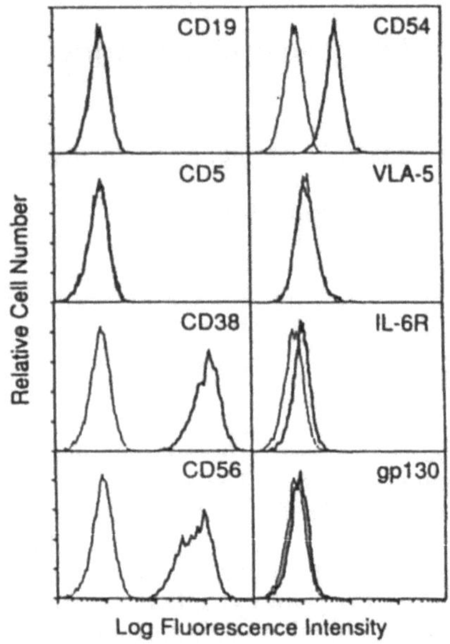

Fig. 1 Surface marker analysis of the freshly isolated human plasmacytoma cells.

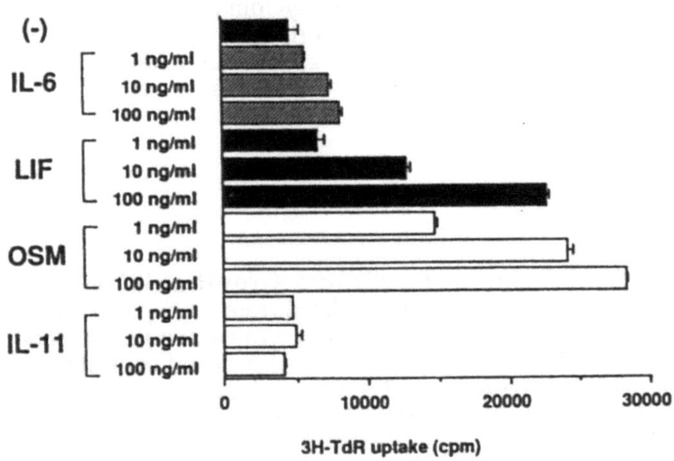

Fig. 2 *In vitro* augmentation of plasmacytoma cell growth by IL-6, LIF, OSM and IL-11 at the density of 0.3X10^6/ml on 96 well plates for 3 days in 0.2ml of RPMI-1640 medium with 10% FCS and 5X10^{-5}M 2ME. DNA synthesis was measured after the last 16 hours pulse lable of a 3 day-culture with ^{3}[H] TdR.

Anti-gp130 mAb Inhibited the IL-6-, LIF- and OSM-Dependent Plasma-cytoma Cell Growth

To test whether this IL-6-, LIF- and OSM-dependent cell growth is through the common signal transducer, gp130, we cultured these plasmacytoma cells with anti-gp130 mAb. Anti-gp130 mAb clearly blocked their growth in the presence or absence of these cytokines (Fig. 3). This indicates that IL-6, LIF and OSM are growth factors for plasmacytoma cells using the common signal transducer, gp130.

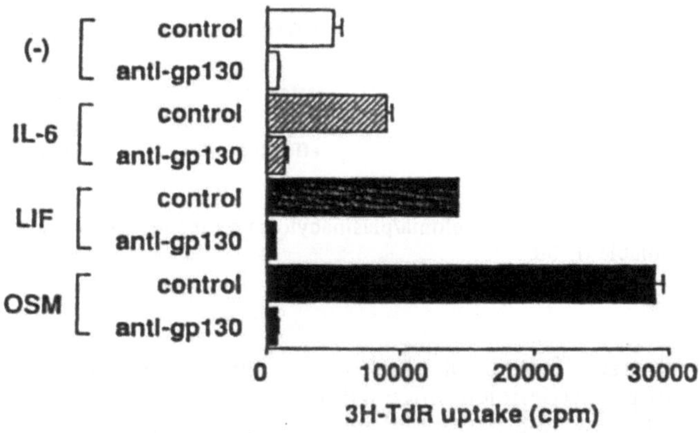

Fig. 3 Inhibition of plasmacytoma cell growth by anti-gp130 mAb. Cells were cultured with 10ng/ml of IL-6, LIF and OSM in the presence of anti-gp130 mAb or control mAb (10μg/ml).

Discussion

The cause of myeloma/plasmacytoma is still unknown and its clinical course widely varies among patients. Especially, the survival of patients with multiple myeloma has not been improved for 30 years (Alexanian & Dimopoulos, 1994). Multiple myeloma is not curable by chemotherapy alone. However, the biological aspects of this disease have recently been made clear and the clinical application of cytokines and their antibodies may be promising. In fact, *in vivo* administration of murine anti-IL-6 mAb to a patient with plasma cell leukemia was therapeutically effective (Klein et al. 1991).

We reported a plasmacytoma case of which the cells proliferated dependently on LIF and OSM as well as IL-6. IL-6 was detected in the culture supernatant of the cells, but LIF or OSM was not, suggesting that these plasmacytoma cells respond to an autocrine growth mechanism mediated by IL-6. But these cells may also respond to a paracrine growth mechanism by LIF or OSM. Although anti-IL-6 mAb and anti-IL-6R mAb inhibited the *in vitro* growth of these cells (data not shown), *in vivo* administration of these antibodies to this patient might not be effective, since the plasmacytoma cells can grow dependently on LIF or OSM. So we would have to block the signal transduction through gp130 with anti-gp130 mAb in this case. We

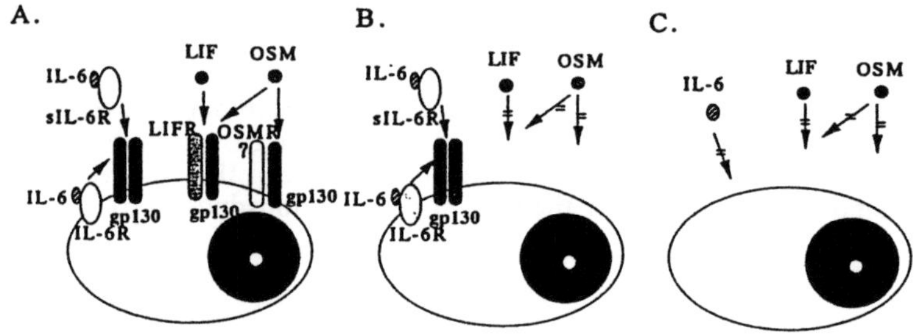

	A.	B.	C.
anti-IL-6 mAb	partially effective	effective	not effective
anti-IL-6R mAb	partially effective	effective	not effective
anti-gp130 mAb	effective	effective	not effective

Fig. 4 New classification of human myeloma/plasmacytoma by the responsiveness of IL-6 related cytokines. sIL-6R : soluble IL-6R

do not know the adverse effects of blockade of the signal transduction through gp130, yet. The analysis of gp130 deficient mice will provide useful information on them. From this point of view, we proposed new classification of human myeloma/ plasmacytoma by the responsiveness of IL-6 related cytokines (Nishimoto et al. 1994). That is, (a) myeloma/plasmacytoma cells whose growth is dependent on IL-6, LIF and OSM, (b) myeloma/plasmacytoma cells whose growth is dependent on IL-6 only, (c) myeloma/plasmacytoma cells whose growth is independent of IL-6, LIF and OSM (Fig. 4). We did not test CNTF yet, but if the CNTF-R is expressed on the cell surface, CNTF may also have the same activity as LIF and OSM (Zhang et al. 1994). IL-11, however, does not induce the growth of freshly isolated myeloma/ plasmacytoma cells, so far (our data and Paul et al. 1992).
Finally, the application of anti-IL-6 mAb, anti-IL-6R mAb and anti-gp130 mAb will be potential therapeutic approaches against group (a) and (b) myeloma/ plasmacytoma. We are now studying for the clinical use of humanized anti-IL-6R mAb for the patients with multiple myeloma.

Acknowledgment

We thank Dr. Hirata for helpful discussion and the editorial assistance.

References

Alexanian R and Dimopoulos M (1994) The treatment of multiple myeloma. New Engl J Med 330: 484-489

Hibi M, Murakami M, Saito M, Hirano T, Taga T and Kishimoto T (1990) Molecular cloning and expression of an IL-6 signal transducer, gp130. Cell 63: 1149-1157

Kishimoto T, Taga T and Akira S (1994) Cytokine signal transduction. Cell 76: 253-262

Kawano M, Hirano T, Matsuda T, Taga T, Horii Y, Iwato K, Asaoku H, Tang B, Tanabe O, Tanaka H, Kuramoto A and Kishimoto T (1988) Autocrine generation and requirement of BSF-2/IL-6 for human multiple myelomas. Nature 332: 83-85

Klein B, Zhang X-G, Jourdan M, Content J, Houssiau F, Aarden L, Piechaczyk M and Bataille R (1989) Paracrine rather than autocrine regulation of myeloma- cell growth and differentiation by interleukin-6. Blood 73: 517-526

Klein B, Wijdenes J, Zhang X-G, Jourdan M, Boiron J-M, Brochier J, Liautard J, Merlin M, Clement C, Morel-Fournier B, Lu Z-Y, Mannoni P, Sany J and Bataille R (1991) Murine anti-interleukin-6 monoclonal antibody therapy for a patient with plasma cell leukemia. Blood 78: 1198-1204

Nishimoto N, Ogata A, Shima H, Yoshizaki K and Kishimoto T (1994) Oncostatin M, leukemia inhibitory factor and interleukin-6 induce the proliferation of human plasmacytoma cells via the common signal transducer, gp130. J Exp Med 179: 1343-1347

Paul SR, Barut BA, Bennet F, Cochran MA and Anderson KC (1992) Lack of a role of interleukin 11 in the growth of multiple myeloma. Leuk Res 16: 247-252.

Sato K, Tsuchiya M, Saldanha J, Koishihara Y, Ohsugi Y, Kishimoto T and Bendig MM (1993) Reshaping a human antibody to inhibit the interleukin 6-dependent tumor cell growth. Cancer Res 53: 851-856

Zhang X-G, Gu J-J, Lu Z-Y, Yasukawa K, Yancopoulos GD, Turner K, Shoyab M, Taga T, Kishimoto T, Bataille R and Klein B (1994) Ciliary neurotrophic factor, interleukin 11, leukemia inhibitory factor and oncostatin M are growth factors for human myeloma cell lines using the interleukin 6 signal transduction. J Exp Med 179: 1337-1342

References

[illegible]

Analysis of Antigen-Driven Positive and Negative Selection of Phosphocholine-Specific Bone Marrow B-Cells

J. J. Kenny[1], P. W. Tucker[2], L. Claflin[3], M. Katsumata[4], M. Green[4], J. Reed[5], and D. L. Longo[6]

[1]B-Cell Development Section, BCDP, PRI/DynCorp, NCI-FCRDC, Frederick, MD 21702
[2]Department of Microbiology, University of Texas, Dallas, TX 75235
[3]Department of Microbiology and Immunology, University of Michigan Medical School, Ann Arbor, MI 48109
[4]Department of Pathology, University of Pennsylvania, Philadelphia, PA 19104
[5]La Jolla Cancer Research Foundation, La Jolla, CA 92037
[6]Biological Response Modifiers Program, NCI-FCRDC, Frederick, MD 21702

1 Introduction

The bone marrow of an adult mouse produces approximately 20 million new immature B cells each day [1,2]. The vast majority of these cells die in the bone marrow with only 3 million entering the mature peripheral B cell pool [2]. Thus, the peripheral B cell pool is comprised predominantly of long-lived $\mu^+\delta^+$ B cells [3] which are selected into this pool via an Ig V region-mediated process that appears to be driven by autoantigens or environmental antigens [4-11]. This receptor-driven selection process is influenced by the anatomical and physiological environment of the B cell [6,8,12,13] as well as the genetic make up of the host [14]. Thus, in Ig-transgenic mice expressing H or H + L chains having specificity for red blood cell (RBC) antigens, the antigen-specific B cells are expressed exclusively in the Ly-1 (CD5) (B1a) B cell subset [11-13,15]. High affinity transgene positive (TG^+) B cells survive only in the peritoneal cavity [12] and undergo apoptosis following i.p. injection of mouse RBC [13]. The overexpression of bcl-2 can prevent this antigen-induced clonal deletion in the peritoneum but not in the bone marrow [16]. On the other hand, lower affinity anti-phosphatidyl choline (PtC)-specific B cells are not clonally deleted in the bone marrow but get selectively expanded in both the spleen and peritoneal cavity by continuous Ig-receptor cross-linking [11,15]. We previously demonstrated that phosphocholine (PC)-specific B cells, which also reside predominantly in the $CD5^+$ B cell subset of normal mice [17], are positively selected into the peripheral lymphoid tissues of V_H1 μ-H-chain transgenic mice that express a normal X-chromosome, but they are clonally deleted in both M167 μ and $\mu\kappa$ transgenic mice which coexpress these M167 anti-PC transgenes and the X-linked immune deficiency gene, *xid* [7,14]. Idiotype and antigen binding analysis of antibodies generated via transfection of variant V_H1 genes in conjunction with the $\kappa 8$, $\kappa 22$, and $\kappa 24$ L chain genes suggest that both the positive and negative selection of idiotype-positive, PC-binding B cells are antigen-mediated and not idiotype-mediated processes [7,18]. The PC-specific B cells that develop in the M167 $\mu\kappa$ 207-4 transgenic mice are present in both the $CD5^-$ and $CD5^+$ B cell subsets [19], and in vitro activation studies [19-21] suggest that these B cells have been previously activated or are being continuously activated via signaling through their Ig-receptor. In Xid mice, this same signal presumably induces the antigen-driven clonal deletion of these PC-specific B cells.

In this paper, we have used several types of transgenic mice to address some of the molecular and biochemical requirements for the positive and negative selection of PC-specific B cells. The data presented: 1) demonstrate that overexpression of the *bcl*-2 gene prevents the clonal deletion of PC-specific B cells in *xid* mice; 2) show that PC-specific B cells which cannot signal through their Ig-receptor are unable to exit the bone marrow; and 3) suggest that the affinity differences in the Ig-receptor of anti-PC B cells may account for the positive and negative selection of PC-specific bone marrow B cells.

2 Results and Discussion

2.1 *Bcl*-2 Prevents the Clonal Deletion of PC-specific B Cells in M167 $\mu\kappa$ Anti-PC Transgenic *xid* Mice

Overexpression of *bcl*-2 is known to inhibit multiple forms of apoptosis [22] including death following treatment of thymocytes with dexamethasone or anti-CD3 antibodies [23-25] or neonatal B cells with anti-μ (G. Nunez, personal communication). Surprisingly, overexpression of *bcl*-2 has minimal or no effect on antigen-induced negative selection of T cells during thymic self-censorship in Bcl-2 transgenic mice [23-25]; however, autoimmune disease due to a wide range of autoantibody production does occur in some Bcl-2 transgenic mice [26]. It is not clear whether the B cells producing these autoantibodies escaped clonal deletion during ontogeny or were generated via somatic mutation following antigen-driven B cell activation.

We have previously shown [14] that the combined expression of the M167 $\mu\kappa$ anti-PC transgenes (TG) with the *xid*, results in an almost total failure to develop B cells in the peripheral lymphoid organs of such mice. Since PC-specific B cells develop normally in the bone marrow of M167 $\mu\kappa$ TG^+ Xid mice, the lack of peripheral PC-specific B cells appears to be due to an Ig-receptor-mediated clonal deletion of these B cells. It was therefore of interest to determine whether or not the overexpression of *bcl*-2 in the anti-PC Xid mice would prevent the clonal deletion of these PC-specific B cells. M167 *μκ* TG$^+$ female mice homozygous for the *xid* gene were produced by first crossing 207-4 *μκ* TG$^+$ male mice to B6.CBA/N female mice and then backcrossing the TG$^+$ F1 males to B6.CBA/N females. The TG$^+$ female progeny of this cross were then mated to Bcl-2 transgenic mice (line 6) [23,27]. The number of cells in the spleen and bone marrow of the 8 phenotypes of F1 progeny derived from this cross were determined, and these cells analyzed by flow cytometry to elucidate whether or not over-expression of *bcl*-2 would prevent the apparent clonal deletion of PC-specific B cells in the Bcl-2^+:V_H1^+ double transgenic male progeny. The data in Table 1 show that the number of splenic B cells increased > 20 fold in V_H1^+ Xid male mice in the presence of *bcl*-2 compared to a 2-fold increase in B cells in the TG$^+$ phenotypically normal female mice. Thus, ~2 x 10^6 B cells were present in the spleens of the Bcl-2^-:V_H1^+ males and this increased to ~ 50 x 10^6 in the Bcl-2^+:V_H1^+ male mice. All of these B cells express both the H chain dependent V_H1-idiotype (id) and M167 binding-site specific 28-5-15 id, and also bind phosphocholine (data not shown). Thus, the overexpression of *bcl*-2 prevented the loss of PC-specific B cells in the TG$^+$ Xid mice making them essentially equivalent to the phenotypically normal Bcl-2^-:V_H1^+ females (41 x 10^6 B cells) even though they still have only half as many B cells as the female mice expressing both the V_H1 and *bcl*-2 transgenes. As previously shown [26-28], the overexpression of *bcl*-2 caused a 3- to 4-fold increase in splenic B cells in both phenotypically normal female and Xid male mice lacking the M167 *μκ* transgenes. *Bcl*-2 had no effect on the bone marrow pre-B cell pool in any of the F1 progeny, which is consistent with the fact that *bcl*-2 is down regulated at this stage of B cell development [29]. The mature, recycling, CD23$^+$ B cells increased 4- to 10-fold in the bone marrow of both Bcl-2^+:V_H1^+ and Bcl-2^+:V_H1^- mice. These mature B cells account for the major increase seen in bone marrow sIgM$^+$ B cells in the presence of *bcl*-2.

The rescue of PC-specific, *xid* B cells by overexpression of *bcl*-2 is most likely due to prevention of antigen-induced, receptor-mediated cell death, although other alternative explanations exist. For example, *bcl*-2 could simply increase the half-life of the B cells so that normal numbers of B cells would eventually accumulate in the peripheral lymphoid organs. Kinetic and Brdu labeling studies are being conducted to address this point. However, it has not, to our knowledge, been directly demonstrated that the selection of immature B cells into the long-lived B cell pool requires a signal through the sIgM receptor and that in the absence of this signal B cells fail to exit the bone marrow and die.

2.2 Signal Transduction via sIg appears to be Required for B Cells to Exit the Bone Marrow

Signal transduction via surrogate L-chain ($V_{pre-B}\lambda_5$) complexes plays a critical role in both the positive selection of μ:$V_{pre-B}\lambda_5$ pre-B cells and the negative selection of pre-B cells expressing Dμ:$V_{pre-B}\lambda_5$ receptor complexes [2,30-32]. To address whether or not a sIg-receptor

Table 1. Comparison of total lymphoid cell numbers, B cells and Pre-B cells in bone marrow and spleen of (B6.Xid.$\mu\kappa$207-4 x Bcl-2)F1 mice

Tissue/Cell[a] Type Analyzed (Total # x 10^{-6})	Sex	Mouse Phenotype[b] Bcl-2 V_H1 - -	Bcl-2 V_H1 - +	Bcl-2 V_H1 + +	Bcl-2 V_H1 + -
Spleen Cells	F	170	80	180	485
	M	107	26	115	224
sIGM$^+$ Spleen Cells	F	89	41	92	369
	M	48	2.3	49	136
Bone Marrow Cells	F	38	45	47	56
	M	52	42	43	62
IgM$^+$B220$^+$ Bone Marrow Cells	F	5.9	1.6	17	23
	M	3.6	3.6	18	28
IgM$^-$B220$^+$ Pre-B Bone Marrow Cells	F	8.6	2.1	1.5	9.3
	M	8.3	1.0	0.4	10.5

[a]The spleen and bone marrow cell numbers were determined after $(NH_4)Cl$ Lysis of the RBC. Bone marrow numbers are based on the cells obtained from two femurs. The number of IgM$^+$ and B220$^+$ cells was calculated from the percent positive cells seen in flow cytometric analysis of the whole spleen and bone marrow cells.
[b]Mice were derived by first crossing B6.CBA/N females with TG$^+$ $\mu\kappa$207-4 males. TG$^+$ F1 males were backcrossed to B6.CBA/N females to produce TG$^+$ *xid/xid* females which were then crossed to males carrying the *bcl*-2 transgene. The 8 phenotypes obtained from this cross were then analyzed by flow cytometry as previously described [7].

mediated signal was required for positive-selection of immature bone marrow B cells and their subsequent survival in the peripheral B cell pool, we produced an Ig-transgenic mouse expressing rearranged T15 (V_H1:κ_{22}) H:L genes in which the spacer, transmembrane and cytoplasmic domains of μ were replaced by I-Aα domains [33]. B cell lines expressing this chimeric PC-specific IgM-(Iaα TM) molecule as their antigen-specific receptor are unable to function in the B cell activation and differentiation events of immediate signal transduction, antigen presentation, or immunoglobulin secretion following binding by anti-id or antigen [33-35]. It was therefore of interest to see if transgenic mice expressing this "dead" chimeric molecule would develop B cells, and whether these B cells would exhibit allelic exclusion or any other sign of altered B cell development. Flow cytometric analysis of spleen cells from TG^+ mice showed that all the B cells expressed the TG-encoded sIgM-Ia*α* chimeric receptor but this molecule was always expressed in association with endogenous $sIgM^b$ (data not shown). Thus, at the level of the spleen, no evidence of allelic exclusion could be seen; however, B cells expressing only the V_H1^+ TG product appear to develop in the bone marrow (data not shown). This suggests that B cells expressing only the IgM-Ia*α* receptor can develop in the absence of endogenous IgM^b but they are not selected into the peripheral lymphoid organs. To test this hypothesis, TG^+ mice were crossed and back-crossed to μ-knock-out (μMT) [36] mice to obtain TG^+ homozygous μMT/μMT mice. The total number of cells in the spleen and bone marrow of the backcross mice were determined and these cells analyzed by flow cytometry for the expression of the transgene. The heterozygous TG^+:μMT/+ mice are phenotypically identical to the original TG^+ founder strain (Table 2, column 2). When one compares these TG^+ mice to the TG^-:μMT/+ controls (columns 1 & 2), one sees that the expression of the transgene has a profound effect on all stages of B cell development. Thus, there is a statistically significant reduction in both splenic and bone marrow $sIgM^+$ B cells and in bone marrow pre-B cells of TG^+ mice. As seen in the founder line, the splenic B cells all co-express the TG-encoded sIgM with endogenous $sIgM^b$, while most of the bone marrow B cells lack endogenous μ^b (data not shown). In addition, greater than 50% of the splenic B cells in the TG^+ mice are $CD23^-$, whereas, the majority of B cells in the TG^- F1 control are $CD23^+$ (data not shown). Thus, the expression of the TG product affects both the number and the type of B cells which develop in the peripheral lymphoid organs. When the TG^+: μMT/+ mice are compared to the TG^+ homozygous μMT/μMT backcross mice (Table 2, columns 2 & 3) one sees that this latter group of mice, which are unable to produce endogenous IgM, produce approximately the same number of bone marrow B cells as the TG^+ heterozygous μMT/+ littermates; however, these B cells do not exit the bone marrow. Thus, there are $< 10^6$ B cells in the spleens of these TG^+ μMT/μMT mice compared to 30×10^6 V_H1^+ B cells in the TG^+ heterozygous μMT/+ mice. These data support the idea that an IgM receptor-mediated signal is required to positively select B cells from the bone marrow into the peripheral lymphoid tissues and that the chimeric IgM-Ia*α* molecule is unable to transduce the required signal.

2.3 The Role of Receptor Affinity in Positive-Selection of B Cells

The affinity limits of sIgM receptors on autoreactive immature B cells which permit positive rather than negative selection of emerging immature bone marrow cells is unknown. Furthermore, nothing is known about the lower limits of sIgM receptor affinity needed for positive selection to occur. Our analysis of the antigen-binding ability of antibodies derived from V1:DFL16.1:J_H1 (V_H1) germ-line and N-region-derived variant H-chains and κ22, κ24, and κ8 L-chains demonstrated that the T15H:κ22L (T15) antibody binds PC at least 20 to 40 times better than other antibodies derived from alternate germ-line forms of the V_H1 H-chain and κ22, κ24, and κ8 L-chains [18]. Thus, the early antigen-driven positive selection of the T15H:κ22L clone during neonatal development could account for the life long dominance of T15-id$^+$ B cells. To achieve affinities in the same range as the T15 antibody, κ24 and κ8 L-chain containing antibodies must have H-chains derived from

Table 2. Comparison of total lymphoid cell numbers, B cells and Pre-B cells in bone marrow and spleen of (PMC-186$\mu\kappa$laαTM x μMT)BC2 mice

Tissue/Cell[a] Type Analyzed (Total No. x 10^{-6})	Mouse Phenotype[b] TG$^-$:μMT/+		TG$^+$:μMT/+		TG$^+$:μMT/μMT		TG$^-$:μMT/μMT
Spleen Cells	120 ± 6	($p<0.001$)[c]	84 ± 4	**($p<0.001$)**	30 ± 3	($p=0.2$)	37 ± 4
sIgM$^+$ Spleen Cells	64 ± 5	($p<0.001$)	30 ± 2	**($p<0.001$)**	0.5 ± 0.1	($p=0.1$)	0.4 ± 0.1
Bone Marrow Cells	41 ± 1	($p=0.4$)	44 ± 3	**($p=0.08$)**	37 ± 2	($p=0.5$)	39 ± 1
IgM$^+$B220$^+$ Bone Marrow Cells	3.4 ± 0.7	($p=0.03$)	1.6 ± 0.2	**($p=0.2$)**	1.2 ± 0.2	($p<0.001$)	0.2 ± 0.1
IgM$^-$B220$^+$ Pre-B Bone Marrow Cells	5.4 ± 0.4	($p<0.001$)	2.8 ± 0.5	**($p=0.5$)**	1.8 ± 0.2	($p=0.004$)	2.6 ± 0.2

[a]The spleen and bone marrow cell numbers were determined after $(NH_4)Cl$ Lysis of the RBC. Bone marrow numbers are based on the cells obtained from two femurs. The number of IgM$^+$ and B220$^+$ cells was calculated from the percent positive cells seen in flow cytometric analysis of the whole spleen and bone marrow cells.

[b]The number of animals analyzed of each phenotype and tissue varied from a low of 8 for the TG$^-$:μMT/μMT spleen and bone marrow to 10 for the TG$^+$:μMT/μMT spleen and bone marrow (the other two groups contained 9 samples). The data shown represent the mean ± SE.

[c]Groups were compared using an unpaired Student's *t*-test. The p values obtained are shown between the groups being compared. The comparison of TG$^+$ normal and TG$^+$:μMT/μMT is in bold face.

variant N-region or somatically mutated V_H1 genes and these variant genes are produced in low frequency only after the higher affinity germ-line encoded T15-id$^+$ B cells have established clonal dominance. The single amino acid differences at the VD-junction of the various germ-line and N-region variant V_H1 H-chains shown in Figure 1 dictate the L-chain which can associate with the H-chain to produce a PC-specific antibody and it appears that antigen rather than an idiotypic network drives the in vivo selection of the PC-specific B cells. Thus, in M167-μ-transgenic mice, from 20 to 75% of the splenic B cells expressing the transgene H-chain product also co-express an endogenous $V_\kappa 24$-$J_\kappa 5$ L-chain and bind PC, whereas no B cells expressing the non-PC binding T15-id formed by the M167H and endogenous κ22L were detected [7].

To demonstrate more conclusively that antigen rather than an idiotype network is responsible for positive selection of PC-specific B cells and to determine the lower affinity limits for antigen-driven expansion of PC-specific B cells, we produced H-chain transgenic mice with each of the variant V_H1 genes shown in Figure 1. The data in Table 3 show that

Nucleotide Sequence of VDJ-genes

```
     VH-1                VD-Junction          DFL16.1                     DJ-Junction          JH1

91  92  93  94  95                        96  97  98  99  100                       100a 100b 100c 100d 101 102
Tyr Cys Ala Arg Asp Ala
TAC TGT GCA AGA GAT GCA CACAGTG   Tyr Tyr Gly Ser Ser Tyr
                AAT     TACTGTG TT TAT TAC TAC GGT AGT AGC TAC CACAGTG Tyr Trp Tyr Phe Asp Val
                Asn                                            CACAGTG TAC TGG TAC TTC GAT GTC
```

H-Chain Amino Acid Sequences

	91	92	93	94	95		96	97	98	99	100		100a	100b	100c	100d	101	102
T15	Tyr	Cys	Ala	Arg	Asp		Tyr	Tyr	Gly	Ser	Ser		Tyr	Trp	Tyr	Phe	Asp	Val
M603-like	Tyr	Cys	Ala	Arg	Asn		Tyr	Tyr	Gly	Ser	Ser		Tyr	Trp	Tyr	Phe	Asp	Val
M511-like	Tyr	Cys	Ala	Arg	Asp	Ala	Tyr	Tyr	Gly	Ser	Ser		Tyr	Trp	Tyr	Phe	Asp	Val
M167	Tyr	Cys	Ala	Arg	Asp	Ala	Asp	Tyr	Gly	Asn	Ser	Tyr	Phe	Gly	Tyr	Phe	Asp	Val

Figure 1. The nucleic acid sequence at the top represents the germ-line sequence for the 3'end of V_H1 gene, the DFL16.1 diversity gene segment, and the J_H1 joining segment. The translated amino acid sequence is shown above the DNA sequence with the T15 H-chain sequence in bold. Alternate splicing, N-region diversification or somatic mutation of the T15 germ-line sequence gives rise to the M603-like, M167-like, or M167 protein sequences shown in the bottom part of this figure. The amino acid numbering was taken from Rudikoff [15].

expression of the T15H and M603-like H-chain genes results in positive selection of PC-specific B cells just as occurred with the M167H chain. However, no PC-specific B cells were seen in mice expressing the M511-like H-chain. This H-chain, when associated with a κ24 L-chain, forms an antibody which binds PC-BSA with an avidity 20 to 40 times less than a T15-IgM antibody [18]. Thus, B cells with receptor affinities for PC below 10^4 may not be capable of being positively selected especially when they have to compete with endogenous T15H. Studies are in progress to show that the PC-specific B cells present in transgenic mice expressing T15H and M603H co-express the expected κ22 and κ8 L-chain genes. Since PC-specific antibodies with affinities much greater than T15 do not exist, it will be difficult to demonstrate the upper affinity limit permitted before B cells become negatively selected.

Table 3. Analysis of antigen-binding cells in V_H1 H-chain transgenic mice

Transgene	%μ^+B220$^+$ B Cells	%PC-ABC	%PC-AB B Cells
T15	8.2	0.24	2.9
M603-like	7.5	0.45	6.0
M511-like	20.2	0.1	0.4
M167	58.0	3.3	5.5

[a]H-chain transgenic mice were produced using the T15-variant genes shown in Figure 1. The M167 mouse (line 243-4) has been described previously [7,37].
[b]Spleen cells were stained with either FITC-conjugated anti-μ or PC-dextran plus PE-conjugated anti-B220 and analyzed using a Becton-Dickenson FaxScan.

References

1. Osmond DG (1986) Population dynamics of bone marrow B lymphocytes. Immunol Reviews 93:103-124
2. Melchers F, Karasuyama H, Haasner D, Bauer S, Kudo A, Sakaguchi N, Jameson B, Rolink A (1993) The surrogate light chain in B-cell development. Immunol Today 14:60-68
3. Forster I, Rajewsky K (1990) The bulk of the peripheral B-cell pool in mice is stable and not rapidly renewed from the bone marrow. Proc Natl Acad Sci USA 87:4781-4784
4. Freitas A, Lembezat M-P, Coutinho A (1989) Expression of antibody V-regions is genetically and developmentally controlled and modulated by the B lymphocyte environment. Int Immunol 1:342-354
5. Gu H, Tarlinton D, Miller W, Rajewsky K, Forster I (1991) Most peripheral cells in mice are ligand selected. J Exp Med 173:1357-1371
6. Freitas AA, Viale A-C, Sundblad A, Huesser C, Coutinho A (1991) Normal serum immunoglobulins participate in the selection of peripheral B cell repertoires. Proc Natl Acad Sci USA 88:5640-5644
7. Kenny JJ, O'Connell C, Sieckmann DG, Fischer RT, Longo DL (1991) Selection of antigen-specific, idiotype-positive B cells in transgenic mice expressing a rearranged M167-μ heavy chain gene. J Exp Med 174:1189-1201
8. Andrade L, Huetz F, Poncet P, Thomas-Vaslin V, Goodhardt M, Coutinho A (1991) Biased V_H gene expression in murine CD5 B cells results from age-dependent cellular selection. Eur J Immunol 21:2017-2023

9. Arnold LW, Spencer DH, Clarke SH, Haughton G (1993) Mechanisms that limit the diversity of antibody: three sequentially acting mechanisms that favor the spontaneous production of germline encoded anti-phosphatidyl choline. Int Immunol 5:1365-1373
10. Clarke SH, McCray SK (1993) V_H CDR3-dependent positive selection of murine V_H12-expressing B cells in the neonate. Eur J Immunol 23:3327-3334
11. Arnold LW, Pennell CA, McCray SK, Clarke SH (1994) Development of B-1 cells: Segregation of phosphatidyl choline-specific B cells to the B-1 population occurs after immunoglobulin gene expression. J Exp Med 179:1585-1595
12. Okamoto M, Murakami M, Shimizu A, Ozaki S, Tsubata T, Kumagai S-I, Honjo T (1992) A transgenic model of autoimmune hemolytic anemia. J Exp Med 175:71-79
13. Murakami M, Tsubata T, Okamoto M, Shimizu A, Kumagai S, Imura H, Honjo T (1992) Antigen-induced apoptotic death of Ly-1 B cells responsible for autoimmune disease in transgenic mice. Nature 357:77-80
14. Kenny JJ, Stall AM, Sieckmann DG, Lamers MC, Finkelman FD, Finch L, Longo DL (1991) Receptor-mediated elimination of phosphocholine-specific B cells in x-linked immune deficient mice. J Immunol 146:2568-2577
15. Hyakawa K, Carmack CE, Shinton SA, Hardy RR (1992) Selection of autoantibody specificities in the Ly-1 B subset. Ann N Y Acad Sci 651:346-353
16. Nisitani S, Tsubata T, Murakami M, Okamoto M, Honjo T (1993) The bcl-2 gene product inhibits clonal deletion of self-reactive B lymphocytes in the periphery but not in the bone marrow. J Exp Med 178:1247-1254
17. Masmoudi H, Mota-Santos T, Huetz F, Coutinho A, Cazenave P-A (1990) All T15 antibodies (but not the majority of V_HT15+ antibodies) are produced by peritoneal CD5+ B lymphocytes. Int Immunol 2:515-520
18. Kenny JJ, Moratz CM, Guelde G, O'Connell CD, George J, Dell C, Penner SJ, Weber JS, Berry J, Claflin JL, Longo DL (1992) Antigen binding and idiotype analysis of antibodies obtained following electroporation of heavy and light chain genes encoding phosphocholine-specific antibodies: a model for T15-idiotype dominance. J Exp Med 176:1637-1643
19. Sieckmann DG, Holmes K, Hornbeck P, Martin E, Guelde G, Bondada S, Longo DL, Kenny JJ (1994) B cells from M167 $\mu\kappa$ transgenic mice fail to proliferate after stimulation with soluble anti-Ig antibodies: a model for antigen-induced B cell anergy. J Immunol 152:4873-4883
20. Hornbeck PV, Donald SP, Kenny JJ (1992) Negative signalling through the B-cell antigen receptor. In Progress in Immunology Vol. VIII, Proceedings of the Eighth International Congress of Immunology, Budapest, J. Gergely, ed. Springer-Verlag, New York, p 358
21. Kenny JJ, Sieckmann DG, Freter C, Hodes R, Hathcock K, Longo DL (1992) Modulation of signal transduction in phosphocholine-specific B cells from $\mu\kappa$ transgenic mice. Curr Top Microbiol Immunol 182:95-103
22. Korsmeyer SJ (1992) Bcl-2:a repressor of lymphocyte death. Immunol Today 13:285-288
23. Siegel RM, Katsumata M, Miyashita T, Louie DC, Greene MI, Reed JC (1992) Inhibition of thymocyte apoptosis and negative antigenic selection in bcl-2 transgenic mice. Proc Natl Acad Sci USA 89:7003-7007
24. Strasser A, Harris AW, Cory S (1991) bcl-2 transgene inhibits T cell death and perturbs thymic self-censorship. Cell 29:889-899
25. Sentman CL, Shutter JR, Hockenbery D, Kanagawa O, Korsmeyer SJ (1991) bcl-2 inhibits multiple forms of apoptosis but not negative selection in thymocytes. Cell 67:879-888

26. Strasser A, Whittingham S, Vaux DL, Bath ML, Adams JM, Cory S, Harris AW (1991) Enforced bcl2 expression in B-lymphoid cells prolongs antibody responses and elicits autoimmune disease. Proc Natl Acad Sci USA 88:8661-8665
27. Katsumata M, Siegel RM, Louie DC, Miyashita T, Tsujimoto Y, Nowell PC, Greene MI, Reed JC (1992) Differential effects of Bcl-2 on T and B cells in transgenic mice. Proc Natl Acad Sci USA 89:11376-11380
28. McDonnell TJ, Deane N, Platt FM, Nunez G, Jaeger U, McKearn JP, Korsmeyer SJ (1989) Bcl-2-immunoglobulin transgenic mice demonstrate extended B cell survival and follicular lymphoproliferation. Cell 57:79-88
29. Li Y-S, Hayakawa K, Hardy RR (1993) The regulated expression of B lineage associated genes during B cell differentiation in bone marrow and fetal liver. J Exp Med 178:951-960
30. Karasuyama H, Rolink A, Shinkai Y, Young F, Alt FW, Melchers F (1994) The expression of V_{preB}/y5 surrogate light chain in early bone marrow precursor B cells of normal and B cell-deficient mutant mice. Cell 77:133-143
31. Gu H, Kitamura D, Rajewsky K (1991) B cell development regulated by gene rearrangement: arrest of maturation by membrane-bound Du protein and selection of Dh element reading frames. Cell 65:47-54
32. Haasner D, Rolink A, Melchers F (1994) Influence of surrogate L chain on D_HJ_H-reading frame 2 suppression in mouse precursor B cells. Int Immunol 6:21-30
33. Webb CF, Nakai C, Tucker PW (1989) Immunoglobulin receptor signaling depends on the carboxyl terminus but not the heavy-chain class. Proc Natl Acad Sci USA 86:1977-1981
34. Parikh VS, Nakai C, Yokota SJ, Bankert RB, Tucker PW (1991) COOH terminus of membrane IgM is essential for an antigen-specific induction of some but not all early activation events in mature B cells. J Exp Med 174:1103-1109
35. Parikh VS, Bishop GA, Liu K-J, Do BT, Ghosh MR, Kim BS, Tucker PW (1992) Differential structure-function requirements of the transmembranal domain of the B cell antigen receptor. J Exp Med 176:1025-1031
36. Kitamura D, Roes J, Kuhn R, Rajewsky K (1991) A B-cell-deficient mouse by targeted disruption of the membrane exon of the immunoglobulin μ chain gene. Nature 350:423-426
37. Storb U, Pinkert C, Arp B, Engler P, Gollahon K, Manz J, Brady W, Brinster R (1986) Transgenic mice with μ and κ genes encoding anti-phosphorylcholine antibodies. J Exp Med 164:627-641

Growth Regulation - C-Myc

Role of the Rel-Family of Transcription Factors in the Regulation of c-myc Gene Transcription and Apoptosis of WEHI 231 Murine B-Cells

Hayyoung Lee, Min Wu, Francis A. La Rosa, Mabel P. Duyao, Alan J. Buckler, and Gail E. Sonenshein

Department of Biochemistry, Boston University Medical School, Boston, MA 02118

Introduction

NF-kB elements mediate control of a number of important growth regulatory genes (rev. in Baeuerle, 1991). We have identified two functional NF-kB elements within the murine c-*myc* gene. The upstream regulatory element or URE is located -1101 to -1091 b.p. upstream of the P1 promoter (Duyao et al. 1990). The internal regulatory element or IRE is at +440 to +459 b.p., within exon 1 (Kessler et al. 1992b). Activation of NF-kB by IL-1 in human dermal FS-4 fibroblasts (Kessler et al. 1992a) or by the *tax* protein of the HTLV-1 virus in T lymphocytes (Duyao et al. 1992) leads to induction of transcription of the c-*myc* promoter through binding to the URE and IRE. NF-kB is now known to be a family of dimeric transcription factors, whose subunits all contain an amino terminal *rel* homology domain, which mediates both dimerization and DNA binding. Classical NF-kB is a heterodimer composed of a p50 and a p65 subunit (Ghosh et al. 1990; Ruben et al. 1991). The c-*rel* gene encodes a 75 kDa protein (Simek and Rice 1988). Various combinations of these Rel family of proteins bind as dimers to NF-kB elements, but their activity has been found to be element specific.

The murine WEHI 231 line (IgM, kappa) had been characterized as an immature B cell on the basis of surface markers and biological properties (Ralph, 1979; Boyd and Schrader, 1981), and is a model system for study of tolerance. Treatment of WEHI 231 cells with an antibody against its surface IgM for 18 hours results in formation of an oligosomal DNA pattern, characteristic of apoptotic fragmentation of chromatin within nuclei (Benhamou et al. 1990). Dramatic changes in expression of the c-*myc* oncogene precede DNA degradation. A 5-to 15-fold increase in c-*myc* RNA levels and rate of transcription between the first and second hours is followed by a rapid decline; by 8 hrs of treatment the levels are well below control values (Levine et al. 1986). Over- or inappropriate time of expression of the c-*myc* gene promotes apoptosis; for example, an antisense c-*myc* oligonucleotide inhibited anti-Ig induced apoptosis of WEHI 231 cells (Fischer et al. 1994). Given our evidence for the role of Rel-related factors in the control of c-*myc* gene transcription, we determined the ability of various members of the Rel-family to transactivate the c-*myc* promoter and characterized their role in control of c-*myc* gene transcription and apoptosis of WEHI 231 cells.

Results

Classical NF-kB but not Murine c-Rel Potently Activates Transcription of the c-*myc* Promoter

The murine c-*myc* promoter CAT construct p1.6 Bgl contains 1.2 k.b.p. of upstream and 0.4 k.b.p. of exon 1 sequences, including both the URE and IRE (Duyao et al. 1992). The p1.6 Bgl construct was co-transfected into N.I.H. 3T3 fibroblasts with increasing amounts of vectors expressing the active regions of either the p50 or p65 subunit (La Rosa et al. 1994) alone or in combination (Fig. 1). Increasing amounts of the p50 and p65 expression vectors stimulated CAT activity up to 7.4 fold. Co-transfection with the p65 expression vector alone was sufficient to stimulate CAT activity approximately 3.0 to 6.2 fold. In contrast, expression of the p50 subunit had no detectable effect on CAT activity. Mutation at both the URE and IRE of p1.6 Bgl, converting the internal two G residues involved in protein-DNA interaction (Duyao et al. 1990; Kessler et al. 1992b) to C residues, totally inhibited activation by p65 (data not shown). Thus classical NF-kB can transactivate the c-*myc* promoter directly by binding to the IRE and URE elements.

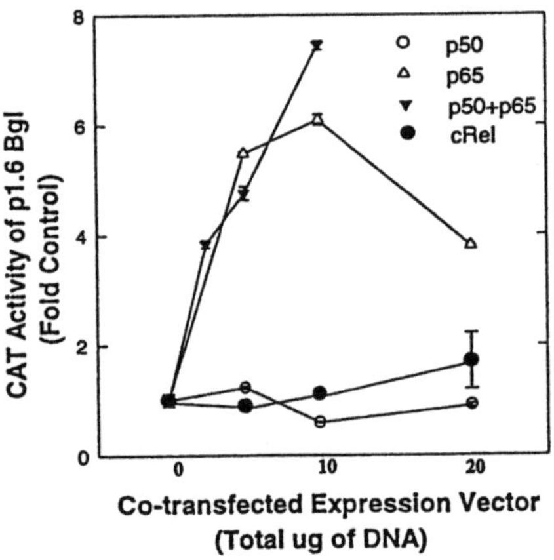

Fig. 1. Effects of p50, p65 and murine c-Rel expression on c-*myc* promoter activity. NIH 3T3 cells were co-transfected via $CaPO_4$ precipitation with 20 ug of p1.6 Bgl and the indicated amounts of Rel-family expression vectors, and parental vector to make up a total of 40 ug DNA.

To measure the effect of murine c-Rel, cultures of N.I.H. 3T3 cells were co-transfected with p1.6Bgl and a vector expressing murine c-Rel. No significant effect was seen with 10 ug of c-*rel* expression vector and only a very modest induction (1.7 +/- 0.4) was seen with 20 ug of vector DNA (Fig. 1). Thus c-Rel, which was found to bind to both the URE and IRE (La Rosa et al. 1994), can only modestly affect c-*myc* gene transcription.

Anti-Ig Treatment Induces Classical NF-kB

Nuclear extracts were prepared from WEHI 231 cells in exponential growth or following 1, 2, 8 and 12 hours of anti-Ig treatment. Mobility shift analysis was performed with the 221 b.p. Bgl II-Acc I fragment that spans -1139 to -921 b.p. relative to the P1 promoter (Fig. 2A). The profile obtained with nuclear extracts from exponentially growing cells (E) is similar to that seen previously (Duyao et al. 1990). Footprinting and competition analysis has shown that formation of the complexes in bands numbered 1 through 6 are due to interaction with the URE element (Duyao et al. 1990). After 1 h of anti-Ig treatment, there was a very significant increase in binding. In particular, a transient increase in intensity of complexes in the region of band 3 was noted, as well as an overall increase in background intensity between bands 2 and 5 (Fig. 2A). The increase in binding was diminished by 2 hours. By 8 hours, the levels of the all of the complexes except for band 2 were below those seen in extracts from control cells. By 12 hours of treatment, there was a general reduction in formation of the larger

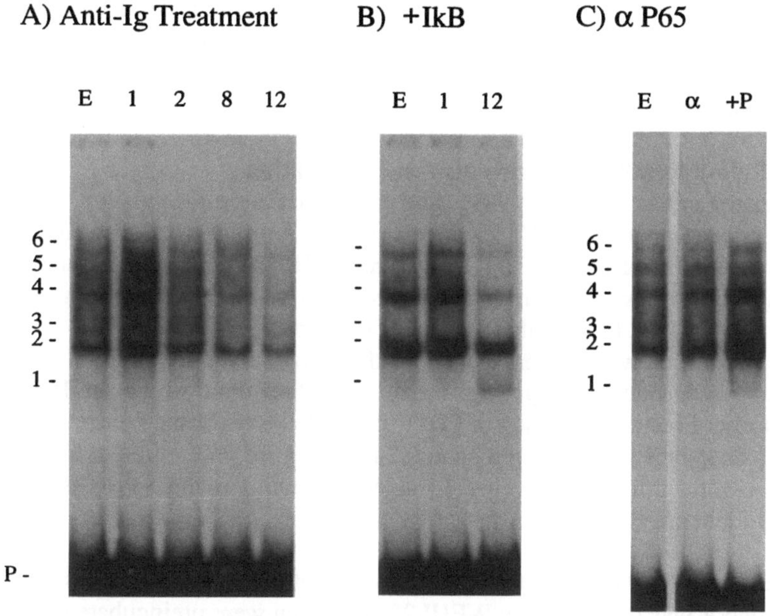

Fig. 2. Anti-Ig induces expression of p65-containing complexes. A) Time course of changes in binding to the URE element; B) Effects of IkB-α; C) Effects of an anti-p65 antibody.

complexes and an increase in formation of band 1. Thus an early transient increase in binding of the NF-kB-family within 1 hour follows anti-Ig treatment and parallels changes seen in the rate of c-*myc* gene transcription (Levine et al. 1986). In particular, binding of band 3 increases significantly.
Previously, we noted that band 3 co-migrated with the major complex induced upon 70Z/3 pre-B to B cell differentiation (Duyao et al. 1990), suggesting this band represents binding of classical NF-kB. Addition of a p65 antibody to binding reactions with an extract from cells in exponential growth resulted in a selective reduction in formation of band 3 (Fig. 2C). Incubation of the p65 antiserum preparation in the presence of its cognate peptide prevented the inhibition of binding, confirming the specificity of the reaction. Addition of low concentrations of purified IkB-α, which selectively competes for interaction with p65 preventing binding, blocked band 3 formation with E, 1 and 12 hour extracts (Fig. 2B), indicating the band 3 complexes contain p65. With the extracts prepared 1 hour post anti-Ig treatment, addition of IkB-α was further found to significantly reduce formation of the increased background binding between bands 2 and 5 (Fig. 2B).

Additional antibody analysis was performed with antibodies that recognized p50 and c-Rel subunits in either homodimer or heterodimer complexes (data not shown). The major complex band 2 represents a heterodimer of p50/c-Rel. Only a slight increase and decrease in expression of this species was detected upon anti-Ig treatment. Band 1, which is induced 8 to 12 hours following anti-Ig treatment is a homodimer of p50. Binding of this species to NF-kB URE element would fail to transactivate the c-*myc* promoter. Most importantly, band 3 represents binding of p50/p65 classical NF-kB. Anti-Ig treatment resulted in a rapid transient activation of its binding, which peaked at 1 hour, consistent with the increase in c-*myc* gene transcription.

Inhibition of NF-kB Induction Prevents c-*myc* Activation
Schreck, Baeuerle and coworkers (1992) have shown that the anti-oxidant pyrrolidinedithiocarbamate (PDTC) selectively inhibits activation of Rel-related factors. To measure the effects of PDTC addition, WEHI 231 cells were incubated for 1 hour in media containing increasing concentrations of PDTC, ranging from 0.01 mM to 1.0 mM, and then treated for 1 hour with anti-Ig. Mobility shift analysis was performed with nuclear extracts and an oligonucleotide probe containing the sequence of the kappa light chain NF-kB element, which similarly binds this family of factors. Addition of PDTC prevented Rel-related factor induction in a dose-dependent fashion up to 0.2 mM (Fig. 3). Activation of all of the complexes were similarly prevented as judged by binding to the c-*myc* 221 b.p. fragment (data not shown).

The effects of PDTC were tested on induction of c-*myc* mRNA levels. Cytoplasmic RNA was isolated from WEHI 231 cells that were preincubated in the absence or presence of 0.2 mM PDTC for 1 hour and then treated with anti-Ig for an additional 1 hour (1+1h PDTC). As seen in the resulting Northern blot (Fig.

4), anti-Ig treatment of WEHI 231 caused a significant increase in c-*myc* mRNA levels, as we have reported previously (Levine et al. 1986), that was inhibited upon pre-incubation with PDTC. As control, RNA was isolated from similarly pre-treated WEHI 231 cells (1h PDTC) or following an additional 1hour without

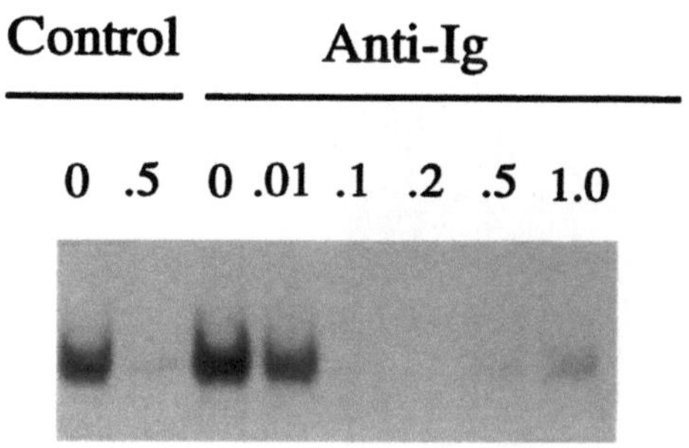

Fig. 3. The anti-oxidant pyrrolidinedithiocarbamate blocks activation of Rel binding.

subsequent anti-Ig treatment (1+1h PDTC). Only a slight effect was seen on the basal expression of c-*myc* mRNA. This result suggest that binding of the family of Rel-related factors plays only a small role in the constitutive level of c-*myc* transcription, which is consistent with identification of p50/c-Rel, a weak transactivator of the c-*myc* promoter, as the major binding complex in exponentially growing cells. To confirm the specificity of the effects on c-*myc* expression and to evaluate the early effects of PDTC on cell growth, the Northern

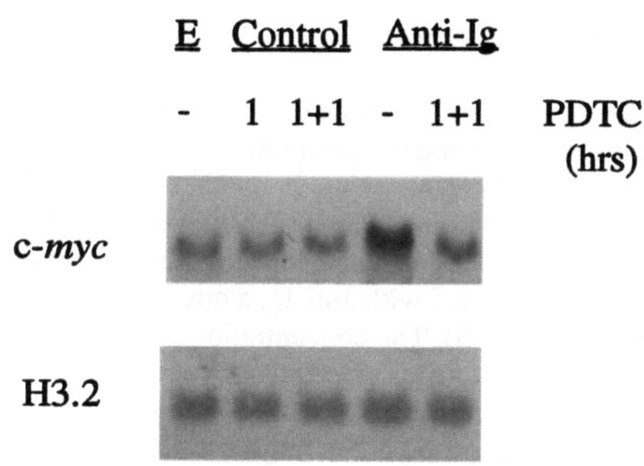

Fig. 4. PDTC blocks induction of c-*myc* mRNA levels following anti-Ig treatment.

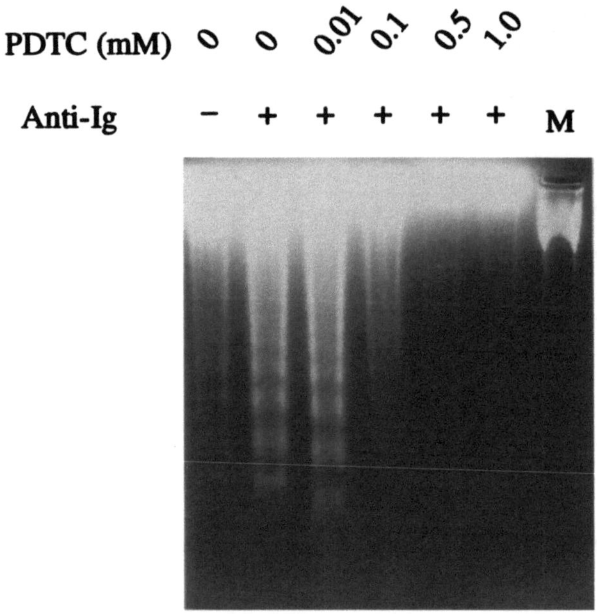

Fig. 5. PDTC inhibits anti-Ig induced apoptosis of WEHI 231 cells.

blot was probed for histone H3.2 mRNA, an S-phased expressed RNA. No changes in expression of this mRNA resulted from the PDTC treatment. Thus PDTC, a selective inhibitor of Rel-related factor activation (Schreck et al. 1992), blocks the increase in c-*myc* mRNA expression in WEHI 231 cells induced by anti-Ig treatment.

The Anti-Oxidant PDTC Prevents Apoptosis

Since expression of c-*myc* has been implicated in induction of apoptosis, the effects of PDTC treatment on apoptosis were measured. We first sought to determine the kinetics of DNA damage under our conditions. Degradation of the DNA was detectable between 8 and 12 h following incubation of WEHI 231 cells with anti-Ig (data not shown), which is somewhat earlier than the 18-24 h times reported previously (Benhamou et al. 1990). When cells that had been treated with PDTC for 1 hour were similarly incubated with anti-Ig, a dose dependent inhibition of apoptosis was noted (Fig. 5). The concentration that effectively blocked apoptosis was the same as that which affected activation of Rel-related factors (Fig. 3). Similar testing of a second anti-oxidant N-acetylcysteine (NAC) demonstrated partial abrogation of apoptotic DNA fragmentation (data not shown). Thus treatment with anti-oxidants known to block activation of Rel-related factors prevents anti-Ig mediated apoptosis of WEHI 231 cells.

Discussion

Here we have identified multiple nuclear complexes containing c-Rel, p50 and p65 *rel*-related subunits in extracts of the murine B cell lymphoma line WEHI 231 (IgM, kappa), and have implicated the changes in their expression in the transient activation of c-*myc* gene transcription and apoptosis induced by anti-Ig treatment of these cells. The predominant complex binding to the c-*myc* URE NF-kB element in extracts of exponential growing cells was surprisingly a p50/c-Rel heterodimer (band 2), which we found to be a weak trans-activator of the c-*myc* promoter. Treatment of WEHI 231 cells with anti-Ig resulted in a transient increase in binding activity of p65-containing complexes, including classical NF-kB (p50/p65) activity (band 3), a potent transactivator of the c-*myc* promoter. Pre-treatment of the cells with the anti-oxidant PDTC totally prevented induction of Rel-factor binding and of c-*myc* mRNA levels following anti-Ig treatment. After 4 hours of treatment, Rel-related binding complex formation and the rate c-*myc* gene transcription were slightly below levels in exponential growth (Levine et al. 1986, and data not shown). These findings suggest that an increase and decrease in competition for binding of induced p65-containing complexes, such as classical NF-kB, for less active Rel-containing complexes upon anti-Ig treatment contribute to the transient increase in the rate of c-*myc* transcription induced by anti-Ig treatment of WEHI 231 cells.

Pre-treatment with PDTC also prevented apoptosis, as judged by the absence of the normal ladder of degraded DNA within 13 hours of anti-Ig treatment and cell morphology (data not shown). These results suggest a possible role for NF-kB in activation of the pathway leading to programmed cell death in WEHI 231 cells. High levels of c-Rel expression have been associated with programmed cell death in the devloping avian embryo and in bone marrow cells in vitro (Abbadie et al. 1993). In WEHI 231 cells, these effects could be mediated via regulation of changes in c-*myc* expression. Overexpression of c-*myc* accelerated apoptosis following IL-3 deprivation of the 32D IL-3 dependent myeloid cell line, and upon growth arrest either by serum deprivation, isoleucine deprivation, or thymidine block of 3T3 fibroblasts (Askew et al. 1991; Evan et al. 1992). Addition of an antisense c-*myc* oligonucleotide has been shown to inhibit receptor mediated apoptosis in immature T cells and some T cell hybridomas (Shi et al. 1992), and more recently the anti-Ig induced apoptosis in WEHI 231 cells (Fischer et al. 1994). Thus inhibition of the normal activation of Rel-related factors and therefore induction of c-*myc* could account for the affects of PDTC. Alternatively, other genes regulated by the NF-kB family or effects of the anti-oxidant PDTC on other genes or signalling pathways could be responsible for the inhibition of apoptosis. This inhibition, however, is not likely to be mediated through affects on the expression of the gene product of bcl-2, which has been found to have anti-oxidant activity (Hockenberry et al. 1993), since overexpression of bcl-2 failed to prevent anti-Ig induced apoptosis of WEHI 231 cells (Cuende et al. 1993).

Acknowledgments

We thank Drs. U. Siebenlist, J. Cleveland, T. Gilmore, S. Ghosh and I. Verma for generously providing constructs, and T. Rothstein for the NF-kB oligonucleotide. This work was supported by N.I.H. grant CA36355.

References

Abbadie C, Kabrun N, Bouali F, Vandenbunder B and Enrietto P (1993) High levels of c-*rel* respression are associated with programmed cell death in the developing avian embryo and in bone marrow cells in vitro. Cell 75:899-912

Askew DS, Ashmun R, Simmons B and Cleveland J (1991) Constitutive c-*myc* expression in IL-3-dependent myeloid cell line suppresses cycle arrest and accelerates apoptosis. Oncogene 6:1915-1922

Baeuerle PA (1991) The inducible transcription activator NF-kB: regulation by distinct protein subunits. Biochem Biophys Acta 1072:63-73

Benhamou L, Cazenave P and Sarthou P (1990) Anti-immunoglobulins induce death by apoptosis in WEHI-231 B lymphoma cells. Eur J Immunol 20: 1405-1407

Boyd A and Schrader J (1981) The regulation of growth and differentiation of a murine B-cell lymphoma. II. The inhibition of WEHI 231 by anti-immunoglobulin antibodies. J Immunol 126:2466-2469

Cuende E, Alez-Martinez J, Ding L, Gonzalez-Garcia M, Martinez-A C and Nunez G (1993) Programmed cell death by *bcl*-2-dependent and independent mechanisms in B lymphoma cells. EMBO J 12:1555-1560

Duyao MP, Buckler AJ and Sonenshein GE (1990) Interaction of an NF-kB-like factor with a site upstream of the c-*myc* promoter. Proc Natl Acad Sci USA 87:4727-4731

Duyao MP, Kessler DJ, Spicer D, Bartholomew C, Cleveland J, Siekevitz M and Sonenshein GE (1992) Transactivation of the c-*myc* promoter by human T cell leukemia virus type I *tax* is mediated by NF-kB. J Biol Chem 267:16288-16291

Evan G, Wyllie A, Gilbert C, Littlewood T, Land H, Brooks M, Waters C, Penn L and Hancock D (1992) Induction of apoptosis in fibroblasts by c-*myc* protein. Cell 69:119-128.

Fischer G, Kent S, Joseph L, Green D and Scott D (1994) Lymphoma models for B cell activation and tolerance. X. anti-u-mediated growth arrest and apoptosis of murine B cell lymphomas is prevented by the stabilization of myc. J Exp Med 179:221-228

Ghosh S, Gifford AM, Riviere L, Tempst P, Nolan G and Baltimore D (1990) Cloning of the p50 DNA binding subunit of NF-kB: homology to *rel* and *dorsal*. Cell 62:1019-1029

Hockenberry D, Oltvai Z, Yin X-M, Milliman C and Korsmeyer S (1993) Bcl-2 functions in an antioxidant pathway to prevent apoptosis. Cell 75:241-251

Kessler DJ, Duyao MP, Spicer, DB and Sonenshein GE (1992a) NF-kB-like factors mediate interleukin 1 induction of c-*myc* gene transcription in fibroblasts. J Exp Med 176:787-792

Kessler, DJ, Spicer D, La Rosa, F and Sonenshein GE (1992b) A novel NF-kB element within exon 1 of the murine c-*myc* gene. Oncogene 7:2447-2453

La Rosa, FA, Pierce, JW and Sonenshein GE (1994) Differential regulation of the c-*myc* oncogene promoter by the NF-kB family of transcription factors. Mol. Cell. Biol. 14:1039-1044

Levine RA, McCormack JE, Buckler AJ and Sonenshein GE (1986) Transcription and posttranscriptional control of c-myc gene expression in WEHI 231 cells. Mol Cell Biol 6:4112-4116

Ralph P (1979) Functional subsets of murine and human B lymphocyte cell lines. Immunological Rev 48:107-121

Ruben SM, Dillon P, Schreck R, Henkel T, Chen C-H, Maher M, Baeuerle PA and Rosen C (1991) Isolation of a *rel*-related human cDNA that potentially encodes the 65 kD subunit of NF-kB. Science 251:490-493

Schreck R, Meier B, Mannel D, Droge W and Baeuerle P (1992) Dithiocarbamates as potent inhibitors of nuclear factor kB activation in intact cells. J Exp Med 175:1181-1194

Shi Y, Glynn J, Guilbert L, Cotter T, Bissonnette R and Green D (1992) Role for c-*myc* in activation induced apoptotic cell death in T-cell hybridoma. Science 257:212-215

Simek S and Rice N (1988) p59$^{v\text{-}rel}$, the transforming protein of reticulendotheliosis virus, is complexed with at least four other proteins in transformed chicken lymphoid cells. Oncogene Res 2:103-119

c-myc Transcription is Initiated from Po in 70 % of Patients with Multiple Myeloma

R. G. Hoover, V. Kaushal, C. Lary, P. Travis and T. Sneed

Departments of Pathology and Medicine (Hematology/Oncology), Arkansas Cancer Research Center, Little Rock, Arkansas 72205

Introduction

The Role of c-myc in Multiple Myeloma

The role of the c-myc gene is well established in Burkitt's lymphoma and in murine plasmacytomas [1]. In these neoplasms the c-myc locus is involved in a reciprocal chromosomal translocation with the immunoglobulin heavy chain or light chain loci, resulting in inappropriate activation of c-myc transcription. In patients with multiple myeloma, however, the role of the c-myc gene is less clear. Less than 1% of patients with multiple myeloma have abnormalities of chromosome 8, and t(8:14) are rare [2]. At the structural level, abnormalities are seen in less than 5% of patients [2]. However, approximately 85-90% of patients show levels of c-myc mRNA and protein, in tumor cells, that are comparable to actively proliferating cells, suggesting a possible role for c-myc in this neoplasm [3]. A number of oncogenes have been shown to be abnormal in multiple myeloma [4]. Some of these genes (eg. ras) act at a proximal level in the signal transduction apparatus, while others subserve regulatory functions (eg. p53, Rb). Thus, although c-myc may be structurally normal in myeloma, its transcription may be inappropriately activated or regulated by mutations in other genes.

Regulation of c-myc transcription

The regulation of c-myc transcription is extremely complicated [5]. The gene is composed of three exons (Figure 1). The first exon is untranslated and contains the two principal promoters, termed P1 and P2. Initiation from these promoters accounts for approximately 85-100% of transcripts in most cells. An additional promoter, termed Po, is located approximately 600bp 5' to P1 and accounts for approximately 5% of transcripts in some cell lines. In addition, a fourth promoter, termed P3, is located in the first intron and accounts for approximately 5% of transcripts in selected tumor lines. Finally, an antisense promoter has been described in the second intron [6]. P1 and P2 produce transcripts of 2.4Kb and 2.2Kb respectively (Figure 1). Each of these transcripts contains a single open reading frame (ORF). Translation of this ORF from two different start sites results in the typically abundant 64Kd and 67Kd c-myc proteins [5] (Figure 1).

Interestingly, transcripts initiated from Po contain two additional ORF that have been shown to produce proteins of 58Kd and 15Kd in some instances [7]. The structural properties of these two additional proteins have not been extensively studied and their role in oncogenesis is unknown.

Multiple, overlapping, positive and negative, cis-acting, regulatory elements have been described 5' of the first exon. In addition, a number of known DNA binding proteins have been shown to bind to upstream elements and regulate c-myc transcription under a variety of circumstances [5].

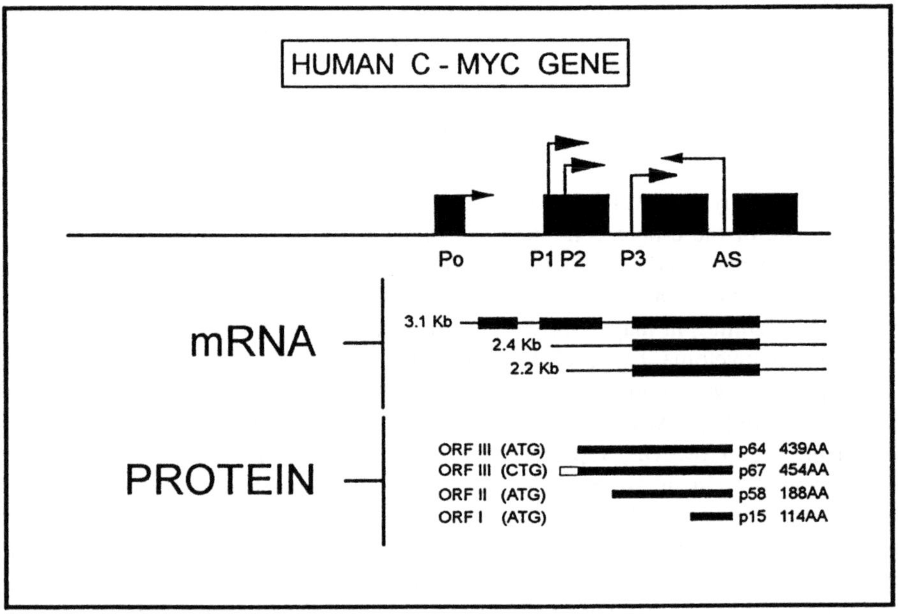

Figure 1. Promoters, transcripts, and proteins of the human c-myc gene.

C-myc is a cytokine responsive gene

C-myc transcription can be stimulated by a number of cytokines including IL2, IL3 and IL6 [8-10]. Interestingly, stimulation by these cytokines appears to be promoter selective. For example, IL2 has been shown to selectively stimulate transcription from the P2 promoter in proliferating human T lymphocytes [9]. Collectively, studies examining the stimulation of c-myc transcription by cytokines have suggested multiple regulatory pathways influencing c-myc transcription and the likely existence of multiple, separate cytokine response elements.

C-myc transcription initiation in multiple myeloma cell lines

Ten human myeloma cell lines were examined (Table I). C-myc transcription was assessed by Northern blotting and S1 nuclease protection analysis. Seven of the ten cell lines analyzed expressed c-myc mRNA constitutively. Three of ten showed no detectable levels of c-myc mRNA in an unstimulated state. Of the seven lines that expressed c-myc, six lines transcribed c-myc from the Po promoter exclusively (Figure 2). One cell line, ARH-77 transcribed c-myc from both P1 and P2, but not from Po.

Table I

CELL LINE	PROMOTER		
	Po	P1	P2
ARH - 77	−	+	+
COL	+	−	−
MER	+	−	−
LER	+	−	−
CLO	+	−	−
ARK	+	−	−
ARD	+	−	−
LES	−	−	−
U266	−	−	−
8226	−	−	−

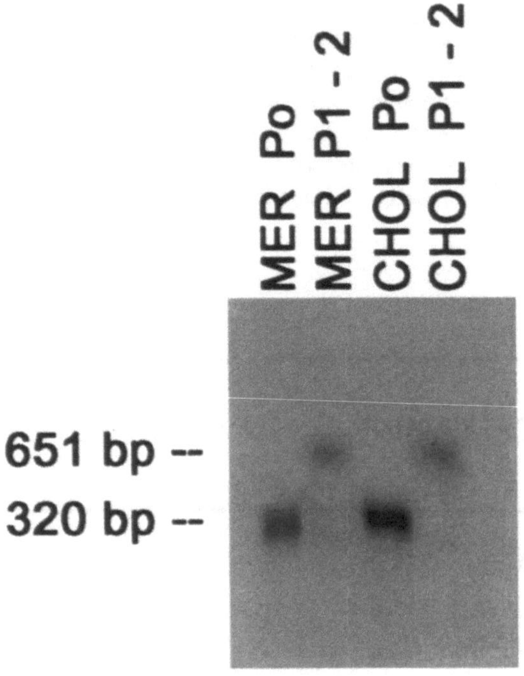

Figure 2: C-myc is transcribed exclusively from Po in some human myeloma cell lines. The cells lines COL and MER were analyzed for c-myc transcription by S1 nuclease protection. Two probes were used. With the Po probe a band in the range of 256 - 385bp indicates Po transcription. With the P1P2 probe a band of 651bp indicates Po transcription while bands of 579bp and/or 399bp would indicate transcription from P1 and/or P2 respectively.

Po Transcription is Stimulated by IL2 and IL6

C-myc is transcribed from P1 and P2 in the ARH-77 cell line. If these cells are washed and recultured at low density c-myc transcription decreases dramatically (Figure 3). If the cells are then stimulated for 6 hours with 100U/ml/10^6 cells recombinant IL2 or IL6, c-myc transcription is stimulated 40 - 50 fold (Figure 3). Interestingly, if ARH-77 cells are stimulated with both IL2 and IL6, c-myc transcription initiates from Po as well as P1 and P2 (Figure 4). This finding suggests that the Po promoter is cytokine responsive and regulated independently of the P1 and P2 promoters.

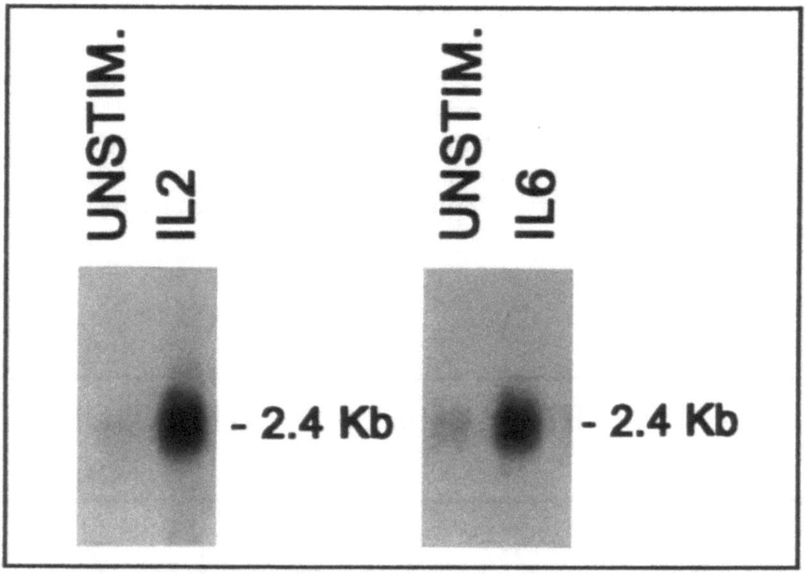

Figure 3: IL2 or IL6 stimulate c-myc transcription in ARH-77 cells. ARH-77 cells were stimulated with 100U IL2 or IL6 for 6 hours. C-myc mRNA levels were assessed by Northern blotting.

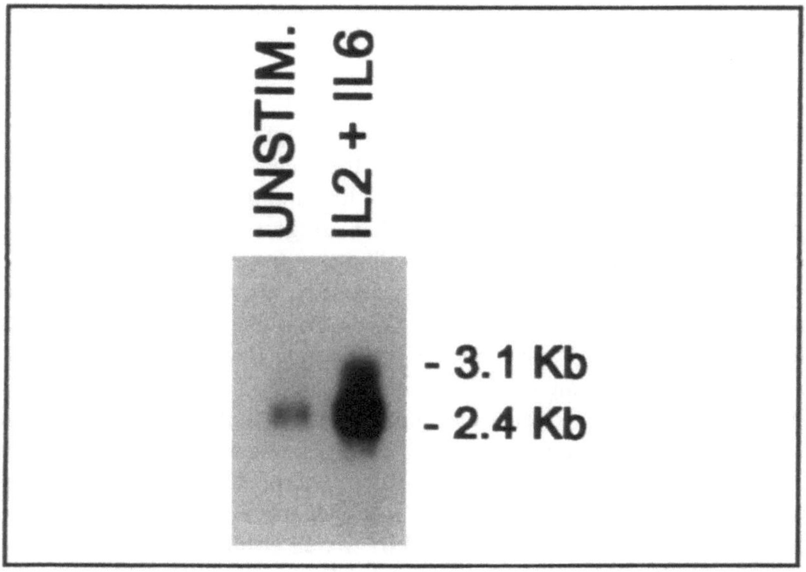

Figure 4: IL2 and IL6 stimulate c-myc transcription from Po as well as P1 and P2 in ARH-77 cells. ARH-77 cells were treated with 100 U IL2 and IL6 for 6 hours. C-myc mRNA levels were assessed by Northern blotting.

Primary Human Myeloma Cells Transcribe c-myc from Po

Bone marrow derived tumor cells from 37 patients with multiple myeloma were analyzed for the expression of c-myc by Northern blotting. 26/37 patients examined showed c-myc transcripts of 3.1Kb in size consistent with transcription initiation from Po (Table II). Tumor cells from 7 patients did not show c-myc transcription in the unstimulated state and 4 patients showed c-myc transcripts of 2.2 - 2.4 Kb in size. Thus, approximately 70% of patients with multiple myeloma transcribe c-myc from the Po promoter.

Table II

c - MYC TRANSCRIPTION INITIATION			
	Po	P1/P2	NONE
PATIENTS N = 37	26	4	7

Discussion

Involvement of the c-myc gene in human multiple myeloma is distinctly unlike that in other B cell neoplasms, in that, the gene is usually not involved in translocations with the immunoglobulin gene loci and gross structural abnormalities are not common [2]. Recent evidence has suggested that point mutations are lacking in the portions of the genes encoding the protein, however, it is currently unknown whether transcription regulatory elements are mutated [11]. Regardless, it has been shown that myeloma tumor cells from the majority of patients with myeloma show levels of c-myc mRNA and protein that are comparable to those seen in actively proliferating cells [3].

Mutations of other oncogenes have been shown in human myeloma with varying prevalence [4]. Genes such as ras, p53, Rb, BCL-1 and BCL-2 have been shown to be mutated, however, their role in the pathogenesis of this disorder is unclear. It is possible that there is no one single pathogenetic event that causes myeloma. However, since c-myc is transcribed and translated at high levels in the majority of patients, it is likely that this gene plays a significant role in this disease, regardless of the mechanism of gene dysregulation.

Multiple myeloma is a cytokine responsive neoplasm [12]. IL6 appears to play a pivotal role in the proliferation and/or differentiation of this tumor. However, other cytokines likely play significant roles as well [12]. The c-myc gene is also a cytokine responsive gene. IL2, IL3 and IL6 as well as other cytokines have been shown to stimulate the transcription of this gene [8-10]. Interestingly, these cytokines appear to have a differential effect on promoter selection by the c-myc gene, suggesting the possibility of separate cytokine regulatory elements within the c-myc locus [8-10]. For the first time, we have shown that the combination of IL2 and IL6 can selectively stimulate transcription from the seldom used Po promoter. The production of this elongated transcript is provocative in that two additional open reading frames are present in this molecule. At this time, it is unknown whether these additional ORF are translated in myeloma cells or what role, if any, these protein products might play in the pathogenesis of this disease.

In this manuscript, we have presented data showing that tumor cells from 70% of patients transcribe the c-myc gene exclusively from Po. The following are offered as possible explanations for this observation.

1) The structure of the c-myc gene in myeloma is entirely normal and Po transcription is driven by exogenous cytokines (ie. IL6 and IL2).
2) Point mutations effecting transcriptional regulatory elements are present and abrogate transcription from P1 and P2, thus permitting transcription from Po.
3) A combination of #1 and #2.
4) The primary lesion occurs in a gene proximal to c-myc and permits transcription from Po and not from other c-myc promoters.

At this time, these hypotheses are mere conjecture. Additional studies need to be performed in order to ascertain the mechanism of our observations.

REFERENCES

1. Spencer, C.A., and M. Groudine (1991) Control of c-myc gene regulation in normal and neoplastic cells. Adv. Cancer Res. 56:1.
2. Barlogie, B., Epstein, J., Selvanayagam, P., and R. Alexanian (1989) Plasma cell myeloma: New biological insights and advances in therapy. Blood 73:865
3. Greil, R., Fashing, B., Loidl, P., and H. Huber (1991) Expression of the c-myc proto-oncogene in multiple myeloma and chronic lymphocytic leukemia: An in situ analysis. Blood 78:180.
4. Durie, B.G.M. (1992) Cellular and molecular genetic features of myeloma and related disorders. Hematology Oncology Clinics of N.A. 6:463.
5. Marcu, K., Bossone, S., and A. Patel (1992) Myc function and regulation Ann Rev. of Biochem. 61:809.
6. Spicer, D., and G. Sonenshein (1992) An antisense promoter of the murine c-myc gene is localized within intron 2 Mol. Cell. Biol. 12:1324.

7. Bentley, D. and M. Groudine (1986) Novel promoter upstream of the human c-myc gene and regulation of c-myc expression in B cell lymphoma. Mol. Cell. Biol. 6:3481.
8. Broome, H.E., Reed, J.C., Godillot, E., and R.G. Hoover (1987) Differential promoter utilization by the c-myc gene in mitogen and interleukin 2 stimulated human lymphocytes. Mol. Cell. Biol. 7:2988.
9. Tanner, J., and G. Tosato (1992) Regulation of B cell growth and immunoglobulin gene transcription by IL6. Blood 79:452.
10. Chang, Y., Spicer, D., and G. Sonenshein (1991) Effects of IL3 on promoter usage, attenuation and antisense transcription of the c-myc oncogene in the IL3 dependent Ba/F3 early pre-B cell line. Oncogene 6:1979.
11. Yano, T., Hoang, A., Lewis, B., Clark, H., Barret, J., Dang, C., and M. Raffeld (1994) Coding region mutations of the c-myc gene in 8Q24 translocated lymphomas and their significance. Proceedings of the Conference on Mechanisms in B Cell Neoplasia, Bethesda, MD. April 1994.
12. Klein, B., and R. Bataille (1992) Cytokine network in human multiple myeloma. Hematology Oncology Clinics of N.A. 6:273.

Clustered Mutations in the Transcriptional Activation Domain of Myc in 8q24 Translocated Lymphomas and their Functional Consequences

M. Raffeld, T. Yano, A.T. Hoang[1], B. Lewis[1], H.M. Clark, T. Otsuki and C.V. Dang[1]
Hematopathology Section, Laboratory of Pathology, National Institutes of Health, Bethesda, MD and [1]Division of Hematology, Departments of Medicine, Cell Biology and Anatomy, Johns Hopkins Oncology Center, The Johns Hopkins University School of Medicine, Baltimore, MD

Introduction

Translocation of the Myc gene on chromosome 8q24 to an immunoglobulin gene segment on chromosomes 14q32, 22q11 or 2p12 is believed to be the central event in the pathogenesis of Burkitt's lymphoma (Dalla-Favera et al. 1982). The molecular heterogeneity of Myc translocations in Burkitt's lymphoma, suggests that there may be several mechanisms that account for the resulting deregulation of Myc gene transcription, and the consequent abnormalities in cell growth regulation that lead to neoplasia. For example, in many cases of Burkitt's lymphoma, the transcriptional enhancer of the immunoglobulin heavy chain gene, Eμ, is located on the same chromosome as the translocated Myc gene, suggesting that it may be responsible for activating Myc transcription (Taub et al. 1982). In other cases, however, Eμ and Myc are located on different chromosomes (Rabbitts et al. 1983), and other mechanisms must operate. In sporadic Burkitt's lymphomas the breakpoints on chromosome 8q24 frequently occur in intron 1 of the Myc gene, separating the oncogene's regulatory elements from its coding sequences (Pelicci et al. 1986). This suggests that loss of normal regulatory elements may also play a role in abnormal regulation of the translocated oncogene (Taub et al. 1984; Bentley and Groudine 1988; Hann et al. 1988). In addition, the translocated Myc gene may be accompanied by mutations or deletions in the first exon sequence that affect regions responsible for transcriptional attenuation (Cesarman et al. 1987), or that result in changes in Myc promoter usage (Spencer et al. 1990). Furthermore, specific mutations within intron 1 have been identified that are associated with decreased binding of a nuclear factor (Zajac-Kaye et al. 1988). Thus, there are a variety of molecular abnormalities identified within the Myc gene locus in Burkitt's lymphoma, and although the precise mechanism of Myc deregulation is debated, the ultimate consequence of the translocation for the development of neoplasia is unquestioned.

Mutation of the Myc gene product itself has not been considered to be a major factor contributing to the tumorigenic activity of the abnormally expressed Myc proteins since early analyses of the first cell lines studied revealed a normal Myc sequence in the translocated allele (Battey et al. 1983; Stanton et al. 1984). Nonetheless, several endemic and sporadic Burkitt's lymphoma cell lines (Rabbitts et al. 1983; Rabbitts et al. 1984; Showe et al. 1985; Care et al. 1986; Murphy et al. 1986; Szajnert et al. 1987) contain mutations in the second or third exons. The location of some of these mutations, particularly those in the transcriptional activation domain (TAD) of Myc (Kato et al. 1990), suggested that they could have functional consequences. We, therefore, decided to survey Myc rearranged primary tumors for coding region mutations and to study the functional consequences of the mutations that might be present.

Results and Discussion

Mutations are frequent in the transcriptional activation domain in Burkitt's lymphoma

We first screened for Myc coding region mutations in 15 cases of Myc-rearranged Burkitt's lymphomas that had previously been characterized at the molecular level (Yano et al. 1992). Screening was performed by PCR-SSCP using 9 overlapping sets of oligonucleotide primers covering the entire second and third exons (Yano et al. 1993). Fourteen of the 15 lymphomas displayed abnormal migration patterns for at least one of the five amplified segments of the second exon. In contrast, only one tumor revealed an abnormal migration pattern in the third exon.

Sequencing of the abnormal conformers revealed a total of 41 mutations in the second exon of the Myc gene in the 14 lymphomas, while only 1 mutation was found in the third exon (Table 1, cases B1-B19). Thirty-three of the 42 mutations were missense mutations. Case B5 had a small insertion of 3 nucleotides coding for an extra leucine between positions 56 and 57. Interestingly, the identical insertion has been reported in the Raji cell line (Rabbitts et al. 1983). The remaining 8 mutations were point mutations that did not change the amino acid residue. As shown in Figure 1, mutations were distributed over the entire second exon from codon 5 to codon 253. Notably, 7 of the 14 tumors had one or two nonsense mutations clustered into a narrow region spanning codons 38-63. Four of these had mutations within one or more proline residues, 3 at codon 57, 1 at codon 60 and 1 at codon 63. Seven tumors possessed mutations in the central portion of the second exon between codons 95 and 138.

Table 1. Coding region mutations in 8q24 translocated lymphomas

Case*	Codon	Mutation	
B1		no mutation	
B2	115	TTC->CTC	Phe->Leu
	240	CTC->GTC	Leu->Val
B3a and B3b	36	CAG->CTG	Gln->Leu
	95	ACG->CCG	Thr->Pro
	183	GCC->GTC	Ala->Val
B3b	57	CCC->TTC	Pro->Phe
	122	GAG->GAA	Glu->Glu
B4	5	GTT->ATT	Val->Ile
	54	GAG->GAA	Glu->Glu
	57	CCC->TCC	Pro->Ser
	60	CCC->CAC	Pro->His
	306	CAG->CAC	Gln->His

B5	56a	CTG trinucleotide insertion	
B7	5	GTT->ATG	Val->Ile
	95	ACG->GCG	Thr->Ala
	115	TTC->CTC	Phe->Leu
	175	AGC->AGT	Ser->Ser
	179	CAG->CAA	Gln->Gln
B8	29	GAG->GAC	Glu->Asp
	244	ACA->AAA	Thr->Lys
B9	137	GGC->GCC	Gly->Ala
	224	CTC->ATC	Leu->Ile
	253	GAG->AAG	Glu->Lys
B10	39	GAG->GAC	Glu->Asp
	44	GCG->GTG	Ala->Val
B11	47	GAG->GAA	Glu->Glu
	57	CCC->ACC	Pro->Thr
	73	TCC->TCA	Ser->Ser
	74	TAC->CAC	Tyr->His
B13	29	GAG->GAA	Glu->Glu
	138	TTC->TCC	Phe->Ser
	174	TCC->TAC	Ser->Tyr
B16	39	GAG->GAC	Glu->Asp
	138	TTC->TGC	Phe->Cys
B17	38	AGC->ACC	Ser->Thr
	57	CCC->TCC	Pro->Ser
	138	TTC->TGC	Phe->Cys
	175	AGC->AGT	Ser->Ser
	176	TTG->ATG	Leu->Met
B18	19	GTG->GTA	Val->Val
	25	TGC->TAC	Cys->Tyr
	172	TCC->TGC	Ser->Cys
B19	44	GCG->ACG	Ala->Thr
	63	CCT->GCT	Pro->Ala
A1	35	CAG->CAA	Glu->Glu
	58	ACC->GCC	Thr->Ala
A2	210	TCG->TCA	Ser->Ser
A4	68	GGG->GGA	Gly->Gly
	87	GAC->GGC	Asp->Gly
	129	ATC->CTC	Ile->Leu
A5	56	CTG->GTG	Leu->Val
	57	CCC->ACC	Pro->Thr
	68	GGG->GGA	Gly->Gly
	223	CTG->CTC	Leu->Leu
	239	GTG->GTA	Val->Val
A6	58	ACC->AAT	Thr->Asp
A9	34	CAG->CAA	Glu->Glu
A12	176	TTG->ATG	Leu->Met
A21	4508**	G->A	
	71	TCG->TGG	Ser->Trp
	82	CTT->CAT	Leu->His
	114-120	18 bp deletion	

*Numbered cases beginning with B refer to classical Burkitt's lymphoma cases. Numbered cases beginning with A refer to AIDS-associated lymphomas.

**Location in bp from 5' Hind III site of Myc gene.

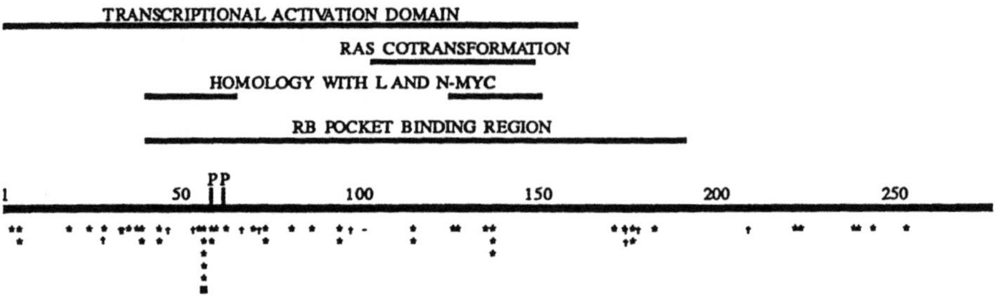

Fig. 1. Map of the second exon of Myc showing the location of the mutations identified in Burkitt's and AIDS-associated lymphomas. ★ represents missence mutations; and †, sense mutations. Properties associated with the second exon are shown at the top of the figure.

Mutations occur in other lymphomas with Myc Involvement

We wanted to know whether these mutations were specific to classical Burkitt's lymphomas or whether they could be found in other lymphomas with 8q24 translocations. To answer this question we extended our mutational analysis to other lymphomas known to have a high frequency of Myc translocation. Twenty-two aggressive AIDS-associated lymphomas, including 11 small non-cleaved cell lymphomas, 10 immunoblastic lymphomas, and 1 diffuse large cell lymphoma were screened. Of the 22 AIDS-NHL studied, nine had Myc rearrangements or cytogenetic translocations involving the 8q24 locus. Eighteen mutations were identified in eight cases. Six of these occurred in the Myc rearranged cases, while only two were found in cases without identifiable Myc rearrangements. The distribution of mutation was similar to the distribution in Burkitt's lymphoma (Table 1, cases A1-A21). In particular, three of the eight cases had missense mutations involving codons 56, 57 or 58. Thus, it appears that mutations of the Myc gene are also found in lymphomas other than classical Burkitt's lymphoma, and that they are primarily associated with lymphomas that possess Myc rearrangement.

Mutations are ongoing in Burkitt's lymphoma

Prior to sequencing we had first subcloned the entire exon 2 fragment from each lymphoma into the pAMP vector (Nisson et al. 1991) and screened for the abnormal conformers found in the genomic DNA. During this screening, we noted the presence of one or two additional populations of subclones that showed unexpected mobilities in the SSCP analysis in three of the Burkitt's lymphoma cases. These additional populations do not represent PCR artifacts since they were reproducible in multiple, separate PCR reactions. Sequence analysis revealed that these subclones possessed the identical mutations as the major subclone, plus one or two additional mutations. An example of such a case is B3 (Fig.2). This case has two major subpopulations (B3a and B3b), one with mutations at codons 36, 95, and 183, accounting for 57% of the population and a second subpopulation with additional mutations at positions 57 and 122, accounting for 29% of the population. There is also a third subpopulation, B3c, that accounts for 14% of the subclones. This minor subpopulation is mutated at codons 8, 36, 95 and 183 and appears directly descended from B3a. We were also able to study a cell line derived from the biopsy, which interestingly, is most closely related to the minor subpopulation B3c, having the same 4 mutations plus an additional codon 57 mutation. This codon 57 mutation is different from the one found in B3b. This type of case clearly establishes that the mutation process is ongoing, and occurs following Myc translocation. This conclusion is consistent with data recently reported by Eick (Albert et al. 1994) who have also described ongoing mutations in the Raji Burkitt's lymphoma cell line.

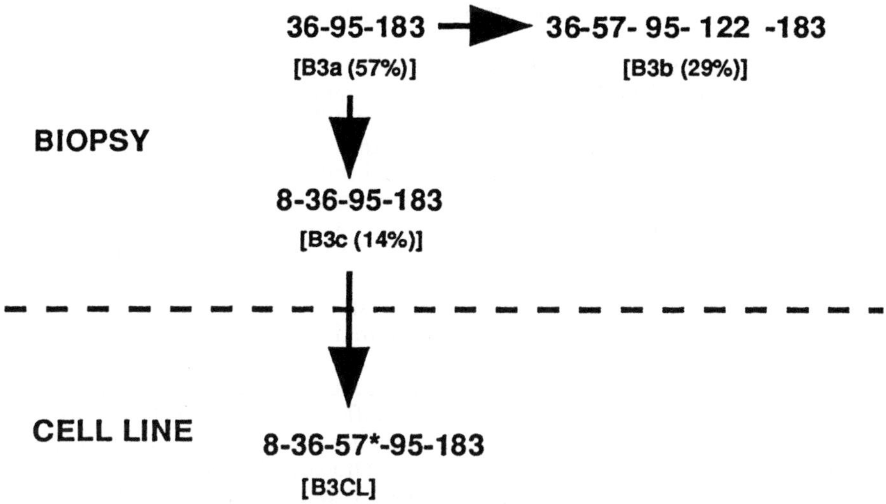

Fig. 2. Ongoing mutations of the Myc gene in a Burkitt's lymphoma. The illustrated case, B3, has two major subpopulations (B3a and B3b) defined by different but related Myc mutations. B3a has mutations at codons 36, 95, and 183, accounting for 57% of the tumor cell population. B3b contains additional mutations at positions 57 and 122, accounting for 29% of the population. A third minor subpopulation, B3c, accounts for 14% of the population. This minor subpopulation is mutated at codons 8, 36, 95 and 183 and appears directly descended from B3a. The cell line B3CL is most closely related to the minor tumor subpopulation B3c, having the same 4 mutations plus an additional codon 57 mutation. This codon 57★ mutation is different from the one found in B3b.

Do the mutations have a functional effect?

It seemed reasonable to speculate that these mutations have been selected for their ability to provide the tumor with a proliferative advantage. The mutations are clonal and they are clustered in the transcriptional activation domain of the Myc protein (Kato et al. 1990). Because the Myc gene is already deregulated in these lymphomas as a result of translocation, we postulated that these mutations might further affect functional regions that would ultimately result in a growth advantage to the tumor subclone.

Lymphoma-derived Myc mutations augment Myc transforming activity

To determine the pathophysiological significance of lymphoma-derived Myc mutations, selected Myc mutants (B3a, B3b, B5, B11, B17, B19 and A1 in Table

1) were tested for neoplastic transforming activity. These mutants were expressed in RAT1a fibroblasts that are able to be transformed by Myc alone. As compared to wild-type Myc produced by a Maloney leukemia virus LTR-driven expression vector, the mutants derived from Burkitt's lymphoma are more potent in transforming RAT1a fibroblasts. Both colony numbers and sizes of transfected RAT1a cells in soft agarose were increased by Myc mutants when compared to wild-type Myc.

Myc mutants resist suppression of Myc-mediated transactivation by p107

Based on observations by Bernards and coworkers (pers. commun.) and work by Gu et al. (1994), we studied the interaction of GAL4-Myc chimeric proteins and the retinoblastoma-related protein, p107. As initially observed by Bernards and coworkers (presented at the Ninth Annual Oncogenes Meeting, Frederick, MD, 1993), p107 was found to bind to the Myc aminoterminal transactivation domain and inhibits Myc-mediated transactivation. We assessed the specificity of the inhibition of Myc-mediated transactivation by p107. We observed that p107 was incapable of suppressing transactivation by three other transactivation domains (VP16, NF1 or CTF1 and USF) at a concentration of p107 that suppresses Myc transactivation. These studies suggest that p107 has a specific effect on the Myc transactivation domain in this limited survey of transactivation domains.

We then tested 7 Myc mutants, which had increased transforming activities, for their susceptibility to p107 suppression, specifically because the transactivation levels of these chimeric GAL4-Myc mutant proteins were not uniformly greater or less than that of wild-type GAL4-Myc. By titration of p107 input plasmid DNA amounts, we observed that wild-type GAL4-Myc transactivation was suppressed by p107 in a dose-dependent fashion. The Myc mutants were uniformly resistant to the suppression by p107. Most remarkable is the mutant (Al, Table 1) that contains only a single missense mutation 58Thr -> Ala. This mutant lacks a critical phosphorylation site that down-regulates Myc transforming activity. The other mutants contain alterations that cluster around 58Thr. These observations suggest that mutations acquired in the Myc transactivation domain somehow confer resistance to suppression by p107. Several possibilities can be envisioned as mechanisms for resistance to suppression by p107. First, the mutations may simply prevent the binding of p107 to the Myc transactivation domain. Second, Myc mutations may enhance the ability of Myc to interact with the basal transcriptional machinery through the TATA-binding protein or TBP (Hateboer et al. 1993; Maheswaran et al. 1994). Third, the Myc mutations may resist a p107-mediated phosphorylation of Myc that inactivates the transactivation function of Myc.

Myc mutations do not appear to alter the binding of Myc to p107 or TBP

To test the hypotheses outlined above regarding the mechanism by which Myc mutants resist suppression by p107, we studied the interaction of the Myc transactivation domain with either p107 or TBP in a mammalian two-hybrid system (termed KISS) that we developed previously (Fearon et al. 1992). p107 or TBP was linked to the GAL4 DNA-binding domain, whereas the Myc wild-type and mutant transactivation domains were linked to the potent Herpes simplex VP16 protein transactivation domain. With appropriate controls, we determined that wild-type Myc transactivation can bind p107 or TBP in this two-hybrid system. With such an interaction, the VP16 transactivation domain is tethered via Myc and p107 or TBP to the GAL4 DNA binding domain, thereby reconstituting an active GAL4 transactivator. Surprisingly, the mutant Myc transactivation domains interacted with p107 as well as the wild-type Myc transactivation domain. Further, interactions of Myc mutants and TBP did not appear to be augmented over that between wild-type Myc and TBP. These observations suggest that Myc mutations do not alter the interaction of Myc with either p107 or TBP as determined by the two-hybrid system.

Based on the prevalence of mutations affecting Myc threonine-58 in lymphomas, we speculate that an alteration in the ability of Myc to be phosphorylated by a p107-associated kinase underlies the resistance of Myc mutants to suppression by p107. As such, mutations that affect 58Thr directly will naturally result in the lack of phosphorylation of this critical residue which is uniformly mutated in v-Myc. Although mutations affecting 58Thr are common in Burkitt's lymphomas, many other mutations cluster around this residue rather than directly affect it. With mutations that cluster around 58Thr, it is reasonable to anticipate that phosphorylation of 58Thr may be absent because the kinase consensus recognition motif is altered by flanking amino acid substitutions. In fact, we have studied an available Burkitt's lymphoma cell line and found (in collaboration with S. Hann), that one such Burkitt's Myc mutant (B7 in Table 1) displays a phosphorylation pattern that is virtually identical to that of mutants with alteration of 58Thr (Lutterbach and Hann 1994). This observation supports our hypothesis that phosphorylation of 58Thr is affected by mutations clustering around this residue. It remains to be determined whether a p107-associated kinase is capable of phosphorylating Myc at 58Thr and abolishing its transactivating function.

Acknowledgments

This work was supported in part by NIH grant CA 51497 (CVD). CVD is a scholar of the Leukemia Society of America.

References

Albert T, Urlbauer B, Kohlhuber F, Hammersen B, Eick D (1994) Ongoing mutations in the N-terminal domain of c-Myc affect transactivation in Burkitt's lymphoma cell lines. Oncogene 9:759-63

Battey J, Moulding C, Taub R, Murphy W, Stewart T, Potter H, Lenoir G, Leder P (1983) The human c-myc oncogene: structural consequences of translocation into the IgH locus in Burkitt lymphoma. Cell 34:779-87

Bentley DL, Groudine M (1988) Sequence requirements for premature termination of transcription in the human c-myc gene. Cell 53:245-56

Care A, Cianetti L, Giampaolo A, Sposi NM, Zappavigna V, Mavilio F, Alimena G, Amadori S, Mandelli F, Peschle C (1986) Translocation of c-myc into the immunoglobulin heavy-chain locus in human acute B-cell leukemia. A molecular analysis. Embo J 5:905-11

Cesarman E, Dalla-Favera R, Bentley D, Groudine M (1987) Mutations in the first exon are associated with altered transcription of c-myc in Burkitt lymphoma. Science 238:1272-5

Dalla-Favera R, Bregni M, Erikson J, Patterson D, Gallo RC, Croce CM (1982) Human c-myc onc gene is located on the region of chromosome 8 that is translocated in Burkitt lymphoma cells. Proc Natl Acad Sci U S A 79:7824-7

Fearon ER, Finkel T, Gillison ML, Kennedy SP, Casella JF, Tomaselli GF, Morrow JS, Dang CV (1992) Karyoplasmic interaction selection strategy: a general strategy to detect protein-protein interactions in mammalian cells. Proc Natl Acad Sci U S A 89:7958-62
Gu W, Bhatia K, Magrath IT, Dang CV, Dalla-Favera R (1994) Binding and suppression of the Myc transcriptional activation domain by p107. Science 264:251-4
Hann SR, King MW, Bentley DL, Anderson CW, Eisenman RN (1988) A non-AUG translational initiation in c-myc exon 1 generates an N-terminally distinct protein whose synthesis is disrupted in Burkitt's lymphomas. Cell 52:185-95
Hateboer G, Timmers HT, Rustgi AK, Billaud M, Van't Veer LJ, Bernards R (1993) TATA-binding protein and the retinoblastoma gene product bind to overlapping epitopes on c-Myc and adenovirus E1A protein. Proc Natl Acad Sci U S A 90:8489-93
Kato GJ, Barrett J, Villa GM, Dang CV (1990) An amino-terminal c-myc domain required for neoplastic transformation activates transcription. Mol Cell Biol 10:5914-20
Lutterbach B, Hann SR (1994) Hierarchical phosphorylation at N-terminal transformation-sensitive sites in c-Myc protein is regulated by mitogens in mitosis. Mol Cell Bio in press:
Maheswaran S, Lee H, Sonenshein GE (1994) Intracellular association of the protein product of the c-myc oncogene with the TATA-binding protein. Mol Cell Biol 14:1147-52
Murphy W, Sarid J, Taub R, Vasicek T, Battey J, Lenoir G, Leder P (1986) A translocated human c-myc oncogene is altered in a conserved coding sequence. Proc Natl Acad Sci U S A 83:2939-43
Nisson PE, Rashtchian A, Watkins PC (1991) Rapid and efficient cloning of Alu-PCR products using uracil DNA glycosylase. PCR Methods Appl 1:120-3
Pelicci PG, Knowles D, Magrath I, Dalla-Favera R (1986) Chromosomal breakpoints and structural alterations of the c-myc locus differ in endemic and sporadic forms of Burkitt lymphoma. Proc Natl Acad Sci U S A 83:2984-8
Rabbitts TH, Forster A, Baer R, Hamlyn PH (1983) Transcription enhancer identified near the human C mu immunoglobulin heavy chain gene is unavailable to the translocated c-myc gene in a Burkitt lymphoma. Nature 306:806-9
Rabbitts TH, Forster A, Hamlyn P, Baer R (1984) Effect of somatic mutation within translocated c-myc genes in Burkitt's lymphoma. Nature 309:592-7
Rabbitts TH, Hamlyn PH, Baer R (1983) Altered nucleotide sequences of a translocated c-myc gene in Burkitt lymphoma. Nature 306:760-5
Showe LC, Ballantine M, Nishikura K, Erikson J, Kaji H, Croce CM (1985) Cloning and sequencing of a c-myc oncogene in a Burkitt's lymphoma cell line that is translocated to a germ line alpha switch region. Mol Cell Biol 5:501-9
Spencer CA, LeStrange RC, Novak U, Hayward WS, Groudine M (1990) The block to transcription elongation is promoter dependent in normal and Burkitt's lymphoma c-myc alleles. Genes Dev 4:75-88
Stanton LW, Fahrlander PD, Tesser PM, Marcu KB (1984) Nucleotide sequence comparison of normal and translocated murine c-myc genes. Nature 310:423-5
Szajnert MF, Saule S, Bornkamm GW, Wajcman H, Lenoir GM, Kaplan JC (1987) Clustered somatic mutations in and around first exon of non-rearranged c-myc in Burkitt lymphoma with t(8;22) translocation. Nucleic Acids Res 15:4553-65
Taub R, Kirsch I, Morton C, Lenoir G, Swan D, Tronick S, Aaronson S, Leder P (1982) Translocation of the c-myc gene into the immunoglobulin heavy chain locus in human Burkitt lymphoma and murine plasmacytoma cells. Proc Natl Acad Sci U S A 79:7837-41
Taub R, Moulding C, Battey J, Murphy W, Vasicek T, Lenoir GM, Leder P (1984) Activation and somatic mutation of the translocated c-myc gene in burkitt lymphoma cells. Cell 36:339-48
Yano T, Sander CA, Clark HM, Dolezal MV, Jaffe ES, Raffeld M (1993) Clustered mutations in the second exon of the Myc gene in sporadic Burkitt's lymphoma. Oncogene 8:2741-8
Yano T, van Krieken JHJM, Magrath IT, Longo DL, Jaffe ES, Raffeld M (1992) Histogenetic correlations between subcategories of small noncleaved cell lymphomas. Blood 79:1282-90
Zajac-Kaye M, Gelmann EP, Levens D (1988) A point mutation in the c-myc locus of a Burkitt lymphoma abolishes binding of a nuclear protein. Science 240:1776-80

Association with C-Myc: An Alternated Mechanism for c-Myc Function

A. Shrivastava[1] and K. Calame[1,2]

[1]Department of Biochemistry and Molecular Biophysics, [2]Department of Microbiology, Columbia University College of Physicians and Surgeons, New York, NY 10032.

Introduction

The *c-myc* proto-oncogene encodes a ubiquitously expressed nuclear phosphoprotein that is involved in the control of cellular proliferation and differentiation (1, 2, 3). Alteration of the *c-myc* locus by chromosomal translocation, amplification or retroviral insertion is often associated with tumorigenesis in different species including humans (1, 3). In normal cells c-MYC plays a crucial role in determining whether cells divide or undergo programmed cell death (4).

The precise function of the c-MYC protein is not known, although the presence of basic helix loop helix leucine zipper (bHLH-Zip) domains seems to suggest that it may act as a DNA-binding transcription factor (5). It has been shown that c-MYC, dimerized with bHLH-Zip protein Max, can bind to specific DNA sequences (6, 7, 8) and activate transcription of artificial promoters with c-MYC binding sites (9, 10, 11). However, few natural genes that are c-MYC responsive in a DNA-site dependent manner have been identified. c-MYC is also known to negatively autoregulate its own transcription but no c-MYC binding site has been found in the *c-myc* promoter (12). This suggests that there may be other ways c-MYC can function. Recently we have found that c-MYC associates with transcription factor Yin -Yang-1 (YY1) and affects its function (13). We suggest that association with YY1 may be one important way c-MYC functions.

YY1 (CF1, δ, NF-E1, or UCRBP) is a 65-kD zinc finger DNA binding protein, belonging to the GLI-Kruppel family of transcription factors (14, 15, 16, 17, 18). YY1 is known to regulate transcription of many genes. Its activity appears to be dependent upon the context of its binding site; in some genes YY1 activates transcription, in others it represses transcription and in some others it is an initiator of transcription. There is a rapidly growing number of YY1-dependent genes which are shown in Table 1. Deletion or mutation of YY1 sites was recently found to be the mechanism by which HPV16 escapes cellular repression in cervical cancers (23). In some genes, YY1 binding sites overlap with activator binding sites and repression is relieved when activators such as serum response factor or AP-1 displace YY1 (19, 21, 22, 24). YY1-dependent repression can also be relieved by adenovirus E1A protein (13).

It is obvious that, in contrast to the paucity of c-MYC-dependent genes, many important cellular and viral genes are YY1-dependent. The c-MYC/YY1 association thus takes on added interest since YY1-dependent genes may be indirect targets of c-MYC. Since c-MYC levels are known to vary in response to mitogens, hormones and other signals, c-MYC may activate or repress transcription of YY1 regulated genes by modulating YY1 activity. This provides an alternative mechanism for c-MYC-dependent transcription regulation.

Table 1

YY1 Regulated Genes	YY1 Functions	References
c-fos promoter	repressor	19,20
Skeletal α-actin promoter	repressor	21
Ig Kappa 3' enhancer	repressor	17
AAV P5 promoter	repressor	14
HPV-18	repressor	22
LCR of HPV-16	repressor	23
LTR of MoMLV	repressor	18
β-casein promoter	repressor	25
ε-globin promoter	repressor	26,27
N-ras promoter*	repressor	22,28
c-myc promoter	activator	29,30
IgH intronic enhancer	activator	29,30
rpL30 promoter	activator	16,31
rpL32 promoter	activator	16,31
rpL7* promoter	activator	32
DHFR(m) promoter	activator	33,34
Leaky-late genes of HSV-1	activator	35,36
IAP Upstream element	activator	37
Human retrotransposon LINE-1	activator	38
AAV P5 promoter	initiator	39
LTR of HIV-1	initiator	40
CK promoter*	initiator	41
MLP of Ad12	ND	42
α-globin promoter	ND	43

Ig, immunoglobulin; AAV, adeno-associated virus; HPV, human papilloma virus; LCR, long control region; LTR, long terminal repeat; MoMLV moloney murine leukemia virus; DHFR (m), mouse dihydrofolate reductase; HSV-1, herpes simplex virus type 1; IAP, intracisternal A-particle; HIV-1, human immuno deficiency virus type 1; CK, creatine kinase; MLP, major late promoter; Ad12, human adenovirus type 12; ND, not determined; asterisks indicate sites on the basis of sequence similarity to YY1 DNA binding site.

YY1 Associates with c-MYC

YY1 association with c-MYC was first detected in a two hybrid (44) screen of a cDNA library from the human monocytic line HL-60. Subsequent experiments with various controls confirmed the specificity of YY1's interaction with c-MYC in this system (13). To further confirm the specificity of the interaction, a glutathione-S-transferase (GST) affinity matrix assay was used (13, 45). c-MYC association with YY1 was disrupted by E1a which competes with c-MYC for association with YY1 and is known to have structural similarity with c-MYC (13, 46). Other bHLH-Zip proteins like USF and Max did not interact with YY1 in this assay, emphasising the specificty of c-MYC/YY1 association (13). Finally, Max was shown to compete with YY1 for association with c-MYC, thus ruling out the possibility of a ternary complex containing c-MYC, Max and YY1 (13).

Regions of YY1 and c-MYC involved in their association

The regions of c-MYC and YY1 which have been shown to be involved in

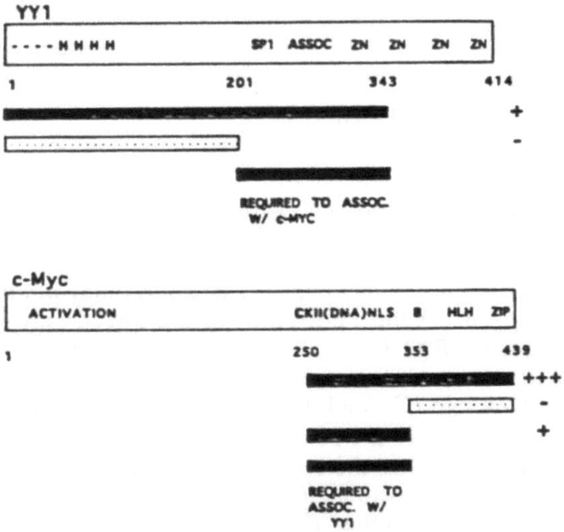

Fig. 1. Regions of YY1 and c-MYC involved in association. 1, 201, 343, 414 in YY1 and 1, 250, 353, 439 in c-MYC, the respective amino acid postions in ful length YY1 and c-MYC protein; ----, negatively charge residues; HHHH, poly histidine stretch; SP1 assoc, domain involved in association with SP1; ZN, zinc fingers; Activation, activation domain of c-MYC; CKII, caesin kinase II phosphorylation sites in c-MYC; DNA, non specific DNA binding domain; NLS, nuclear localization signal; B, basic domain; HLH, helix loop helix domain; ZIP, leucine zipper domain.

their association are summarized in Fig. 1. c-MYC requires amino acids 250-439 for maximal association with YY1 although some association occurs with amino acids 250-353. Amino acids 353-439 include the bHLHZip domain of c-MYC. There is evidence that amino acids 250-353 are also important for some functions of c-MYC. i) Amino acids 290 to 318 are conserved among species and among N-MYC, L-MYC, and c-MYC (1). ii) Removal of amino acids 265 to 353 decreases (but does not abolish) the ability of c-MYC to cooperate with Ras in transforming rat embryo fibroblasts (47). iii) This region is required for transformation of Rat1 cells (48).

YY1 requires amino acids 201-343 for its association with c-MYC but it has not been shown that this region is sufficient for its binding (13, Fig 1). The same region of YY1 is shown to be involved in interaction with another transcription factor SP1 (49, 50).

Effect of c- MYC association on YY1 functions

We investigated the effect of c-MYC on YY1's ability to activate and repress transcription. We showed that cotransfected c-MYC inhibited the YY1-dependent activation of *c-myc* promoter and also inhibited the GAL4-YY1-dependent repession of a thymidine kinase promoter with upstream GAL4 binding sites (13). c-MYC dependent inhibition of YY1 provides one explanation for the negative autoregulation of the *c-myc* promoter by c-MYC protein.

How does c-MYC effect YY1 functions?

c-MYC-dependent inhibition of YY1 function can be explained by at least three different mechanisms (Fig. 2). c-MYC association may block the ability of YY1 to interact with other transcription factors or the basal transcription machinery components (Fig. 2A). Alternatively c-MYC association may inhibit the ability of YY1 to bind DNA (Fig. 2B). A third possibility is that c-MYC may sequester YY1 to c-MYC-Max binding sites (Fig. 2C).

As mentioned above, GST assays suggested that YY1 and Max can not simultaneously associate with c-MYC. To further investigate this question we added unlabeled c-MYC and ^{35}S-labeled YY1 to a GST-Max column; no retention of YY1 was observed (not shown). Since there is no evidence for YY1/c-MYC/Max complexes, the model C appears to be unlikely.

Next we investigated whether YY1/c-MYC complexes can bind to YY1 sites. Increasing amounts of purified GST-c-MYC (259-439 aa) were added to recombinant YY1 in an electrophoretic mobility shift assay (Fig. 3). Low mobility complexes which were specifically dependent upon GST-c-MYC (lanes 1-3) and which did not

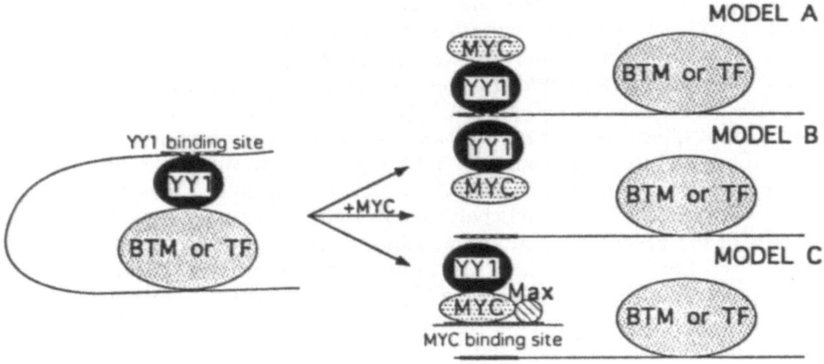

Fig.2. Possible mechanisms to explain the c-MYC dependent inhibition of YY1 functions. BTM, basal transcription maschinery; TF, transcription factors.

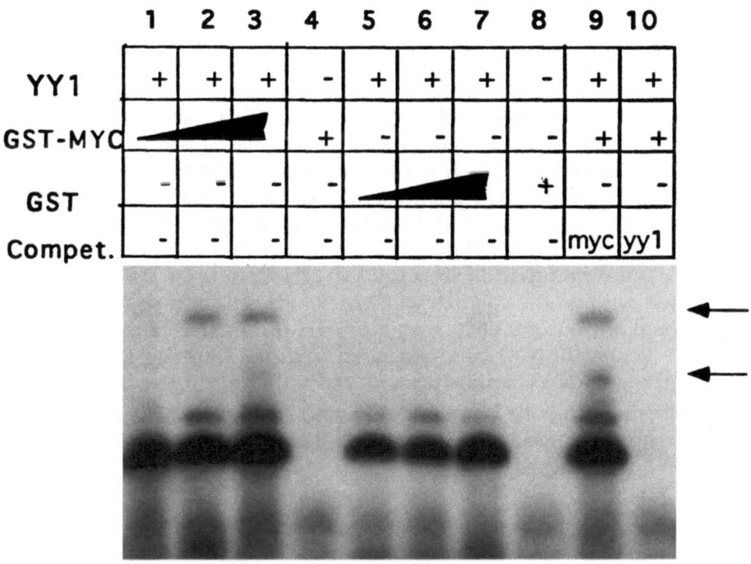

Fig.3. Effect of c-MYC-YY1 association on DNA binding ability of YY1. ^{32}P labeled mE1 oligo from IgH enhancer was used as a probe in these gel-shift assays. In the gel-shift reactions in lane 1-3, increasing amount of purified GST-MYC along with constant amount of YY1 was used; lane 4, only GST-MYC (same amount as in lane 3) was used witout any YY1; lane 5-7, increasing amount of purified GST was used along with constant amount of YY1; lane 8, GST alone without any YY1 was used; unlabeled c-MYC and YY1 DNA binding site oligos was used as a competitor in lane 9 and 10, arrows indicate low mobility complexes.

appear with control GST (lanes 5-7) were observed. The complexes could be competed by YY1 binding sites (lane 10) and not by c-MYC binding sites (lane 9). When YY1 was omitted no complexes were observed (lane 4). These data show that YY1/c-MYC complexes can bind YY1 sites although we can not rule out the possibility that association with c-MYC alters the affinity of YY1 for its site. These data, along with the finding that c-MYC inhibits GAL4-YY1 which binds via GAL4 sites, makes the model B also seem unlikely. Although we currently have no direct data to support model A, we favor the possility that c-MYC inhibits YY1 activity by interfering with important contacts between YY1 and other transcription factors or components of the basal transcription machinery. Recently it has been shown that YY1, TFII B and RNA polymerase II are sufficient to initiate transcription on a supercoiled DNA template carrying the YY1 initiator element from the AAV P5 promoter (51). It has also been shown that YY1 directly interacts with TFII B (51, A. Berrier, A. Shrivastava - unpublished). Thus c-MYC might disrupt functionally important YY1-TFII B interactions.

c-MYC Associates with YY1 *in vivo*

To confirm that c-MYC and YY1 associate in a physiologically relevant context, we have used coimmunoprecipitation (coIP). First, recombinant YY1 and c-MYC were mixed *in vitro* and coIPed using a polyclonal antiserum to murine c-MYC for the immunoprecipitation and a monoclonal antibody against YY1 to develop an immunoblot of the precipitated proteins. This assay confirmed the association between the two proteins and allowed appropriate conditions to be found for subsequent coIPs from cell extracts. As shown in Fig. 4, we over-expressed YY1 in 293 T (Bosc) cells and made whole cell extracts from these cells. These extracts were mixed with whole cell extract from Daudi cells and used for coIP of YY1 with polyclonal antisera made against a COOH-terminal 13 amino acid peptide of human c-MYC. A specific YY1 band was seen from the IP using MYC antiserum but not from the IP using MYC antiserum blocked with peptide or preimmune serum. These data show that YY1 and c-MYC associate in cells. We have subsequently shown similar association in a murine B cell line, M12, which was not overexpressing YY1 (not shown).

Fig. 4. Co-immunoprecipitation of YY1 with c-MYC. lane P, M, P+M represent immunoprecipitation with preimmune, α-MYC and α-MYC + peptide; p.cl., polyclonal.

Conclusion

c-MYC, a potent regulator of growth and differentiation, can associate with YY1, a versatile transcription regulator *in vitro* and *in vivo*. This association can happen in the absence of any other protein or DNA. We have also shown that in transient transfection assays c-MYC can modify the functions of YY1 as a transcriptional regulator. These data suggest that c-MYC may regulate transcription of many genes by modulating YY1 (Fig. 5). Experimental evidence suggests that inhibition of YY1 by c-MYC may involve disruption of YY1's interactions with other transcription factors or basal transcription machinery components.

It appears that c-MYC inhibition of transcriptional proteins extends beyond YY1. c-MYC association also inhibits the activity of TFII-I (52), another transcriptional initiator. c-MYC also associates with TBP (53,54) although the functional consequences of this association are not known. In addition, c-MYC associates with Rb (55) and p107 (56); these interactions may affect c-MYC activity in a cell-cycle dependent manner. Overall, the current data indicate that c-MYC may exert its effects on cell growth by multiple mechanisms involving association not only with Max, but also association with many other proteins. Further studies on these interactions may be expected to increase our understanding of MYC's role in cell growth regulation and in carcinogenesis.

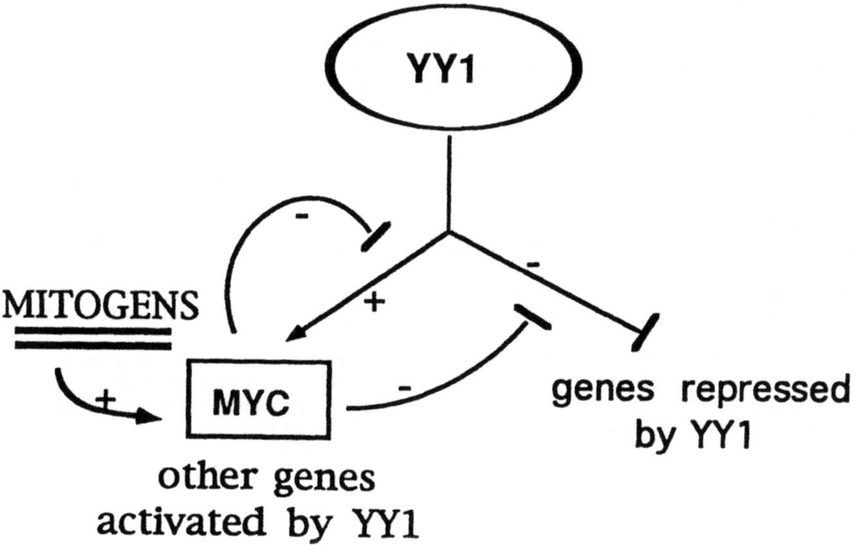

Fig.5. How c-MYC can regulate transcription of various genes regulated by YY1 including c-*myc*. +, activation; -, repression.

References

1. Marcu K. J., Bossone S. A., Patel A. J. (1992). *myc* Function and regulation. Annu. Rev. Biochem., **61**, 809.
2. Luscher B., Eisenman R. N. (1990). New light on Myc and Myb. Genes Dev., **4**, 2025.
3. R. Dalla-Favera (1991) in Origins of human cancer, J. Brugge, T. Curran, E. Harlow, F. Mccormick, Cold Spring Harbor Laboratory Press, pp.543-551.
4. Evan G. I. and Littlewood T. D. (1993) Curr Opin Genet Dev., **3**, **44**.
5. Blackwell, T. K. *et al.* (1990). Sequence-specific DNA-binding by the c-Myc protein. Science, **250**, 1149.
6. Blackwood E. M. and Eisenman R. N. (1991). Max: A helix-loop-helix zipper protein that forms a sequence-specific DNA binding complex with Myc. Science, **251**, 1211.
7. Blackwood E. M. *et al.* (1992). Myc and Max associate in vivo. Genes Dev., **6**, 71 (1992).
8. Prendergast G. C. *et al.* (1991). Association of Myn, the murine homolog of Max, with c-Myc stimulates methylation-sensitive DNA binding and Ras cotransformation. Cell, **65**, 395.
9. Kretzner, L. Blackwood, E. M. and Eisenman, R. N. (1992) Nature, **359**, 426.
10. Amati, B. *et al.* (1992) Nature, **359**, 423.
11. Gu W. *et al.* (1993). Opposite regulation of gene transcription and cell proliferation by c-Myc and Max. Proc. Natl. Acad. Sci. USA, **90**, 2935.
12. Grignani F. *et al.* (1990). Negative autoregulation of c-myc gene expression is inactivated in transformed cells. EMBO J., **9**, 3913.
13. Shrivastava A. *et al.* (1993). Inhibition of transcription regulator Yin-Yang-1 by association with c-Myc. Science, **262**, 1889.
14. Shi Y. *et al.* (1991). Transcriptional repression by YY1, a human GLI-Kruppel-related protein, and relief of repression by adenovirus E1A protein Cell, **67**, 377.
15. Kakkis E. *et al.* (1989). Yin-Yang 1 activates the c-myc promoter. Nature, **339**, 718.
16. Hariharan N., Kelly D. E., Perry R. P. (1991). Delt, a transcription factor that binds to downstream elements in several polymerase II promoters, is a functionally versatile zinc finger protein. Proc. Natl. Acad. Sci. U.S.A., **88**, 9799.
17. Park K., Atchison, M., L. (1991). Isolation of a candidate repressor/activator, NF-E1 (YY1-1, delta), that binds to the immunoglobulin heavy-chain μ-E1 site. Proc. Natl. Acad. Sci. U.S.A., **88**, 9804.
18. Flanagan J. R. *et al.* (1992). Cloning of a negative transcription factor that binds to the upstream conserved region of Moloney murine leukemia virus. Mol. Cell. Biol., **12**, 38.
19. Gualberto A. *et al.* (1992). Functional antoganism between YY1 and the serum response factor. Mol. Cell. Biol., **12**, 4209.
20. Natesan S., Gilman M. Z. (1993). DNA bending and orientation-dependent function of YY1 in the c-fos promoter. Genes Dev., **7**, 2497.
21. Lee T. C., Shi Y., Schwartz R. J. (1992). Displacement of BrdU-induced YY1 by serum response factor activates skeletal alpha actin transcription in embryonic myoblasts. Proc. Natl. Acad. Sci. U.S.A., **89**, 9814.
22. Bauknecht T. *et al.* (1992). Identification of a negative regulatory domain in the human papillomavirus type 18 promoter: interaction with the transcriptional repressor YY1. EMBO J., **11**, 4607.
23. May M. *et al.* (1994). The E6/E7 promoter of extrachromosomal HPV16 DNA in cervical cancers escapes from cellular repression by mutation of target sequences for YY1. EMBO J., **13**, 1460.
24. Vincent C. K. *et al.* (1993). Different regulatory sequences control creatine kinase-M gene expression in directly injected skeletal and cardiac muscle. Mol. Cell. Biol., **13**, 1264.

25. Meier V. S., Groner B. (1994). The nuclear factor YY1 participates in repression of the β–casein gene promoter in mammary epithelial cells and in counteracted by mammary gland factor during lactogenic harmone induction. Mol. Cell. Biol., **14**, 128.
26. Peters B. *et al.* (1993). Protein-DNA interact ion in the epsilon-globin gene silencer. J. Biol. Chem., **268**, 3430.
27. Gumucio D. L. *et al.* (1993). Phylogenetic footprinting reveals unexpected complexity in transfactor binding upstream from the epsilon-globin gene. Proc. Natl. Acad. Sci. USA, **90**, 6018.
28. Paciucci R. and Pellicer, A. (1991). Dissection of the mouse N-ras gene upstream regulatory sequences and identification of the promoter and a negative regulatory element. Mol. Cell. Biol., **11**, 1334.
29. Riggs K. J. *et al.* (1991). Common factor 1 is a transcriptional activator which binds in the c-myc promoter, the skeletal alpha-actin promoter, and the immunoglobulin heavy chain enhancer. Mol. Cell. Biol. **11**, 1765.
30. Riggs K. J. *et al.* (1993). Yin-Yang 1 activates the c-myc promoter. Mol. Cell. Biol. **13**, 7487.
31. Hariharan N. *et al.* (1989). Equipotent mouse ribosomal protein promoters have a similar architecture that includes internal sequence elements. Genes Dev., **3**, 1789.
32. Meyuhas O. and Klein A. (1989). The mouse ribosomal protein L7 gene. J. Biol. Chem., **265**, 11465.
33. Farnham P. J. and Means A. L. (1990). Sequences downstream of the transcription initiation site modulate the activity of the murine dihydrofolate reductase promoter. Mol. Cell. Biol., **10**, 1390.
34. Azizkhan J. C. *et al.* (1993). Transcription from TATA-less promoters: dihydrofolate reductase as a model. Crit. Rev. Eukaryot. Gene Expr., **3**, 229.
35. Chen S. *et al.* (1992). Transactivation of the major capsid protein gene of herpes simplex virus type I requires a cellular transcription factor. J. Virol., **66**, 4304.
36. Mills L. K., Shi Y., Millette R. L. (1994). YY1 is the cellular factor shown previously to bind to regulatory regions of several leaky-late (βγ, γ1) genes of herpes simplex virus type I. J. virol., **68**, 1234.
37. Satyamoorthy K. *et al.* (1993). The intracisternal A-particle upstream element interacts with transcription factor YY1 to activate transcription: pleiotropic effects of YY1 on distinct DNA promoter elements. Mol. Cell. Biol., **13**, 6621.
38. Becker K. J. *et al.* (1993). Binding of the ubiquitous nuclear transcription factor YY1 to a cis regulatory sequence in the human LINE-1 transposable element. Hum. Mol. Genet., **2**, 1697.
39. Seto E., Shi Y., Shenk T. *et al.* (1991). YY1 is an initiator sequence-binding protein that directs and activates transcription in vitro. Nature, **354**, 241.
40. Margolis D. M., Somasundaran M., Green M. R. (1994). Human transcription factor YY1 represses human immunodeficiency virus type 1 transcription and virion production. J. Virol., **68**, 905.
41. Mitchell M. T. and Benfield P. A. (1990). Two different RNA polymerase II initiation complexes can assemble on the rat brain creatine kinase promoter. J. Biol. Chem., **265**, 8259.
42. Zock C., Iselt A. and Doerfler W. (1993). A unique mitigator sequence determines the speciec specificty of the major late promoter in adenovirus type 12 DNA. J. Virol., **67**, 682.
43. Yost S. E., Shewchuk B., Hardison R. (1993). Nuclear protein-binding sites in a transcriptional control region of the rabbit alpha-globin gene. Mol. Cell. Biol., **13**, 5439.
44. Fields S., Song O.-K. (1989). A novel genetic system to detect protein-protein interactions. Nature, **340**, 245.
45. Artandi S. and Calame K. (1993). Glutathione S-transferase fusion protein affinity chromatography to assess biochemical interactions between DNA-binding proteins. Methods Mol. Genet. **1**, 267.
46. Ralston R. and Bishop J. M. (1983). The protein products of the myc and myb oncogenes and adenovirus E1a are structurally related. Nature, 306, 803.
47. Penn L. et al. (1990). Domains of human c-myc protein required for autosuppression and cooperation with ras oncogenes are overlapping. Mol. Cell. Biol., 10, 4961.
48. Stone J. et al. (1987). Definition of regions in human c-myc that are involved in transformation and nuclear localization. Mol. Cell. Biol., 7, 1697.
49. Seto E., Lewis B., Shenk T. (1993). Interaction between transcription factors Sp1 and YY1. Nature, **365**, 462.

50. Lee J. S. , Galvin K. M., Shi Y. (1993). Evidence for physical interaction between the zinc-finger transcription factors YY1 and Sp1. Proc. Natl. Acad. Sci. USA, 90, 6145.
51. Usheva A. and Shenk T. (1994). TATA-binding protein-independent initiation: YY1, TFIIB, and RNA polymerase II direct basal transcription on supercoiled template DNA. Cell, 76, 1115.
52. Roy A. *et al.* (1993). Direct role for Myc in transcription initiation mediated by interactions with TFII-I. Nature, 365, 359.
53. Hateboer G. *et al.* (1993). TATA-binding protein and the retinoblastoma gene product bind to overlapping epitopes on c-Myc and adenovirus E1a protein. Proc. Natl. Acad. Sci. USA, 90, 8489.
54. Maheswaran S., Lee H., Sonenshein G. E. (1994). Intracellular association of the protein product of the c-myc oncogene with the TATA-binding protein. Mol. Cell. Biol., 14, 1147.
55. Rustgi, A. K., Dyson, N. andBernards, R. (1991). Amino-terminal domains of c-myc and N-myc proteins mediate binding to the retinoblatoma gene product. Nature, 352, 541.
56. Gu W. *et al.* (1994). Binding and suppression of the Myc transcriptional activation domain by p107. Science, 264, 251.

The Role of Ornithine Decarboxylase in c-Myc-Induced Apoptosis

G. Packham[1] and J.L. Cleveland[1,2].
1, Department of Biochemistry, St. Jude Children's Research Hospital, 332 N. Lauderdale, Memphis, TN 38105, and 2, Department of Biochemistry, University of Tennessee, Memphis, TN 38163.

Translocations or amplifications which activate members of the *myc* gene family are common in human cancer. c-Myc is a central regulator of normal cell proliferation and differentiation (Luscher and Eisenman, 1991; Marcu et al. 1992) and can induce apoptosis (Askew et al. 1991, 1993; Evan et al. 1992; Bissonette et al. 1992; Shi et al. 1992). c-Myc is a transcription factor (Amati et al. 1992; Kretzner et al. 1992), however, the targets regulated by c-Myc which mediate its activities are not known. Here, we discuss the role of the ornithine decarboxylase (ODC) gene as a direct transcriptional target of c-Myc and demonstrate a role for ODC enzyme activity in c-Myc-induced apoptosis.

Introduction

Enforced c-*myc* expression in cells reveals several biological activities of c-Myc consistent with its role as an oncogene. Specifically, c-Myc is sufficient to promote cell cycle progression (Askew et al. 1991; Evan et al. 1992), inhibit differentiation (Coppola and Cole 1986; Prowchownik and Kukowska 1986; Freytag 1988) and to cause morphological transformation of some fibroblast cell lines (Eilers et al. 1989). Our laboratory has utilized the murine myeloid cell line 32D.3 to analyze c-Myc function in hematopoietic cells. 32D.3 cells are absolutely dependent on IL-3 for their growth and viability (Valtieri et al. 1987). Expression of c-*myc* is tightly regulated by this ligand (Cleveland et al. 1989) and following removal of IL-3, c-*myc* expression is rapidly down-regulated and cells accumulate in the G1 phase of the cell cycle (Askew et al. 1991). After extended periods in the absence of IL-3, 32D.3 cells eventually lose viability through the induction of apoptosis (Askew et al. 1991). By contrast, enforced expression of c-*myc*, in the absence of IL-3, continues to drive cells into S phase (Askew et al. 1991). However, this is associated with a marked acceleration in the induction of apoptosis (Askew et al. 1991). Similar observations have been made following hormone-mediated induction of a c-Myc estrogen receptor (ER) chimeric protein in serum-deprived fibroblasts (Evan et al. 1992).

Considering the association between activation of *myc* genes and human cancer, it is surprising that c-Myc is sufficient to induce apoptosis as well as cell cycle progression. This observation implies that activation of c-Myc is not sufficient for tumor cell growth as additional genetic lesions are required to specifically suppress c-Myc-induced apoptosis. Several genes which cooperate with c-*myc* in tumorigenesis, e.g., *bcl2*, *pim1*, v-*raf*, have been shown to suppress apoptosis (Vaux et al. 1988; Nunez et al. 1990; Moroy et al. 1993; Cleveland et al.

1994), suggesting that c-Myc can induce apoptosis *in vivo* and that suppression of this activity contributes to tumorigenesis.

An attractive rationale for the treatment of tumors which carry activated *myc* alleles would therefore be to specifically modulate c-Myc functions to promote cell death. For example, it might be possible to inhibit c-Myc-induced cell cycle progression while leaving the ability of c-Myc to induce apoptosis intact, or, alternatively, to activate apoptosis without affecting cell cycle progression. To achieve this goal it is necessary to identify targets regulated by c-Myc and to understand their roles in c-Myc-induced apoptosis and cell cycle progression.

Recently, our laboratory has analyzed the role of ornithine decarboxylase (ODC) as a mediator of c-Myc function. ODC catalyzes the decarboxylation of ornithine to putrescine and is a key regulator of polyamine biosynthesis (Tabor and Tabor 1984; Pegg 1986). Polyamines are highly abundant components of prokaryotic and eukaryotic cells, although their exact function in cells is not known. A number of parallels between c-Myc and ODC indicate that ODC is a good candidate c-Myc-responsive gene. First, c-Myc and ODC levels (and those of polyamines) are highly regulated in cells and are associated with cell growth (Bowlin et al. 1986; Luscher and Eisenman 1991; our unpublished observations). Second, similar to c-Myc (Heikkila et al. 1987), ODC activity is required for cell cycle progression into S phase as a specific inhibitor of ODC enzyme activity arrests cells in G1 (Pegg 1986; Bowlin et al. 1986; our unpublished observations). Third, ODC, like c-Myc, is sufficient (Eilers et al. 1989; Auvinen et al. 1992), and cooperates with activated *ras* (Land et al. 1983; Hibshoosh et al. 1991) in transformation of fibroblasts. Finally, ODC enzyme activity is required for the transformation of cells by tyrosine kinase oncogenes (Auvinen et al. 1992), a function which is c-Myc-dependent (Sawyers et al. 1992).

Our laboratory, and others, have demonstrated that the ODC gene is likely a direct transcriptional target of c-Myc. ODC expression is normally rapidly down-regulated following withdrawal of growth factors from cells. By contrast, enforced c-*myc* expression in cells results in the constitutive, growth factor independent expression of ODC RNA, demonstrating that c-Myc is sufficient to induce expression of the endogenous ODC gene (Dean et al. 1987; Askew et al. 1991; Wagner et al. 1993; Tavitigian et al. 1994). Furthermore, c-Myc is a potent activator of ODC gene-reporter fusion constructs and site-directed mutagenesis demonstrated that this effect was mediated by two conserved CACGTG sequence motifs (the identified c-Myc:Max binding sequence [Blackwood and Eisenman 1992]) located in ODC intron 1 (Bello-Fernandez et al. 1993). Each site contributed to the efficient activation of ODC by c-Myc, and each was sufficient to confer c-Myc-responsiveness to a heterologous promoter (Bello-Fernandez et al. 1993). A larger genomic fragment which included both binding sites and the natural intervening sequences was activated by c-Myc to approximately the same extent as the intact promoter, suggesting that all c-Myc-responsive sequences lie within this region (Packham and Cleveland, submitted). Furthermore, a single binding site from the ODC promoter could confer c-Myc-responsiveness independent of position and orientation and over distances of 1.7kb, suggesting that the interaction between the c-Myc transactivation domain and the basic transcription machinery is remarkably flexible (Packham and Cleveland, submitted). More recently, c-Myc has been shown to regulate the endogenous ODC gene without requiring *de novo* protein synthesis (Wagner et al. 1993).

Results

Enforced Expression of c-Myc or ODC Augments ODC Enzyme Activity in 32D.3 Cells.

Enforced c-*myc* expression in 32D.3 cells accelerates apoptosis following IL-3 withdrawal and this is associated with the constitutive, growth-factor independent expression of ODC RNA (Askew et al. 1991). To determine whether enforced c-*myc* expression also has effects on ODC enzyme activity, 32D.3 -derived clones with constitutive c-*myc* expression (*myc* clones; Askew et al. 1991) were analyzed for ODC enzyme activity and compared to control cell lines grown in IL-3. *myc* clones overexpressed ODC activity (3 to 8 fold) relative to control lines (Packham and Cleveland 1994). Therefore, c-*myc* expression has effects on ODC enzyme activity as well as ODC RNA. Similar findings have been reported following activation of a hormone-regulated MycER chimeric protein in fibroblasts (Wagner et al. 1993).

To determine the consequence of enforced ODC expression in 32D.3 cells following IL-3 withdrawal, cells were transfected with a murine ODC expression construct and individual clones expressing exogenous ODC transcripts were identified (ODC clones). Analysis of ODC enzyme activity in these cells, revealed that the clones also had augmented ODC enzyme activity (2.4 to 12 fold) relative to control cells (Packham and Cleveland 1994).

Enforced ODC Activity is Sufficient to Induce Apoptosis.

To address whether enforced expression of ODC, like c-*myc*, is sufficient to compromise cell survival, the ODC clones were deprived of IL-3 and their viability examined with time. ODC clones showed accelerated loss of viability following IL-3 withdrawal, relative to control cells (Fig. 1; Packham and Cleveland 1994). Analysis of genomic DNA from IL-3-deprived ODC clones revealed a characteristic pattern of DNA degradation resulting from internucleosomal nicking of DNA, typical of cells undergoing apoptosis (Packham and Cleveland 1994). In agreement with this observation, morphological analysis of cells revealed changes such as nuclear condensation and formation of apoptotic bodies (Fig. 2). Therefore, enforced expression of ODC, like c-*myc*, is sufficient to induce accelerated apoptosis in 32D.3 cells.

ODC is an Effector of c-Myc-induced Apoptosis.

To directly test the role of ODC enzyme activity in c-Myc-induced apoptosis, we have used α-difluoromethylornithine (DFMO) (Pegg 1986), an ODC enzyme substrate analog that acts as a specific irreversible inhibitor of ODC enzyme activity. Treatment of 32D.3 cells with DFMO rapidly and effectively inhibited ODC enzyme activity (Packham and Cleveland 1994). As expected, DFMO was an effective inhibitor of apoptosis of ODC clones following IL-3 withdrawal, as DFMO-treated ODC clones lost cell viability at rates similar to parental 32D.3 cells (Packham and Cleveland 1994). Importantly, DFMO also inhibited rates of cell death of *myc* clones (Packham and Cleveland 1994). DFMO was an effective inhibitor for early times following IL-3 withdrawal, although at later times inhibition was less complete. By contrast, the prolonged cell death of parental 32D.3 cells, which is likely ODC and c-Myc-independent, was not inhibited by DFMO (Packham and Cleveland 1994).

Discussion

We have demonstrated that ODC, like c-Myc, is sufficient to induce accelerated apoptosis in 32D.3 cells following IL-3 withdrawal. Since c-*myc* is not expressed in IL-3-deprived ODC clones, induction of apoptosis by ODC is c-*myc* independent. Furthermore, DFMO, a specific inhibitor of ODC, delays c-Myc-induced apoptosis. Therefore, ODC is downstream of c-Myc and is a mediator of c-Myc in the induction of apoptosis. Since ODC is also likely a direct transcriptional target of c-Myc, c-Myc regulation of transcription is, therefore, relevant to its observed biological effects.

Expression of ODC, like c-*myc*, is tightly associated with cell growth (Bowlin et al. 1986; Pegg 1986; our unpublished observations). More recently, however, the precise regulation of ODC has been proposed to prevent toxicity due to inappropriate expression of ODC enzyme activity (Morris 1991) and our observation that enforced ODC expression induces apoptosis is clearly consistent with this concept. Therefore, ODC, like c-Myc, is associated with both cell proliferation and cell death.

Our observations also support the concept that apoptosis is induced by multiple pathways. c-Myc-induced apoptosis is apparently mediated by both ODC-dependent and ODC-independent pathways, as DFMO only partially inhibits c-Myc-induced cell death. Furthermore, the death of parental 32D.3 cells is c-Myc- (and ODC) independent as DFMO has no effects on cell survival and commitment to death of 32D.3 cells occurs after down-regulation of ODC and c-*myc* expression. Similar c-Myc-dependent and independent pathways have been described in T-cells (Shi et al. 1992). Therefore, multiple pathways can induce apoptosis, although in each case the morphological and biochemical changes associated with apoptosis are conserved, suggesting that each pathway ultimately activates a common effector molecule.

The observation that ODC is a mediator of c-Myc-induced apoptosis, and the relatively good correlation between the activity of ODC and c-Myc in other biological systems, suggests that ODC may be a global mediator of c-Myc function. We have also analyzed the role of ODC in cell cycle progression. In contrast to c-*myc*, which promotes continued entry of cells into S phase in the absence of IL-3, ODC clones withdraw from cycle and accumulate in G1, like parental cells. Therefore, ODC is not sufficient to induce cell cycle progression in the absence of IL-3 (Packham and Cleveland, in preparation). Importantly, the ability of ODC to induce apoptosis but not S phase entry demonstrates that we have separated cell cycle progression and apoptosis. It has been suggested that c-Myc-induced apoptosis is a default process in response to inappropriate cell cycle progression. This is clearly not the case for ODC (and ODC-dependent c-Myc-induced apoptosis) as ODC induces apoptosis without any demonstrable effects on cell cycle progression. Therefore, ODC seems to be a direct inducer of apoptosis and may modulate other c-Myc functions.

Acknowledgments

We thank Marion Merrill Dow Research Laboratories, Merrill Dow Inc., for generously providing DFMO. We are especially indebted to Hui Yang and Elsie White for excellent technical assistance. This work was supported by National

Institute of Diabetes and Digestive and Kidney Diseases Grant DK44158 (J.L.C.), National Cancer Institute Center Support Grant PO CA21765, and by the American Lebanese Syrian Associated Charities. G.P. is the recipient of the Martin Morrison Endowed postdoctoral fellowship.

References

Amati B, Dalton S, Brooks MW, Littlewood TD, Evan GI and Land H (1992) Transcriptional activation by the human c-Myc oncoprotein in yeast requires interaction with Max. Nature (London) 359:423-425.

Askew DS, Ashmun RA, Simmons BC and Cleveland JL (1991) Constitutive c-*myc* expression in an IL-3 dependent myeloid cell line suppresses cell cycle arrest and accelerates apoptosis. Oncogene 6:1915-1922.

Askew DS, Ihle JN and Cleveland JL (1993) Activation of apoptosis associated with enforced Myc expression in myeloid progenitor cells is dominant to the suppression of apoptosis by interleukin-3 or Erythropoietin. Blood 82:2079-2087.

Auvinen M, Paasinen A, Andersson LC and Holtta E (1992) Ornithine decarboxylase activity is critical for cell transformation. Nature (London) 360:355-358.

Bello-Fernandez C, Packham G and Cleveland JL (1993) The ornithine decarboxylase gene is a transcriptional target of c-Myc. Proc. Natl. Acad. Sci. USA 90: 7804-7809.

Bissonnette RP, Echeverri F, Mahboubi A and Green DR (1992) Apoptotic death induced by c-*myc* is inhibited by bcl-2. Nature (London) 359:552-555.

Blackwood EM and Eisenman RN (1992) Max: A helix-loop-helix zipper protein that forms a sequence specific DNA-binding complex with Myc. Science 251:1211-1217.

Bowlin TL, McKnown BJ and Sunkara PS (1986) Ornithine decarboxylase induction and polyamine biosynthesis are required for the growth of interleukin-2- and interleukin-3-dependent cell lines. Cell Immunol. 98:341-350 (1986).

Cleveland JL, Dean M, Rosenberg N, Wang JYI and Rapp UR (1989) Tyrosine kinase oncogenes abrogate interleukin-3 dependence of murine myeloid cell through signaling pathways involving c-*myc*: Conditional regulation of c-*myc* transcription by temperature-sensitive v-abl. Mol. Cell. Biol. 9:5685-5695.

Cleveland JL, Troppmair J, Packham G, Askew DS, Lloyd P, Gonsalez-Garcia M, Nunez G, Ihle JN and Rapp UR (1994) v-*raf* suppresses apoptosis and promotes growth of interleukin-3-dependent myeloid cells. Oncogene, in press.

Coppola JA and Cole MD (1986) Constitutive c-*myc* oncogene expression blocks mouse erythroleukemia cell differentiation but not commitment. Nature (London) 320: 760-763.

Dean M, Cleveland JL, Rapp UR and Ihle JN (1987) Role of *myc* in the abrogation of IL-3 dependence of myeloid FDC-P1 cells. Oncogene Res. 1:61-76.

Eilers M, Dicard P, Yamamoto KR and Bishop JM (1989) Chimaeras of Myc oncoprotein and steroid receptors cause hormone-dependent transformation of cells. Nature (London) 340:66-68.

Evan GI, Wylie AH, Gilbert CS, Littlewood TD, Land H, Brooks M, Waters CM, Penn LZ and Hancock DC (1992) Induction of Apoptosis in Fibroblasts by c-myc protein. Cell 69:119-128.

Freytag S (1988) Enforced expression of the c-*myc* oncogene inhibits cell differentiation by precluding entry into a distinct pre-differentiation state in G0/G1. Mol. Cell. Biol. 8:1614-1624.

Heikkila R, Schwab G, Wickstrom E, Lohie SL, Pliznik DH, Watt R and Nacres LM (1987) A c-*myc* antisense oligonucleotide inhibits entry into S phase but not progression from G0 to G1. Nature (London) 328:445-449.

Hibshoosh H, Johnson M and Weinstein IB (1991) Effects of overexpression of ornithine decarboxylase (ODC) on growth control and oncogene-induced cell transformation. Oncogene 6:739-743.

Kretzner L, Blackwood E and Eisenman RN (1992) Myc and Max proteins possess distinct transcriptional activities. Nature (London) 359:426-429.
Land H, Parada LF and Weinberg RA (1983) Tumorigenic conversion of primary embryo fibroblasts requires at lest two cooperating oncogenes. Nature (London) 304: 596-598.
Luscher B and Eisenman RN (1991) New light on Myc and Myb. Part I: *myc*. Genes & Dev. 4:2025-2035.
Macu KB, Bossone SA and Patel AJ (1992) Myc function and regulation. Annu. Rev. Biochem. 61:809-860.
Morris DR (1991) A new perspective on ornithine decarboxylase regulation: Prevention of polyamine toxicity is the overriding theme. J. Cell. Biochem. 46:102-105.
Moroy T, Grzeschiczek A, Petzold S and Hartmann KU (1993) Expression of a Pim-1 transgene accelerates lymphoproliferation and inhibits apoptosis in lpr/lpr mice. Proc. Natl. Acad. Sci. USA. 90: 10734-10738.
Nunez G, London L, Hockenberry D, Alexander M, McKeran JP and Korsmeyer SJ (1990) Deregulated Bcl-2 expression selectively prolongs survival of growth factor-deprived hemopoietic cell lines. J. Immunol. 144:3602-3610.
Packham G and Cleveland JL (1994) Ornithine decarboxylase is a mediator of c-Myc-induced apoptosis. Mol. Cell Biol. in press.
Pegg A.E (1986) Recent advances in the biochemistry of polyamines in eukaryotes. Biochem. J. 234:249-262.
Prowchownik E.V and Kukowska J (1986) Deregulated expression of c-*myc* in murine erythroleukemia cells prevents differentiation. Nature (London) 322:848-850.
Sawyers CL, Callahan W and Witte ON (1992) Dominant negative Myc blocks transfromation by Abl oncogenes. Cell 70:901-910.
Shi Y, Glynn JM, Guilbert LJ, Cotter TG, Bissonnette RP and Green DR (1992) Role for c-*myc* in activation-induced apoptotic cell death in T cell hybridomas. Science 257:212-214.
Tabor CW and Tabor H (1984) Polyamines. Annu Rev. Biochem. 53:749-790.
Tavitigian SV, Zabludoff S and Wold BJ (1994) Cloning of mid-G1 serum response genes and identification of a subset regulated by conditional myc expression. Mol. Biol. Cell 5:375-388.
Valtieri M, Tweardy DJ, Caracciolo D, Johnson K, Mavilio F, Altmann S, Santoli D and Rovera G (1987) Cytokine-dependent granulocytic differentiation. Regulation of proliferative and differentiative responses in a murine progenitor cell line. J. Immunol. 138:3829-3835.
Vaux DL, Cory S and Adams JM (1988) Bcl-2 gene promotes haemopoietic cell survival and cooperates with c-*myc* to immortalize pre-B cells. Nature (London) 335:440-442.
Wagner AJ, Meyers C, Lamins LA and Hay N (1993) c-Myc induces the expression and activity of ornithine decarboxylase. Cell Growth and Differentiation 4:879-883.

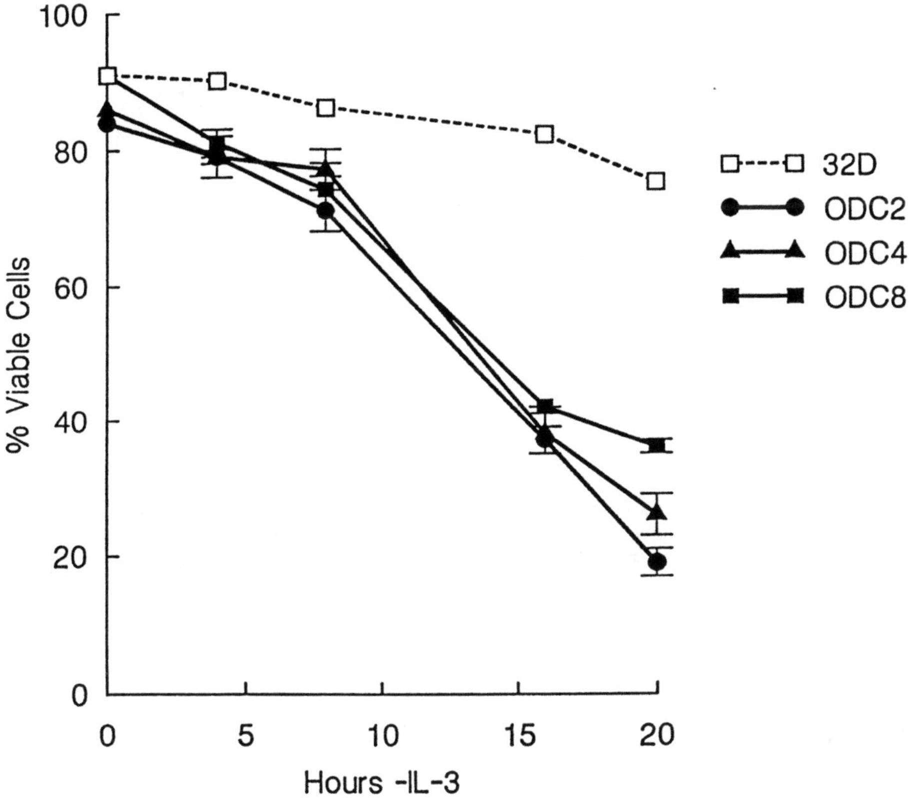

Fig.1. Enforced ODC expression accelerates cell death in 32D.3 cells following IL-3 withdrawal. Parental 32D.3 cells and ODC clones were thoroughly washed to remove IL-3 and resuspended in medium without IL-3. Cell viability was determined using trypan blue dye exclusion at the indicated time points following IL-3 withdrawal. Error bars are derived from the average of duplicate viability determinations.

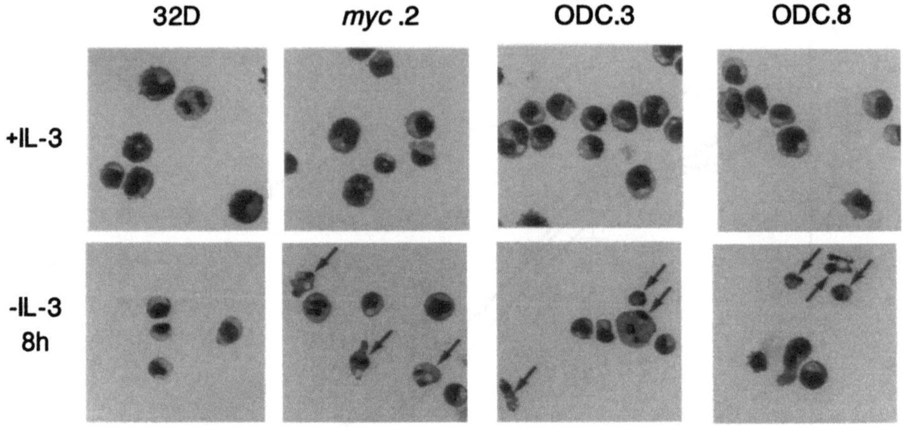

Fig.2. Enforced expression of ODC and c-*myc* in 32D.3 cells induces apoptosis following IL-3 withdrawal. Parental 32D.3 cells, ODC clones and a c-*myc* clone (Askew et al. 1991) were thoroughly washed to remove IL-3 and resuspended in medium without IL-3. Cytospins were prepared eight hours after removal of IL-3, and cells stained with Wright-Geimsa. Cells not washed of IL-3 were analyzed as controls. Arrows indicate cells undergoing apoptosis.

Growth Factor Regulation of Cell Cycle Progression and Cell Fate Determination

Thomas W. Beck[1], Nancy S. Magnuson[2] and Ulf R. Rapp[3]

[1]Program Resources Inc./Dyncorp, Frederick Cancer Research and Development Center, Frederick MD 21702-1201; [2]Department of Microbiology, Washington State University, Pullman, WA 99164-4233; [3]Laboratory of Viral Carcinogenesis, National Cancer Institute, Frederick Cancer Research Facility, Frederick, MD 21702-1201

The interaction between growth factors and their cell surface receptors initiates the cell cycle which under different physiological conditions can lead to cell growth, differentiation, survival or apoptosis (programmed cell death) (Figure 1). The sequence of events leading to cellular proliferation induced by growth factors involves activation of an intracellular signaling cascade resulting in induction of the expression of specific target genes in the nucleus which are sequentially and temporally expressed in defined classes, such as immediate early, delayed early and late. These gene products encode a large number of proteins which collaborate and indeed some are essential for driving cells through Gap1 (G1) of the cell cycle and into DNA synthesis (S phase). For successful entry into S phase, the growth factor must be in continuous contact with the cells. If however, the growth factor is removed prior to late G1 (R), cells will not commit to S phase and instead arrest in G1. This commitment step is controlled at least in part by certain members of a family of gene products, the cyclins, which are differentially regulated during the course of the cell cycle (Figure 1). The cyclins undergo specific protein/protein interactions with another family of proteins called cyclin dependent protein kinases (cdks), which are serine/threonine protein kinases. In general, complex formation with cyclin activates cdk kinase activity, but regulation of cyclin dependent kinase (cdk) activity appears to be more complicated involving both phosphorylation events catalyzed by other protein kinases (eg. CAKs) and kinase inhibitors (eg. p16 and p21/Cip-1/Waf-1). Once activated, cdks phosphorylate and inactivate retinoblastoma (Rb) and p107 proteins which are negative regulators for progression into S phase. As a result of Rb inactivation, transcription factors in the E2F family are activated and induce the expression of late genes encoding proteins involved in DNA synthesis and replication. The study of the mechanism of action of negative growth effectors is a major focus of ongoing research in several laboratories. TGF-b, for example, has been proposed to block cell cycle progression by inhibiting cyclin E-Cdk2 complexes (Koff, A et al., 1993) thus keeping RB in the active form and preventing accumulation of free E2F. Another proposed mechanism involves the downregulation of*myc* expression by TGF-b (Munger et al. 1992), an interesting view considering the recent demonstration that Myc complexes with p107 *in vivo* (Gu et al., 1994). In any case, it is not yet clear how TGF-b works.

How does growth factor/growth factor receptor signaling work? In general, the ligand/receptor interaction triggers intrinsic or associated phosphotyrosine kinase (PTK) activity which can stimulate several pathways as indicated in Figure 2. However, we know that at least two are essential and sufficient in immortalized cell lines for growth. One of these pathways involves the Raf/MAPK and the other involves c-Myc. Of these two pathways, the Raf/MAPK pathway is better understood. The major events following growth factor binding to receptor starts with the small GTPase proto-oncogene, Ras, which is charged with GTP. Activated Ras translocates Raf-1 from the cytosol to the membrane via a direct physical interaction where activation occurs by an unknown mechanism. In turn, activated Raf-1 then initiates a cascade of phosphorylation events involving protein kinases which are highly conserved evolutionarily. Raf-1 directly phosphorylates and activates MEK (MAP kinase kinase) which in turn phosphorylates and activates MAP kinase (mitogen-activated protein kinases)/ERK (extracellular signal-regulated kinase) isoforms. Activated MAP kinases then catalyze the phosphorylation of multiple oncogene class transcription factors (eg. TCF/Elk-1, NF-kb, Myc) leading to changes in gene transcription. In addition, activated MAP kinase also catalyzes the phosphorylation and activation of the Rsk isozymes, Rsk-1 and Rsk-2/MAPKAP-2 kinase and PLA2. The Rsk and MAP kinases have a broad and completely unrelated substrate specificity, and therefore might contribute differentially to the ultimate physiological responses that result from cell stimulation by growth factors.

In contrast, the sequence of events leading to activation of the c-Myc pathway is not completely clear. It is known that the *myc* expression is tightly controlled at multiple levels, including transcription initiation and elongation and mRNA stability (K. B. Marcu et al., 1992) and recent findings indicate that NF-kB plays an important role in the transcriptional activation of the *c-myc* promoter (La Rosa et al., 1994). On the other hand, details concerning the function of c-Myc in cell cycle regulation are only beginning to emerge. It was recently shown that the Rb-related protein called p107 associates with the amino-terminal domain c-Myc *in vitro* and *in vivo* (Gu et al, 1994). Moreover, when p107 was overexpressed in NIH3T3 cells, cell proliferation was blocked prior to S phase of the cell cycle. Unlike the situation with Rb, however, the p107-mediated growth arrest could not be rescued by cyclin A, cyclin E or E2F1 suggesting that p107 targets other cell cycle progression factors such as Myc-Max transcriptional complexes by binding to Myc and inhibiting its transcriptional activity. Alternatively, p107 may switch Myc from a positive to a negative regulator, analogous to the action of Rb on E2F (Herschback and Johnson, 1993).

As previously stated, the Raf/MAPK pathway is much better understood than most other signaling pathways. In fact, recent biochemical and genetic studies from a variety of model systems have continued to identify other homologs belonging to the families forming the five tiers of the Raf/MAPK pathway as illustrated in Figure 3. Additional homologs in this pathway may yet be identified. The rapid activation of the Raf/MAPK pathway by receptor tyrosine kinases as well as G-protein coupled receptors occur through pathways dependent on Ras as well as Raf (Troppmair et al., 1992; Thomas et al., 1992, Crespo et al., 1994). More recent studies of signaling resulting from cellular stress or treatment with tumor necrosis factor (TNF) alpha have provided evidence for the existence of yet another homolog of the MAPK family (reviewed in Kolesnick and Golde, 1994; Kyriakis et al., 1994). These protein kinases have been designated stress-activated protein kinases (SAPKs) and appear to be activated by sphingomyelinase-mediated responses to TNF-alpha. Therefore, SAPKs define a new TNF-alpha and stress-activated signaling pathway, possibly initiated by sphingomyelin-based second messages. Although the precise mechanisms coupling TNF-alpha signaling at the cell surface to the cytoplasm remains unknown, recent studies suggest Raf-1 may play a role in some circumstances. Consistent with this involvement we have found that TNF-alpha stimulates Raf-1 kinase activity and dominant negative versions of Raf-1 block TNF-induced NF-kB activation in cells where it induced growth but not in cells in which it blocks growth (Rapp and Pfzermeir, unpublished results; Finco and Baldwin, 1993).

In addition to multiple interconnecting pathways involving Raf-1 now being identified and the physical association between Raf and activated Ras *in vivo* and *in vitro* (Van Aelst et al., 1993; Vljtek et al., 1993; Moodie et al., 1993; Zang et al., 1993 Warne et al., 1993), more detail is being uncovered about the mechanism of activation of Raf-1. Attempts to activate Raf-1 by GTP-Ras *in vitro* have not been successful (reviewed in Magnuson et. al. 1994). We had originally proposed based on our observations that 1) dominant negative versions of Raf-1 blocked cell proliferation, 2) oncogenic Ras was insufficient for Raf-1 activation in the absence of serum, and 3) Raf-1 was a substrate for PKC that Ras would function as a shuttle to translocate inactive Raf-1 in the cytosol to activating Raf-1 kinase kinases in the membrane (Reed et al., 1991, Bruder et al., 1992, Troppmair et al., 1992; Troppmair et al., 1994; Rapp et al., 1994). One group has recently demonstrated that Raf-1 appears to translocate together with Ras to the plasma membrane where it is found to be enzymatically active (Traverse et al., 1993). One potential activator of Raf-1 kinase may be a product(s) of phosphatidylcholine-phospholipase C (PC-PLC) as this enzyme was demonstrated to activate Raf kinase *in situ* as well as trigger cell growth in a Ras-independent yet Raf-dependent manner (Cai et al., 1993). Most recent findings have demonstrated that when Raf-1 is targeted to the plasma membrane by the addition of the COOH-terminal membrane localization signal of K-Ras, it is enzymatically activated to the same extent as Raf coexpressed with the oncogenic mutant Ras (Stokoe et al., 1994; Leevers et al., 1994). This supports our suspicion that Ras functions as a regulated, membrane-bound anchor for Raf-1, and that something else contributes to Raf-1 activation. What exactly the activator might be is not known, although incubation of GTP-bound Ras and Raf with a particular membrane fraction results in Raf activation (Kumata, T. and U.R. Rapp, unpublished). Nevertheless, we hypothesize that more than one factor may be responsible. For example, Raf-1 has been shown to be associated with the T cell receptor-CD3 and IL-2 beta chain receptor complexes in unstimulated cells and this suggests that Raf activators may be nearby tyrosine kinases (Maslinski et al., 1992; Loh et al., 1994). This predicts that Raf activation through these receptors should be Ras-independent.

It must also be remembered that in addition to the existence of multiple potential activators of Raf-1, there is accumulating evidence that there may be alternate substrates for Raf-1. As found in 3T3 L1 cells, which undergo adipocytic differentiation when chronically exposed to insulin, all three kinases, MAPK, RSK, and Raf-1 kinase, are activated. However, the activation of Raf-1 by insulin is totally dissociated from that of MAPK and RSK indicating that separate parallel signals emerge from Ras and that there may be other substrates of activated Raf-1 (Porras et al., 1994). Similarly, *v-raf* has been shown to confer CSF-1 independent growth to a macrophage cell line and induce immediate early gene expression without MAPK activation (Busher et al., 1993). Obviously something other than MAPK must be activated in these cells to facilitate the transcriptional activation of the immediate early gene expression. It seems plausible then that other, as of yet undescribed, substrates for Raf-1 and/or MEK may mediate these MAPK-independent responses.

What is the contribution of Raf-1 to the mitogenic events brought about by growth factor stimulation? One factor that Raf-1 shares in common with the small integral membrane protein, Bcl-2, is the ability to act synergistically with c-Myc to induce IL-3 independent growth and survival of IL-3 dependent hematopoietic cells (Cleveland et al., in press; Cleveland et al, 1986; Vaux et al., 1988). When Bcl-2 alone is overexpressed, it can suppress apoptosis in a wide variety of situations including the cell death resulting from growth factor deprivation. Since both Raf-1 and Bcl-2 might be considered members of the same oncogene complementation group based on their ability to cooperate with c-Myc in rendering hematopoietic cells independent of IL-3 for continuous proliferation and survival, it is surprising that Bcl-2 and Raf-1 can act synergistically to suppress apoptosis (Miyashita et al., in press). One explanation for which there is evidence is that Bcl-2 directly interacts with Raf-1 (Sato et al., submitted). This interaction may lead to the phosphorylation of protein(s) associated with Bcl-2 who's apoptotic activity may thereby be regulated by Raf-1 by phosphorylation. This brings up the question of why Raf-1 and c-Myc both appear to be necessary in proliferation, survival and differentiation? Raf-1 seems to be necessary for promoting cell cycle progression. When Raf-1 is over expressed, for example, in IL-3 dependent hematopoietic cells, it can prolong cell survival, but by itself, it cannot abrogate IL-3-dependence. Over expression of c-Myc, on the other hand, only commits such IL-3-dependent cells to the S phase of the cell cycle but does not provide cells with the proper preparation to go through the S phase and, therefore, also cannot abrogate IL-3 dependence (Heikkila et al., 1987; Troppmair et al., 1988). Without a second signal to promote cell cycle progression, apoptosis occurs (Askew et al, 1991; Askew et al., in press). Therefore, as expected, when IL-3-dependent cells are made to overexpress both c-Myc and Raf-1, growth factor independent cells result (Cleveland et al., in press). Thus, presumably at least part of the explanation for the ability of the *myc* and *raf* oncogenes to collaborate in the development of IL-3 independence relates to the apoptosis suppressing activity of Raf-1 and its possible role in regulating the cell cycle.

Acknowledgements

The content of this publication does not necessarily reflect the views or policies of the Department of Health and Human Services, nor does mention of trade names, commercial products, or organizations imply endorsement by the U.S. government.

References

Askew, D.S., R.A. Ashmun, B.C. Simmons and J.L. Cleveland. 1991. Constitutive *c-myc* expression in an IL-3 dependent myeloid cell line suppresses cell cycle arrest an accelerates apoptosis. Oncogene 6:1915-1922.

Askew, D.S., J.N. Ihle and J.L. Cleveland. Activation of apoptosis associated with enforced Myc expression in myeloid progenitor cells in dominant to the suppression of apoptosis by interleukin-3 or Erythropoietin. Blood (in press).

Bruder, J.T., G. Heidecker, and U. R. Rapp UR. 1992. Serum-, TPA- and Ras-induced expression from Ap-1/Ets-driven promoters requires Raf-1 kinase. Genes & Dev 6:545-556.

Buscher, D., P. Dello Sbarba, R.A. Hipskind, U.R. Rapp, E.R. Stanley and M. Baccarini. 1993. v-raf confers CSF-1 independent growth to a macrophage cell line and leads to immediate early gene expression without MAP-kinase activation. Oncogene 8:3323-3332.

Cai, H., P. Erhardt, J. Troppmair, M.T. Diaz-Meco, G. Sithanandam, U.R. Rapp, J. Moscat, G.M. Cooper. 1993. Hydrolysis of phosphatidylcholine couples Ras to activation of Raf protein kinase during mitogenic signal transduction. Mol. Cell. Biol. 13:7645-7651.

Cleveland, J., M. Jansen, K. Bister, T. Fredrickson, H. Morse, J. Ihle, and U. Rapp. 1986. Interaction between raf and myc oncogenes in transformation *in vivo* and *in vitro.* J. Cell Biochem. 30:185-218.

Cleveland, J.L.,M. Dean, N. Rosenberg, J. Y-J. Wang and U.R. Rapp. 1989. Tyrosine kinase oncogenes abrogate interleukin-3 dependence of murine myeloid cells through signalling pathways involving *c-myc*: Conditional regulation of *c-myc* transcription by temperature sensitive v-abl. Mol. Cell. Biol. 9:5685-5695.

Cleveland, J.L., J. Troppmair, D.S. Askew, P. Lloyd, J.N. Ihle and U.R. Rapp. 1994. Raf-1 is required for optimal growth of interleukin-3-dependent myeloid cells and synergizes with Myc to promote autonomous growth by suppressing apoptosis.

Crespo, P., N. Xu, J. Daniotti, J. Troppmair, U.R. Rapp, and J.S. Gutkind. Signalling through transforming G protein-coupled receptors in NIH3T3 cells involves c-Raf activation: evidence for a protein kinase C independent pathway. (Submitted).

Finco, T.S. and A.S. Baldwin. 1993. KB site dependent induction of gene expression by diverse inducers of nuclear factor KB requires Raf-1. J. Biol. Chem. 268:17676-17679.

Gu, W., K. Bhatia, I.T. Magrath, C.V. Dang, and R. Dalla-Favera. 1994. Binding and suppression of the Myc transcriptional activation domain by p107. Science 264:251-254 .

Heikkila, R., G. Schwab, E. Wickstrom, S.L. Lohie, D. H. Pliznik, R. Watt, and L. M. Nacres. 1987. A *c-myc* antisense oligonucleotide inhibits entry into S-phase but not progression from Go to G1. Nature (London) 328:445-449.

Herschback, B.M. and A.D. Johnson. 1993. Transcriptional repression in eukaryotes. Annu. Rev. Cell Biol. 9:479-509.

Koff, A., M. Ohtsuki, K., Polyak, J. Roberts, and J. Massague. 1993. Negative regulation of G1 in mammalian cells: inhibition of cyclin-dependent kinase by TGF-beta. Science 260: 536-539.

Kolesnick, R. and D.W. Golde. 1994. The sphingomyelin pathway in tumor necrosis factor and interleukin-1 signaling. Cell, 77:325-328.

Kolch, W., G. Heidecker, and U. R. Rapp. 1991. Raf-1 protein kinase is required for growth of induced NIH/3T3 cells. Nature 349:426-428.

Kyriakis, J.M., P. Banerjee, E. Nikolakaki, T. Dal, E.A. Rube, M.F. Ahmad, J. Avruch and J.R. Woodgett. 1994. The stress-activated protein kinase subfamily of c-Jun kinases. Nature 369:156-160.

La Rosa, F.A., J.W. Pierce and G.E. Sonenshein. 1994. Differential regulation of the c-myc oncogene promoter by the NF-kB/rel family of transcription factors. Mol. Cell. Biol. 14:1039-1044.

Leevers, S.J., H.G. Paterson and C. J. Marshall. 1994. Requirement for Ras in Raf activation is overcome by targeting Raf. Nature 369:411-414.

Loh, C., C. Romeo, B. Seed, J.T. Bruder, U. Rapp and A. Rao. 1994 . Association of Raf with the CD3 delta and gamma chains of the T cell receptor-CD3 complex. J. Biol. Chem. 269:8817-8825.

Magnuson, N.S., T. Beck, H. Vahidi, H. Hahn, U. Smola and U.R. Rapp. The raf-1 serine/threoine protein kinase. In: Herrmann, R. and R. Mertelsmann (eds): Hematopoietic growth factors in clinical applications. 2nd ed. (in press).

Marcu,K. B., S.A. Bossone, and A.J. Patel. 1992. Myc function and regulation. Ann. Rev. Biochem. 61:809-860.

Maslinski, W., B. Remillard, M. Tsudo and T.B. Strom. 1992. Interleukin-2 (IL-2) induces tyrosine kinase-dependent translocation of active raf-1 from the IL-2 receptor into the cytosol. J. Biol. Chem. 267:15281-15284.

Miyashita, T., L. Hovey, T. Torigoe, S. Krajewsky, J. Troppmair, U.R. Rapp and J.C. Reed. Novel form of oncogene cooperation: synergistic suppression of apoptosis by combination of *bcl-2* and *raf* oncogenes. Oncogene (in press).

Moodie SA, B.M. Willumsen, M.J. Webber, and A.Wolfman. 1993. Complexes of Ras-GTP with Raf-1 and mitogen-activated protein kinase kinase. Science 260:1658-1661.

Munger, K., J. Pietenpol, M. Pittelkow, J. Holt, and H. Moses. 1992. Transforming growth factor beta-1 regulation of c-myc expression, pRB phosphorylation, and cell cycle progression in keratinocytes. Cell Growth Differ. 3:291-298.

Porras, A., K. Muszynski, U.R. Rapp and E. Santos. 1994. Dissociation between activation of Raf-1 kinase and the 42-kDa mitogen-activated protein kinase/90-kDa S6 kinase (MAPK/RSK) cascade in the insulin/Ras pathway of adipocytic differentiaion of 3T3 L1 cells. J. Biol. Chem. 269:12741-12748.

Rapp U.R., J. T. Bruder , and J. Troppmair . 1993. Role of the Raf signal transduction pathway in fos/jun regulation and determination of cell fates. CRC, Crit. Rev. (in press).

Reed, J.C. S. Yum, M.P. Cuddy, B.C. Turner, and U.R. Rapp. 1991. Differential regulation of the p72-74 Raf-1 kinase in 3T3 fibroblasts expressing ras or src oncogenes. Cell Growth Diff. 2:235-243.

Sato, T., M. Hanada, S. Bodrug, S. Irie, N. Iwama, L. Boise, C. Thompson, E. Golemis, L. Fong, H-G. Wang and J.C. Reed. Investigations of interactions between members of the Bcl-2 protein family using yeast two-hybrid system. (submitted).

Stokoe, D., S.G. Macdonald, K. Cadwallader, M. Symons, and J. Hancock. 1994. Activation of Raf as a result of recruitment to the plasma membrane. Science, 264:1463-1467.

Thomas, S.M., M. De Marco, G. D'Arcangelo, S. Halegoua, J.S. Brugge. 1992. Ras is essential for nerve growth factor- and phorbol ester-induced tyrosine phosphorylation of MAP kinases. Cell 68:1031-1046.

Troppmair, J., J.T. Bruder, H. App, H. Cai, L. Liptak, J. Szeberenyi, G.M. Cooper and U.R. Rapp. 1992. Ras controls coupling of growth factor receptors and protein kinase C in the membrane to Raf-1 and B-Raf protein serine kinases in the cytosol. Oncogene 7: 1867-1873.

Troppmair, J., J. T. Bruder, H. Munoz, P.A. Lloyd, J. Kyriakis, P. Banerjee, J. Avruch and U.R. Rapp. 1994. Mitogen-activated protein kinase/extracellular signal-regulated protein kinase activation by oncogenes, serum and 12-*O*-tetradecanoylphorbol-13-acetate requires raf and is necessary for transformation. J. Biol. Chem. 269:7030-7035.

Troppmair, J., M. Huleihel, J. Cleveland, J.F. Mushinski, J. Kurie, H.C. Morse, III, J.S. Was, M. Potter and U.R. Rapp. 1988. Plasmacytoma induction by J series of v-myc recombinant retroviruses: Evidence for the requirement of two (raf and myc) oncogenes for transformation. Current Topics in Micro. Immun. 141:110-114.

Van Aelst L, M. Barr, S. Marcus, A. Polverino, and M. Wigler. 1993. Complex formation between RAS and RAF and other protein kinases. Proc Natl Acad Sci USA 90:6213-6217.

Vaux, D., S. Cory, and J. Adams. 1988. *Bcl-2* gene promotes hematopoietic cell survival and cooperates with c-myc to immortalize pre-B cells. Nature 335:440-442.

Vojtek AB, S.M. Hollenberg, J.A. Cooper. 1993. Mammalian Ras interacts directly with the serine/threonine kinase Raf. Cell 74:205-214.

Warner PH, P.R. Viciana, and J. Downward. 1993. Direct interaction of Ras and the amino-terminal region of Raf-1 *in vitro.* Nature (Lond) 364:352-355.

Zhang X-f, J. Settleman, J.M. Kyriakis, E. Takeuchi-Suzuki, Elledge, M.S. Marshall, J.T. Bruder, U.R. Rapp,and J. Avruch. 1993. Normal and oncogenic p21ras proteins bind to the amino-terminal regulatory domain of c-Raf-1. Nature. 364:308-313.

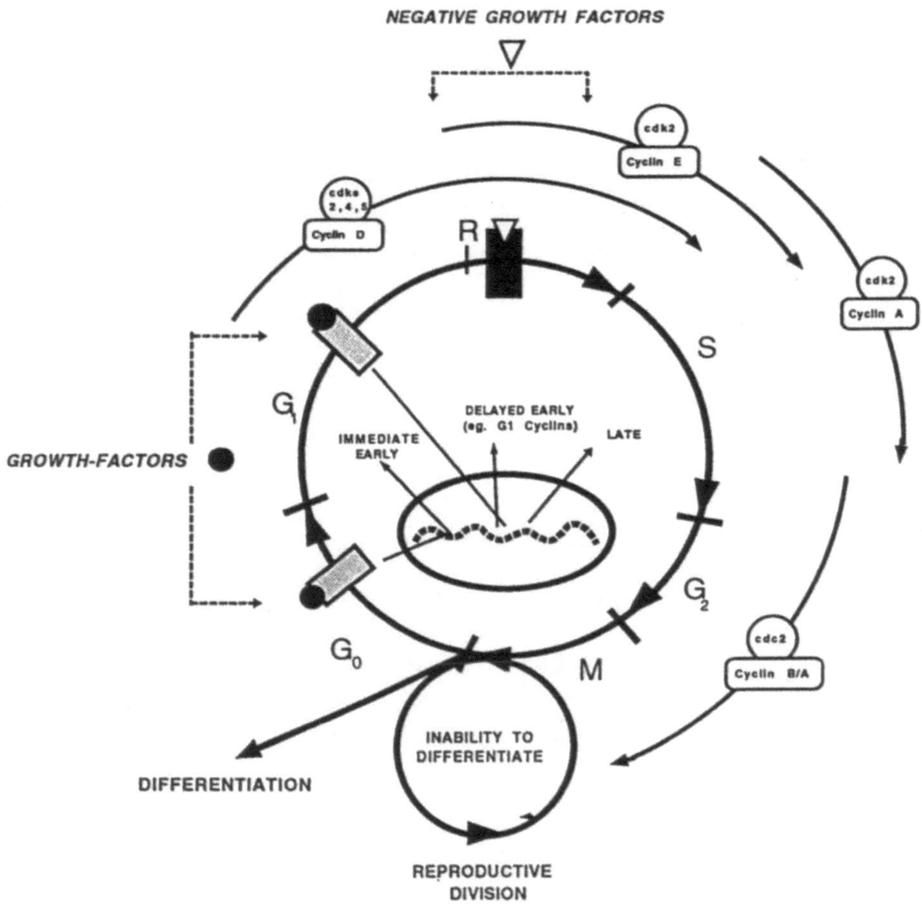

Figure 1. A model of key regulatory events in the mammalian cell cycle mediated by growth factors and negative growth effectors.

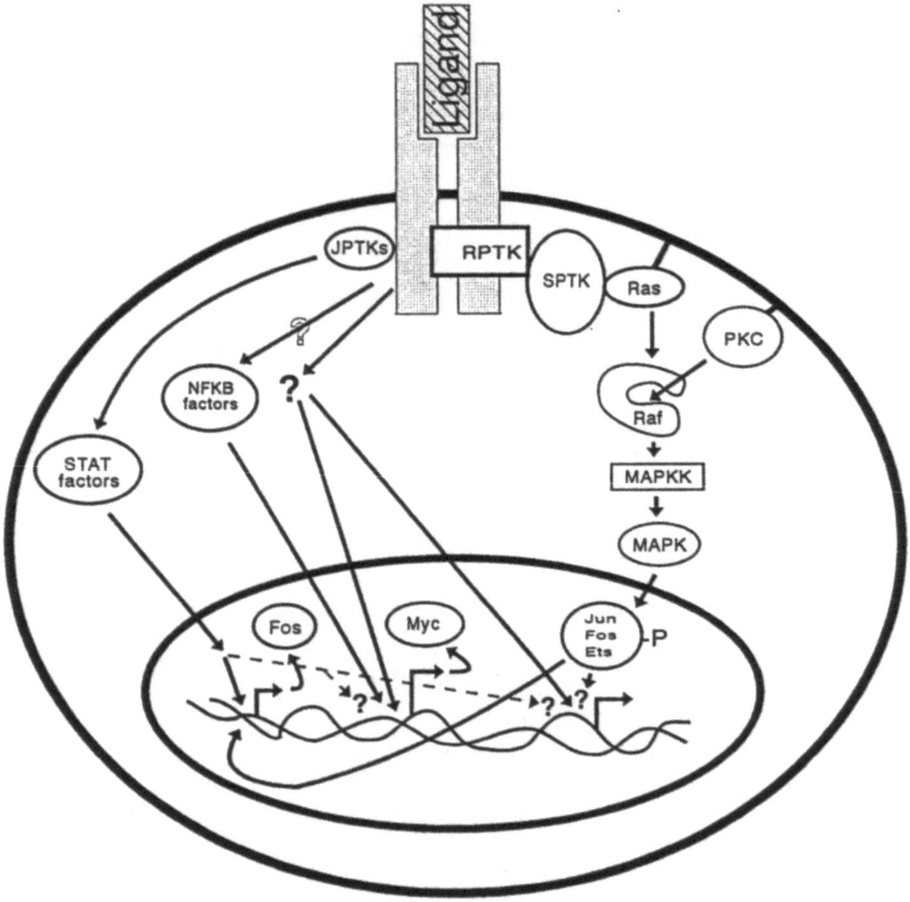

Figure 2. Growth factor receptor activation of Raf-1. These receptor systems include those with intrinsic or associated with phosphotyrosine kinase activity. Ligand-dependent Raf-1 kinase activation can be triggered by both receptor groups. Abbreviations used: RPTK, receptor phosphotyrosine kinase; SPTK, Src phosphotyrosine kinase; JPTK, JAK phosphotyrosine kinase; STAT factors, signal transducers/activators of transcription; PKC, protein kinase C; MAPKK, mitogen-activated protein kinase kinase; and MAPK, and mitogen-activated protein kinase. Fos, Jun, Myc and Ets represent various transcription factors.

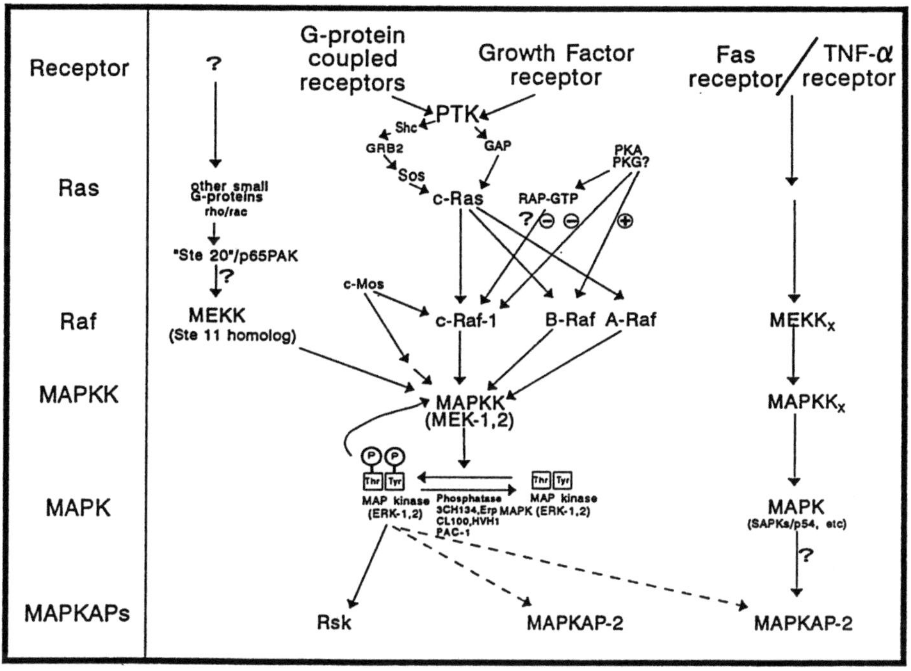

Figure 3. The Raf-1/MAPK pathway is presented in this schematic representation as having five major tiers. Each tiers contains one or more homologs which have been identified to date. It is possible there are other components of each tier that have not yet been identified. These are indicated with a subscript “x”.

Growth Regulation: Apoptosis, BCL-2, Dormancy

Defective Apoptosis due to Bcl-2 Overexpression May Explain Why B-CLL Cells Accumulate in G0

D. Gottardi, A. Alfarano, A.M. De Leo, A. Stacchini, L. Bergui and F. Caligaris-Cappio.
Dipartimento di Scienze Biomediche e Oncologia Umana, Università di Torino, Torino, Italy

B-chronic lymphocytic leukemia (B-CLL) is an accumulative disease of resting B lymphocytes (1). As the clonal expansion of B-CLL appears to reflect an extended survival of monoclonal B cells rather than an acceleration of their proliferative activity, the crucial problem of B-CLL biology becomes why these cells accumulate in the G0 phase of the cell cycle. Several data (reviewed in ref. 2) lead to conclude that B-CLL cell extended survival may be due both to a number of abnormalities that prevent an adequate mitogenic response and to the malignant cell's inability to undergo apoptosis.

It has been suggested (3) that the expression of Bcl-2 gene may be central to B-CLL escape from apoptosis. Bcl-2 gene product is overexpressed in B-CLL cells (4, 5). The mechanisms underlying the increased expression are still poorly understood, as Bcl-2 levels do not reflect a rearrangement of Bcl-2 gene, which is a rare event in CLL cells (6), neither a gene amplification, nor a Bcl-2 mRNA prolonged half-life (7). Recent findings suggest the possibility that Bcl-2 overexpression might be the result of Bcl-2 gene hypomethylation (5). The expression of Bcl-2 mRNA and protein can be modulated in B-CLL cells induced to activation, proliferation and/or differentiation by protocols of B cell stimulation which include cytokines and thioredoxin (4). These findings indicate that proliferation and Bcl-2 expression are inversely related in malignant B-CLL cells from peripheral blood, as it is observed in normal B lymphocytes from secondary follicles (8, 9). However, the magnitude of stimuli required to downregulate Bcl-2 indicates that the ability to downregulate Bcl-2 is miniaturized in B-CLL cells in comparison with normal B lymphocytes.

To investigate the role of Bcl-2 overexpression in B-CLL pathophysiology we have used fludarabine, a drug with a marked clinical activity in some B-CLL cases refractory to conventional treatment (10) which has been shown to act by inducing apoptosis of target cells (11). The aim was to define whether fludarabine is able to induce apoptosis of fresh B-CLL samples and whether apoptosis might be related to Bcl-2 modifications.

Material and Methods

Peripheral blood malignant B cells from 7 patients were separated on Ficoll-Hypaque gradient. Cells were cultured at the concentration of $2x10^6$/ml in RPMI 1640 + FCS 10% in presence of: a) fludarabine at the concentration of 0,6 μg/ml, 6 μg/ml and 30 μg/ml; b) fludarabine 6 μg/ml + methilprednisolone 1mM.

After 4, 8, 24, 48, 72 hours of culture, cells were harvested and analysed. Total RNA was extracted and analyzed by RT-PCR and by Northern Blot for Bcl-2 expression using the MBR probe (kind gift of Dr. C. Croce, Jefferson University,

Philadelphia). The controls were a β2 microglobulin probe for RT-PCR and an actin probe (mouse alfa-actin cDNA clone, p91) for Northern Blot experiments. Bcl-2 protein levels were evaluated with a cytofluorograph using an anti-Bcl-2 antibody (kindly provided by Dr David Mason, John Radcliffe Hospital, Headington, UK). DNA was extracted and electrophoresed on 2% agarose gel to establish if internucleosomal DNA cleavage had occurred. Apoptosis was also evaluated by cytofluorograph analysis and by morphology on May Grunwald Giemsa stained cytospins.

All patients were treated with fludarabine according to standard protocols (10): 40 mg/square meter for 5 days associated with prednisolone 50 mg/ square meter. *In vitro* studies have been compared to the clinical consequences of the *in vivo* administration of the drug.

Results

Fludarabine May Induce Bcl-2 Downregulation in Vitro

In 2/7 cases studied, a significant downregulation of Bcl-2 mRNA was evident at 24h and maximal at 48h (fig. 1). A downregulation of Bcl-2 protein was prominent after 48 h of in vitro culture in presence of fludarabine (Fig. 2). Both events took place in the samples treated with fludarabine 30 μg/ml or with fludarabine 6 μg/ml + methilprednisolone 1mM. No Bcl-2 mRNA downregulation nor Bcl-2 protein modification were observed in the remaining five cases irrespective of the concentration of fludarabine and/or the addition of methiprednisolone.

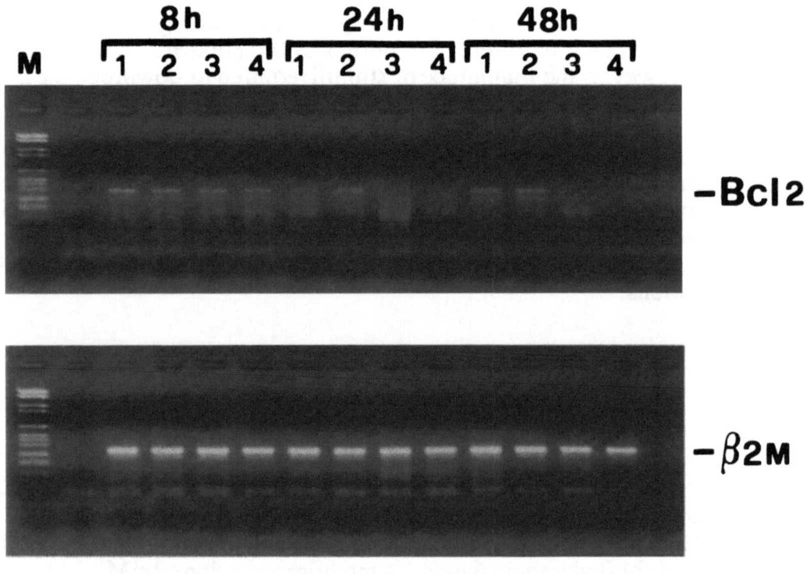

Fig. 1: RT-PCR of B-CLL cells cultured for different times in the following conditions: lane 1: control; lane 2: methiprednisolone 1mM; lane 3: fludarabine 30

μg/ml; lane 4: fludarabine 30 μg/ml + methilprednisolone 1 mM.

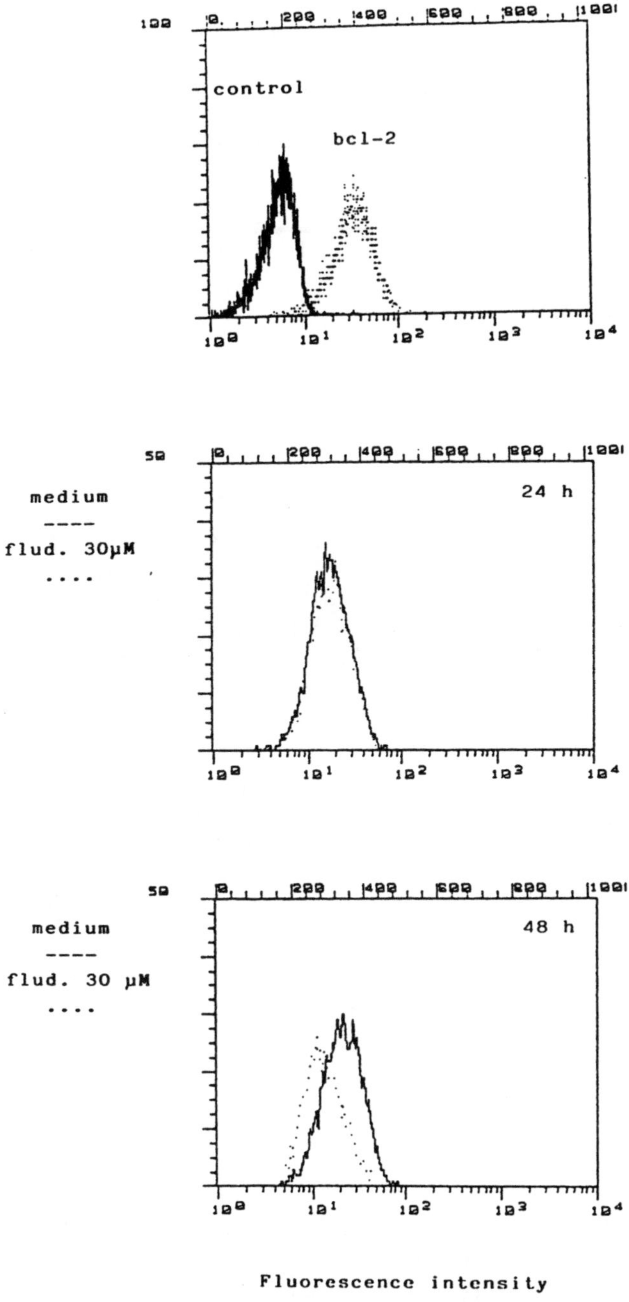

Fig. 2: cytofluorograph assessment of Bcl-2 protein downregulation

Fludarabine-Induced Bcl-2 Downregulation is Followed by Massive in Vitro Apoptosis

In the 2 cases where fludarabine treatment *in vitro* had been associated with Bcl-2 downregulation a significant degree of apoptosis could be subsequently revealed by morphology, cytofluorometric analysis and DNA ladder (Fig. 3). Apoptosis was maximal after 72h of culture in presence of fludarabine 30 µg/ml or fludarabine 6 µg/ml + methiprednisolone 1mM. On the contrary, the remaining five cases, where no Bcl-2 downregulation had been observed after *in vitro* exposure to the drug, also showed either low or absent evidence of apoptosis.

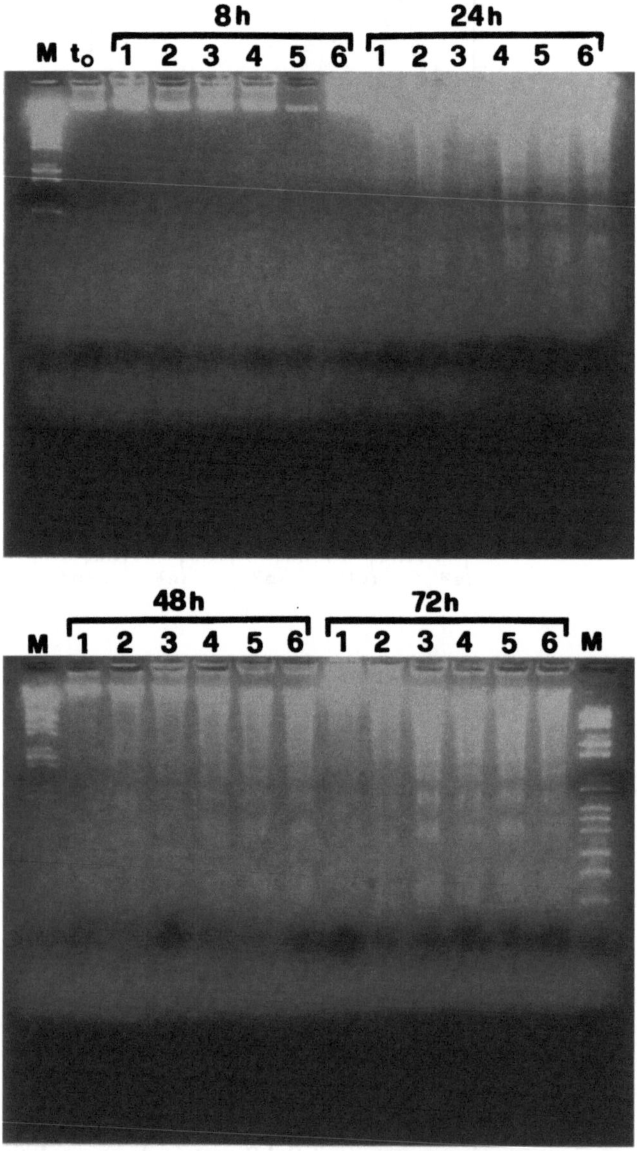

Fig. 3: DNA ladder from B-CLL cells cultured for different times in the following conditions: lane 1: control; lane 2: methilprednisolone 1mM; lane 3: fludarabine 6 μg/ml; lane 4: fludarabine 6 μg/ml + methilprednisolone 1mM; lane 5: fludarabine 30 μg/ml; lane 6: fludarabine 30 μg/ml + methilprednisolone 1mM. M: markers; t0: time 0.

B-CLL and Fludarabine: in Vitro / in Vivo Correlations

The *in vitro* data correlated with the *in vivo* outcome. The cases who had shown in vitro Bcl-2 downregulation and prominent apoptosis after the cell exposure to fludarabine also exhibited a markedly conspicuous clinical response. As for the remaining 5 patients, three had a clinical response that was slow and lagged behind, while the other two did not show any *in vivo* clinical response.

DISCUSSION

A plausible conclusion from these findings is that Bcl-2 overexpression inhibits apoptosis and leads to the relentless accumulation of malignant cells frozen in G0. It is presently unclear whether B-CLL cells do not easily downregulate Bcl-2 because they are unable to be activated by the mitogenic signals that activate normal B lymphocytes or whether they cannot be activated because they are unable to switch off Bcl-2 expression. Nevertheless, the data here presented document that, irrespective of the mechanisms that lead to Bcl-2 overexpression in B-CLL cells, Bcl-2 certainly might prevent apoptosis and favour the malignant cell blockage in G0. It might be asked why fludarabine has been able to downregulate Bcl-2 only in 2/7 cases. This question brings in the issue of why Bcl-2 antisense oligodeoxynucleotides are effective in promoting apoptosis only in a limited number of cells (7). At least two reasons may be advocated and are at present experimentally tested. First, Bcl-2 is a member of a family of genes whose known members, Bcl-xL, Bcl-xS and Bax, either synergize with Bcl-2 or counteract its activity (12, 13). We have to learn which is the pattern in fresh cells from different patients and how this pattern may relate to the in vitro and in vivo response both to fludarabine and to antisense oligodeoxynucleotides. Second, it is still unclear which role the deregulation of Bcl-2 has in the mechanisms that control apoptosis and involve the cooperation between Bcl-2, APO1/Fas and c-myc (14, 15). Notwithstanding that, it is clear that by downregulating Bcl-2 we are able to cause massive apoptosis of B-CLL cells: this observation not only underlines the central role of Bcl-2 in preventing B-CLL cell apoptosis and promoting their G0 accumulation but also may help developing new and more promising approaches to the treatment of the disease.

Acknowledgements: this work was supported by A.I.R.C. and by Progetto Finalizzato CNR - A.C.R.O. D.G. is recipient of an A.I.R.C. fellowship; A.M.D.L. is recipient of a G. Ghirotti fellowship. The helpful secretarial assistance of Mrs. Giuliana Tessa is greatly acknowledged.

References

1. Dameshek W (1967) Chronic lymphocytic leukemia - an accumulative disease of immunologically incompetent lymphocytes. Blood 29: 566-584.
2. Caligaris-Cappio F, Gottardi D, Alfarano A, Stacchini A, Gregoretti MG, Ghia P, Bertero M, Novarino A, Bergui L (1993) The nature of the B lymphocyte in B-chronic lymphocytic leukemia. Blood Cells 19: 601-613.
3. Schena M, Gottardi D, Ghia P, Larsson LG, Carlsson M, Nilsson K, Caligaris-Cappio F (1993) Role of bcl-2 in the pathogenesis of B-chronic lymphocytic leukemia. Leukemia and Lymphoma 11:173-179.
4. Schena M, Larsson LG, Gottardi D, Gaidano GL, Carlsson M, Nilsson K, Caligaris-Cappio F (1992) Growth and differentiation-associated expression of bcl-2 in B-chronic lymphocytic leukemia cells. Blood 79: 2981-2989.
5. Hanada M, Delia D, Aiello A, Stadtmauer E, Reed JC (1993) bcl-2 gene hypomethylation and high level expression in B-cell chronic lymphocytic leukemia. Blood 82: 1820-1828.
6. Adachi M, Cossman J, Longo D, Croce CM, Tsujimoto Y (1989) Variant translocation of the bcl-2 gene to immunoglobulin l light chain gene in chronic lymphocytic leukeemia. Proc. Natl. Acad. Sci. USA 86: 2771-2774.
7. Caligaris-Cappio F, Ghia P, Gottardi D, Parvis G, Gregoretti MG, Nilsson K, Schena M (1992) The role of BCL-2 in the natural history of B-chronic lymphocytic leukemia. Current Topics Microbiol Immunol 182: 279-286.
8. Liu YJ, Johnson GD, Gordon J, Maclennan ICM (1992) Germinal centres in T-cell-dependent antibody responses. Immunol Today 13: 17-21.
9. Liu YJ, Mason DY, Johnson GD, Abbot S, Gregory CD, Hardie DL, Gordon J, Maclennan ICM (1991) Germinal center cells express bcl-2-protein after activation by signals which prevent their entry to apoptosis. Eur J Immunol 21: 1905-1910.
10. O'Brien S, Kantarjian H, Beran M, Smith T, Koller C, Robertson LE, Lerner S, Keating M (1993) Results of fludarabine and prednisone therapy in 264 patients with chronic lymphocytic leukemia with multivariate analysis-derived prognostic model for response to treatment. Blood 82: 1695-1700.
11. Robertson LE, Chiubb S, Meyn RE, Story M, Ford R, Hittelman WN, Plunkett W (1993) Induction of apoptotic cell death in chronic lymphocytic leukemia by 2-Chloro-2'-deoxyadenosine and 9-b-D-arabynosyl-2-fluoroadenine. Blood 81: 143-150.
12. Boise LH, Gonzalez-Garcia M, Postema CE, Ding L, Lindsten T, Turks LA, Mao X, Nunez G, Thompson CB (1993) bcl-x, a bcl-2-related gene that functions as a dominant regulator of apoptotic death. Cell 74: 597-608.
13. Oltvai ZN, Milliman CL, Korsmeyer SJ (1993) Bcl-2 heterodimerizes in vivo with a conserved homolog, Bax, that accelerates programmed cell death. Cell 74: 609-619.
14. Mapara MY, Bargon R, Zugck C, Dohner H, Jonker RR, Krammer PH, Dorken B (1993) APO-1 mediated apoptosis or proliferation in human chronic B lymphocytic leukemia: correlation with bcl-2 oncogene expression. Eur J Immunol 23: 702-708.
15. Larsson LG, Gray HE, Totterman T, Pettersson U, Nilsson K (1987) Drastically increased expression of MYC and FOS protooncogenes during in vitro differentiation of chronic lymphocytic leukemia cells. Proc Natl Acad Sci USA 84: 223- 225.

Lyn Tyrosine Kinase Signals Cell Cycle Arrest in Mouse and Human B-Cell Lymphoma

R. H. Scheuermann[1], E. Racila[2] and J. W. Uhr[2]
Laboratory of Molecular Pathology and Department of Pathology[1], Department of Microbiology[2], University of Texas Southwestern Medical Center, Dallas, TX 75235.

Tumor dormancy is an operational term used to describe a prolonged quiescent state in which tumor cells are present, but tumor progression is not clinically apparent. In order to study the mechanisms underlying tumor dormancy, we have utilized two murine models of dormancy with an aggressive murine B-cell lymphoma (BCL_1) [1-3]. In the first model, BALB/c mice are immunized to the idiotype (Id) of the BCL_1 immunoglobulin (Ig) before challenge with BCL_1. In the second model, antibody to various epitopes on the BCL_1 Ig (an IgM-λ) are injected into SCID mice before or after challenge with BCL_1. In naive BALB/c or SCID mice, the tumor grows primarily in the spleen and splenomegaly is detected approximately one month after challenge with 3 x 10^4-10^6 BCL_1 cells. In Id-immune mice injected with 10^6 BCL_1, 70% do not develop splenomegaly by day 60. We have used this time period as an arbitrary cut-off and have considered such mice to harbor dormant tumor. When the spleens of Id-immunized BALB/c mice that display tumor dormancy are examined, a population of dormant lymphoma cells (DLC) can be isolated by multiparameter cell sorting. The rarity of λ light chain in BALB/c mice (usually less than 0.5%) has facilitated isolation of the DLC. Examination of these cells shows that they are physiologically different from those of BCL_1 cells growing in non-immune BALB/c. Thus DLC are smaller in size, have a less "aggressive" morphology, a different pattern of oncogene expression and a portion of the cells appear to be in cell-cycle arrest (CCA) as evidenced by an increased proportion of cells in the G_0 or G_1 phases of the cell cycle. The population of DLC was relatively stable for the 200 days of observation (about 0.5 x 10^6 DLC/spleen). Since there still are cycling DLC, and the size of the population is stable, we presume that cell death must be occurring and balancing cell replication. Almost 100 dormant BALB/c mice have been followed for up to two years of age. Regrowth of tumor occurs at a steady rate suggesting a stochastic process. In 10 mice in which regrowth of tumor occurred, the "escapee" tumor cells were reinjected into Id-immune mice. In 8/10 mice, dormancy was not induced suggesting that the cells had undergone a further mutation which had made them non-susceptible to the immune response.

In an effort to determine the cellular changes responsible for the phenomenon of dormancy and to study the underlying mechanisms of dormancy induction, we have evaluated a cell line, BCL_1.3B3, derived from BCL_1 which behaves like BCL_1 with regard to induction of dormancy in both Id immune and SCID-antibody mice. Importantly, this cell line allows *in vitro* analyses that are not possible with BCL_1 which is passaged only *in vivo*. Results from the SCID-antibody model indicate that anti-Ig antibodies must be the major contributors to dormancy induction. Therefor, we have examined the effects of anti-Ig treatment on cell growth of BCL_1.3B3 [4]. A variety of different polyclonal and monoclonal

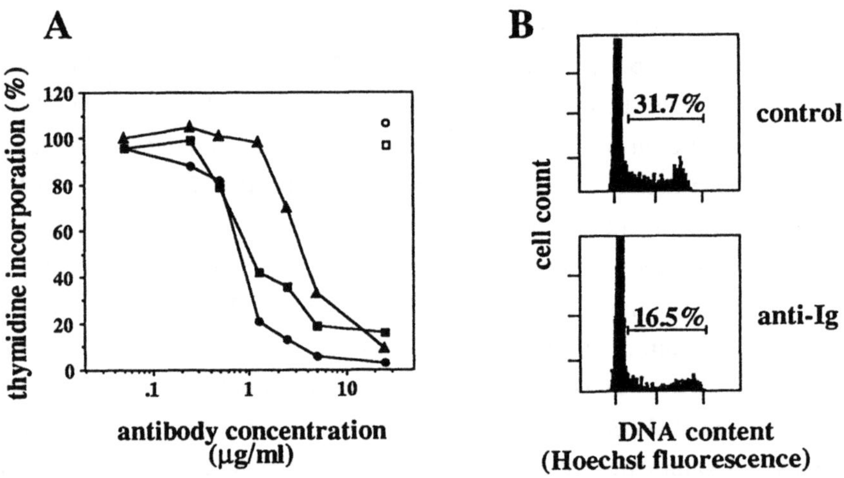

Fig. 1. Anti-Ig antibodies induce cell cycle arrest in BCL_1.3B3. A. Induction of growth arrest by different anti-Ig antibodies. Cells (3 x 10^4) were plated in triplicate in 0.1 ml of medium containing antibodies at the indicated concentrations. After 8 hr, 1 µCi ^{3}H-thymidine was added; cells were harvested and analyzed for incorporation 16 hr later. Thymidine incorporation is given as the percent of incorporation in cells treated with medium alone. Anti-Ig antibodies used were goat anti-mouse µ (closed squares), rabbit anti-mouse µ (closed triangles), and rabbit anti-BCL1 idiotype (closed circles); control antibodies were rabbit anti-ovalbumin (open circles) and goat anti-ovalbumin (open squares). All antibodies were affinity purified against their cognate antigen. B. Anti-Ig induces cell cycle arrest. Cells were treated with either rabbit anti-ovalbumin (control Ab) or rabbit anti-BCL_1 idiotype (anti-Ig) at 25 µg/ml for 24 hr and analyzed for DNA content by Hoechst fluorescence and for viability by 7-amino actinomycin D (7-AAD) exclusion. In the histograms of Hoechst fluorescence gating on viable cells (7-AAD$^-$), cells with a 1n DNA content are in the G_0 or G_1 phases of the cell cycle, cells with $1<n<2$ DNA content are in the S phase, and cells with a 2n DNA content are in G_2 or M. The percent of cells in the combined S, G_2 and M phases is indicated in each panel as a measure of cell cycle progression.

anti-Ig reagents cause growth arrest of BCL$_1$.3B3 as measured by ^{3}H-thymidine incorporation (Fig. 1A). Growth arrest induced by anti-Ig consists of two different components - apoptosis and cell cycle arrest. Apoptosis in treated cells has been demonstrated (data not shown) by classical DNA laddering, the presence of cells on flow cytometry that have a reduced content of DNA but impermeable plasma membranes and, finally, detection of DNA-strand breaks using the APO-tag methodology. Cell cycle arrest is indicated by a reduction in the proportion of viable cells in the S, G_2 and M phases of the cell cycle (Fig. 1B).

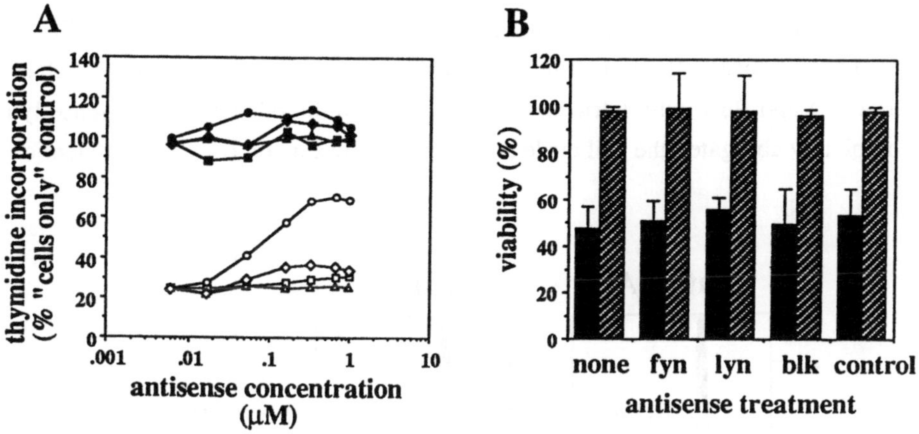

Fig. 2. Antisense oligonucleotides to the lyn tyrosine kinase prevents growth arrest but not cell death induced by anti-Ig in BCL$_1$.3B3. A. Prevention of growth arrest by antisense lyn. Cells were pre-treated with antisense (AS) oligonucleotides, either control AS (diamonds), ASfyn (squares), ASblk (triangles) or ASlyn (circles) for 24 hrs. Either control antibodies (closed symbols) or anti-Ig antibodies (open symbols) were added and thymidine incorporation measured. Thymidine incorporation is given as the percent of incorporation in cells treated with medium alone. Results are a pool of five different experiments. B. Anti-Ig effect on cell viability is not abrogated by ASlyn. BCL$_1$.3B3 cells were pre-incubated for 24 hr with AS oligonucleotides at 0.67 μM as indicated, treated with rabbit anti-ovalbumin (hatched bars) or rabbit anti-BCL$_1$ idiotype (solid bars) antibodies, and analyzed 24 hr later for viability by trypan blue exclusion. Results are an average of five different experiments. Error bars represent ± standard deviation.

Anti-Ig antibodies that are effective in inducing dormancy are also able to induce the cell cycle arrest and apoptosis responses *in vitro*, whereas other antibodies that bind to different cell surface molecules which are ineffective *in vivo* have no effect on BCL$_1$.3B3 cells *in vitro*. We therefor conclude that dormancy is induced by the capacity of anti-Ig to cause apoptosis and cell cycle arrest through the activation of signaling pathways initiating with membrane Ig (mIg) expressed by BCL_1.3B3. Several src-family tyrosine kinases, Lyn, Fyn, Blk and Lck, have been found to associate with mIg and are thought to be important in the signaling cascades proceeding from mIg [5-7]. To determine if any of these kinases are essential for signaling growth arrest following mIg engagement, BCL_1.3B3 cells have been treated with antisense oligonucleotides targeted at the translation initiation regions of each kinase gene and the effects on anti-Ig-mediated growth inhibition measured by ^{3}H-thymidine incorporation (Fig. 2A) [4]. Antisense reagents targeted at the blk or fyn genes had no effect on anti-Ig-mediated growth inhibition. On the other hand, antisense lyn abrogated much of the inhibitory effect. Evaluation of apoptosis by measuring cell viability in antisense treated cells revealed that antisense lyn had no effect on the ability of anti-Ig to induce cell death (Fig. 2B). These results were confirmed by DNA laddering and flow cytometric analysis as well. On the other hand, antisense lyn completely abrogated the cell cycle arrest induced by anti-Ig (Fig. 3). Prevention

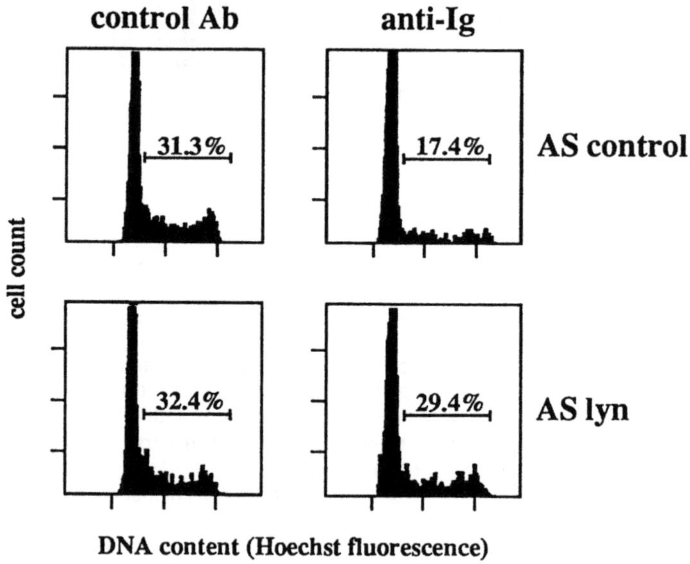

Fig. 3. ASlyn abrogates anti-Ig induced cell cycle arrest. BCL_1.3B3 cells were pre-incubated for 24 hr with either AS control (top row) or ASlyn (bottom row) at 0.67 µM, treated with rabbit anti-ovalbumin (left column) or rabbit anti-BCL_1 idiotype (right column), and analyzed 24 hr later for cell cycle progression as described for Figure 2. The percent of viable cells in the combined S, G_2 and M phases of the cell cycle 24 hr later is indicated.

of cell cycle arrest but not apoptosis is probably the explanation for the partial reversal in growth inhibition seen in the ^{3}H-thymidine assay. These results suggest that the signaling pathways responsible for these two responses must be partially distinct and must separate shortly after the signal enters the cell.

In order to determine if these signaling pathways are a common component of B cell neoplasms, the response to mIg engagement was examined in the Daudi Burkitt's lymphoma (human) cell line. As in BCL_1.3B3, Daudi cells treated with anti-Ig were induced into apoptosis (data not shown) and cell cycle arrest (Fig. 4) [4]. These two signaling pathways must be very similar in the two cell lines since antisense lyn treatment abrogated cell cycle arrest induced by mIg engagement in Daudi (Fig. 4), but had no effect on apoptosis (data not shown).

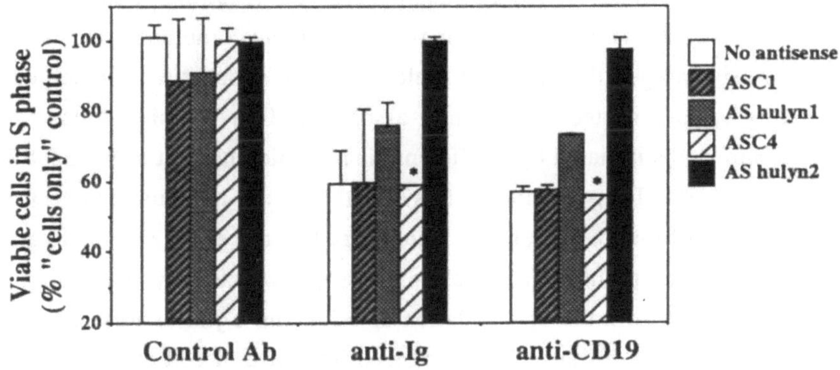

Fig. 4. ASlyn abrogates cell cycle arrest induced by anti-Ig or anti-CD19 in Daudi Burkitt's lymphoma. Daudi cells were suspended in growth medium to 0.7 x 10^6 cells/ml and pre-treated with or without AS oligonucleotides (as indicated) at a final concentration of 5 µM for 24 hrs. AS hulyn1 and AS hulyn2 target the lyn translation initiation region and an internal lyn sequence, respectively; ASC1 and ASC4 are two different control AS reagents. Cells were treated with either goat anti-human IgM, HD-37 (anti-CD19), or goat anti-ovalbumin and incubation continued for an additional 24 hrs. before analysis of DNA content by 7-AAD and Hoechst fluorescence. As a measure of cell cycle progression the percent of viable cells in the S phase of the cell cycle was calculated and compared with the percent of cells in S phase for cells in medium alone ("cells only"). Values represent the average of two experiments ± one standard deviation, except that values indicated with * which represent a single determination.

In Daudi, the engagement of a different cell surface receptor, CD19, has been found to induce cell cycle arrest, but differs from mIg engagement in that little apoptosis is observed [4, 8]. Since Lyn/CD19 interactions have recently been reported [9], the importance of Lyn in the cell cycle arrest signaling initiating from

CD19 was examined. Once again, treatment of cells with antisense lyn abrogated the cell cycle arrest response induced by anti-CD19 (Fig. 4).

Discussion

Based on these results we have developed a model for the cellular changes associated with tumor dormancy induction in BCL_1 (Fig. 5). In Id-immune mice, anti-idiotype antibodies bind to the cell surface of the tumor cells. However, rather than initiating cellular destruction through classical immunological effector mechanisms like ADCC or complement-mediated effects, binding of these antibodies to mIg activates signal transduction cascades resulting in the inhibition of tumor cell growth. Growth inhibition results from the combined action of two independent signaling pathways - cell cycle arrest and apoptosis. The cell cycle arrest pathway is initiated by the mIg-associated Lyn tyrosine kinase; the apoptosis pathway is initiated by a different kinase which has yet to be identified (possibly Blk [7]). Transmission of this signal into the nucleus results in the down-regulation of c-myc expression, which may be critical for cell cycle arrest induction, the down-regulation of bcl2 expression, which may be critical for apoptosis induction, and the up-regulation of c-fos and c-jun expression which may be critical for one or both. The combined influence of cell cycle arrest and apoptosis is sufficient to maintain the population of tumor cells stable for long periods of time resulting in tumor dormancy. Inactivation of any component of these two pathways would then allow the tumor to escape dormancy and regrow.

These findings demonstrate that even highly malignant neoplastic cells contain signaling pathways that, when activated, can arrest the growth of the transformed cells. Indeed, tumor growth is governed by a complex interaction between growth stimulatory, inhibitory and survival pathways. The transformed phenotype of any particular neoplastic clone is dictated by the pattern of which pathways remain active and which have been inactivated during the transformation process.

The cellular response to mIg or CD19 engagement in the Daudi Burkitt's lymphoma includes the induction of cell cycle arrest in some of the neoplastic cells. This finding is particularly relevant to cancer therapy since antibody reagents targeting these two cell surface molecules are being evaluated in clinical

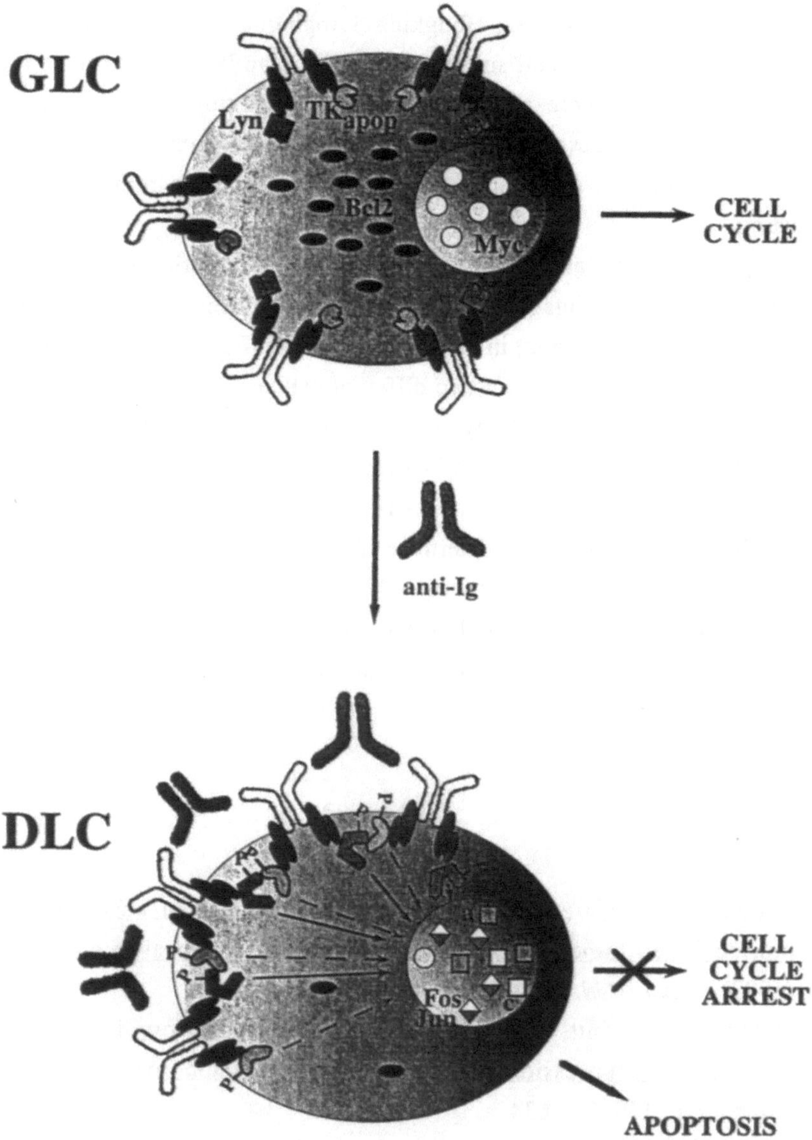

Fig. 5. Model for anti-Ig-mediated growth control in B cell lymphomas. Growing lymphoma cells (GLC) are converted into dormant lymphoma cells (DLC) through the action of anti-Ig antibodies. See text for details.

trials for the treatment of non-Hodgkin's lymphoma [10,11]. The fact that engagement of these two cell surface receptors by antibody induces cell cycle arrest, and that cell cycle arrest is associated with tumor dormancy in the BCL_1.3B3 model suggests that tumor regression in patients given this type of treatment, may be due to the induction of tumor dormancy rather than tumor eradication.

Our results indicate that the Lyn tyrosine kinase is essential for signaling cell cycle arrest following engagement of mIg in two different cell lines from two different species. In addition, in Daudi, Lyn is essential for signaling cell cycle arrest initiated by engagement of two different cell surface molecules - mIg and CD19. Although Lyn was initially characterized as a B cell-specific kinase it is now clear that Lyn is also expressed in a variety of other normal and transformed cell types, including macrophages, platelets, T cell leukemias and neuroblastomas [12-15]. It will be important to determine if the Lyn kinase is a mediator of cell cycle arrest originating from a variety of cell surface receptors in a variety of different cell types. If so, this would make Lyn an ideal candidate for therapeutic targeting; agonistic activation of Lyn might arrest cell growth in a variety of cancers.

References

1. Krolick, K.A., Uhr, J.W., Slavin, S. & Vitetta, E.S. *In vivo* therapy of a murine B cell tumor (BCL1) using antibody-ricin A chain immunotoxins. (1982) *J. Exp. Med.* **155**, 1797-1809.
2. George, J.J.T., Tutt, A.L. & Stevenson, F.K., Anti-idiotypic mechanisms involved in suppression of a mouse B cell lymphoma *BCL1*., (1987) *J.Immunol.* **138**, 628-634.
3. Yefenof, E., Picker, L.J., Scheuermann, R.H., Tucker, T.F., Vitetta, E.S. & Uhr, J.W. Cancer dormancy: isolation and characterization of dormant lymphoma cells. (1993) *Proc.Natl.Acad.Sci.U.S.A.* **90**, 1829-1833.
4. Scheuermann, R.H., Racila, E. Tucker, T., Yefenof, E., Street, N.E., Vitetta, E.S., Picker, L.J. & Uhr, J.W. Lyn tyrosine kinase signals cell cycle arrest but not apoptosis in B-lineage lymphoma cells (1994) *Proc.Natl.Acad.Sci.U.S.A.* **91**, 4048-4052.

5. Cambier, J.C. & Campbell, K.S. Membrane immunoglobulin and its accomplices: new lessons from an old receptor. (1992) *FASEB J.* **6**, 3207-3217.
6. Kim, K.-M., Alber, G., Weiser, P. & Reth, M. Signaling function of the B-cell antigen receptors. (1993) *Immunol.Rev.* **132**, 125-145.
7. Yao, X.-R. & Scott, D.W. Expression of protein tyrosine kinases in the Ig complex of anti-mu-sensitive and anti-mu-resistant B-cell lymphomas: role of the p55blk kinase in signaling growth arrest and apoptosis. (1993) *Immunol. Rev.* **132**, 163-186.
8. Ghetie, M.-A., Picker, L.J., Richardson, J.A., Tucker, K., Uhr, J.W. & Vitetta, E.S. Anti-CD19 inhibits the growth of human B-cell tumor lines *in vitro* and of Daudi cells in SCID mice by inducing cell cycle arrest. (1994) *Blood* **83** , 1-9
9. van Noesel, C.J.M., Lankester, A.C., van Schijndel, G.M.W. & Van Lier, R.A.W. The CR2/CD19 complex on human B cells contains the src-family kinase Lyn, (1993) *Interntl.Immunol.* **5**, 699-705.
10. Brown, S.L., Miller, R.A. & Levy, R. Anti-idiotype antibody therapy of B-cell lymphoma (1989) *Seminars in Oncology* **16**, 199-210.
11. Vitetta, E. S., Thorpe, P.E. & Uhr, J. W. Immunotoxins: magic bullets or misguided missiles? (1993) *Immunol.Today* **14**, 252-259.
12. Boulet, I., Ralph, S., Stanley, E., Lock, P., Dunn, A.R., Green, S.P. & Phillips, W.A. Lipopolysaccharide- and interferon-gamma-induced expression of hck and lyn tyrosine kinases in murine bone marrow-derived macrophages (1992) *Oncogene* **7**, 703-710.
13. Cichowski, K., McCormick, F. & Brugge, J.S. p21rasGAP association with Fyn, Lyn, and Yes in thrombin-activated platelets (1992) *J.Biol.Chem.* **267**, 5025-5028.
14. O'Connor, R., Torigoe, T., Reed, J.C. & Santoli, C. Phenotypic changes induced by interleukin-2 (IL-2) and IL-3 in an immature T-lymphocytic leukemia are associated with regulated expression of IL-2 receptor beta chain and of protein tyrosine kinases LCK and LYN (1992) *Blood* **80**, 1017-1025
15. Bielke, W., Ziemieki, A., Kappos, L. & Miescher, G.C. Expression of the B cell-associated tyrosine kinase gene Lyn in primary neuroblastoma tumours and its modulation during the differentiation of neuroblastoma cell lines (1992) *Biochem.Biophys.Res.Comm.* **186**, 1403-1409.

A Link Between the Antioxidant Defense System and Calcium: A Proposal for the Biochemical Function of Bcl-2.

G.W. Bornkamm[1] and C. Richter[2]

[1]Institut für Klinische Molekularbiologie und Tumorgenetik, Hämatologikum der GSF, 81377 München, Germany

[2]Laboratorium für Biochemie I, ETH, CH 8092 Zürich, Switzerland

Introduction

Tumorigenesis has long been viewed as a problem of perturbed regulation of cell proliferation. It has, however, become increasingly apparent during the last years that disturbance of the equilibrium between cell survival and cell death may equally contribute to the development of a tumor. Elimination of cells has first been described by morphologists as an important physiological process in developmental biology for which the term programmed cell death (PCD) or apoptosis has been coined, and has since then been recognized as a generally important phenomenon in many different areas of biology (Wyllie et al., 1980). Apart from morphological criteria, apoptosis is only poorly defined and discrimination from other forms of cell death is often difficult. Regardless of the definition of apoptosis, it is apparent that susceptibility versus resistance to toxic conditions or death inducing signals is an extremely important property of a cell determining its fate in its environmental context.

Bcl-2 has become the paradigm of a growing list of (onco)genes which regulate cell survival rather than proliferation (Vaux et al., 1988; for review see Korsmeyer, 1992). The gene is activated in follicular lymphoma by chromosomal translocation and juxtaposition to one of the immunoglobulin gene loci (Tsujimoto et al., 1984). Constitutive expression of Bcl-2 in B cells of transgenic mice leads to prolonged survival of B cells after antigenic stimulation and a long lasting immune response resulting in follicular hyperplasia which predisposes to the development of B cell malignancies (McDonnell et al., 1989; McDonnell and Korsmeyer, 1991). The biochemical function of Bcl-2 protein is still elusive. Bcl-2 is a 24 KD protein with a hydrophobic carboxy terminus which targets it to the nuclear envelope, the endoplasmic reticulum and the mitochondrial membrane (Chen-Levy et al., 1989; Hockenbery et al., 1990; Monaghan et al., 1992). There is some controversy about the localization of Bcl-2 at the inner or outer mitochondrial membrane (Hockenbery et al., 1990, 1993). Deletion of the membrane anchor only partially inactivates the survival protecting function of Bcl-2 (Hockenbery et al., 1993).

Role of Bcl-2 in an antioxidant pathway

Several lines of evidence have implicated Bcl-2 as part of the cellular antioxidant defense system:

- Bcl-2 protects thymocytes against the damaging effect of γ-irradiation (Strasser et al., 1991),
- antioxidants and glutathione peroxidase promote survival of FL5.12 cells (pro B-lymphocytes) after Il-3 withdrawal similarly to Bcl-2 (Hockenbery et al., 1993),
- Bcl-2 protects FL5.12 cells against H_2O_2 toxicity (Hockenbery et al., 1993),
- Bcl-2 protects 2B4 (T cell hybridoma) cells against the toxicity of the redox cycling quinone menadione (Hockenbery et al., 1993),
- Bcl-2 does not interfere with the formation of hydroperoxides in 2B4 cells but blocks their damaging effect (lipid peroxidation) (Hockenbery et al., 1993),
- Bcl-2 protects neural cells against the damaging effect of t-butyl-hydroperoxide, menadione, and depletion of glutathione (GSH) (Zhong et al., 1993),
- protection is associated with an increase in GSH level and a decrease in the formation of free radicals (Kane et al., 1993),
- the intracellular GSH level is instrumental in protecting Burkitt lymphoma cells from being killed at low cell density and/or serum concentration (Falk et al., submitted),
- TNFα toxicity is mediated by superoxide anion production in mitochondria (Wong et al., 1989; Schulze-Osthoff et al., 1992; Hennet et al., 1993b),
- Bcl-2 protects L929 cells from TNFα toxicity without interfering with superoxide anion production (Hennet et al., 1993a),
- the phenotype of bcl-2 knock-out mice (fulminant lymphoid apoptosis, polycystic kidney disease, and hypopigmented hair) suggests that damage through oxygen free radicals is a common denominator in the develpment of the various symptoms (Veis et al., 1993).

Role of Bcl-2 in intracellular Ca^{2+} homeostasis and toxicity

In addition to the role of Bcl-2 as part of the antioxidant defense system, evidence is mounting which implicates Bcl-2 in Ca^{2+} homeostasis and toxicity:

- Bcl-2 protects thymocytes against the toxic effect of the Ca^{2+} ionophore ionomycine (Strasser et al., 1991),
- Bcl-2 inhibits cell death induced by Ca^{2+} ionophore A23187 in neural cells without altering the intracellular free Ca^{2+} concentration (Zhong et al., 1993),
- withdrawal of Il-3 from Il-3-dependent immature myeloid 32D cells induces redistribution of intracellular Ca^{2+} from the endoplasmic reticulum and nuclear envelope to the mitochondria. Bcl-2 inhibits redistribution of intracellular Ca^{2+} upon Il-3 withdrawal (Baffy et al., 1993),
- TNFα toxicity is blocked by Ruthenium Red, an inhibitor of Ca^{2+} uptake into mitochondria (Hennet et al., 1993b),
- Bcl-2 increases the mitochondrial membrane potential $\Delta\psi$ (Hennet et al., 1993a),

- toxicity of oxidants is blocked by inhibitors of "Ca^{2+} cycling" in mitochondria (i.e. by Ca^{2+} uptake as well as Ca^{2+} release inhibitors) and/or by preventing opening of the "Ca^{2+} dependent mitochondrial inner membrane permeability transition pore" (see below).

The interplay of oxidants and Ca^{2+}

Oxidants may increase the intracellular free Ca^{2+} level directly by inactivating the ATP driven Ca^{2+} pumps in the endoplasmic reticulum and plasma membrane (for review see Orrenius et al., 1989). An increase in the cytosolic free Ca^{2+} level is associated with an increase in Ca^{2+} import into mitochondria. Mitochondria have low physiological free Ca^{2+} levels but can accomodate large amounts of Ca^{2+} ions, thus acting as a safety device for the maintenance of cytosolic free Ca^{2+} homeostasis (for review see Richter, 1992). Ca^{2+} import into mitochondria is directly dependent on the cytosolic free Ca^{2+} concentration and is driven by the electrogenic gradient which is formed by expulsion of protons from mitochondria into the cytosol by the respiratory chain. Another direct action of oxidants may be an induction of a conformotional change in the mitochondrial Ca^{2+} release channel. Ca^{2+} ions are released from mitochondria in exchange to Na^+ and H^+, the contribution of Na^+ and H^+ to Ca^{2+} antiport varying in different tissues. The nature of the H^+/Ca^{2+} antiporter is still a matter of debate. It is regarded as a specific H^+/Ca^{2+} antiporter by some authors (Richter and Kass, 1991; Schlegel et al., 1992) and as a "Ca^{2+} dependent permeability transition pore" by others (Gunter and Pfeiffer, 1990).

Oxidants may perturb Ca^{2+} regulation in mitochondria also in an indirect fashion. Glutathione (GSH) is the main substrate for hydroperoxides and, by reduction of substrates, is itself converted to its oxidized disulfide (GSSG) by glutathione peroxidase (GSHPx). Glutathione is regenerated by glutathione reductase (GR) and NADPH yielding GSH and $NADP^+$, and NADPH itself is regenerated through the mitochondrial energy-linked transhydrogenase system at the expense of NADH. As the net reaction, oxidants are reduced thereby increasing the level of NAD^+. Under phyisological conditions, i.e. at low mitochondrial free Ca^{2+} concentration, NADH is regenerated from NAD^+ by oxidation of substrates such as fatty acids. NADH again drives the respiratory chain, the generation of the membrane potential $\Delta\psi$ and ATP production. At increased mitochondrial free Ca^{2+} concentration, however, NAD^+ , instead of being converted to NADH by substrate oxidation, is hydrolysed into ADP-ribose and nicotinamide by NAD glycohydrolase. Mitochondrial Ca^{2+} release was proposed to be linked to mitochondrial pyridine nucleotide oxidation by Lehninger et al. (1978) and was subsequently shown to be coupled to hydrolysis of NAD^+ (Lötscher et al., 1979, 1980), even though the precise biochemical mechanism involved has not been understood. Opening of the Ca^{2+} release channel has been proposed to be triggered by protein-mono-ADP-ribosylation (Frei and Richter, 1988), it is, however, also conceivable that cyclic ADP-ribose is involved as a second messenger. The recent discovery of an inositol trisphosphate (IP_3)-independent Ca^{2+} release pathway which is induced by cyclic ADP-ribose (Clapper et al., 1987; Lee et al., 1989; Koshiyama et al., 1991; Meszaras et al., 1993) might favor the latter possibility. Oxidative stress and Ca^{2+} thus synergize in the depletion of mitochondrial NADH and NAD^+ and in the stimulation of the Ca^{2+} release pathway

pathway in mitochondria. If, because of an inefficient extrusion or compartmentalization of Ca^{2+}, the cytosolic Ca^{2+} concentration remains untolerably high, Ca^{2+}, exported from the mitochondria, is reimported through the Ca^{2+} import channel thus leading to a futile Ca^{2+} import/export cycle ("Ca^{2+} cycling", Carafoli, 1987) dissipating the mitochondrial membrane potential and energy production. Inhibitors of Ca^{2+} uptake into mitochondria (ruthenium red) as well as inhibitors of Ca^{2+} release (cyclosporin A, L-carnitine) (Crompton et al., 1988; Broekemeier et al., 1989; Richter et al., 1990; Galli and Fratelli, 1993; Gogvadze and Richter, 1993) disrupt Ca^{2+} cycling and prevent the irreversible damage caused by oxidative stress.

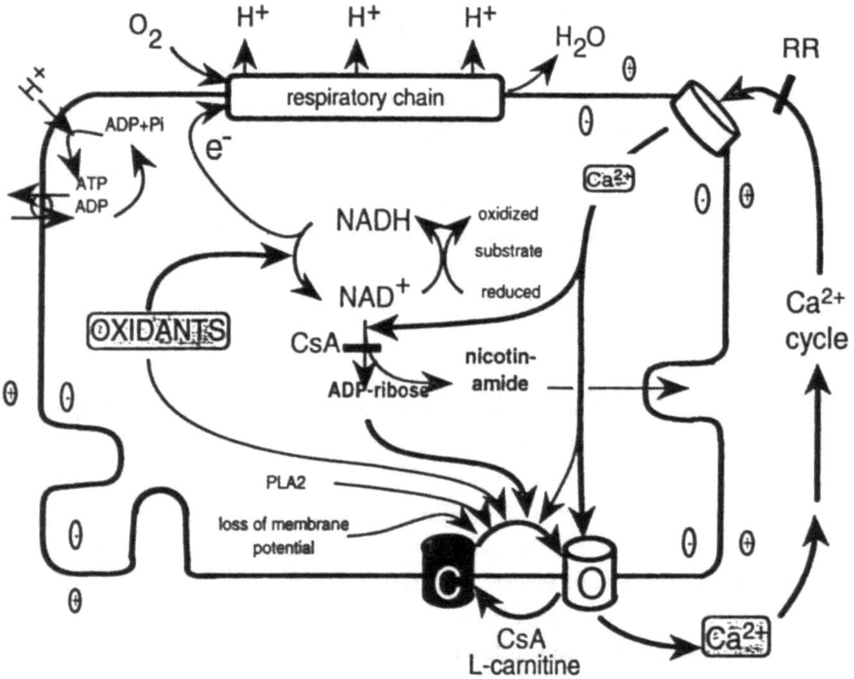

Figure 1:
Schematic representation of mitochondrial pyridine nucleotide and Ca^{2+} metabolism. C and O represent the Ca^{2+} release channel in closed and open configuration, respectively.
Abbreviations: CsA cyclosporin A, PLA2: phospholipase A2, RR: ruthenium red.

Proposal for a biochemical function of bcl-2

We propose that Bcl-2 is involved in the regulation of Ca^{2+} release in endoplasmic reticulum, nuclear envelope and mitochondria. This model is compatible with (i) the subcellular localization of Bcl-2, (ii) the synergistic action of oxidants and Ca^{2+} in inducing irreversible cell damage, (iii) the fact that Bcl-2 interferes with the damaging effect of reactive oxygen species without acting as a scavenger of free radicals, and (iv) the fact that Bcl-2 prevents intracellular Ca^{2+} redistribution from intracellular stores to mitochondria.

The following targets of Bcl-2 for regulating Ca^{2+} release may be envisaged:

- cyclophilins (peptidyl-prolyl-cis-trans-isomerase) as targets of cyclosporin A, which may regulate NAD glycohydrolase activity and/or the opening of the Ca^{2+} release channel,
- NAD glycohydrolase,
- intermediates of cyclic ADP-ribose signalling,
- protein-mono-ADP-ribosylation,
- phospholipase A2,
- the Ca^{2+} release channel, or the "Ca^{2+} dependent inner mitochondrial permeability transition pore".

Perspective

A particular interesting property of Bcl-2 is its ability to exert a protective or partially protective function against a variety of different conditions which are detrimental to the cell such as γ-irradiation (Strasser et al., 1991), high temperature (Cuende et al., 1993), cytostatic drugs (Walton et al., 1993), virus infection (Alnemri et al., 1992), TNFα (Hennet et al., 1993a) and Fas/APO-1 treatment (Itoh et al., 1993), growth factor deprivation, and overexpression of genes inducing apoptosis such as p53 and c-myc (Fanadi et al., 1992; Wang et al., 1993). A notable exception is the death induced by antigen specific cytotoxic T cells, which is not counteracted by Bcl-2 (Vaux, 1993; Vaux et al., 1994). Most importantly, the protective potential of Bcl-2 defines a step on the pathway to cell death - presumably the final step -, which is shared among different toxic agents or conditions. Unterstanding the step on the pathway, which is defined by Bcl-2, will therefore help in designing rational strategies for protecting cells against damage induced by a variety of conditions.

The model outlined here suggests an important role of NADH/NAD^+ and NADPH/$NADP^+$ as crossover points of energy and redox metabolism in the regulation of Ca^{2+} homeostasis and the cell's susceptibility or resistance to apoptosis. Remarkably, the cell surface antigen CD38 has recently been shown to exhibit NAD glycohydrolase activity capable also of generating cyclic ADP-ribose (for review, see Malavasi et al., 1994). Cyclic ADP-ribose, again, was shown to act as a second messenger inducing Ca^{2+} release from intracellular stores through an IP_3- independent pathway (Clapper et al., 1987; Lee et al., 1989) . The localization of CD38 on the cell surface and its nature as an ectoenzyme might be difficult to reconcile with a function of CD38 in the regulation of intracellular Ca^{2+} release and apoptosis. However, NAD glycohydrolase may also be localized at other sites than the outer cell membranes. It will be interesting to see whether overexpression of NAD glycohydrolase is capable of increasing the susceptibility of a cell to undergo apoptosis when exposed to oxidative stress.

The Ca^{2+} release channel of mitochondria and other subcellular sites is biochemically still ill defined. Provided that Bcl-2 is in fact involved in regulation of Ca^{2+} release from various subcellular sites, Bcl-2 might be instrumental in defining biochemically and molecularly this important cellular structure. It might be particularly intriguing to see whether Bcl-2 is able to interfere with IP_3-independent Ca^{2+} signalling when it is ectopically expressed in, e.g., endocrine secretory cells or muscle cells.

References

Alnemri, E.S., Robertson, N.M., Fernandes, T.F., Croce, C.M. and Litwack, G. (1992) Overexpressed full-length human BCL2 extends the survival of baculovirus-infected Sf9 insect cells. Proc. Natl. Acad. Sci. USA 89: 7295-7299

Baffy, G., Miyashita, T., Williamson, J.R., Reed, J.C. (1993) Apoptosis Induced by Withdrawal of Interleukin-3 (IL-3) from an IL-3-dependent Hematopoietic Cell Line Is Associated with Repartitioning of Intracellular Calcium and Is Blocked by Enforced Bcl-2 Oncoprotein Production. J. Biol. Chem. 268: 6511-6519

Broekemeier, KM, Dempsey, ME and Pfeiffer, DR. (1989) Cyclosporine A is a potent inhibitor of the inner membrane permeability transition in liver mitochondria. J Biol Chem 264: 7826-7830

Carafoli, E. (1987) Intracellular Ca^{2+} homeostasis. Ann. Rev. Biochem. 56: 395-433

Chen-Levy, Z., Nourse, J. and Cleary, M. (1989) The bcl-2 Candidate Proto-Oncogene Product Is a 24-Kilodalton Integral-Membrane Protein Highly Expressed in Lymphoid Cell Lines and Lymphomas Carrying the t(14;18) Translocation. Mol. Cell. Biol. 9: 701-710

Clapper, D.L., Walseth, T.F., Dargie, P.J. and Lee, H.C. (1987) Pyridine Nucleotide Metabolites Stimulate Calcium Release from Sea Urchin Egg Microsomes Desensitized to Inositol Trisphosphate. J. Biol. Chem. 262: 9561-9568

Crompton, M., Ellinger, H. and Costa, A. (1988) Inhibition by cyclosporin A of a Ca^{2+}-dependent pore in heart mitochondria activated by inorganic phosphate and oxidative stress. Biochem. J. 255: 357-360

Cuende, E., Alés-Martínez, J.E., Ding, L., Gónzalez-García, M., Martinez-A, C. and Nunez, G. (1993) Programmed cell death by bcl-2 dependent and independent mechanisms in B lymphoma cells. EMBO J. 12: 1555-1560

Fanidi, A., Harrington, E.A. and Evan, G.I. (1992) Cooperative interaction between c-myc and bcl-2 proto-oncogenes. Nature 359: 554-556

Frei, B. and Richter, C. (1988) Mono(ADP-ribosylation) in Rat Liver Mitochondria. Biochemistry 27: 529-535

Galli, G. and Fratelli, M. (1993) Activation of Apoptosis by Serum Deprivation in a Teratocarcinoma Cell Line: Inhibition by L-Acetylcarnitine. Exp. Cell Res. 204: 54-60

Gogvadze, V. and Richter, C. (1993) Cyclosporine A protects mitochondria in an *in vitro* model of hypoxia/reperfusion injury. FEBS Lett. 333: 334-338.

Gunter, T.E. and Pfeiffer, D.R. (1990) Mechanisms by which mitochondria transport calcium. Am. J. Physiol. C755-786.

Hennet, T., Bertoni, G., Richter, C. and Peterhans, E. (1993a) Expression of BCL-2 Protein Enhances the Survival of Mouse Fibrosarcoid Cells in Tumor Necrosis Factor-mediated Cytotoxicity. Cancer Res. 53: 1456-1460

Hennet, T., Richter, C. and Peterhans, E. (1993b) Tumour necrosis factor-α induces superoxide anion generation in mitochondria of L929 cells. Biochem. J. 289: 587-592

Hockenbery, D.M., Nunez, G., Milliman, C., Schreiber, R.D. and Korsmeyer, S.J. (1990) Bcl-2 is an inner mitochondrial membrane protein that blocks programmed cell death. Nature 348: 334-336.

Hockenbery, D.M., Oltval, Z.N., Yin, Y.M., Milliman, C.L., Korsmeyer, S.J. (1993) Bcl-2 Functions in an Antioxidant Pathway to Prevent Apoptosis. Cell 75: 241-251

Itoh, N., Tsujimoto, Y., and Nagata, S. (1993) Effect of bcl-2 on Fas Antigen-Mediated Cell Death. J. Immunol. 151: 621-627

Kane, D.J., Sarafian, T.A., Anton, R., Hahn, H., Gralla, E.B., Valentine, J.S., Örd, T., and Bredesen, D.E. (1993) Bcl-2 Inhibition of Neural Death: Decreased Generation of Reactive Oxygen Species. Science 262: 1274-1277

Korsmeyer, S.J. (1992) Bcl-2 initiates a new category of oncogenes: regulators of cell death. Blood 80: 879-886

Koshiyama, H., Lee, H.C. and Tashjian, A.H. (1991) Novel Mechanism of Intracellular Calcium Release in Pituitary Cells. J. Biol. Chem. 266: 16985-16988

Lee, H.C., Walseth, T.F., Bratt, G.T., Hayes, R.N. and Clapper, D.L. (1989) Structural Determination of a Cyclic Metabolite of NAD^+ with Intracellular Ca^{2+}-mobilizing Activity. J. Biol. Chem. 264: 1608-1615

Lehninger, A.L., Vercesi, A. and Bababunmi, E.A. (1978) Regulation of Ca^{2+} release from mitochondria by the oxidation-reduction state of pyridine nucleotides. Proc. Natl. Acad. Sci. USA 75: 1690-1694

Lötscher, H.R., Winterhalter, K.H., Carafoli, E., Richter, C. (1979) Hydroperoxides can modulate the redox state of pyridine nucleotides and the calcium balance in rat liver mitochondria. Proc. Natl. Acad. Sci. USA 76: 4340-4344

Lötscher, H.R., Winterhalter, K.H., Carafoli, E. and Richter, C. (1980) Hydroperoxide-induced Loss of Pyridine Nucleotides and Release of Calcium from Rat Liver Mitochondria. J. Biol. Chem. 255: 9325-9330

Malavasi, F., Funaro, A., Roggero, S., Horenstein, A., Calosso, L. and Mehta, K. (1994) Human CD38: a glycoprotein in search of a function. Immunology Today 15: 95-97

McDonnell, T.J., Deane, N., Platt, F.M., Nunez, G., Jaeger, U., McKearn, J.P. and Korsmeyer, S.J. (1989) bcl-2-immunoglobulin transgenic mice demonstrate extended B cell survival and follicular lymphoproliferation. Cell 57: 79-88

McDonnell, T.J. and Korsmeyer, S.J. (1991) Progression from lymphoid hyperplasia to high-grade malignant lymphoma in mice transgenic for the t(14;18). Nature 349: 254-256

Meszaros, L.G., Bak, J. and Chu, A. (1993) Cyclic ADP-ribose as an endogenous regulator of the non-skeletal type ryanodine receptor Ca^{2+} channel. Nature 364: 76-79

Monaghan, P., Robertson, D., Amos, T.A.S., Dyer, J.J.S., Mason, D.Y. and Greaves, M.F. (1992) Ultrastructural Localization of BCL-2 Protein. J. Histochem. Cytochem. 40: 1819-1825

Orrenius, S., McConkey, D.J., Bellomo, G. and Nicotera, P. (1989) Role of Ca^{2+} in toxic cell killing. Trends Pharmacol. Sci. 10: 281-285

Richter, C., Theus, M. and Schlegel, J. (1990) Cyclosporine A Inhibits Mitochondrial Pyridine Nucleotide Hydrolysis and Calcium Release. Biochem. Pharmacol. 40: 779-782

Richter, C. and Kass, G.E.N. (1991) Oxidative stress in mitochondria: its relationship to cellular Ca^{2+} homeostasis, cell death, proliferation and differentiation. Chem Biol Interact 77: 1-23

Richter, C. (1992) Mitochondrial calcium transport. In L. Ernster (Ed.) Molecular Michanisms in Bioenergetics, Elsevier Science Publishers, 349-358

Schlegel, J., Schweizer, M. and Richter, C. (1992) "Pore" formation is not required for the hydroperoxide-induced Ca^{2+} release from rat liver mitochondria. Biochem. J. 285: 65-69

Schulze-Osthoff, K., Bakker, A.C., Vanhaesebroeck, B., Beyaert, R., Jacob, W.A. and Fiers, W. (1992) Cytotoxic Activity of Tumor Necrosis Factor Is Mediated by Early Damage of Mitochondrial Functions. J. Biol. Chem. 267: 5317-5323

Strasser, A., Harris, A.W. and Cory, S. (1991) bcl-2 Transgene Inhibits T Cell Death and Perturbs Thymic Self-Censorship. Cell 67: 889-899

Tsujimoto, Y., Finger, L.R., Yunis, J., Nowell, P.C. and Croce, C.M. (1984) Cloning of the chromosome breakpoint of neoplastic B cells with the t(14;18) chromosome translocation. Science 226, 1097-1099

Vaux, D.L., Cory, S. and Adams, J.M. (1988) bcl-2 gene promotes haemopoietic cell survival and co-operates with c-myc to immortalize pre-B cells. Nature 335: 440-442

Vaux, D.L. (1993) Toward an understanding of the molecular mechanisms of physiological cell death. Proc. Natl. Acad. Sci. USA 90: 786-789

Vaux, D.L., Haecker, G., and Strasser, A. (1994) An Evolutionary Perspective on Apoptosis. Cell 76: 777-779

Veis, D.J., Sorenson, C.M., Shutter, J.R. and Korsmeyer, S.J. (1993) Bcl-2-Deficient Mice Demonstrate Fulminant Lymphoid Apoptosis, Polycystic Kidneys, and Hypopigmented Hair. Cell 75: 229-240

Wang, Y., Szekely, L., Okan, I., Klein, G. and Wiman, K.G. (1993) Wild-type p53-triggered apoptosis is inhibited by bcl-2 in a v-myc-induced T-cell lymphoma line. Oncogene 8: 3427-3431

Walton, M.I., Whysong, D., O'Connor, P.M., Hockenbery, D., Korsmeyer, S.J. and Kohn, K.W. (1993) Constitutive Expression of Human Bcl-2 Modulates Nitrogen Mustard and Camptothecin Induced Apoptosis. Cancer Research 53: 1853-1861

Wong, G.H.W., Elwell, J.H., Oberley, L.W., Goeddel, D.V. (1989) Manganous Superoxide Dismutase Is Essential for Cellular Resistance to Cytotoxicity of Tumor Necrosis Factor. Cell 58: 923-931

Wyllie, A.H., Kerr, J.F.R. and Currie, A.R. (1980) Cell death: the significance of apoptosis. Int. Rev. Cytol. 68: 251-306

Zhong, L.T., Sarafian, T., Kane, D.J., Charles, A.C., Mah, S.P., Edwards, R.H. and Bredesen, D.E. (1993) bcl-2 inhibits death of central neural cells induced by multiple agents. Proc. Natl. Acad. Sci. USA 90: 4533-4537

Heterodimerization with Bax is Required for Bcl-2 to Repress Cell Death

Xiao-Ming Yin, Zoltán N. Oltvai, and Stanley J. Korsmeyer
Howard Hughes Medical Institute, Division of Molecular Oncology, Departments of Medicine & Pathology, Washington University School of Medicine, St. Louis, MO 63110

The t(14;18)(q32;q21) found in the follicular B cell lymphomas is the most common translocation associated with human lymphoid malignancies (Fukuhara 1979, Yunis 1987). Molecular cloning of the t(14;18) chromosomal breakpoint revealed a proto-oncogene Bcl-2 at 18q21 (Tsujimoto 1985, Bakhshi 1985, Cleary, 1985). Translocation creates a Bcl-2-Ig fusion gene that is markedly deregulated resulting in inappropriately elevated levels of Bcl-2-Ig RNA and the 25 kd Bcl-2 protein (Graninger 1987, Seto 1988). Bcl-2 is a novel oncogene which extends cell survival by inhibiting a variety of apoptotic deaths (Vaux 1988, Hockenbery 1990, Nunez 1990). Deregulated expression of Bcl-2 in either Bcl-2-Ig (McDonnell 1990) or lck^{pr}-Bcl-2 (Sentman 1991, Strasser 1991) transgenic models induce B or T cell lymphomas by blocking apoptosis and extending cell survival. Conversely, bcl-2 gene ablation results in fulminant lymphocyte death, as well as polycystic kidney disease and hair hypopigmentation. This documents a normal physiological role for Bcl-2 in maitaining homeostasis of lymphocytes (Veis 1993). Recently, a family of Bcl-2-related proteins has been defined which includes Bcl-x_L, another mammalian gene that represses apoptosis (Boise 1993), and Ced-9, a Bcl-2 homolog found in C. elegans that inhibits programmed cell death in the worm (Hengartner 1994a). Interestingly, a new member found in mammalian cells, Bax, does not protect cells from apoptosis. In contrast, it counteracts Bcl-2's effect when over-expressed in cells bearing Bcl-2 (Oltavi 1993). Bax forms heterodimers with Bcl-2 and homodimers with itself. It has been proposed that the competing dimerization between Bcl-2/Bax heterodimers and Bax/Bax homodimers may determine whether a cell should die or not (Oltavi 1993). Other members in this family include MCL-1 (Kozopa, 1993), A1(Lin 1993), BHRF1 of Epstein-Barr virus (Baer 1984) and LWM5-HL of African swine fever virus (Neila 1993). The most homologous region among all family members is within two highly conserved areas, denoted as <u>B</u>cl-2 <u>h</u>omology 1 and 2 (BH1 and BH2) domains (Fig. 1).

To determine the functional importance of these conserved domains, site-directed mutagenesis of the most conserved amino acids in both domains was undertaken

(Fig. 1A,B). Mutant Bcl-2 proteins were expressed in a murine IL-3 dependent cell line, FL5.12 (Nunez 1990) and the glucocorticoid sensitive murine T cell hybridoma, 2B4 (Ucker 1989). All stable clones selected for further analysis expressed equivalent levels of Bcl-2 protein by flow cytometry and Western blot analysis (Yin 1994).

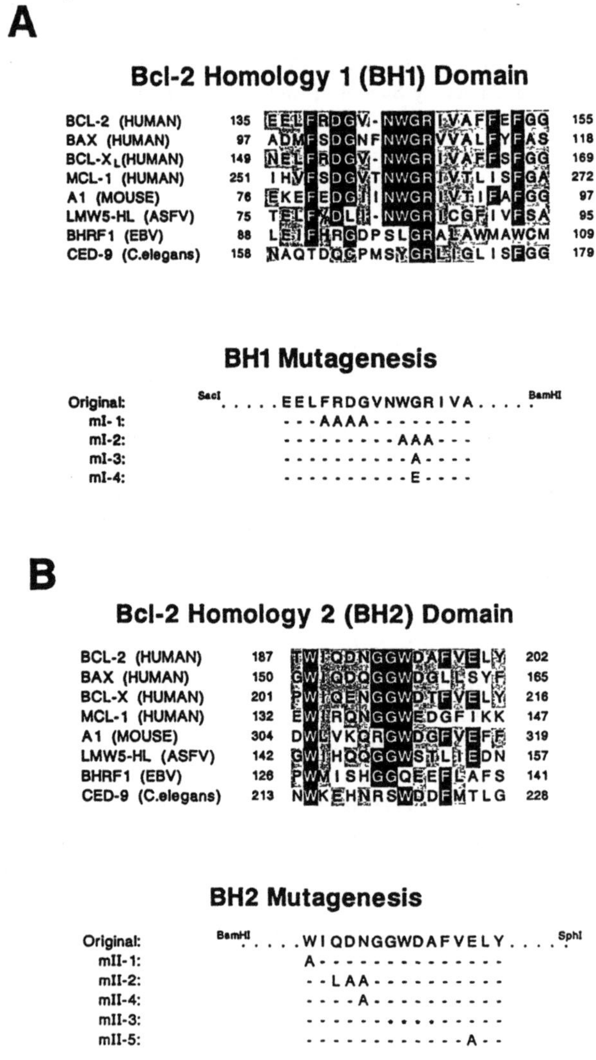

Fig. 1. Schematic representation of the BH1 (A) and BH2 (B) domains of Bcl-2 family proteins and mutagenesis in each domain. The alignment was maximized by introducing insertions marked by dashes. Identical amino acids are in black, conserved residues are shaded. Mutagenesis was conducted by either amino acid substitutions or deletions. Dashes indicate no change. Stars indicate deletion of the residues.

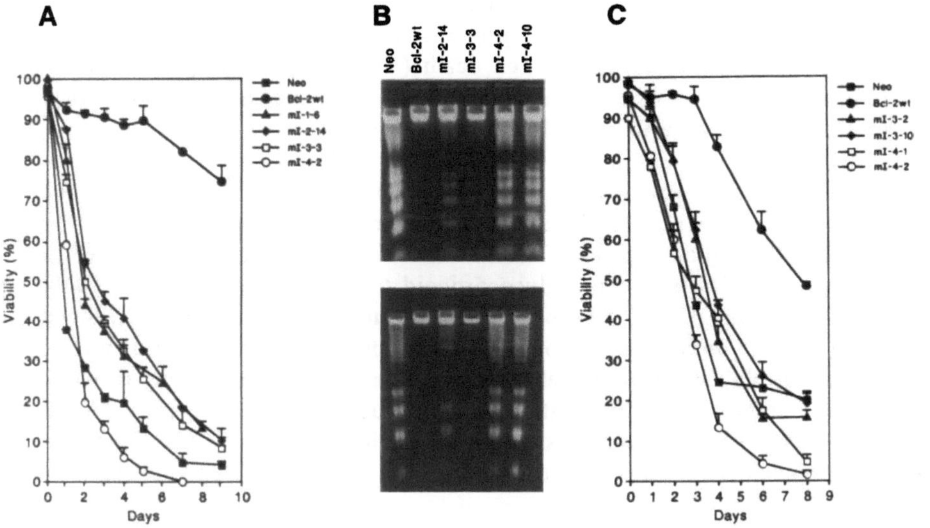

Fig. 2. Cell death assays of BH1 clones. A, Viability of representative FL5.12 clones bearing Neo control, wild-type or BH1 mutant Bcl-2 after IL-3 deprivation. B, DNA fragmentation assay of representative BH1 mutant FL5.12 clones following IL-3 deprivation for 24 or 48 hours. C, Viability assay of representative 2B4 clones bearing Neo control, wild-type, or BH1 mutant Bcl-2 after 225 rad of γ-irradiation.

FL5.12 clones possessing wild-type or mutant Bcl-2 were assessed in a cell death assay following IL-3 deprivation. Within BH1 domain, alanine replacement of FRDG (mI-1) or WGR (mI-2) markedly decreased the death repressor activity of Bcl-2 in multiple clones tested (Fig. 2A). Clones bearing mutant proteins underwent an apoptotic death with oligosomal-length DNA fragmentation and DNA release (Wyllie 1980, Fig. 2B).

The frequent structural importance of glycine residues (Creighton 1984) prompted the single amino acid substitution of gly^{145} with alanine (mI-3). All five mI-3 clones tested lacked death repressor activity (Fig. 2A). Of note this same glycine position was found to be changed to a glutamic acid in the gain of function mutation in ced-9 (Hengartner 1994b). Ced-9 is a Bcl-2 homolog in C. elegans (Hengartner 1994a). Consequently, the same glutamic acid substitution was created in Bcl-2 (mI-4), but proved to be loss of function when assessed in mammalian cells (Fig. 2A). To assess whether these subtle modifications would eliminate Bcl-2 activity in other death pathways, constructs were introduced into 2B4 cells. Neither mI-3 or mI-4 Bcl-2 proteins could block γ-irradiation induced (Fig. 2C) or glucocorticoid induced cell death (Yin 1994). All clones of each series of mutants died with similar kinetics, and mI-4 mutants displayed somewhat increased death compared to control cells (Neo)(Fig. 2).

Within BH2 domain Trp (position 188 of Bcl-2) is universally present in all family members (Fig. 1B). Replacement of this amino acid with alanine (mII-1) abrogated Bcl-2's ability to protect FL5.12 cells when IL-3 was withdrawn (Fig. 3A) or to protect 2B4 cells from glucocorticoid treatment (Fig. 3B). The QDN motif (190-192) and Glu^{200} are conserved in many family members. Clones expressing Bcl-2 with substitutions at these positions (mII-2 and mII-5, respectively) displayed approximately half of the death-repressor activity of wild-type Bcl-2 in either the IL-3 deprivation assay or glucocorticoid assay (Fig. 3A,B). However, a single amino acid substitution of alanine for Asn^{192} (mII-4) had no effect on Bcl-2's death repressor activity (Fig. 3A). The $Trp^{195}Asp^{196}$ position of

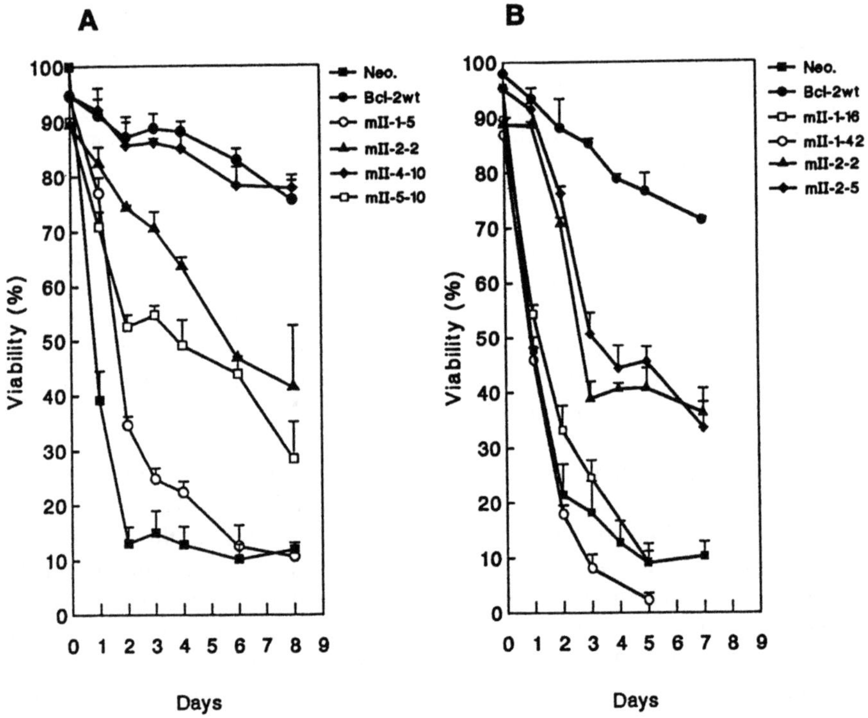

Fig. 3. Cell death assay of BH2 clones. A. Viability of representative FL5.12 clones bearing Neo control, wild-type or BH2 mutant Bcl-2 after IL-3 deprivation. B. Viability assay of representative 2B4 clones after treatment with 10^{-6}M dexamethasone.

Bcl-2 corresponds to the junction of exon II and exon III. Surrounding amino acids are conserved and represent a conserved exon junction in Bax (Oltvai 1993), Bcl-x_L (Boise 1993), Ced-9 (Hengartner 1994a), and possibly of other family members. A Bcl-2 mutant (mII-3) which deleted the GWDA motif (194-197) also completely eliminated Bcl-2's ability to block apoptosis (data not shown). However, the levels of the mII-3 protein in stable transfectants were consistently lower than that in Bcl-2 wildtype clones, suggesting that this mutation may also have affected stability of the protein (see below). Consistent with the above results, the mII-1 and mII-2 Bcl-2 proteins behave similarly in γ-irradiation-induced death (data not shown).

Since Bax counters Bcl-2 activity and forms heterodimers with Bcl-2 (Oltvai 1993) we examined each Bcl-2 mutant for association with Bax. Immunoprecipitation of human Bcl-2 wt protein with the 6C8 MAb (Hockenbery 1990) co-precipitated murine Bax from FL5.12 cells (Fig. 4). However, all mutants which eliminated Bcl-2's death repressor activity failed to interact with Bax. This includes the single substitution of gly^{145} (Yin 1994) or Trp^{188} (Fig. 4A). In addition, mII-3 Bcl-2 protein which eliminated the GWDA motif, while present in lower amounts, also failed to associate with Bax (Fig. 4D). The relationship

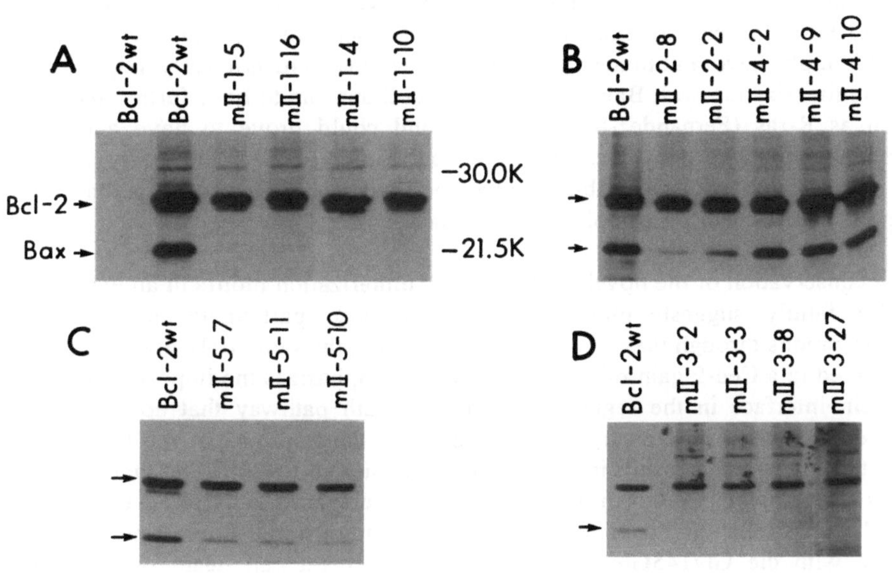

Fig. 4. Immunoprecipitation of Bcl-2 wild-type (wt) and BH2 mutant proteins. FL5.12 clones expressing wt, mII-1(A), mII-2 or mII-4(B), mII-5(C) or mII-3(D) of BH2 mutant Bcl-2 were either immunoprecipitated with a control hamster antibody (lane 1 of panel A) or anti-Bcl-2 MAb (6C8).

of heterodimerization and function was further strengthened by examination of the mII-2 and mII-5 Bcl-2 proteins that had decreased death repressor activity and also demonstrated a diminished interaction with Bax (Fig. 4B,C). In contrast, mII-4 protein which fully protected cells from death interacted with Bax to a similar extent as Bcl-2 wt (Fig. 4B). The same series of Bcl-2 mutants expressed in 2B4 cells showed the same pattern of heterodimerization (not shown). Thus, the same amino acids required for Bcl-2 function were also needed for heterodimerization with Bax, and the ability of Bcl-2 to protect cells from apoptosis seemed to correlate with its ability to heterodimerize with Bax.

Since Bax can form homodimers, it was possible that Bcl-2 would also form homodimers. To explore this possibility, several approaches were taken, which include using crosslinking agents or introducing a second copy of HA-tagged wild-type Bcl-2 into a Bcl-2 expressing cell line for a co-immunoprecipitation assay. Results indicate that bcl-2 homodimers exist, but neither the BH1 nor BH2 mutants described here disrupted Bcl-2 homodimerization. (Yin 1994).

The reciprocal ability of Bcl-2 and Bax to repress or promote death, prompted analysis of Bcl-2 mutants for function and dimerization. The BH1 and BH2 domains proved critical for Bcl-2's function and the formation of Bcl-2/Bax heterodimers. Select mutations in BH1 or BH2 domains still enabled Bcl-2 homodimerization, yet the Bcl-2 homodimers were insufficient to protect cells from death. We can not exclude that conserved Bcl-2 residues such as Gly^{145}, or Trp^{188} might have additional roles, but each amino acid is clearly required for heterodimerization with Bax. In addition, Bcl-2 appears to have further partners, such as R-ras (Fernandez-Sarabia 1993) and could prove to have additional functional domains. However, the striking correlation between the ability of Bcl-2 to repress cell death and its ability to heterodimerize with Bax suggests that Bcl-2 represses cell death by complexing with Bax.

The conservation of the novel BH1 and BH2 dimerization motifs in an expanding Bcl-2 family suggests other members may also participate in competing dimerizations through these domains. The fact that the same BH1 glycine residue is altered in a Ced-9 gain-of-function mutation emphasizes the importance of this protein interface in the regulation of a cell death pathway that appears to be universal to multicellular organisms. However, the finding that the Gly145Glu substitution in Bcl-2 results in loss-of-function coupled with the fact that Bcl-2 wt does not completely compensate for Ced-9 in C. elegans (Vaux 1992, Hengartner 1994a) may indicate a difference in their protein interactions. Interestingly, human Bcl-2 with the Gly145Glu mutation failed to protect cell death in the worm (Hengartner 1994b), suggesting that mammalian Bcl-2 may have to interact with a worm homolog of Bax to prevent cell death. While both BH1 and BH2 domains are required for Bcl-2 to bind with Bax, their importance to other family members is under investigation. Overall, the current data favors a model in which Bcl-2 must bind Bax to exert its death-repressor activity and suggest that strategies to disrupt Bcl-2/Bax heterodimers might be expected to promote cell death.

Acknowledgement

Work was supported in part by NIH CA 49712-05. Z. N. O. is supported by NIH physicians training grant T32 HL07088.

References

Baer B, Bankier AT, Biggin MD, Deininger PL, Farrell PJ, Gibson TJ, Hatfull G, Hudson GS, Satchwell SC, Seguin C, Tuffnell PS, Barrell BG (1984) DNA sequence and expression of the B95-8 Epstein-Barr virus genome. Nature 310:207-211.

Bakhski A , Jensen JP, Goldman P, Wright JJ, McBride OW, Epstein AL, Korsmeyer SJ (1985) Cloning of the chromosomal breakpoint of t(14;18) human lymphomas: clustering around J_H on chromosome 14 and near a transcriptional unit on 18. Cell 41:899-906.

Boise LH, Gonzalez-Garcia M, Postems CE, Ding L, Lindsten T, Turka LA, Mar X, Nunez G, Thompson CB (1993) Bcl-x, a Bcl-2 related gene that functions as a dominant regulator of apoptotic cell death. Cell 74:597-608.

Cleary ML, Sklar J (1985) Nucleotide sequence of a t(14;18) chromosomal breakpoint in follicular lymphoma and demonstration of a breakpoint cluster region near a transcriptionally active locus on chromosome 18. Proc. Natl. Acad. Sci. 82:7439-7443.

Creighton TE (1984) Proteins, *Structures and Molecular Principles* Chap. 1, 5, & 6., WH Freeman and Comp., New York.

Fernandez-Sarabia MJ, Bischoff JR (1993) Bcl-2 associates with the *ras*-related protein R-*ras* p23. Nature 366:274-275 (1993).

Fukuhara S, Rowley JD, Varrakojis D, Golomb HM (1979) Chromosome abnormalities in poorly differentiated lymphocytic lymphoma. Cancer Res 39:3119-3128.

Graninger WB, Seto M, Boutain B, Goldman P, Korsmeyer SJ (1987) Expression of Bcl-2 and Bcl-2-Ig fusion transcripts in normal and neoplastic cells. J Clin Invest 80:1512-1515.

Hengartner MO. Horvitz HR (1994a) C. elegans cell survival gene ced-9 encodes a functional homolog of the mammalian proto-oncogene bcl-2. Cell,76:665-676.

Hengartner MO, Horvitz HR (1994b) Activation of *C. elegans* cell death protein CED-9 by an amino-acid substitution in a domain conserved in Bcl-2. Nature 369:318-320

Hockenbery DM, Nunez G, Milliman C, Schreiber RD, Korsmeyer SJ (1990) Bcl-2 is an inner mitochondrial membrane protein that blocks programmed cell death. Nature 348:334-336.

Kozopas KM, Yang T, Buchan HL, Zhou P, Craig RW (1993) MCL-1, a gene expressed in programmed myeloid cell differentiation, has sequence similarity to Bcl-2. Proc. Natl. Acad. Sci. 90:3516-3520.

Lin EY, Orlofsky A, Berger MS, Prystowsky MB (1993) Characterization of A1, a novel hemopoietic-specific early-response gene with sequence similarity to Bcl-2. J. Immunol., 151:1979-1988.

McDonnell TJ, Deane N, Platt FM, Nunez G, Jaeger U, McKearn JP, Korsmeyer SJ (1989) Bcl-2 immnoglobulin transgenic mice demonstrate extended B cell survival and follicular lymphoproliferation. Cell 57:79-88.

Neilan, JG, Lu Z, Afonso CL, Kutish GF, Sussman MD, Rock DL (1993) An african swine fever virus gene with similarity to the proto-oncogene bcl-2 and the Epstei-Barr virus gene BHRF1. J. Virol. 67:4391-4394.

Nunez G, London L, Hockenbery D, Alexander M, McKearn, JP, Korsmeyer SJ (1990) Deregulated Bcl-2 gene expression selectively prolongs survival of growth factor-deprived hemopoietic cell lines. J. Immunol. 144:3602-3610.

Oltvai ZN, Milliman CL, Korsmeyer SJ, (1993) Bcl-2 heterodimerizes *in vivo* with a conserved homolog, Bax, that accelerates programmed cell death. Cell 74:609-619.

Sentman CL, Shutter JR, Hockenbery D, Kanagawa O, Korsmeyer SJ (1991) Bcl-2 inhibits multiple forms of apoptosis but not negative selection in thymocytes. Cell 67:879-888.

Seto M, Jaeger U, Hockett RD, Graninger W, Bennet S, Goldman P, Korsmeyer SJ (1988) Alternative promoters and exons, somatic mutation and transcriptional deregulation of the bcl-2-Ig fusion gene in lymphoma. EMBO J. 7:123-131.

Strasser A, Harris AW, Cory, S (1991) Bcl-2 transgene inhibits T cel death and perturbs thymic self-censorship. Cell 67:889-899.

Tsujimoto Y, Gorham J, Cossman J, Jaffe E, Croce CM (1985) The t(14;18) chromosome translocations involved in B-cell neoplasms result from mistakes in VDJ joining. Science 229:1390-1393.10.

Ucker DS, Ashwell JD, Nickas G (1989) Activation-driven T cell death I. Requirements for de novo transcription and translation and association with genome fragmentation. J. Immunol. 143:3461-3469.

Vaux DL, Cory S, Adams JM (1988) Bcl-2 gene promotes hemopoietic cell survival and cooperates with c-myc to immortalize pre-B cells. Nature 335:440-442.21.

Vaux DL, Weissman IL, & Kim SK (1992) Prevention of programmed cell death in *Caenorhabditis elegans* by human bcl-2. Science 258:1955-1957.

Veis DJ, Sorenson CM, Shutter JR, Korsmeyer SJ (1993) Bcl-2 deficient mice demonstrate fulminant lymphoid apoptosis, polycystic kidneys and hypopigmented hair.

Wyllie AH, Kerr JFR, Currie AR (1980) Cell death: the significance of apoptosis. Int. Rev. Cytol. 68:251-306.

Yin X-M, Zoltan ON, Korsmeyer SJ (1994) BH1 and BH2 domains of Bcl-2 are required for inhibition of apoptosis and heterodimerizatin with Bax. Nature 369:321-323.

Yunis JJ, Frizzera G, Oken MM, McKenna J, Theologides A, Arnesen M (1987) Multiple recurrent genomic defects in follicular lymphoma: a possible model for cancer. N Engl J Med 316:79-84.

Growth Regulation: Cyclin D1, ABL

Mantle Cell/Centrocytic Lymphoma: Molecular and Phenotypic Analysis Including Analysis of the bcl-1 Major Translocation Cluster by PCR

M. E. Williams[1], L. R. Zukerberg[2,3], N. L. Harris[2], W-I. Yang[2], A. Arnold[3], S. D. Finkelstein[4], and S. H. Swerdlow[4]

[1]Departments of Internal Medicine and Pathology, University of Virginia School of Medicine, Charlottesville, VA;
[2]Department of Pathology and [3]Endocrine Oncology Unit, Massachusetts General Hospital and Harvard Medical School, Boston, MA;
[4]Department of Pathology, University of Pittsburgh School of Medicine, Pittsburgh, PA.

Introduction

Mantle cell/centrocytic lymphoma (MCL) is a biologically and clinically distinct subtype of non-Hodgkin's lymphoma (NHL) characterized by small lymphocytes with irregular nuclei and a diffuse, vaguely nodular, or mantle zone growth pattern [1,2]. The neoplastic cells are typically positive for CD5, the pan-B cell antigens CD19 and CD20, and surface IgM and IgD, but negative for CD10 and CD23. Clinical features include male predominance, a median age of onset of 64 years, and presentation with advanced stage disease in the majority of patients. Peripheral blood, bone marrow, and extranodal disease involving the gastrointestinal tract or Waldeyer's ring is common [1,2]. The median survival is approximately 30-40 months. MCL is estimated to comprise about 4-7% of all NHL.

The observation that the majority of MCL contain the chromosomal translocation t(11;14)(q13;q32) or rearrangements of the 11q13 bcl-1/cyclin D1 loci stimulated much of the recent interest in MCL, and indeed was instrumental in verifying it as a distinct subtype of NHL [1, 3-7]. It is now clear that this translocation leads to the aberrant expression of cyclin D1 (CCND1; previously designated bcl-1 or PRAD1), a G1 cyclin involved in cell cycle regulation [8-11]. MCL may thus serve as an important model for determining the role of cell cycle deregulation in lymphomagenesis, which may also be relevant to other types of human neoplasia [12,13]. Additional mutations in MCL, such as the p53 point mutations recently demonstrated in transformed follicular NHL, remain to be identified [14,15].

Although Southern blot analysis has been useful for characterizing known or suspected MCL, the technique is limited by the need to analyze multiple 11q13 breakpoints and the need for fresh or frozen cells or tissue [7,16]. We therefore developed a polymerase chain reaction (PCR) assay for the 11q13 bcl-1 major translocation cluster (MTC) breakpoint [17]. More recently we have utilized a polyclonal anti-cyclin D1 antibody immunohistochemical stain to determine the frequency of expression in mantle cell and related lymphomas, and have correlated this expression with bcl-1/cyclin D1 rearrangements.

Results and Discussion

Southern Blot Analysis

The demonstration of chromosome 11q13 bcl-1 and cyclin D1 rearrangements provided important confirmatory evidence that MCL represented a distinct subtype of NHL. These rearrangements can provide supportive evidence for a diagnosis of MCL, especially in cases with atypical morphologic or phenotypic features, as they occur rarely in other low- and intermediate-grade NHL [1,2,18]. Recent reports, however, have documented the occurrence of bcl-1 rearrangements and/or cyclin D1 expression in splenic lymphoma with villous lymphocytes [D. Jadayel, personal communication (19)] and multiple myeloma [20-22].

We have utilized a panel of bcl-1 and cyclin D1 translocation breakpoint probes and the immunoglobulin heavy chain joining gene probe JH to detect rearrangements and to determine (by comigration of rearranged fragments) the t(11;14) in human lymphomas and lymphocytic leukemias [3,4,7]. Including previously reported cases, we have now studied 41 morphologically and phenotypically diagnosed MCL. Twenty-eight (68%) demonstrated bcl-1 or cyclin D1 rearrangements. Of these, 14 occur at the bcl-1 MTC locus and 9 at the p94PS locus; none have been identified with the p210 and p11EH probes [4]. Another 3 showed rearrangement approximately 15 kb upstream of cyclin D1 exon I, while 3 were rearranged within 1-2 kb of exon I without apparent disruption of the coding region [7,23]. One of the latter was rearranged with both the p94PS probe on multiple restriction enzyme digests and with Bam HI only with probes flanking the 5′ side of cyclin D1 exon I [23]. It is uncertain whether these Bam HI rearrangements represent a point mutation or a Bam HI polymorphism. No rearrangements were identified with these probes in follicular small cleaved cell NHL (n=20), large cell lymphoma (n=23), small lymphocytic lymphoma/chronic lymphocytic leukemia (n=31), or monocytoid B cell lymphoma (n=7). All cases showed clonal JH rearrangements. Thus, bcl-1 or cyclin D1 rearrangements are strongly associated with MCL and can provide supportive evidence for that diagnosis. However, the need for multiple probe and enzyme combinations and fresh or frozen material for DNA extraction limits the diagnostic utility of this technique. In addition, other breakpoints lie outside the regions detected by these probes, as indicated by two cases in our series which contain karyotypically demonstrated t(11;14)(q13;q32) but which lack detectable rearrangements, and as recently reported by Raynaud et al [10]. Thus, alternative assays are necessary to more clearly define the diagnostic utility of this molecular marker.

Polymerase Chain Reaction (PCR)

Translocations at the bcl-1 MTC locus are tightly clustered and can be amplified by PCR utilizing MTC and concensus JH primers. The MTC breakpoints of six MCL and four t(11;14)-containing B cell lines fell within an 82 base pair span [17]. Nine of these translocations involved the immunoglobulin heavy chain J4 locus. The PCR products ranged from approximately 200 bp to 600 bp in size (depending upon the MTC primer utilized and the precise MTC and JH breakpoints) and can be detected on agarose gels [17]. PCR can thus provide a rapid assay for the most common of the 11q13 breakpoints, although it is present in only about half of all MCL. More recently, Rimokh et al [24] and Molot et al [25] have also reported PCR results for MTC translocations.

Immunohistochemical Staining

The cyclin D1 gene is expressed at the mRNA level in MCL, whereas it appears not to be expressed in other NHL or in normal B cells [8-11,26]. Such expression is postulated to be driven by the juxtaposed immunoglobulin regulatory elements. Expression at the protein level, especially in clinical samples, has been difficult to assess due to the lack of sensitive and specific anti-cyclin D1 antibodies.

Recently Yang et al [27] reported that each of 15 MCL showed nuclear staining with a polyclonal rabbit anti-cyclin D1 antibody. Non-MCL and reactive hyperplasia cases were uniformly negative with the exception of a single case of B-CLL which showed an unusual "dot-like" pattern of nuclear staining. Utilizing this same antibody, we analyzed cases from our series of morphologically and phenotypically typical and atypical MCL previously characterized by Southern blot. The preliminary results confirm the findings of Yang et al [27], and, importantly, revealed cyclin D1 staining in some cases which lacked detectable bcl-1 or cyclin D1 rearrangements [S. H. Swerdlow et al, manuscript in preparation]. Recent data also suggests that cyclin D1 is an important expressed component of a chromosome 11q13 amplicon in some human solid tumors, and that it may be prognostically relevant in head and neck squamous cell carcinomas and in breast adenocarcinomas [12,13]. Thus, assessment of cyclin D1 expression in these tumors may provide a useful molecular marker for assessing prognosis and for treatment stratification.

Immunostaining for p53 was also performed utilizing the D07 monoclonal antibody (DAKO Corp., Carpinteria, CA) in 21 MCL specimens. Preliminary analysis showed three were positive, six were weakly positive and twelve were negative. Analysis of p53 point mutations in these cases is in progress.

Conclusions

MCL represents a distinct clinical and pathologic entity characterized by the chromosomal translocation t(11;14)(q13;q32), bcl-1 and cyclin D1 rearrangements, and overexpression of cyclin D1 at both the mRNA and protein levels. The use of PCR to detect this translocation, or the use of anti-cyclin D1 antibodies to determine cyclin D1 expression, can help confirm a diagnosis of MCL. These techniques also will facilitate ongoing studies to assess the spectrum of NHL and other neoplasms with altered cyclin D1 expression and should enhance our understanding of the role of cell cycle deregulation in cancer pathogenesis.

References

1. Swerdlow SH, Williams ME (1993) Centrocytic lymphoma: A distinct clinicopathologic, immunophenotypic, and genotypic entity. Pathol Annual 28:171-197
2. Zucca E, Stein H, Coiffier B (1994) European Lymphoma Task Force (ELTF): Report of the workshop on mantle cell lymphoma (MCL). Annals Oncol, in press
3. Williams ME, Westermann CD, Swerdlow SH (1990) Genotypic characterization of centrocytic lymphoma: Frequent rearrangement of the chromosome 11 bcl-1 locus. Blood 76:1387-1391
4. Williams ME, Meeker TC, Swerdlow SH (1991) Rearrangement of the chromosome 11 bcl-1 locus in centrocytic lymphoma: Analysis with multiple breakpoint probes. Blood 78:493-498
5. Rimokh R, Berger F, Cornillet P, Wahbi K, Rouault JP, Ffrench M, Bryon PA, Gadoux M, Gentilhomme O, Germain D, Magaud JP (1990) Break in the bcl1 locus is closely associated with intermediate lymphocytic lymphoma subtype. Genes, Chromosomes and Cancer 2:223-226
6. Medeiros LJ, Van Krieken JH, Jaffe ES, Raffeld M (1990) Association of bcl-1 rearrangements with lymphocytic lymphoma of intermediate differentiation. Blood 76:2086-2090

7. Williams ME, Swerdlow SH, Rosenberg CL, Arnold A (1992) Characterization of chromosome 11 translocation breakpoints at the bcl-1 and PRAD1 loci in centrocytic lymphoma. Cancer Res 52:5541s-5544s
8. Rosenberg CL, Wong E, Petty EM, Bale AE, Tsujimoto Y, Harris NL, Arnold A (1991) PRAD1, a candidate BCl1 oncogene: Mapping and expression in centrocytic lymphoma. Proc Natl Acad Sci USA 88:9638-9642
9. Rimokh R, Berger F, Delsol G, Charrin C, Bertheas MF, Ffrench M, Garoscio M, Felman P, Coiffier B, Bryon PA, Rochet M, Gentilhomme O, Germain D, Magaud JP (1993) Rearrangement and overexpression of the bcl-1/PRAD-1 gene in intermediate lymphocytic lymphomas and in t(11q13)-bearing leukemias. Blood 81:3063-3067
10. Raynaud S, Bekri S, Leroux D, Grosgeorge J, Klein B, Bastard C, Gaudray P, Simon MP (1993) Expanded range of 11q13 breakpoints with differing patterns of cyclin D1 expression in B-cell malignancies. Genes, Chromosomes and Cancer 8:80-87
11. Williams ME, Swerdlow SH (1994) Cyclin D1 overexpression in non-Hodgkin's lymphoma with chromosome 11 bcl-1 rearrangement. Annals Oncol 5(Suppl.1):S71-S73
12. Williams ME, Gaffey MJ, Weiss LM, Wilczynski C, Schuuring E, Levine P (1993) Chromosome 11q13 amplification in head and neck squamous cell carcinoma. Arch Otolaryngol Head Neck Surg 119:1238-1243
13. Gaffey MJ, Frierson HF, Williams ME (1993) Chromosome 11q13, c-erbB-2, and c-myc amplification in invasive breast carcinoma: Clinicopathologic correlations. Mod Path 6:654-659
14. Sander CA, Yano T, Clark HM, Harris C, Longo DL, Jaffe ES, Raffeld M (1993) p53 mutation is associated with progression in follicular lymphomas. Blood 82:1994-2004
15. LoCoco F, Gaidano G, Louie DC, Offit K, Chaganti RSK, Dalla-Favera R (1993) p53 mutations are associated with histologic transformation of follicular lymphoma. Blood 82:2289-2295
16. deBoer CJ, Loyson S, Kluin PM, Kluin-Nelemans HC, Schuuring E, van Krieken JHJM (1993) Multiple breakpoints within the BCL-1 locus in B-cell lymphoma:Rearrangements of the cyclin D1 gene. Cancer Res 53:4148-4152
17. Williams ME, Swerdlow SH, Meeker TC (1993) Chromosome t(11;14)(q13;q32) breakpoints in centrocytic lymphomas are highly localized at the bcl-1 major translocation cluster. Leukemia 7:1437-1440
18. Raffeld M, Jaffe ES (1991) bcl-1, t(11;14), and mantle cell-derived lymphomas. Blood 78:259-263
19. Jadayel D, Matutes E, Dyer MJS, Brito-Babapulle V, Khokar MT, Oscier D, Catovsky D (1994) Splenic lymphoma with villous lymphocytes: Analysis of bcl-1 rearrangements and expression of the cyclin D1 gene. Blood, in press
20. Fiedler W, Weh HJ, Hossfeld DK (1992) Comparison of chromosome analysis and bcl-1 rearrangement in a series of patients with multiple myeloma. Brit J Haematol 81:58-61
21. Seto M, Yamamoto K, Iida S, Akao Y, Utsumi KR, Kubonishi I, Miyoshi I, Ohtsuki T, Yawata Y, Namba M, Motokura T, Arnold A, Takahashi T, Ueda R (1992) Gene rearrangement and overexpression of PRAD1 in lymphoid malignancy with t(11;14)(q13;q32) translocation. Oncogene 7:1401-1406
22. Travis P, Sawyer J, Lary C, Hoover R, Barlogie B (1993) Translocation 11;14 and BCL-1 gene abnormalities in multiple myeloma. Blood 82 (Suppl.1):261a
23. Williams ME, Swerdlow SH, Rosenberg CL, Arnold A (1993) Chromosome 11 translocation breakpoints at the PRAD1/cyclin D1 gene locus in centrocytic lymphoma. Leukemia 7:241-245.

24. Rimokh R, Berger F, Delsol G, Digonnet I, Rouault JP, Tigaud JD, Gadoux M, Coiffier B, Bryon PA, Magaud JP (1994) Detection of the chromosomal translocation t(11;14) by polymerase chain reaction in mantle cell lymphomas. Blood 83:1871-1875
25. Molot RJ, Meeker TC, Wittwer CT, Perkins SL, Segal GH, Masih AS, Braylan RC, Kjeldsberg CR (1994) Antigen expression and polymerase chain reaction amplification of mantle cell lymphomas. Blood 83:1626-1631
26. Withers DA, Harvey RC, Faust JB, Melnyk O, Carey K, Meeker TC (1991) Characterization of a candidate bcl-1 gene. Mol Cell Biol 11:4846-4853
27. Yang WI, Zukerberg LR, Motokura T, Arnold A, Harris NL (1994) Pradl/Cyclin D1 (bcl-1) protein expression in low grade B-cell lymphomas and reactive hyperplasia. Am J Pathol, in press

Acknowledgements

This work was supported in part by an American Cancer Society (ACS) Research Grant and by NCI Cancer Center Support Grant P30 CA44579 (to M.E.W.), PHS grant 5f32CA09260-2 (to L.R.Z.), and NIH grants DK 11794 and CA 5590, and and ACS Faculty Research Award (to A.A.). We also thank P. Ennis, S. Likowski and M. Whitefield for expert technical assistance, Dr. T. Meeker for providing probes and PCR primers, and Dr. Toru Motokura for his involvement in developing and producing the anti-cyclin D1 antibody used in these studies.

Cyclin D1 as the Putative bcl-1 Oncogene

A. W. Harris, S. E. Bodrug, B. J. Warner, M. L. Bath, G. J. Lindeman and J. M. Adams
The Walter and Eliza Hall Institute of Medical Research, PO Royal Melbourne Hospital, Melbourne 3050, Australia

Introduction

The junction points of recurring chromosome translocations in hematopoietic tumors have yielded a rich harvest of oncogenes by molecular cloning (Korsmeyer 1992). One such karyotype alteration, the t(11;14) (q13;q32) in certain human B-lymphoid neoplasms, was first described 15 years ago by Vandenberghe et al. (1979), but has only recently yielded a candidate oncogene, cyclin D1. In this paper, we briefly review the path by which the gene was discovered, the clues to its role in progression of the G_1 phase of the cell cycle, and the recently published evidence for its tumorigenic activity, including our own analysis of cyclin D1 transgenic mice. For a comprehensive review of cyclins and their possible roles in oncogenesis, see Sherr (1993) and Motokura and Arnold (1993).

The *bcl*-1 Locus in B-cell Lymphomas

Ten years ago, Tsujimoto et al. (1984) cloned part of a locus joined to the immunoglobulin heavy chain gene complex by the 11;14 translocation in a human chronic lymphocytic leukemia cell line. The 11q13 locus was designated *bcl*-1 (B-cell lymphoma/leukemia 1) and it was postulated to contain an oncogene that could be activated by its juxtaposition to immunoglobulin enhancer sequences in B-lymphoid tumor cells bearing the translocation. For nearly a decade, however, the postulate remained untestable because no gene could be found in the vicinity of the translocation breakpoints. Thus, molecular scrutiny of the flanking regions gradually moved the putative oncogene further away from the initially detected translocation breakpoints. While these breakpoints defined a major translocation cluster (Tsujimoto et al. 1985; Ince et al. 1988; Koduru et al. 1989), some B-cell tumors were later found to have an 11q13 breakpoint at 36 kb (Rabbitts et al. 1988) or even 63 kb (Meeker et al. 1989) downstream. Remarkably, the relevant oncogene eventually proved to be about 120 kb away.

The frequency of the *Igh-bcl*-1 translocation among all human B-cell non-Hodgkin lymphomas is very low, less than 5% (Athan et al. 1991). It is almost entirely confined to histologically low-grade tumors and has been identified in occasional cases of chronic lymphocytic leukemia and plasma cell leukemia, but not in follicular center B-cell lymphomas, which typically instead have activated the *bcl*-2 gene by a 14:18 chromosome translocation (Korsmeyer 1992). The *bcl*-1 rearrangement has proven to be characteristic of lymphomas whose cells seem to derive from primary lymphoid follicles or the mantle zones of secondary follicles (Weisenburger et al. 1987; Williams et al. 1990, 1991; Medeiros et al. 1990; Rimokh et al. 1990). These tumors, designated diffuse small cleaved-cell, centrocytic or mantle-zone lymphomas, probably are equivalent to those that have been described as diffuse lymphocytic lymphomas of intermediate differentiation. While commonly diffuse in histological character, they can also be somewhat nodular, as might befit cells with residual mantle cell character. Their cell surface phenotype, which is distinct from that of follicular center lymphomas, is usually $IgM^{+(high)}$ $CD5^{+}$ $CD10^{-}$ $CD23^{-}$. Accordingly, it has been proposed that these tumors be unified under the designation mantle cell lymphoma (Raffeld and Jaffe, 1991; Banks et al. 1992). The argument for a common biological origin has been reinforced by the recent finding that tests for 11q13 rearrangements with hybridization probes covering 110 kb of the *bcl*-1 locus showed about 70% of such tumors to be positive (Williams et al. 1992).

The *bcl*-1 Locus Yields a Putative Oncogene

Other kinds of genetic alteration to the *bcl*-1 locus have been found in several non-hematopoietic tumors, indicating a wider role for the putative oncogene in cellular transformation. In up to 20% of cases of mammary carcinoma and squamous cell carcinomas of the head, neck and lung, variable portions of the locus have been amplified 2- to 10-fold (Zhou et al. 1988; Berenson et al. 1989, 1990; Ali et al. 1989). Furthermore, occasional cases of benign parathyroid adenoma harbor an inversion within chromosome 11 that joins the parathyroid hormone gene to a transcribed DNA sequence initially designated *PRAD1* in the chromosome band bearing *bcl*-1 (Arnold et al. 1989). This finding was the key that unlocked the oncogene problem. The *PRAD1* sequence was found to be overexpressed in tumor cells bearing the amplified or translocated locus and located within 130 kb of the *bcl*-1 major translocation cluster (Lammie et al. 1991; Rosenberg et al. 1991). When its nucleotide sequence was determined (Motokura et al. 1991; Withers et al. 1991), it proved to encode a new member of the cell cycle-regulating family of cyclin proteins, cyclin D1, isolated at the same time by other workers through independent strategies (Xiong et al. 1991; Matsushime et al. 1991).

Control of the cellular mitotic cycle is currently being defined in terms of the activity of protein kinases that comprise catalytic subunits of the cyclin-dependent kinase (cdk) family complexed with obligate positive regulators, the cyclins (Reed 1992). Passage through the first gap phase between mitosis or quiescence and DNA replication seems to be promoted by the G_1 cyclins D1, D2, D3 and E (and perhaps also C) (Sherr 1993). Whereas most cyclins oscillate with the cycle but are found in all cell types, the D cyclins are produced in response to growth factors and are differentially expressed in various cell types. Cyclin D1 is required for progression through G_1 in rat fibroblasts (Baldin et al. 1993) as well as in various human tumor cells, including those that exhibit gene amplification (Lukas et al. 1994). Its enforced overexpression shortened that phase of the cell cycle in both continuously proliferating cells and in cells emerging from quiescence (Quelle et al. 1993; Jiang et al. 1993). However, cyclin D1 is not expressed by proliferating normal lymphoid cells (Ajchenbaum et al. 1993; Bodrug et al. 1994), nor by lymphoid tumor cells lacking alterations to the *bcl*-1/cyclin D1 locus (Palmero et al. 1993; Lukas et al. 1994). These cells presumably control their G_1 phase through the actions of the other D cyclins. Thus, expression of cyclin D1 in lymphomas bearing the 11;14 translocation is ectopic, further implicating its deregulation in the origin of those tumors. The activation of this putative oncogene seems to be solely by deregulation. No changes to the coding region were found in the overexpressed transcripts from a parathyroid adenoma or a centrocytic lymphoma (Rosenberg et al. 1993).

Some circumstantial evidence also points towards a role for cyclin D1 in a few mouse tumors. An association with tumorigenesis by the Friend mouse leukemia virus has been indicated by the location of an occasionally used proviral integration site called *Fis*-1 close to the cyclin D1 gene. A myeloid leukemia and a T-cell tumor with a provirus at this site were found to overexpress the cyclin mRNA (Lammie et al. 1992), presumably under the influence of the strong enhancer in the retroviral long terminal repeat sequence. This situation is analogous to that in *bcl*-1 translocations in which the immunoglobulin heavy chain enhancer is thought to act on the cyclin D1 gene.

Transformation Tests on Cultured Cells

Attempts to induce malignant transformation by engineered overexpression of the cyclin D1 gene in cultured fibroblasts have produced somewhat divergent results. Although transfected cells of the Rat6 cell line did not assume a transformed growth pattern in culture, they did become capable of generating fibrosarcomas in nude mice (Jiang et al. 1993), but no neoplastic changes were evident in two other fibroblast lines (Quelle et al. 1993). With primary baby rat kidney cells, which show a transformed phenotype in vitro when transfected with cooperating pairs of known oncogenes, combinations of highly expressed cyclin D1 with *myc*, activated *ras*, adenovirus E1A or mutant p53 did not produce foci of transformation (Hinds et al. 1994). Some transforming activity was demonstrable in a three-way combination of the cyclin with *ras* and a partially defective mutant of E1A (Hinds et al. 1994), but these transfected cells were not tested for tumorigenesis. By contrast, rat embryo fibroblasts have recently been reported to acquire both transformed growth characteristics in vitro and tumorigenic activity in vivo when co-transfected with cyclin D1 and activated H-*ras* (Lovec et al. 1994). Neither the cyclin alone, nor cyclin combined with *myc*, produced transformation in this test system. While most of these studies suggest that overexpression of cyclin D1 can contribute to malignancy, its effects are more subtle than those of the well-studied oncogenes. Perhaps the extent of its contribution depends critically on the level of expression achieved and the nature of the cells under test.

Eμ-cyclin D1 Transgenic Mice

To determine whether enforced cyclin D1 expression, mimicking the 11;14 chromosome translocation, can produce lymphomas requires a test in lymphoid cells in a whole animal. We have attempted to do this by constructing transgenic mice, which have previously proven to be useful in analyzing the lymphomagenic activity of oncogenes (Adams and Cory 1991). The results have been fully documented elsewhere (Bodrug et al. 1994).

The transgene construct was designed to express its mouse cyclin D1 cDNA preferentially in lymphoid cells. It contained the enhancer from the 5' intron of the mouse immunoglobulin heavy chain locus (Eμ) and the potent hybrid SRα promoter. The prospects for lymphomagenic activity initially looked promising when three of the 11 primary transgenic animals developed thymic T-cell lymphomas. However, one of these did not express the transgene and subsequent monitoring of over 200 transgene-expressing descendants of several founder animals up to one year of age yielded only two more tumors. Since the background incidence of thymoma in non-transgenic mice of the same strain was around 1%, the transgene was not making a significant contribution to spontaneous lymphomagenesis. One transgenic strain (Cyd 76), which showed high expression of cyclin D1 protein in both the B- and T-cell lineages, was studied further in detail.

Despite the lack of spontaneous tumorigenic activity, the cyclin D1 transgene did prove to function cooperatively with the c-*myc* gene in transforming lymphoid cells to malignancy. Doubly transgenic mice from crosses between the Cyd 76 and Eμ-*myc* strains developed lymphomas an average of eight weeks earlier (7 weeks vs 15 weeks) than the littermates that inherited only the Eμ-*myc* transgene. Most of the cyclin D1/*myc* tumors were pre-B- and B-cell lymphomas like those that occur spontaneously in Eμ-*myc* mice, but two of the 14 tumors characterized were T-cell lymphomas, implying that cyclin D1 can be oncogenic for T as well as B lymphocytes. We are not aware of any extensive screen of human T-cell neoplasms for rearrangements in the *bcl*-1/cyclin D1 locus. It might be expected that some would have translocated the cyclin D1 gene into a T-cell receptor locus, although a translocation of this kind is not among the recognized recurrent karyotypic abnormalities in T-cell tumors (Rowley 1990).

We also tested whether cyclin D1 could cooperate with an activated *ras* gene in T-cell transformation, using a transgenic mouse strain that carries a mutant N-*ras* gene and spontaneously develops thymic T-cell lymphomas (Harris et al. 1988). Adding the Cyd 76 transgene to this strain produced a small, marginally significant acceleration of T-cell tumor onset, indicating weak cooperation between cyclin D1 and *ras*. Thus, the strong cooperation reported for fibroblasts (Lovec et al. 1994) does not seem to apply to lymphocytes. On the other hand, the collaborative action of the cyclin with *myc* seen here in lymphoid cells does not seem to apply to fibroblasts (Hinds et al. 1993; Lovec et al. 1994). Thus, cell type may strongly influence the oncogenic action of cyclin D1.

Since the Eμ-cyclin D1 transgenic mice did not spontaneously develop a significant incidence of tumors, they could be assayed for non-malignant alterations to their lymphoid cells. In adult mice, however, no alterations were detected. The cellular composition of their hemato-lymphoid organs assessed by surface marker analysis was normal. There was no change in the proportion of lymphoid cells actively in cycle; neither were the lymphocytes more responsive to mitogenic stimuli. Clearly, expression of the G_1 cyclin, by itself, is not sufficient to drive lymphocytes to proliferate. In addition, it did not noticeably shorten the G_0-S phase interval in mitogen-activated T or B cells. On the other hand, the transgene did cause a change in lymphocyte development in very young animals. They contained fewer mature B cells and T cells, indicating a reduced rate of production. This effect was greatly accentuated by coexpression of the Eμ-*myc* transgene. Thus, the two genes cooperate in inhibiting lymphocyte differentiation as they do in transforming these cells to malignancy.

Concluding Remarks

The long search for the *bcl*-1 oncogene appears to have reached its goal. Our findings with Eμ-cyclin D1 mice (Bodrug et al. 1994) greatly strengthen the case that cyclin D1 is the relevant oncogene in all the B-lymphoid tumors that carry the 11:14 translocation. By extrapolation, these observations also raise the odds that cyclin D1 contributes to the development of the carcinomas where the gene is amplified. Surprisingly, however, the cyclin D1 transgene had negligible tumorigenic impact on its own. Perhaps its oncogenic activity can only be realized in the context of another genetic change. Impressive synergy in lymphomagenesis was observed here with the *myc* gene. Whether a similar oncogenic partnership holds for the human tumors remains to be determined.

How cyclin D1 acts as an oncogene remains unclear. The most obvious possibility would be that it augments the activity of a Cdk operative in the G_1 phase, but no change in cell cycle behavior was evident. The only pre-neoplastic effect observed was a partial block to lymphoid development in the young mice and this effect was accentuated by a co-expressed *myc* gene. Presumably the Myc protein, which is a transcription factor, provides a complementary function. It might, for example, induce expression of cyclins, such as cyclin E or A, that control other rate-limiting steps in G_1 and early S phase. Thus, concerted expression of cyclin D1 and Myc may keep most of the cells in cycle and thereby impede differentiation.

Acknowledgments

We thank Dr CJ Sherr for mouse cyclin D1 cDNA and monoclonal anti-mouse cyclin D1 antibody. This work was supported by the National Health and Medical Research Council (Canberra) and by the US National Cancer Institute (CA43140).

References

Adams JM, Cory S (1991) Transgenic models of tumor development. Science: 254:1161-1167

Ajchenbaum F, Ando K, DeCaprio JA, Griffin JD (1993) Independent regulation of human D-type cyclin gene expression during G_1 phase in primary human T lymphocytes. Proc Natl Acad Sci USA 268:4113-4119

Ali IU, Merlo G, Callahan R, Lidereau R (1989) The amplification unit on chromosome 11q13 in aggressive primary human breast tumors entails the *bcl*-1, *int*-2 and *hst* loci. Oncogene 4:89-92

Arnold A, Kim HG, Gaz RD, Eddy RL, Fukushima Y, Byers MG, Shows TB, Kronenberg HM (1989) Molecular cloning and chromosomal mapping of DNA rearranged with the parathyroid hormone gene in a parathyroid adenoma. J Clin Invest 83:2034-2040

Athan E, Foitl DR, Knowles DM (1991) *BCL*-1 rearrangement. Frequency and clinical significance among B-cell chronic lymphocytic leukemias and non-Hodgkin's lymphomas. Am J Pathol 138:591-599

Baldin V, Lukas J, Marcote MJ, Pagano M, Draetta G (1993) Cyclin D1 is a nuclear protein required for progression in G_1. Genes Dev 7:812-821

Banks PM, Chan J, Cleary ML, Delsol G, De Wolf-Peeters C, Gatter K, Grogan TM, Harris NL, Isaacson PG, Jaffe ES, Mason D, Pileri S, Ralfkiaer E, Stein H, Warnke RA (1992) Mantle

cell lymphoma. A proposal for unification of morphologic, immunologic, and molecular data. Am J Surg Pathol 16:637-640
Berenson JR, Yang J, Mickel RA (1989) Frequent amplification of the *bcl*-1 locus in head and neck squamous cell carcinomas. Oncogene 4:89-92
Berenson JR, Koga H, Yang J, Pearl J, Holmes EC, Figlin R, the Lung Cancer Study Group (1990) Frequent amplification of the *bcl*-1 locus in poorly differentiated squamous cell carcinoma of the lung. Oncogene 5:1343-1348
Bodrug SE, Warner BJ, Bath ML, Lindeman GJ, Harris AW, Adams JM (1994) Cyclin D1 transgene impedes lymphocyte maturation and collaborates in lymphomagenesis with the *myc* gene. EMBO J 13:2124-2130
Harris AW, Langdon WY, Alexander WS, Hariharan IK, Rosenbaum H, Vaux D, Webb E, Bernard O, Crawford M, Abud H, Adams JM, Cory S (1988) Transgenic models for hematopoietic tumorigenesis. Curr Top Microbiol Immunol 141:82-93
Hinds PW, Dowdy SF, Eaton EN, Arnold A, Weinberg RA (1994) Function of a human cyclin gene as an oncogene. Proc Natl Acad Sci USA 91:709-713
Jiang W, Kahn SM, Zhou P, Zhang Y-J, Cacace AM, Infante AS, Doi S, Santella RM, Weinstein IB (1993) Overexpression of cyclin D1 in rat fibroblasts causes abnormalities in growth control, cell cycle progression and gene expression, Oncogene 8:3447-3457
Koduru PR, Offit K, Filippa DA (1989) Molecular analysis of breaks in BCL-1 proto-oncogene in B-cell lymphomas with abnormalities of 11q13. Oncogene 4:929-934
Korsmeyer SJ (1992) Chromosomal translocations in lymphoid malignancies reveal novel proto-oncogenes. Annu Rev Immunol 10:785-807
Lammie GA, Fantl V, Smith R, Schuuring E, Brookes S, Michalides R, Dickson C, Arnold A, Peters G (1991) D11S287, a putative oncogene on chromosome 11q13, is amplified and expressed in squamous cell and mammary carcinomas and linked to BCL-1. Oncogene 6:439-444
Lammie GA, Smith R, Silver J, Brookes S, Dickson C, Peters G (1992) Proviral insertions near cyclin D1 in mouse lymphomas: a parallel for BCL1 translocations in human B-cell neoplasms. Oncogene 7:2381-2387
Lovec H, Sewing A, Lucibello FC, Müller R, Möröy T (1994) Oncogenic activity of cyclin D1 revealed through cooperation with Ha-*ras*: link between cell cycle control and malignant transformation. Oncogene 9:323-326
Lukas J, Pagano M, Staskova Z, Draetta G, Bartek J (1994) Cyclin D1 protein oscillates and is essential for cell cycle progression in human tumor cell lines. Oncogene 9:707-718
Matsushime H, Roussel MF, Ashmun RA, Sherr CJ (1991) Colony-stimulating factor 1 regulates novel cyclins during the G1 phase of the cell cycle. Cell 65:701-713
Medeiros LJ, Van Krieken JH, Jaffe ES, Raffeld M (1990) Association of *bcl*-1 rearrangements with lymphocytic lymphoma of intermediate differentiation. Blood 76:2086-2090
Meeker TC, Grimaldi JC, O'Rourke R, Louie E, Juliusson G, Einhorn S (1989) An additional breakpoint region in the *BCL-1* locus associated with the t(11;14)(q13;q32) translocation of B-lymphocytic malignancy. Blood 74:1801-1806
Motokura T, Bloom T, Kim HG, Jüppner H, Ruderman JV, Kronenberg HM, Arnold A (1991) A novel cyclin encoded by a *bcl1*-linked candidate oncogene. Nature 350:512-515
Motokura T, Arnold A (1993) Cyclins and oncogenesis. Biochim Biophys Acta 1155:63-78
Palmero I, Holder A, Sinclair AJ, Dickson C, Peters G (1993) Cyclins D1 and D2 are differentially expressed in human B-lymphoid cell lines. Oncogene 8:1049-1054
Quelle DE, Ashmun RA, Shurtleff SA, Kato J-y, Bar-Sagi D, Roussel MF, Sherr CJ (1993) Overexpression of mouse D-type cyclins accelerates G_1 phase in rodent fibroblasts. Genes Dev 7:1559-1571
Rabbitts PH, Douglas J, Fischer P, Nacheva E, Karpas A, Catovsky D, Melo JV, Baer R, Stinson MA, Rabbitts TH (1988) Chromosome abnormalities at 11q13 in B cell tumours. Oncogene 3:99-103
Raffeld M, Jaffe ES (1991) bcl-1, t(11;14), and mantle cell-derived lymphomas. Blood 78:259-263
Reed SI (1992) The role of p34 kinases in the G1 to S-phase transition. Annu Rev Cell Biol 8:529-561
Rimokh R, Berger F, Cornillet P, Wahbi K, Rouault J-P, Ffrench M, Bryon P-A, Gadoux M, Gentilhomme O, Germain D, Magaud J-P (1990) Break in the *BCL1* locus is closely associated with intermediate lymphocytic lymphoma subtype. Genes Chromos Cancer 2:223-226

Rosenberg CL, Wong E, Petty EM, Bale AE, Tsujimoto Y, Harris NL, Arnold A (1991) *PRAD1*, a candidate oncogene: mapping and expression in centrocytic lymphoma. Proc Natl Acad Sci USA 88:9638-9642

Rosenberg CL, Motokura T, Kronenberg HM, Arnold A (1993) Coding sequence of the overexpressed transcript of the putative oncogene PRAD1/cyclin D1 in two primary human tumors. Oncogene 8:519-521

Rowley JD (1990) Molecular cytogenetics: Rosetta Stone for understanding cancer - twenty-ninth G. H. A. Clowes Memorial Award Lecture. Cancer Res 50:3816-3825

Sherr CJ (1993) Mammalian G1 cyclins. Cell 73:1059-1065

Tsujimoto Y, Yunis J, Onorato-Showe L, Erickson J, Nowell PC, Croce CM (1984) Molecular cloning of the chromosomal breakpoint of B-cell lymphomas and leukemias with the t(11;14) chromosome translocation. Science 224:1403-1406

Tsujimoto Y, Jaffe E, Cossman J, Gorham J, Nowell PC, Croce CM (1985) Clustering of breakpoints on chromosome 11 in human B-cell neoplasms with the t(11;14) chromosome translocation. Nature 315:340-343

Van Den Berghe H, Parloir C, David G, Michaux JL, Sokal, G (1979) A new characteristic karyotypic anomaly in lymphoproliferative disorders. Cancer 44:188-195

Weisenburger DD, Sanger WG, Armitage JO, Purtilo DT (1987) Intermediate lymphocytic lymphoma: immunophenotypic and cytogenetic findings. Blood 69:1617-1621

Williams ME, Westermann CD, Swerdlow SH (1990) Genotypic characterization of centrocytic lymphoma: frequent rearrangement of the chromosome 11 *bcl*-1 locus. Blood 76:1387-1391

Williams ME, Meeker TC, Swerdlow SH (1991) Rearrangement of the chromosome 11 bcl-1 locus in centrocytic lymphoma: analysis with multiple breakpoint probes. Blood 78:493-498

Williams ME, Swerdlow SH, Rosenberg CL, Arnold A (1992) Characterization of chromosome 11 translocation breakpoints at the *bcl*-1 and *PRAD1* loci in centrocytic lymphoma. Cancer Res 52(Suppl):5541s-5544s

Withers DA, Harvey RC, Faust JB, Melnyk O, Carey K, Meeker TC (1991) Characterization of a candidate *bcl-1* gene. Mol Cell Biol 11:4846-4853

Xiong Y, Connolly T, Futcher B, Beach D (1991) Human D-type cyclin. Cell 65: 691-699

Zhou DJ, Casey G, Cline MJ (1988) Amplification of human *int*-2 in breast cancer and squamous carcinomas. Oncogene 2:279-282

Pre-B-Cells Transformed by ts Abelson Virus Rearrange κ and γ Genes in Early G1

L. C. Wang, Y. Y. Chen, and N. Rosenberg
Immunology Graduate Program and Department of Pathology, Tufts University School of Medicine, Boston, MA 02111

Introduction

Abelson murine leukemia virus (Ab-MLV) induces a rapidly progressive, invariably fatal lymphoma in mice and transforms pre-B lymphocytes *in vitro* (reviewed in Rosenberg and Witte 1988). Like normal pre-B cells, the transformants have undergone V(D)J joining at the immunoglobulin heavy chain locus but most of them have not recombined the κ or λ light chain genes. Consistent with this pattern, the cells express a constellation of differentiation markers normally associated with pre-B lymphocytes. These phenotypic properties have always suggested that Ab-MLV transformation arrests B lymphocyte differentiation at the pre-B cell stage. Recent work using temperature sensitive (*ts*) Ab-MLV transformation mutants has confirmed this hypothesis (Chen et al. 1994). Pre-B cells transformed by *ts* mutants undergo light chain gene rearrangement at a high frequency after shift to the nonpermissive temperature. Differentiation arrest is mediated in part by the ability of the virus to suppress the expression of molecules that are critically important for pre-B cell differentiation. The RAG-1 and RAG-2 proteins and NF-κB are two of these targets (Chen et al. 1994; Klug et al. 1994).

Ab-MLV transformation is mediated by the protein tyrosine kinase activity of the v-*abl* protein, the single product of Ab-MLV (Rosenberg and Witte 1988). Similar to other protein tyrosine kinase oncoproteins, expression of v-*abl* protein stimulates tyrosine phosphorylation of a variety of cellular molecules. These events allow the cell to bypass normal cell cycle controls and provide a constitutive growth signal. Although the pathways involved are far from understood, signalling through the *ras* and *myc* pathways is almost certainly involved (Sawyers et al. 1992; Pendergast et al. 1993). In Ab-MLV-transformed lymphoid cells, the critical growth stimulatory signals from the v-*abl* protein are delivered during the G1 phase of the cell cycle. Cells transformed by *ts* Ab-MLV mutants undergo G1 arrest when shifted to the nonpermissive temperature (Chen and Rosenberg 1992). In the absence of other stimulatory signals such as overexpression of the *bcl*-2 gene, the cells undergo apoptosis.

Both uncontrolled cell cycle progression and differentiation arrest are important aspects of Ab-MLV-induced tumor development. These two features of the transformation process combine to subvert the normal regulatory circuits that control pre-B lymphopoeisis. Continued growth allows survival of lymphoid precursors with nonfunctional immunoglobulin gene rearrangements. Because differentiation is arrested, the cells do not express functional immunoglobulin proteins on their surface and are not subject to normal selection and growth control mediated by the interaction of membrane immunoglobulin with environmental signals. Recent data from several groups (Schlissel et al. 1993; Lin and Desiderio 1994) have suggested that normal lymphocytes undergo V(D)J joining during the G1 phase of the cell cycle, perhaps because levels of RAG-2 protein are highest during this phase. The work described in this communication addresses the stage during G1 that is affected by the v-*abl* protein and examines the relationship between cell cycle arrest and light chain gene rearrangement.

Results and Discussion

Earlier work (Chen and Rosenberg 1992) demonstrated that lymphoid cells transformed with *ts* Ab-MLV undergo G1 arrest following shift to the nonpermissive temperature. To define when during G1 the cells are arrested, the expression of genes important for the G1/S transition was examined. One such group of genes is the G1 cyclin genes, D1, D2, and D3 (reviewed in Sherr 1993). Cyclin D mRNAs and proteins appear during early to mid G1 and reach maximal levels at the G1/S boundary. Northern analyses of RNAs prepared from lymphoid cells transformed with wild type (*wt*) Ab-MLV revealed that most clones expressed either the D1 or D2 cyclin genes (Fig. 1 and data not shown). Similar levels of these RNAs were observed in cells maintained at either the permissive (34.0°C) or nonpermissive (39.5°C) temperature. Similar results were observed when *ts* transformants were grown at the permissive temperature. However, RNAs prepared from *ts* transformants that had undergone G1 arrest following shift to the nonpermissive temperature did not contain G1 cyclin sequences. In addition, *ts* transformants that were arrested in G1 at the nonpermissive temperature produced large amounts of cyclin RNAs 6 and 12 hours after shift back to the permissive temperature (Fig. 1, lanes H→L). These increases coincided with G1 progression (data not shown).

A second marker of the G1 phase is the retinoblastoma protein (pRB). This molecule is active at the G1/S transition, a process that is regulated by phosphorylation of pRB. Cells in G1 contain unphosphorylated and hypophosphorylated pRB; when cells undergo the G1/S transition, pRB is

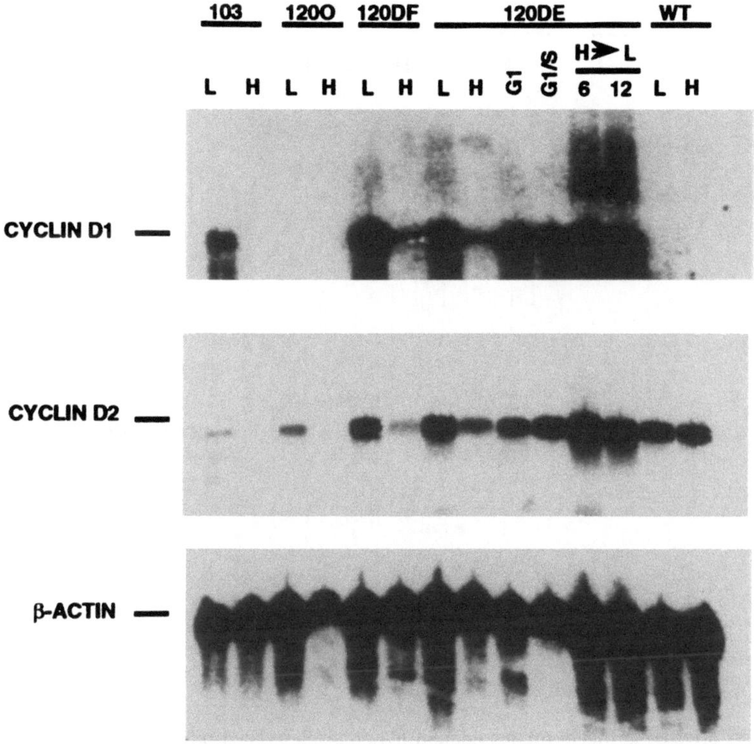

Fig. 1. Expression of D cyclins in the *ts* transformants. Cells transformed with *wt* (WT) and *ts* (103, 120O, 120DF, 120DE) Ab-MLV were maintained at the permissive (L) and nonpermissive (H) temperature and RNAs were harvested. G1 and S populations were prepared using nocodazole and hydroxyurea, respectively, as described elsewhere (Chen and Rosenberg 1992). For lanes H→L, cells were maintained at the nonpermissive temperature for 20 hours and shifted back to the permissive temperature for 6 or 12 hours prior to harvesting the RNAs. Northern blots were analyzed with D1 and D2 cyclin probes and a β-actin probe.

hyperphosphorylated (Chen et al. 1989). When the phosphorylation state of pRB was examined in *ts* transformants, cells maintained at the permissive temperature contained mostly hyperphosphorylated pRB (Fig. 2). A similar pattern was observed in *wt* transformants maintained at either the permissive or nonpermissive temperature. However, when *ts* transformants were incubated at the nonpermissive temperature for 20 hours, the majority of pRB was in the hypophosphorylated form. There also was a species of pRB that migrated around 98 kD (Fig. 2, arrow head) that appeared only in *ts* cells maintained at the nonpermissive temperature. A similar form of pRB has been noted in growth-arrested, terminally differentiated human hematopoietic cell lines (Mihara et al. 1989).Taken together, the experiments examining G1 cyclin expression and pRB phosphorylation suggest that the *ts* transformants are arrested in the early phase of G1, prior to committment to S phase. These data are consistent with cell synchronization and cell cycle experiments reported previously (Chen and Rosenberg 1992). The early phases of G1 transition are usually regulated by

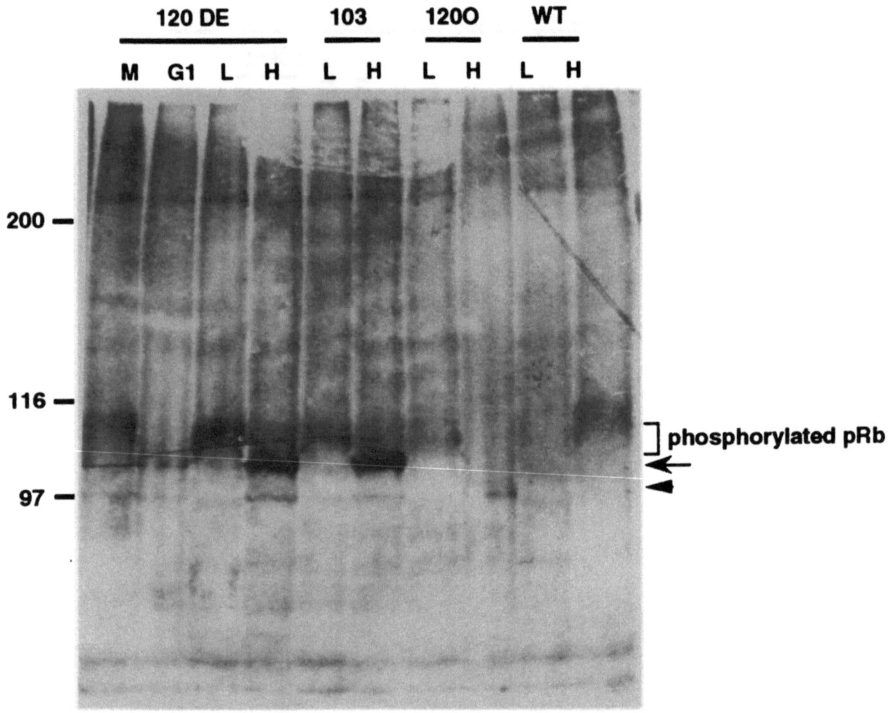

Fig. 2. The G1-arrested *ts* transformants contain hypophosphorylated pRB. Cell lysates from *ts* (120DE, 103, 120O) and *wt* (WT) transformants were analyzed on western blots probed with anti-pRB antibody. L, cells grown at the permissive temperature; H, cells maintained at the nonpermissive temperature for 20 hours; M, cells synchronized at M phase; G1, cells synchronized in early G1. Cell synchronization conditions were identical to those described elsewhere (Chen and Rosenberg 1992). Positions of protein standards (kD), the dephosphorylated form of pRB (arrow) and a species of pRB of 98 kD (arrow head) are indicated.

growth factors in normal cells. Interleukins such as IL-7 probably act at this point in normal B lineage precursors. Despite this, the function of v-*abl* in regulating cell cycle progression cannot be replaced by IL-7 or other interleukins (Y. Y. Chen and N. Rosenberg, unpublished data).

Earlier experiments have shown that a high frequency of *ts* transformants rearrange κ and λ light chain genes after shift to the nonpermissive temperature (Chen et al. 1994). In these studies, *ts* transformants were shifted to the nonpermissive temperature, then shifted back to the permissive temperature 4 days later and subcloned. This type of analysis does not evaluate the time required at the nonpermissive temperature to activate the gene rearrangement process. The protein tyrosine kinase activity of the *ts* v-*abl* protein decreases by 4 hours after temperature shift to the nonpermissive temperature, and increases in RAG-1 and RAG-2 RNAs can be detected at this time (Chen et al. 1994). However, significant proportions of the *ts* cells do not arrest in G1 until 8 to 12 hours after shift to the nonpermissive temperature. To determine how these events correlate

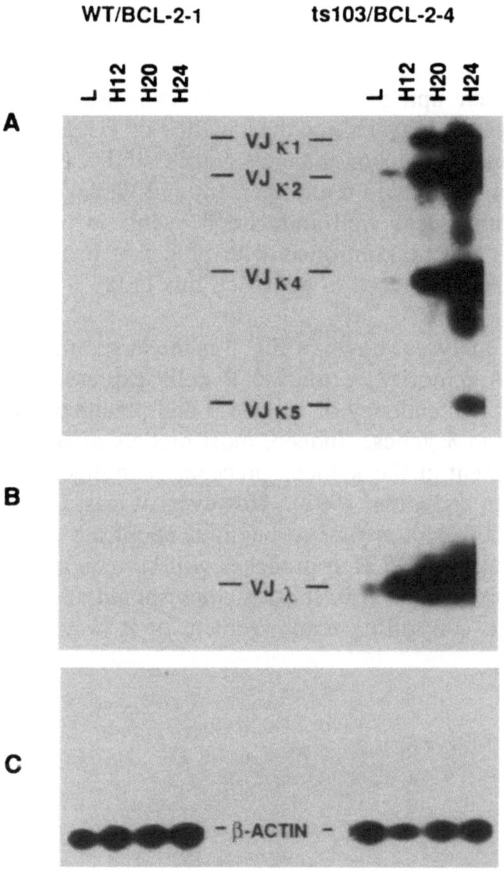

Fig. 3. κ and λ rearrangements can be detected about 12 hours after shift to the nonpermissive temperature. PCR assays were used to detect the products of V(J) joining at the κ and λ loci. (A) For κ, a Jκ(5) primer and a degenerate V(κ) primer were used as described elsewhere (Schissel et. al. 1989). (B) For λ, a Vλ(1,2), a J(λ1) and a J(λ2,3) were used (Zou et al. 1993. (C) The β-actin PCR reactions were performed as described by Hardy et al. (1991). WT/BCL-2-1 is a *wt* transformant that overexpresses *bcl*-2; ts103/BCL-2-4 is a *ts* transformant that overexpresses *bcl*-2. L, permissive temperature; H, nonpermissive temperature. The numbers following the H indicate hours at the nonpermissive temperature.

to activation of gene rearrangement, a PCR assay was used to detect rearranged sequences. When *ts* transformants were shifted to the nonpermissive temperature for as little as 12 hours, rearrangement of both κ and λ sequences could be detected (Fig. 3). As expected, only low levels of rearrangement were detected in *ts* cells grown at the permissive temperature and in *wt* cells maintained at either the permissive or nonpermissive temperature. These experiments suggest that rearrangements coincide with accumulation of cells in the G1 phase of the cell cycle.

Increases in the levels of RAG RNAs (Chen et al. 1994) and proteins (M. S. Huang and N. Rosenberg, unpublished data) can be detected as early as 4 hours after shift to the nonpermissive temperature. Coupled with the PCR analyses, these data suggest that both v-*abl* mediated effects on gene products required for gene rearrangement and on cell growth are important for suppression of pre-B cell differentiation. The time lag between upregulation of RAG expression and detection of rearranged light chain genes may reflect the kinetics of G1 arrest. Alternatively, expression of other gene products or changes in accessibility of the light chain genes may be required for light chain rearrangement and the lag may reflect the time required for these other, less well-understood events to occur. Because the timing of rearrangement can be controlled with precision in the *ts* system, these cells should provide a valuable tool to approach this issue.

A second striking feature of the PCR analyses shown in Fig. 3 is the large amount of V(J)λ rearrangement detected. A minority of murine B cells express λ, a phenomenon that is thought to reflect the tendency of murine B cell precursors to rearrange κ genes more frequently than λ genes. Indeed, most models of B cell differentiation predict that rearrangement at the κ locus precedes rearrangement at the λ locus (reviewed in Langerman and Cohn 1992). However, at least in the *ts* transformants, rearrangements at both loci appear to begin at about the same time. Further studies with more quantitative PCR approaches will be required to confirm these results. The *ts* transformants used in these studies should allow a careful assessment of the mechanisms controlling rearrangement of κ vs λ light chain genes.

Acknowledgements

This work was supported by CA 24220 and AI 35721 from the National Institutes of Health.

References

Chen PL, Scully P, Shew JY, Wang JYJ, Lee WH (1989) Phosphorylation of the retinoblastoma gene product is modulated during the cell cycle and cellular differentiation. Cell 58:1193-1198

Chen YY, Rosenberg N (1992) Lymphoid cells transformed by Abelson virus require the v-*abl* protein tyrosine kinase only during early G1. Proc Natl Acad Sci USA 89:6683-6687

Chen YY, Wang LC, Huang MS, Rosenberg N (1994) An active v-*abl* protein tyrosine kinase blocks immunoglobulin light chain gene rearrangement. Genes & Dev 8:688-697

Hardy RR, Carmack CE, Shinton SA, Kemp JD, Hayakawa K (1991) Resolution and characterization of pre-B and pre-pro-B cell stages in normal mouse bone marrow. J Exp Med 173:1213-1225

Klug CA, Gerety SJ, Chen YY, Rice NR, Rosenberg N, Singh, H (1994) The v-*abl* tyrosine kinase negatively regulates NF-κB and blocks kappa gene transcription in pre-B lymphocytes. Genes & Dev 8:678-687

Langerman RE, Cohn M (1992) What determins the κ/λ ratio? Res Immunol 143:803-839

Lin WC, Desiderio SV (1994) Cell cycle regulation of V(D)J recombination activator protein RAG-2. Proc Natl Acad Sci USA 91:2733-2737

Mihara K, Cao XR, Yen A, Chandler S, Driscoll B, Murphree AL, T'ang A, Fung YT (1989) Cell cycle-dependent regulation of phosphorylation of the human retionblastoma gene product. Science 246:1300-1303

Pendergast AM, Quilliam LA, Cripe LD, Bassing CH, Dai Z, Li N, Batzer A, Rabun KM, Der CJ,Schlessinger J, Gishizky ML (1993) BCR-ABL-induced oncogenesis is mediated by direct interaction with the SH2 domain of the GRB-2 adaptor protein. Cell 75:175-185

Rosenberg N, Witte ON (1988) The viral and cellular forms of the Abelson (*abl*) oncogene. Adv Virus Res 35:39-81

Sawyers CL, Callahan W, Witte ON (1992) Dominant negative MYC blocks transformation by ABL oncogenes. Cell 70:901-910

Schlissel MS, Baltimore D (1989) Activation of immunoglobulin kappa gene rearrangement correlates with induction of germline kappa gene transcription. Cell 58:1001-1007

Schlissel M, Constantinescu A, Morrow T, Baxter M, Peng A (1993) Double-strand signal sequence breaks in V(D)J recombination are blunt, 5'phosphorylated, RAG-dependent, and cell cycle regulated. Genes & Dev 7:2520-2532

Sherr CJ (1993) Mammalian G_1 cyclins. Cell 73:1059-1065

Zou YR, Takeda S, Rajewksy K (1993) Gene targeting in the Igκ locus: efficient generation of λ chain-expressing B cells, independent of rearrangements in Igκ. EMBO J 12:811-820

Development of *btk* Transgenic Mice

E.A. Faust[1], D.J. Rawlings[1], D.C. Saffran[1], and O.N. Witte[1,2]

[1]Department of Microbiology and Molecular Genetics, University of California, Los Angeles, CA 90024.
[2]Howard Hughes Medical Institute, University of California, Los Angeles, CA 90024.

Introduction

B cell development is characterized by the orderly expression of cell surface markers and responses to specific activation signals [1]. Tyrosine kinases are involved in signalling pathways that regulate these events [2]. Bruton's tyrosine kinase (Btk) is a tyrosine kinase expressed in B and myeloid cells [3,4]. Disruption of Btk activity in either humans or mice leads to the related immunodeficiencies XLA and XID [3,4,5,6]. The involvement of Btk in these diseases indicates that Btk plays a critical role in B cell development. The specific functions of Btk in B cell signaling are not yet known.

Our goal is to replace the *btk* gene in XID animals in order to confirm the role of *btk* in this disease, address the potential dominant negative interaction of mutant endogenous Btk on exogenous Btk, and to develop models to study the roles of Btk in B cell development. We are developing *btk* transgenic mice to address these questions.

Bruton's Tyrosine Kinase

Btk is a 77kd protein that contains SH1, SH2, and SH3 domains [3], and Btk shares strong amino acid homology with several other cytoplasmic tyrosine kinases including Dsrc28c and ITK which comprise a new subfamily of tyrosine kinases [7,8,9,10]. Members of this subfamily contain a pleckstrin homology (PH) domain in their highly conserved amino terminal unique regions [11]. PH domains are present in a broad array of signalling proteins including several serine-threonine kinases, GTPases, GTP activating proteins, nucleotide exchange factors and phospholipases [12, 13]. The function of the PH domain is not understood, but it is likely to function in interactions with other signalling molecules.

XLA and XID

XLA results in dramatic reduction in B cells and severe humoral immunodeficiency. Pre-B cells fail to thrive and undergo the normal clonal expansion that results in the production of immature B lineage cells. In XLA, proliferation of cytoplasmic-μ^+ pre-B cells is significantly reduced and fewer cytoplasmic-μ^+ cells enter S phase [14,15]. For these reasons, it has been

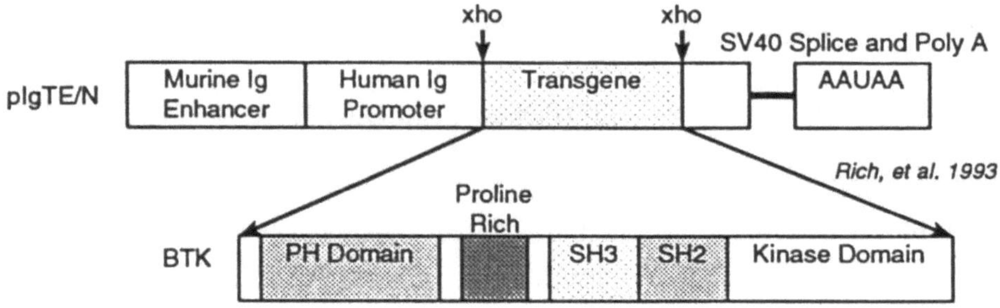

Fig. 1: Structure of *btk* transgene.

proposed that the XLA defect might interfere with B lineage-specific signal transduction pathways critical to early B lineage growth and clonal expansion.

The XID phenotype is less severe than XLA. Mice have 30-50% the normal number of B cells, but these fail to become phenotypically and functionally diverse [16,17]. B cells from XID mice do not respond to thymus independent type 2 antigens such as TNP-Ficoll and have an immature surface phenotype suggestive of disordered maturation. XID B cells exhibit abnormal responses to IgM X-linking, B cell mitogens, IL-10, and do not form colonies in response to LPS in the colony forming unit B cell assay (CFU-B) [16,17,18,19].

XLA patients exhibit a wide range of mutations of the *btk* gene. These mutations can result in a complete lack of gene expression or expression of a variety of mutant proteins [3,4,20]. Since XLA is the result of disruption of a single gene expressed in hematopoetic cells, the disease is a potential candidate for gene therapy.

The XID phenotype results from a single point mutation in the PH domain of *btk* [5,6]. A mutant protein with normal kinase activity is expressed in XID animals [5,6]. It is possible that the endogenous mutant Btk protein expressed in XID and XLA B cells could act in a dominant negative manner by interfering with the function of exogenously added Btk. The introduction of a *btk* transgene into XID mice addresses this issue.

Creation of *btk* Transgenic Mice and Mating onto XID Background

We generated a line of transgenic mice carrying a *btk* cDNA fused to an immunoglobulin (Ig) heavy (H) chain promoter and enhancer (Figure 1). This vector has been used previously to create the IL-7 transgenic mouse (Figure 1). Expression was detected in the bone marrow, lymph nodes, spleen, thymus and skin [21]. The *btk* transgene was inserted into the XHO-1 site of pIgTE/N (Figure 1). pIgTE/N contains a murine H chain Ig enhancer and a human H chain Ig promoter 5' of the XHO-1 site and SV40 sequences containing an intron and polyadenylation signal 3'of the XHO-1 site [21]. A 4.5 kb fragment of DNA containing *Btk* was excised from the plasmid with Sal-I and Bam-H1 and purified by agarose gel electrophoresis.

Transgenic animals were generated by standard technique at the UCLA transgene facility [22]. Transgenic animals were identified by hybridization to DNA prepared from tail biopsies. A fragment derived from the *btk* cDNA

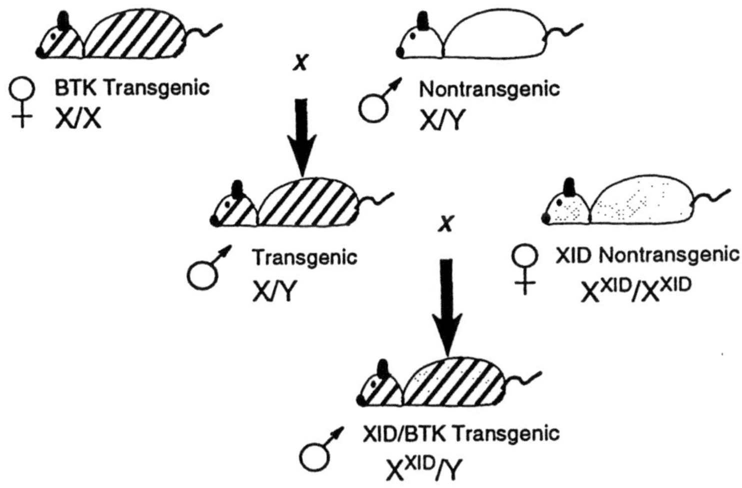

Fig. 2: Mating strategy to cross the *btk* transgene onto an XID background.

was radiolabeled and used as a hybridization probe. Four positive founders were identified. In one animal, the *btk* gene had rearranged. One of the other animals was propagated by repeated crossing with CBA/N (XID) animals (Figure 2).

The female founder was mated with males to put the transgene on a male background. Male transgenic offspring were then mated with a female XID mouse. The resulting male progeny that carried the transgene are XID animals with a *btk* transgene (XID/*btk* transgenic). Litter mates of these animals serve as controls for assays. Male animals without the transgene are negative (XID) controls. All the female progeny express the wildtype X allele (the XID allele is not expressed due to nonrandom X inactivation [23]) thus serving as normal positive controls.

Analysis of XID/*btk* Transgenic Animals

The analysis of the XID/*btk* transgenic animals is depicted in Figure 3. First genomic DNA was prepared from tails and analyzed by Southern blot. Greater than 50% of the animals in any litter inherit the transgene. Next, a group of animals, which included all controls mentioned above, was immunized with 10μg TNP-Ficoll, a T-independent antigen to which XID animals cannot mount an antibody response [16,17]. Seven to ten days later, the animals were sacrificed and spleen, bone marrow, thymus and serum were collected. Expression of the transgene was assessed by Northern blot analysis of the spleen and autokinase assays of the bone marrow and thymus [5]. Btk autokinase activity in the thymus can only result from expression from the transgene since endogenous Btk is not expressed in the thymus [3]. Serum was analyzed by ELISA for levels of IgM and IgG_3, since XID animals have decreased levels of each [16,17]. In addition, the serum was analyzed for antibodies to TNP-Ficoll. Bone marrow cells were transformed by a v-ABL containing retrovirus

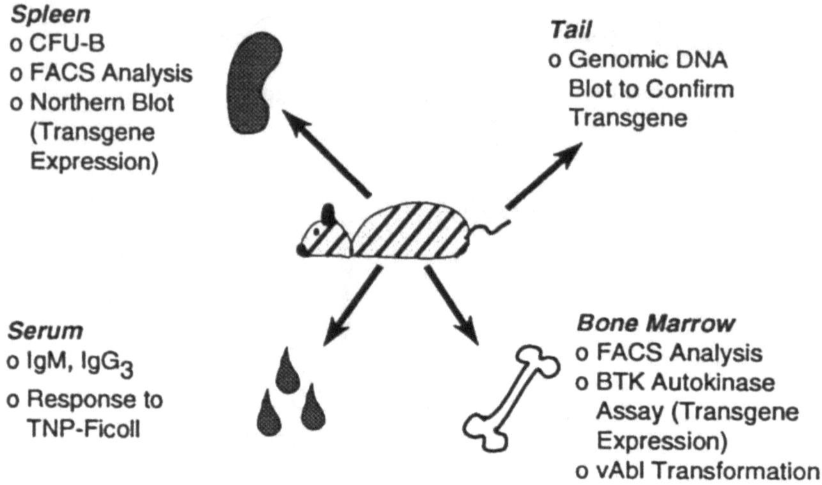

Fig. 3: Analysis of XID/*btk* transgenic mice.

to establish a renewable source of cells [24]. FACS analysis was performed on all tissues to determine if normal ratios of B and T cells exist in these tissues.

Results

FACS profiles of spleen, bone marrow and thymus were indistinguishable from those of litter mate controls. The spleen and bone marrow were analyzed for B220 and Mac-1 expressing cells, while the thymus was analyzed for Thy-1 expressing cells.

Preliminary results indicate that the XID/*btk* transgenic animals exhibited a partial to full rescue of the CFU-B phenotype. Table 1 depicts the results of three separate experiments. In the analysis of the first litter (Exp. 1) the CFU-B phenotype was completely rescued. However, the analysis of the second and third litters (Exp. 2 and Exp. 3) revealed only a partial rescue. This discrepancy may have to do with the age of the litters at the time of the analysis[1]

Only a minimal increase in serum levels of IgG_3 and IgM was observed although there was reconstitution of IgM to near normal levels in Exp. 1 (Table 2). Control X^{XID}/Y male littermates which did not express the transgene were not available for comparison in either Exps. 1 or 3. However, these animals are expected to retain the XID phenotype on the transgenic background strain as supported in Exp. 2 (Table 2). Similarly, partial to no antibody production in response to the TNP-Ficoll challenge was present as compared to XID animals. While the size of the endogenous *btk* transcript and the *btk* transcript from the transgene are indistinguishable by Northern blot analysis, we expected to see an increase in the level of *btk* transcript in transgenic animals because those animals had multiple copies of the transgene. Northern blot analysis of spleen cells from transgenic and nontransgenic animals, how-

[1]The litter analyzed in Exp. 1 was four weeks old whereas the subsequent experiments were done on older animals.

Table 1: Splenic CFU-B colonies of XID/BTK transgenic and control animals.

	Animals	No. Animals	No. CFU-B Colonies[4]
Experiment 1	XID/Y + BTK[1] XID/Y[2] XID/X[3]	2 ND[6] 4	359–TNTC[5] ND 260–TNTC
Experiment 2	XID/Y + BTK XID/Y XID/X	2 2 2	53–153 (90) 0–3 (2) 164–178 (170)
Experiment 3	XID/Y + BTK XID/Y XID/X	4 ND 2	21–330 (115) ND TNTC

[1]XID/BTK transgenic animals.
[2]Negative controls. Male litter mates, nontransgenic.
[3]Positive control. Female litter mates, transgenic and nontransgenic.
[4]The range of the number of CFU-B colonies with the average in parenthesis.
[5]Too numerous to count (greater than 500 colonies per plate).
[6]Not determined.

ever, revealed no detectable difference in the level of *btk* expression. Low levels of Btk autokinase activity were detected in the thymus of transgenic animals, but not in nontransgenic animals. The pIgTE/N vector allows expression in thymus as demonstrated previously in IL-7 transgenic mice [21]. Therefore, the low level *btk* expression seen in the thymus is probably derived from the transgene.

Conclusions

There are at least two alternative explanations for the partial rescue of the CFU-B phenotype and lack of recovery of antibody production in XID/*btk* transgenic animals. First, the endogenous mutant Btk in XID animals acts as a dominant negative, interfering with rescue of function by the exogenous Btk expressed from the transgene. For a mutant allele to act as a complete dominant negative it must be expressed in excess of the wild type allele [25]. The intermediate phenotype seen in these experiments could result from only partial rescue in the presence of this dominant negative effect. However, we would predict that in this situation, with low level expression of the transgene, the endogenous allele would exert a complete dominant negative effect. Alternatively, the *btk* transgene is not expressed at high enough levels to completely rescue the XID phenotype. This explanation is more likely because the transgene mRNA expression was undetectable in the spleen and only low levels of protein expression were detected in the thymus.

Table 2: Serum analysis of XID/BTK transgenic and control animals.

	Animals	No. Animals	Optical Density (260 nm)[1] IgM (Dilut.[2] =1/2000)	IgG3 (Dilut. =1/200)	αTNP-Ficoll (Dilut. =1/100)
Exp. 1	XID/Y+BTK	2	0.89–0.92 (0.91)	0.21–0.25 (0.23)	0.56–0.70 (0.63)
	XID/Y	ND[3]	ND	ND	ND
	XID/X	4	1.03–1.21 (1.15)	0.71–1.31 (1.01)	> 2.0
Exp. 2	XID/Y+BTK	2	0.42–0.50 (0.46)	0.38–0.47 (0.43)	0.14
	XID/Y	2	0.37–0.57 (0.42)	0.30–0.32 (0.31)	0.03–0.10 (0.07)
	XID/X	2	1.5	0.80–1.07 (0.94)	1.5
Exp. 3	XID/Y+BTK	4	0.16–0.25 (0.20)	0.16–0.29 (0.25)	0.00–0.03 (0.02)
	XID/Y	ND	ND	ND	ND
	XID/X	2	0.67–0.93 (0.80)	0.78–0.81 (0.80)	0.96–1.32 (1.10)

[1]Range of the optical density reading at a wavelength of 260 nm with average in parenthesis.
[2]Dilution of serum sample.
[3]Not determined.

Future Experiments

Analyses of additional litters of XID/*btk* transgenic animals are ongoing. In addition, we have created another line of XID/*btk* transgenic mice which may have a higher level of Btk expression. To obtain a line with higher expression, a transgene driven by the SV40 promoter and Ig H enhancer [26] has been constructed.

The long term goal of these studies includes the creation of transgenic animals that express mutant *btk* genes. These mice will be mated to a *btk* knockout mouse whose hematopoietic tissues would be analyzed for abnormalities. Experiments of this type will lead to a better understanding of the role that *btk* plays in B lymphopoiesis.

Acknowledgements

E.A.F. is a recipient of a Biotechnology Training Grant, AI07126-17. D.C.S. is a Fellow of the Leukemia Society of America. D.J.R. is supported by grants AR36834 and AR01912. O.N.W. is an Investigator of the Howard Hughes Medical Institute.

References

1. Rolink A, Melchers F (1991) Molecular and cellular origins of B lymphocyte diversity. Cell 66:1081-1094.

2. DeFranco AL (1992) Tyrosine phosphorylation and the mechanism of signal transduction by the B-lymphocyte antigen receptor. Eur J Biochem 210:381- 388.

3. Tsukada S, Saffran DC, Rawlings DJ, Parolini O, Allen RC, Klisak I, Sparkes RS, Kubagawa H, Mohandas T, Quan S, Belmont JW, Cooper MD, Conley ME, Witte ON (1993) Deficient expression of a B cell cytoplasmic tyrosine kinase in human X-linked agammaglobulinemia. Cell 72:279-290.

4. Vetrie D, Vorechovsky I, Sideras P, Holland J, Davies A, Flinter F, Hammarström L, Kinnon C, Levinsky R, Bobrow M, Smith CIE, Bentley DR (1993) The gene involved in X-linked agammaglobulinaemia is a member of the *src* family of protein-tyrosine kinases. Nature 361:226-233.

5. Rawlings DJ, Saffran DC, Tsukada S, Largaespada DA, Grimaldi JC, Cohen L, Mohr RN, Bazan JF, Howard M, Copeland NG, Jenkins NA, Witte ON (1993) Mutation of the amino-terminal unique region of bruton's tyrosine kinase in murine X-linked immunodeficiency. Science 358:358-361.

6. Thomas JD, Sideras P, Smith CIE, Vorechovsky I, Chapman V, Paul WE (1993) Colocalization of X-linked agammaglobulinemia and X-linked immunodeficiency genes. Science 261:355-358.

7. Gregory RJ, Kammermeyer KL, Vincent WS III, Wadsworth SG (1987) Primary sequence and developmental expression of a novel *Drosophila melanogaster src* gene. Mol Cell Biol 7:2119-2127.

8. Mano H, Mano K, Tang B, Koehler M, Yi T, Gilbert DJ, Jenkins NA, Copeland NG, Ihle JN (1993) Expression of a novel form of *Tec* kinase in hematopoietic cells and mapping of the gene to chromosome 5 near *Kit*. Oncogene 8:417-424.

9. Siliciano JD, Morrow TA, Desiderio S V (1992) *itk*, a T-cell-specific tyrosine kinase gene inducible by interleukin 2. Proc Natl Acad Sci USA 89:11194-11198.

10. Heyeck SD, Berg LJ (1993) Developmental regulation of a murine T-cell- specific tyrosine kinase gene, *Tsk*. Proc Natl Acad Sci USA 90:669-673.

11. Musacchio A, Gibson T, Rice P, Thompson J, Saraste M (1993) The PH domain is a common piece in the structural patchwork of signalling proteins. TIBS 18:343-348.

12. Haslam RJ, Kolde HB, Hemmings BA (1993) Pleckstrin domain homology. Nature 363:309-310.

13. Mayer BJ, Ren R, Clark KL, Baltimore D (1993) A putative modular domain present in diverse signaling protein. Cell 73:629-630.

14. Campana D, Farrant J, Inamdar N, Webster ADB, Janossy G (1990) Phenotypic features and proliferative activity of B cell progenitors in X-linked agammaglobulinemia. J Immunol 145:1675-1680.

15. Pearl ER, Vogler LB, Okos AJ, Crist WM, Lawton AR III, Cooper MD (1978) B lymphocyte precursors in human bone marrow: An analysis of normal individuals and patients with antibody-deficiency states. J Immunol 120:1169-1175.

16. Scher I (1982) The CBA/N mouse strain: an experimental model illustrating the influence of the X-chromosome on immunity. Adv Immunol 33:1-71.

17. Wicker LS, Scher I (1986) X-linked immune deficiency (*xid*) of CBA/N mice. Curr Top Microbiol Immunol 124:87-101.

18. Hitoshi Y, Sonoda E, Kikuchi Y, Yonehara S, Nakauchi H, Takatsu K (1993) IL-5 receptor positive B cells, but not eosinophils, are functionally and numerically influenced in mice carrying the X-linked immune defect. Int Immunol 5:1183-1190.

19. Go NF, Castle BE, Barret R, Kastelein R, Dang W, Mosmann TR, Moore KW, Howard M (1990) Interleukin 10, a novel B cell stimulatory factor: Unresponsiveness of X chromosome-linked immunodeficiency B cells. J Exp Med 172:1625-1631.

20. Saffran DC, Parolini O, Fitch-Hilgenberg ME, Rawlings DJ, Afar DEH, Witte ON, Conley ME (1994) A point mutation in the SH2 domain of Bruton's tyrosine kinase resulting in protein instability and atypical X-linked agammaglobulinemia. N Engl J Med 330:1488-1491.

21. Rich BE, Campos-Torres J, Tepper RI, Moreadith RW, Leder P (1993) Cutaneous lymphoproliferation and lymphomas in interleukin 7 transgenic mice. J Exp Med 177:305-316.

22. Kronenberg M (1992). Trangenic Core Facility UCLA Pamphlet.

23. Conley ME (1985) B cells in patients with X-linked agammaglobulinemia. J Immunol 134:3070-3074.

24. Whitlock C, Denis K, Robertson D, Witte O (1985) In vitro analysis of murine B-cell development. Ann Rev Immunol 3:213-235.

25. Herskowitz I (1987) Functional inactivation of genes by dominant negative mutations. Nature 329:219-222.

26. Rosenbaum H, Webb E, Adams JM, Cory S, Harris AW (1989) N-myc transgene promotes B lymphoid proliferation, elicits lymphomas and reveals cross- regulation with c-myc. EMBO J 8:749-755.

Genomic Instability: General Topics

Genomic Instability in B-Cells and Diversity of Recombinations That Activate c-myc

S. Janz, G. M. Jones, J. R. Müller and M. Potter
Laboratory of Genetics, NCI, NIH

Abstract

Genetic rearrangements activating the proto-oncogene c-*myc* comprise a mandatory oncogenic step in plasma cell tumor development in BALB/cAnPt mice. In the majority of plasmacytomas, c-*myc* activating rearrangements take the form of reciprocal chromosomal translocations t(12;15) that juxtapose c-*myc* to the immunoglobulin heavy chain α locus (IgHα) in particular the switch α region (Sα).The genetic basis for the prevalence of Sα/c-*myc* recombinations in BALB/cAnPt plasmacytomas is not known but may be related to a hypothetical regional genomic instability of the c-*myc* and IgHα loci in BALB/cAnPt mice. We wished to test whether the genomic instability of both loci might be revealed by the diversity of genetic recombinations that can be observed in IgHα and c-*myc*. We employed PCR methods to detect new recombinations of c-*myc* and IgHα in the preneoplastic stage of plasma cell tumor development and found that c-*myc* can be joined to more genes or genomic regions than known before. This is indicative but does not formally prove a particular genomic instability of c-*myc* and IgHα in BALB/cAnPt B cells. Since defective DNA repair provides a mechanistic explanation for genomic instability, we measured the efficiency of repair in IgHα and c-*myc* using an assay that quantitates the removal of UV-induced pyrimidine dimers within specific genomic regions. We used plasmacytoma XRPC 24 as a model system and found that both IgHα and c-*myc* were poorly repaired, whereas c-*abl*, a proto-oncogene not related to conventional pristane-induced plasmacytomagenesis, was efficiently repaired.

Introduction

The contribution of genomic instability to malignant cell transformation, cancer and metastatic progression is a hypothesis that enjoys increasing acceptance. The concept of genomic instability essentially suggests that genetic changes leading to decreased stability of DNA are important components of neoplastic development and provide factors that somehow control the susceptibility of individual humans or inbred strains of mice to develop cancer. The mechanisms of genomic instability, however, remain difficult to define, and information on the genetic and biochemical basis of genomic instability is only currently emerging. On the genetic level, certain DNA sequence and structural motifs such as dispersed repetitive elements, transposable elements, tandem repeats, inverted repeats, and inverted telomere-like sequences have been related to genomic instability. On the biochemical level, errors in DNA metabolism such as slippage in DNA replication, untimely release of DNA endonuclease and alterations in the CpG methylation pattern or deficiencies in DNA repair including gene-specific repair of induced damage in tumor target genes have all been implied to determine genomic instability. The molecular indications of genomic instability are as diverse as the genetic model systems studied and include deletions, inversions, duplications, and amplifications of genes, DNA fingerprint polymorphisms, loss of heterozygosity, and chromosomal rearrangements. Because of the importance of chromosomal rearrangements, we decided to test whether the genomic instability of a given genetic locus might be revealed by the diversity of genetic recombinations that can be observed in that locus. Since our main interest is the pathogenesis of malignant plasma cell tumor development in BALB/cAnPt mice, we chose to look for a potential diversity of rearrangements in two genetic loci that play a prominent role in murine plasmacytomagenesis, the proto-oncogene c-*myc* and the IgHα locus. We performed a PCR analysis, the results of which will be presented here.

Results and Discussion

Genetic recombinations that activate c-*myc* comprise a critical pathogenetic event in plasmacytomagenesis in BALB/cAnPt mice (reviewed in Potter and Wiener, 1992). Rearrangements of c-*myc* typically take the form of non-random reciprocal chromosomal translocations t(12;15) and t(6;15) that juxtapose c-*myc* to an IgH switch region and the *pvt*-1 locus to Igκ, respectively. The predominant chromosomal translocation in BALB/c plasmacytomas (PCTs) is the t(12;15) which occurs in 90% of the tumors. In about 80% of all PCTs with t(12;15) the breaksite in chromosome 12 is in Sα, and only a few PCTs utilize the Sμ or S_γ region (Fig. 1a).

Fig. 2 shows a number of additional recombinations between Sα and c-*myc* in preneoplastic B cells and primary PCTs. The c-*myc* portions of the recombinational fragments obtained in that analysis are positioned along the c-*myc* gene according to the chromosomal breakpoints. The PCR products comprise - from top to bottom - 1 recombinational fragment that has been detected twice in 2 different mesenteric sectors of a BALB/c mouse 50 days post-pristane (#15, 54), 2 fragments in 2 different BALB/c mice 33 days post-pristane (#B73, 82), 2 fragments (#D1, 18) in 2 congenic C.D2-Idh-Pep3 mice (Potter et al. 1994), 1 recombinational fragment with Sγ1 that was found in 3 mesenteric sectors of a BALB/c mouse (#E11, 40, 70) and another recombination product with Sα that was also detected in 3 mesenteric sectors in another mouse (#E61, 83, 90; both #E mice were exceptional insofar as the mesenteric tissue was divided into 50 randomized sectors instead of the usual 5 - 7 prior to PCR analysis), 1 fragment probably indicating an expanded clone since it was reconfirmed in 6 other mesenteric sectors of a C.D2-Idh-Pep3 mouse (#F107, 108, 111, 115, 146, 156, 159), 1 fragment that was detected only once (#F59) and another product that indicated a recombination with Sμ (#F221), 2 fragments in peritoneal exudate cells of a congenic C.D2-Es-Hba mouse (#ES-1, -0.35), 6 different fragments in 6 primary PCTs (#5886, 5890, 5893, 5904, 5905, 5910), 5 fragments in mesenteric omenta (#K7, 15, 16, 21), one of which present in two omental pieces (#K12a, 12b), and 1 fragment that was found in 3 different mesenteric sectors of a BALB/c mouse (#2, 4, 5).

Most of the recombinations appear to be clean double strand breaks in c-*myc* but some occasional alterations were observed. These include a base substitution mutation near the breakpoint (#E61, 83, 90), a 5-bp deletion next to the breaksite (#5890), a 1-bp insertion exactly at the breaksite (#K15), another 8-bp insertion at the breaksite (#F59) and, interestingly, a 41-bp duplication and a 122-bp deletion in two primary PCTs (#5890 and #5904, respectively). The latter observation was made possible by analysis of the reciprocal product of the translocation, chr. 15-.

Fig. 1b indicates that besides the common recombinations between IgH switch regions and c-*myc*, atypical c-*myc* activating rearrangements can also occasionally be seen in murine plasmacytomas. These uncommon recombinations include retroviral insertions in PCTs ABPC 22 and RFPC 2782 (Shaughnessy et al., 1993), an unusual recombination between Sα and *pvt*-1 in ABPC 60 (Shaughnessy et al., 1994), recombinations with repetitive genetic elements such as the insertion of an intracisternal type A particle (IAP) ~2 kbp 3' of c-*myc* in PCT J 558 (Greenberg et al., 1985), a single translocation t(15;16) recombining the Igλ locus with an unknown locus about 30 kbp 3' of c-*myc* (Axelson et al., 1991), and recombinations with the intronic IgH enhancer in ABPC 45 (Fahrlander et al., 1985) and DCPC 21 (Ohno et al., 1991).

We wished to test whether these uncommon recombinations would forebode an even larger diversity of recombinations involving c-*myc*, that in turn might reflect genomic instability of c-*myc*. Therefore we selected DNAs from oil granulomatous tissues that tested positive for Sα/c-*myc* recombinations in previous experiments and designed new primers that anneal in short and long interspersed elements; i. e., SINE and LINE elements, respectively. SINE elements are mainly represented by the B1 and B2 families of 140-bp or 190-bp repeats, whereas LINE elements are mostly dispersed as 3' truncated ends. The selection of dispersed repeats as priming sites to detect new and meaningful types of recombinational fragments of c-*myc* by PCR analysis appeared to be promising since SINE elements, for example, are known to be in close proximity to protein encoding sequences. The repeat PCR primers were paired with PCR primers annealing in c-*myc* in an attempt to detect PCR fragments indicating recombination events between c-*myc* and some known or unknown sequence that is flanked by a repetitive element (Fig. 1c).

Even though this approach turned out to be plagued by a substantial number of artifacts, it did result in 5 recombinational fragments between c-*myc* and unknown sequence (#4 and #21), a minor satellite (#22), a repetitive element of unclear nature (#23) and a LINE-1 element (#2-21-2). Although the importance of these fragments *per se* is unclear, the detection of two more unconventional PCR products that were obtained in unrelated experiments by using specific Sα and Sμ primers together with myc primers suggests that they may be significant. In these two fragments which contain switch regions and therefore appear to have originated in B cells, c-*myc* is joined with the gene for ornithine decarboxylase (ODC, #5904) and a mouse microsattelite (#OG9a), respectively. We conclude that early B cells infiltrating the oil granulomatous tissue of BALB/c mice may carry a diversity of recombination products involving the c-*myc* gene (Fig. 1c).

Fig. 3 illustrates the IgHα portions of the recombinational fragments between Sα and c-*myc* in primary PCTs and preneoplastic B cells in oil granulomas that complement the c-*myc* portions shown in Fig. 2. In two cases, #3 and #6a (at the bottom of Fig. 3), the c-*myc* portions have not been fully determined yet and are therefore missing in Fig. 2. The recombinational fragments are positioned along the IgHα locus according to the recombinational breakpoints. Some of the IgHα sequences flanking the recombination breakpoints with c-*myc* are characterized by deletions and inversions that are denoted by gaps within the fragments and arrows

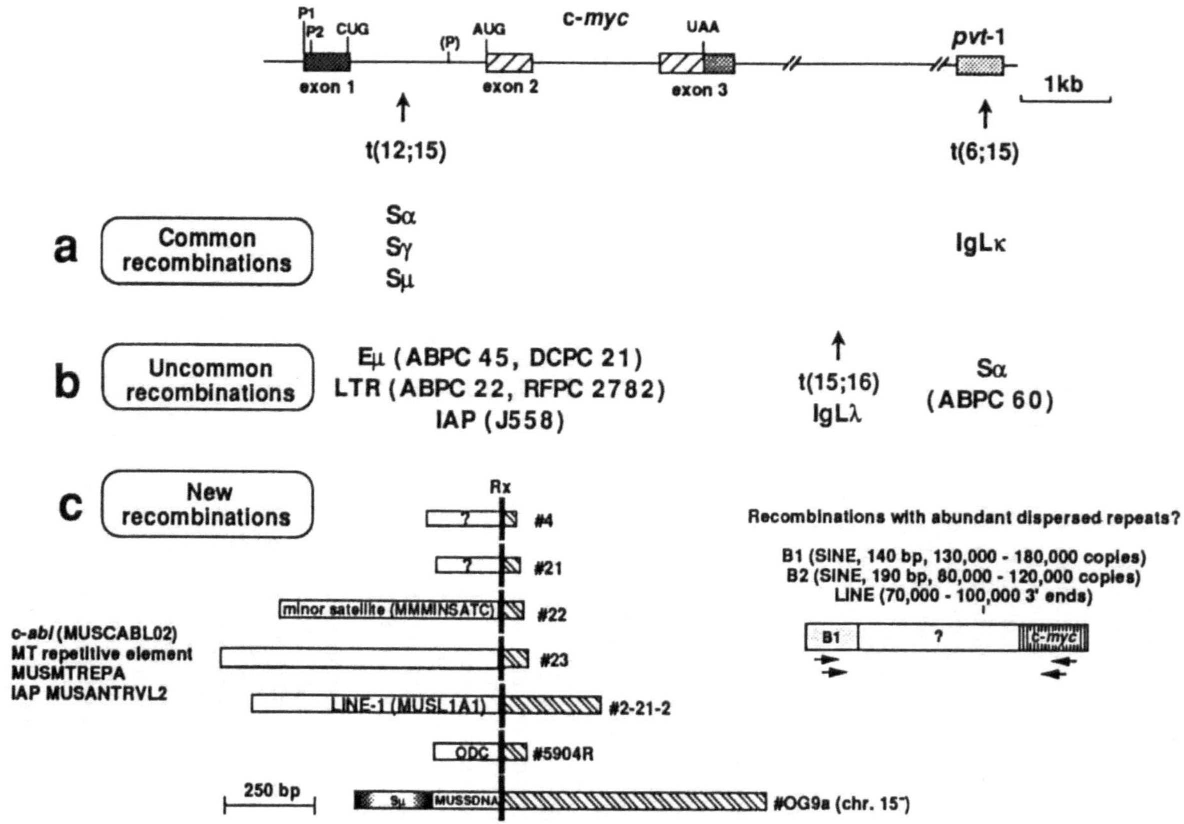

Fig. 1. Diversity of c-*myc* activating genetic recombinations

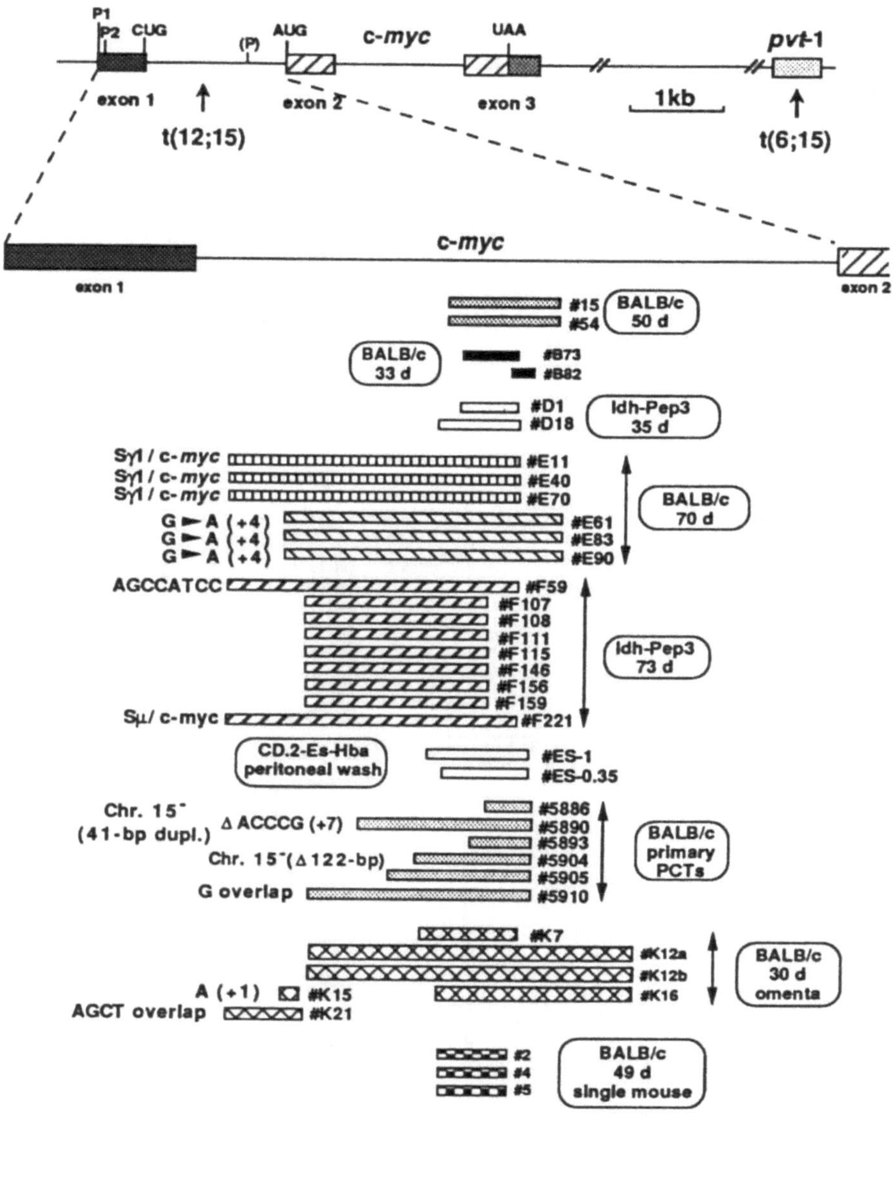

Fig. 2. Recombinations between c-*myc* and IgHα in plasmacytomas and preneoplastic B cells: c-*myc* portions

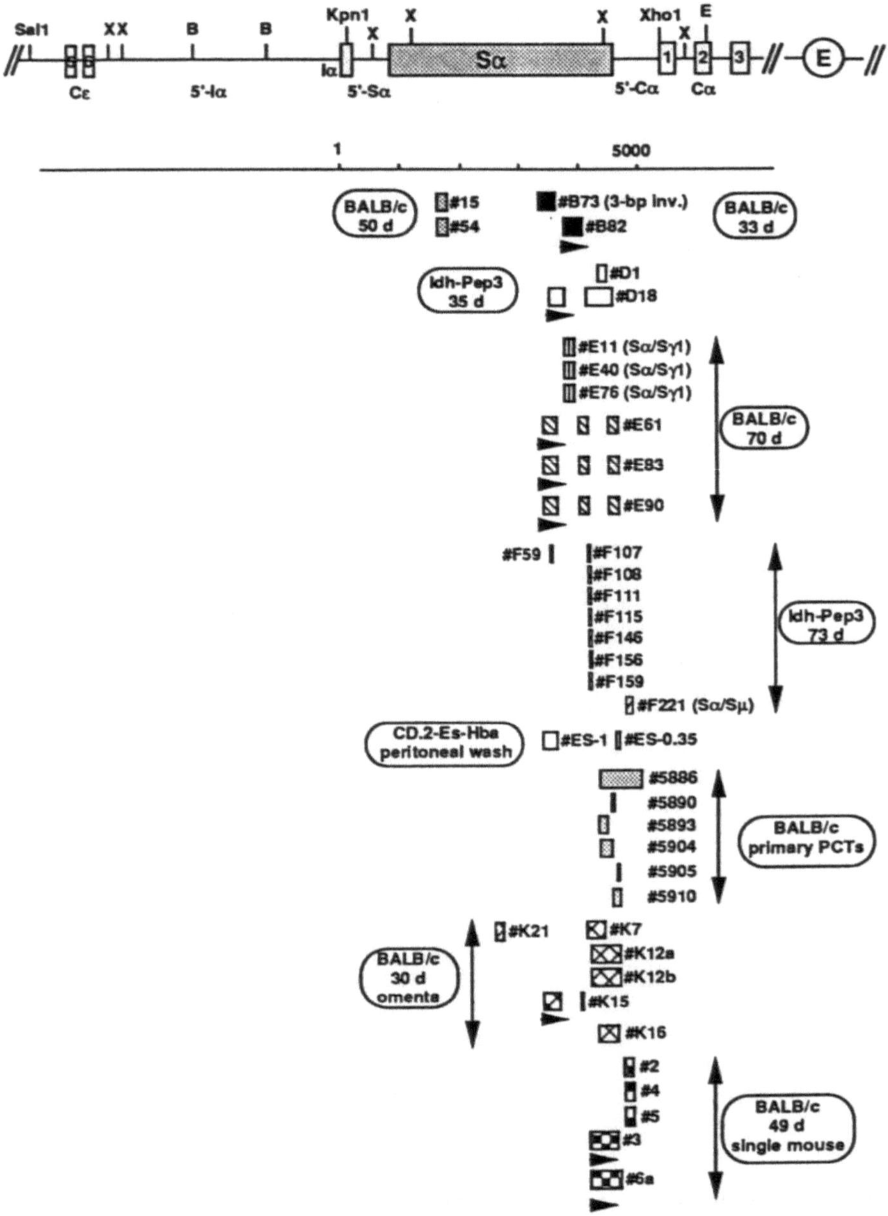

Fig. 3. Recombinations between IgHα and c-*myc* in plasmacytomas and preneoplastic B cells: IgHα portions

pointing to the right, respectively (#B82, D18, E61, 83, 90, K15, 3 and 6a). Since deletions and inversions were also found in intralocus IgHα recombination products (data not shown), it seems feasible that their frequent occurrence might reflect the inherent instability of the IgHα region in BALB/c mice. We also detected two examples of new PCR recombinational products between IgHα and c-*myc* that originated in the 5'-Sα region; i. e., upstream of the known translocation breakpoint region in IgHα (data not shown). We conclude that the recombination region between c-*myc* and IgHα appears to stretch further upstream than thought before. Since we ultimately wish to determine the 5' limit of recombinations in the IgHα locus, we sequenced the region between Iα and Cε (GenBank accession #U08933).

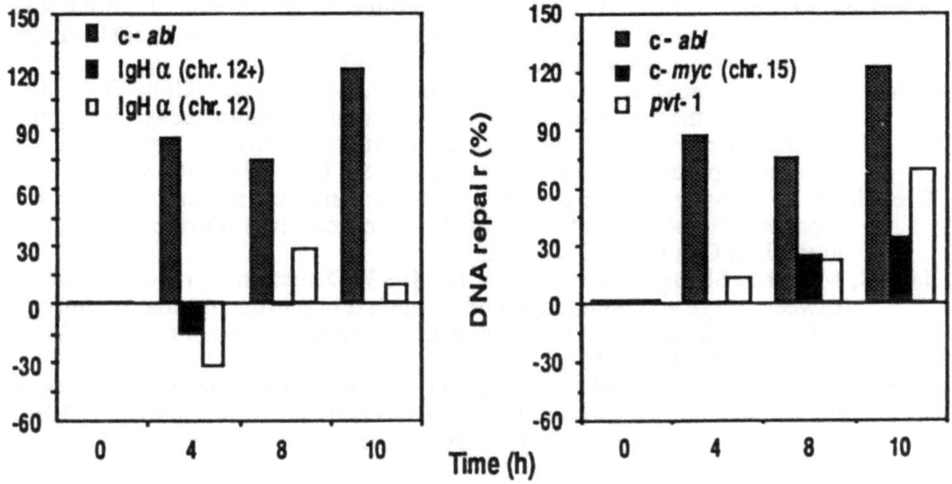

Fig. 4. Gene-specific repair of UV induced pyrimidine dimers in XRPC 24 plasmacytoma cells

Defective DNA repair provides one mechanistic explanation for genomic instability. Therefore we measured the efficiency of repair in IgHα and c-*myc* by employing an assay that quantitates the removal of UV-induced pyrimidine dimers within specific genomic regions. It has been shown that normal splenic B lymphoblasts from BALB/c mice repair the c-*myc*, *pvt*-1, IgHα and Igκ loci poorly (Beecham et al., 1991; 1994). We extended that work using the BALB/c plasmacytoma XRPC 24 as a model system and found that the IgHα locus tested on both the translocated chromosome 12+ and the normal chromosome 12 (left diagram in Fig. 4) and the c-*myc* and *pvt*-1 loci (right diagram in Fig. 4) were poorly repaired, whereas c-*abl*, an oncogene not related to conventional pristane-induced plasmacytomagenesis, was efficiently repaired. We conclude that deficient DNA repair in distinct loci, regional genomic instability, and illegitimate recombinations of c-*myc* might be causally linked in PCT susceptible BALB/c mice. Taken together, the results presented in this study are indicative of the existence of genomic instability in BALB/c mice but, at the same time, do not provide formal proof of it. New, quantitative assays for determining genomic instability need to be developed for future studies.

References

Axelson H, Panda CK, Silva S, Sugiyama H, Wiener F, Klein G, Sumegi J (1991) A new variant 15;16 translocation in mouse plasmacytoma leads to the juxtaposition of c-*myc* and immunoglobulin lambda. Oncogene 6:2263-2270

Beecham EJ, Mushinski JF, Shacter E, Potter M, Bohr VA (1991) DNA repair in the c-*myc* proto-oncogene locus: possible involvement in susceptibility or resistance to plasmacytoma induction in BALB/c mice. Mol Cell Biol 11:3095-3104

Beecham EJ, Jones GM, Link C, Huppi K, Potter M, Mushinski JF, Bohr VA (1994) DNA repair defects associated with chromosomal translocation breaksite regions. Mol Cell Biol 14:1204-1212

Fahrlander PD, Sumegi J, Yang JQ, Wiener F, Marcu KB, Klein G (1985) Activation of the c-*myc* oncogene by the immunoglobulin heavy-chain gene enhancer after multiple switch region-mediated chromosome rearrangements in a murine plasmacytoma. Proc Natl Acad Sci U S A 82:3746-3750

Greenberg R, Hawley R, Marcu KB (1985) Acquisition of an intracisternal A-particle element by a translocated c-*myc* gene in a murine plasma cell tumor. Mol Cell Biol 5:3625-3628

Janz S, Müller J, Shaughnessy J, Potter M (1993) Detection of recombinations between c-*myc* and immunoglobulin switch α in murine plasma cell tumors and preneoplastic lesions by polymerase chain reaction. Proc Natl Acad Sci U S A 90:7361-7365

Ohno S, Migita S, Murakami S (1991) c-*myc* gene in a murine plasmacytoma without visible chromosomal translocations moves to chromosome 12F1 with *Pvt*-1 and rearranges with IgH enhancer-Sμ sequences. Int J Cancer 49:102-108

Potter M, Wiener F (1992) Plasmacytomagenesis in mice: model of neoplastic development dependent upon chromosomal translocations. Carcinogenesis 13:1681-1697

Potter M, Mushinski EB, Wax JS, Hartley J, Mock BA (1994) Identification of two genes on chromosome 4 that determine resistance to plasmacytoma induction in mice. Cancer Res 54:969-975

Shaughnessy JD Jr, Owens JD Jr, Wiener F, Hilbert DM, Huppi K, Potter M, Mushinski JF (1993) Retroviral enhancer insertion 5' of c-*myc* in two translocation-negative mouse plasmacytomas upregulates c-*myc* expression to different extents. Oncogene 8:3111-3121

Shaughnessy J, Wiener F, Huppi K, Mushinski JF, Potter M (1994) A novel c-*myc*-activating reciprocal T(12;15) chromosomal translocation juxtaposes Sα to *Pvt*-1 in a mouse plasmacytoma. Oncogene 9:247-253

Broken-Ended DNA and V(D)J Recombination

M. Schlissel and T. Morrow
Departments of Medicine and Molecular Biology & Genetics
The Johns Hopkins University School of Medicine
Baltimore, Maryland 21205

Introduction

Antigen receptor genes are assembled during lymphocyte development by a highly regulated site-specific DNA recombination mechanism known as V(D)J recombination [1]. There are hundreds of variable (V_H), 12 diversity (D_H), and four joining (J_H) gene-segments in the mouse immunoglobulin (Ig) heavy-chain locus, for example. The recombinase juxtaposes V, D, and J gene segments to produce complete Ig or T cell receptor (TCR) genes. This novel mechanism allows for a tremendous diversity of potential antigen receptor structures with a minimal investment of genetic resources. The price of this elegant solution to the diversity problem, however, is the occasional aberrant recombination event resulting in chromosomal translocation and activation of a proto-oncogene [2].

Despite failure thus far to duplicate the V(D)J recombination reaction in a cell-free system, much has been learned about the reactants and products of this novel reaction (Fig. 1). All genes which rearrange in the immune system are flanked by highly conserved recombination signal sequences (RSSs) known as the heptamer and nonamer--7 and 9 nucleotide regions of conserved DNA separated by either 12 or 23 nucleotides of non-conserved "spacer" DNA. Gene-segments can only recombine with one another if their RSSs have dissimilar spacer lengths (the 12/23 rule) [1]. The RSSs have been shown to be the only DNA sequences necessary to target V(D)J recombination on a reporter construct transfected into a pre-B cell line. The essential nucleotides within these signals have been mapped [3].

V(D)J recombination produces two types of DNA joints, a coding joint and a signal joint (Fig. 1). Coding joints are characterized by imprecise gene-segment joining, with frequent loss or addition of nucleotides. In contrast, signal joints, the reciprocal product of V(D)J recombination, are almost invariably precise head-to-head fusions of the signal heptamers without loss or gain of nucleotides.

The most frequently proposed reaction mechanism for V(D)J recombination consists of RSS recognition, double-strand breakage, and ligation of the appropriate broken ends. Consistent with this proposition, Roth and Gellert first reported the existence of double-stranded DNA breaks in the vicinity of genes within the TCR δ locus in thymocytes undergoing Dδ-to-Jδ rearrangement [4, 5]. In order to study the mechanism and biological regulation of V(D)J recombination, we devised a sensitive PCR-based assay for broken-ended RSS DNA associated with any rearranging locus. We have used this assay to determine the DNA sequence of the broken-ended DNA, to determine the precise structure of the free ends, and to probe the regulation of V(D)J recombination during the cell cycle. In addition, we have examined the effect of perturbations to chromatin structure on the regulation of the recombinase.

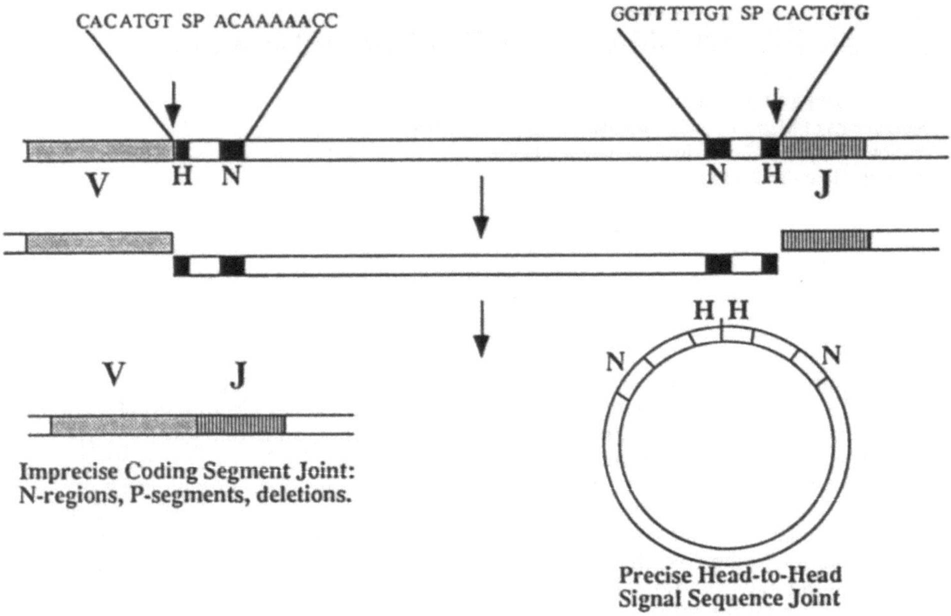

Fig. 1. A proposed reaction pathway for V(D)J recombination. Hypothetical V and J genes with their flanking RSSs are shown. The consensus heptamer and nonamer sequences are indicated above the top line with invariant nucleotides in boldface. SP indicates non-conserved "spacer" DNA. The DNA is cleaved at the heptamer-coding segment junctions yielding four broken DNA ends. The ends are then healed by precise fusion of the signal heptamers and imprecise joining of the coding ends.

Materials and Methods

Detection of double-strand breaks in genomic DNA.

We recently described an LMPCR assay for detecting, mapping, and sequencing double-stranded breaks in genomic DNA [6]. In brief, a synthetic DNA linker is ligated to purified high molecular DNA using T4 DNA ligase. Sites of linker ligation are then assessed relative to any genomic locus by using a PCR assay with a linker-primer and a locus-specific primer. Aliquots of a single linker-ligated genomic DNA sample can be used to assess breaks associated with multiple loci. In the experiment shown in Fig. 3, nuclei were purified by hypotonic lysis [7] and added to the ligation reaction in numbers corresponding to 3 μg total DNA. These reactions were subsequently processed as described [6].

RT-PCR assays.

RT-PCR assays for RAG-1 transcripts, unrearranged κ locus transcripts, and the control H2 transcript were performed exactly as described [8, 9]. RNA was purified by the guanidinium acid-phenol method [7].

Cells, cell culture, and flow cytometry.

The Abelson virus-transformed pre-B cell line 220-8 was grown in RPMI supplemented with 10% heat-inactivated fetal calf serum, 50μM β-mercaptoethanol and antibiotics. LPS (Difco) was added from a 10 mg/ml stock to a final concentration of 30 ug/ml. Sodium butyrate was used at a final concentration of 4mM. Spleen, thymus, bone marrow and fetal liver were obtained from Balb/c mice. Flow cytometry was performed on a FACScan (Becton-Dickinson) on cells stained with monoclonal anti-B220-biotin, avidin-PE, and monoclonal anti-IgM-FITC reagents as described previously [10]..

Results and Discussion

A linker-ligation PCR assay for broken-ended RSS DNA.

We modified the technique of ligation mediated PCR (LMPCR), first devised for genomic sequencing and in vivo footprinting [11], to map the sites of double-stranded breaks in genomic DNA from any characterized locus [6]. Purified total genomic DNA from a tissue undergoing V(D)J recombination is ligated to an asymmetric linker DNA fragment using T4 DNA ligase. The synthetic linker, blunt on one end and containing an 11nt 5' overhang on the other end, will ligate only to blunt-ended, 5' phosphorylated DNA (Fig. 2). Sites of linker ligation are then mapped using a nested PCR strategy. By altering the choice of primers, multiple genes can be studied from a single 1 to 3 ug ligated DNA sample.

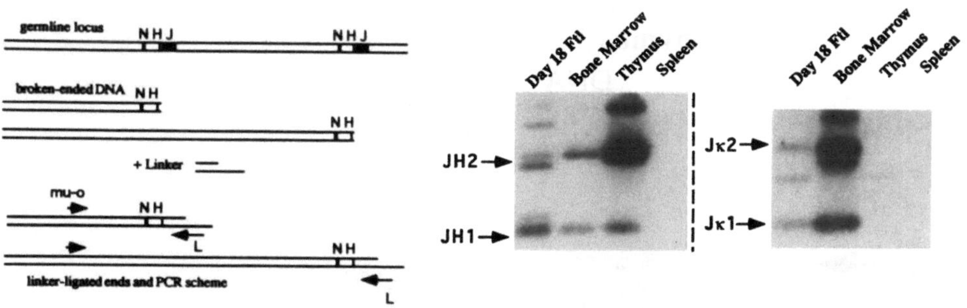

Fig. 2. A linker-ligation PCR assay for broken DNA associated with recombination signal sequences. The diagram outlines the logic of the LMPCR assay. An asymmetric synthetic linker with one blunt end is ligated to genomic DNA. Sites of ligation are mapped by PCR using a locus specific primer and a linker primer as described in the text. DNA purified from day 18 fetal liver, bone marrow, thymus and spleen were analyzed by LMPCR for breaks associated with J_H and Jκ loci. Autoradiograms of Southern blots are shown. Bands indicated by the arrows were shown by DNA sequence analysis to be fusions of the linker precisely to the end of the RSS heptamer [6].

Using this assay we looked for double-stranded DNA breaks in association with the Ig heavy-chain J_H gene-segments and light-chain Jκ gene-segments (Fig. 2). We found fragments corresponding in size to breaks at the signal sequence/coding sequence junction of the J_H genes in d18 fetal liver, adult bone marrow, and newborn thymus, tissues all known to undergo D-to-J_H rearrangement. We found Jκ-associated RSS breaks in d18 fetal liver and bone marrow but not in thymus or adult spleen, again matching the presumed pattern of recombination activity. In a separate study, we reported that sequence analysis of these broken-ended DNA molecules revealed them to be precisely at the heptamer-coding sequence junction [6].

Properties of broken-ended RSS DNA.

We used this LMPCR assay to probe the structure of broken RSS ends. The nature of the assay is such that it only detects blunt, 5' phosphorylated DNA breaks. It was possible that the ends we detected (Fig. 2) represent a fraction of the total broken ended RSS DNA in a population of cells with the remainder being 5' or 3' overhanging or blunt but not phosphorylated. To address this issue, we treated genomic DNA with DNA modifying enzymes designed to "polish" 5' overhanging (Klenow polymerase) or 3' overhanging (T4 DNA polymerase) ends or to phosphorylate free 5' OH groups (T4 polynucleotide kinase) prior to linker ligation. None of these treatments altered the intensity of the broken-end signal we observed [6]. We conclude from this that RSS breaks are blunt and 5' phosphorylated.

The fact that these broken RSS ends are subsequently repaired by precise fusion to one another leads us to propose that protein-protein interaction is responsible for aligning these ends prior to their ligation in vivo to form a signal joint. By analogy with DNA topoisomerases and prokaryotic recombinases, it is possible that these broken signal ends are intimately (? covalently) associated with the recombinase itself. To address this issue we performed the linker ligation assay on nuclei purified from murine thymus (Fig. 3). Detection of RSS breaks from a nuclear template was only moderately less efficient than from purified genomic DNA, and the linker ligation was still dependent on exogenous T4 DNA ligase. We conclude from this experiment that the ends of broken RSS DNA are not protected by protein and are not reactive for ligation in vitro in the absence of exogenous DNA ligase.

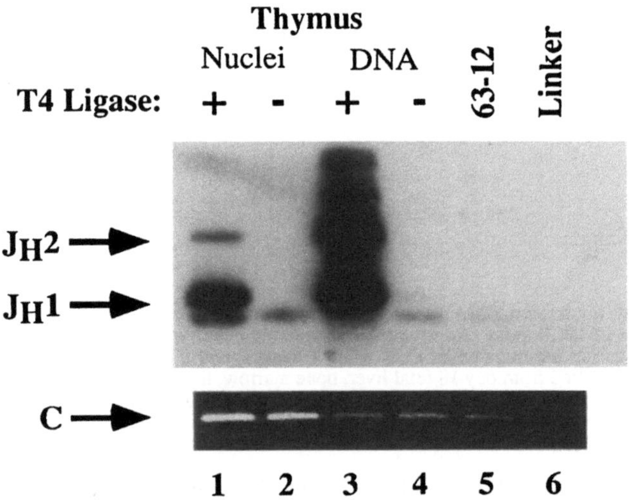

Fig. 3. Broken-ended RSS DNA is accessible to linker ligation in purified thymic nuclei. Similar amounts of DNA in the form of purified nuclei or deproteinized DNA were subjected to the LMPCR assay and analyzed for J_H-associated RSS breaks. T4 DNA ligase was either added (lanes 1, 3, 5, and 6) or omitted (lanes 2 and 4) from the linker ligation step of the assay. Lane 5 contains DNA from a RAG-2-deficient transformed cell line 63-12 [12]. PCR was then performed with the linker primer and J_H locus primers as described [6]. An autoradiogram of a Southern blot is shown with the position of specific J_H1 and J_H2 breaks indicated by arrows. The control signals (labeled C) are PCR assays of a non-rearranging genomic locus using the same DNA samples.

Broken-ended RSS DNA is found only during the G_0/G_1 phase of the cell cycle.

In a recent set of experiments characterizing the properties of broken-ended RSS DNA we reported that our ability to detect RSS breaks is limited to the G_0/G_1 phase of the cell cycle [6]. We used preparative flow cytometry to sort cells into various fractions dependent on their DNA content. We found that nearly all the broken-ended RSS DNA was contained in the fraction of cells with a 1C DNA content.

Earlier assays used to assess the regulation of the V(D)J recombinase relied on PCR-based quantitation of V(D)J recombinant molecules [9, 13, 14]. Since cells undergoing Ig gene rearrangement also undergo division and cell death at various rates, measuring the frequency of rearranged alleles only indirectly reflects the regulation of the recombinase. Since the LMPCR assay for broken-ended DNA detects a recombination reaction intermediate only found in G_0/G_1 cells, detection of broken-ended RSS DNA is a more direct measurement of the locus-specific activity of the recombinase at the time the cells were harvested. We have used this assay to study the developmental regulation of V(D)J recombinase activity.

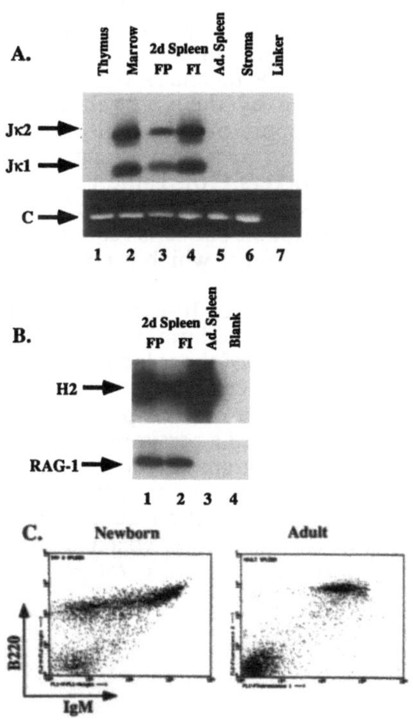

Fig. 4. The generation of Jκ-associated RSS breaks requires activity of the RAG genes and other lineage specific factors. A. Jκ-associated RSS breaks were assayed by LMPCR in DNA purified from various tissues. Samples labeled FP and FI were two different fractions of lymphocytes isolated from splenocytes from a 2 day old animal and thymus (lane 1) was from that same animal. Adult spleen (lane 5) was from a 9 month old animal. Stroma (lane 6) was from a bone marrow stromal cell line (S17) and linker (lane 7) was a reaction without added genomic DNA. C indicates a series of control PCR assays using the identical DNA samples to amplify a non-rearranging genetic locus. B. RT-PCR assays for the RAG-1 mRNA and a control transcript (H2) on RNA purified from some of the same cells as in A. C. Flow cytometric analysis of newborn and adult spleen for B220 and IgM.

The generation of broken-ended RSS DNA requires RAG-1 and RAG-2, reflects the state of lymphocyte development.

In order to explore the relationship between broken-ended RSS DNA and the regulation of V(D)J recombination, we tested DNA purified from various lymphoid tissues for broken-ends associated with the Jκ locus (Fig. 4A). We were able to detect Jκ associated broken-ended RSSs in newborn spleen and adult bone marrow. These molecules were absent, however, from adult spleen and newborn thymus. Newborn spleen contains a significant fraction of immature B cells which are $B220^{+}$ $sIgM^{-}$ (Fig. 4C). These cells express RAG-1 (Fig. 4B) and RAG-2 (not shown) mRNA whereas B cells from adult spleen do not. Newborn thymocytes express very high levels of RAG-1 and RAG-2 mRNA (not shown) but lack Jκ-associated broken-ended RSS DNA. Furthermore, we were unable to detect any broken-ended RSS DNA in lymphoid precursors from RAG-1-deficient mice [6]. We conclude from these studies that RAG gene expression is necessary but not sufficient for the generation of RSS breaks and that the pattern of these breaks reflects the developmental regulation of V(D)J recombinase activity.

Template accessibility and the regulation of V(D)J recombinase activity.

Despite the high degree of sequence conservation between the RSSs which flank all rearranging loci in both B and T cells, the activity of the recombinase is tightly regulated. Ig κ light chain genes, for example, only rearrange in the B cell lineage, most often subsequent to successful Ig heavy-chain gene rearrangement. Yancopoulos and Alt [15] originally proposed that V(D)J recombination is regulated by the accessibility of rearranging loci in chromatin to the recombinase. In support of this notion, they demonstrated that rearranging genes are often transcribed prior to their rearrangement. These initial observations have been extended by many workers [16]. Little has been learned, however, about the nature of the relationship between transcription and gene rearrangement. In particular, it is not known whether transcription per se is required for recombination or whether transcription reflects a chromatin structure which is competent for rearrangement.

In an attempt to address this issue, we studied the induction of Jκ-associated RSS breaks in a transformed pre-B cell line treated with either LPS or sodium butyrate. LPS was shown previously to activate transcription and rearrangement of the κ locus in pre-B cells [14]. Sodium butyrate, an inhibitor of histone deacetylase, alters chromatin structure by affecting the nucleosome core. Actively transcribed genes are often packaged in hyperacetylated nucleosome structures [17]. As expected, LPS caused a persistent elevation in the level of unrearranged κ gene transcripts (Fig. 5). We found a transient increase in the frequency of Jκ-associated RSS breaks in these LPS treated cells. In contrast, sodium butyrate treatment only modestly affected unrearranged κ gene transcription, but produced a persistent increase in Jκ-associated broken-ended RSS DNA (Fig. 5). We conclude from these preliminary observations that alterations in chromatin structure rather than transcription itself is required for the targeting of the V(D)J recombinase.

Conclusion

V(D)J recombination is the only known site-specific recombination system operative in vertebrates. Analogous systems exist in prokaryotes (flagellar antigen switching in salmonella) and lower eukaryotes (mating type switching in yeast). Elucidating the mechanism and regulation of V(D)J recombination is central to understanding normal lymphocyte development and aberrant development resulting in immunodeficiency or malignancy.

An ideal way to approach this problem would be to identify all the components required to carry out V(D)J recombination in a cell-free system. The role of each individual factor could then be studied. Detailed knowledge of the reaction pathway would then lead to hypotheses regarding the regulation of this reaction. Unfortunately, cell-free V(D)J recombination remains elusive. Significant progress, however, has come from analyzing reactants, products and most recently a presumed V(D)J recombination reaction intermediate.

In this and a related paper [6] we have reported the properties of broken-ended RSS DNA. We have demonstrated that it is blunt, 5' phosphorylated, dependent on RAG-1 and RAG-2, and cell-cycle regulated. In addition, we have used the LMPCR assay described above to assess the targeting of the recombinase during lymphoid development and to study the role of chromatin accessibility and unrearranged Ig gene transcription in the regulation of V(D)J recombination. Current efforts are aimed at devising a cell-free system capable of generating broken-ended RSS DNA, signal joints or V(D)J recombinants.

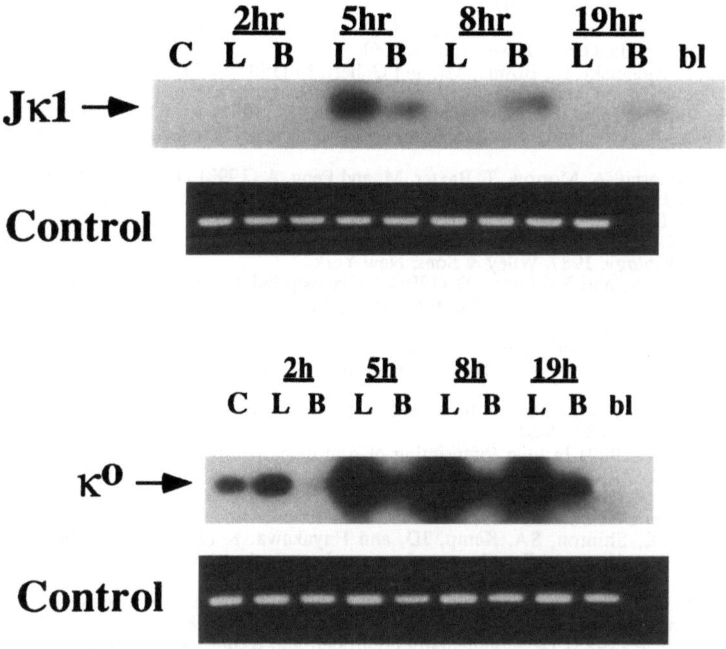

Fig. 5. The effects of LPS and sodium butyrate on unrearranged κ gene transcription and Jκ-associated broken ended RSS DNA in a pre-B cell line. 220-8 cells were cultured in the presence of LPS or sodium butyrate. DNA and RNA were harvested from cells at the indicated times and subjected to LMPCR analysis for broken ends (labeled Jκ1) and RT-PCR assay of unrearranged κ gene transcription (labeled κ^o). The control reactions were PCR assays of an non-rearranging genetic locus (top) or an invariant transcript (bottom). C indicates cells prior to drug treatment, L indicates LPS treated cultures, and B indicates sodium butyrate treated cultures.

Acknowledgements

This work was supported by awards from the Cancer Research Institute (M.S.), and The Culpeper Foundation (M.S.), and an Immunology Training Grant from NIAID (T.M.).

References

1. Tonegawa, S (1983) Somatic generation of antibody diversity. Nature 302:575-581.
2. Korsmeyer, SJ (1992) Chromosomal translocations in lymphoid malignancies reveal novel proto-oncogenes. Ann. Rev. Immunol. 10:785-807.
3. Hesse, JE, Lieber, M, Mizuuchi, K, and Gellert, M (1989) V(D)J recombination: a functional definition of the joining signals. Genes. Devel. 3:1053-1061.
4. Roth, DB, Nakajima, P, Menetski, JP, Bosma, MJ, and Gellert, M (1992) V(D)J recombination in mouse thymocytes: double-strand breaks near T cell receptor δ rearrangement signals. Cell 69:41-53.
5. Roth, DB, Menetski, JP, Nakajima, PB, Bosma, MJ, and Gellert, M (1992) V(D)J recombination: Broken DNA molecules with covalently sealed (hairpin) coding ends in scid mouse thymocytes. Cell 70:1-9.
6. Schlissel, MS, Constantinescu, A, Morrow, T, Baxter, M, and Peng, A (1993) Double-strand signal sequence breaks in V(D)J recombination are blunt, 5' phosphorylated, RAG-dependent and cell cycle regulated. Genes & Devel. 7:2520-2532.
7. Ausubel, F, Brent, R, Kingston, R, Moore, D, Seidman, J, Smith, J, and Struhl, K, ed. *Current Protocols in Molecular Biology*. 1987, Wiley & Sons: New York.
8. Schlissel, MS, Voronova, A, and Baltimore, D (1991) Helix-loop-helix transcription factor E47 activates germ-line immunoglobulin heavy-chain gene transcription and rearrangement in a pre-T cell line. Genes & Devel. 5:1367-1376.
9. Schlissel, MS, Corcoran, LM, and Baltimore, D (1991) Virally-transformed pre-B cells show ordered activation but not inactivation of immunoglobin gene rearrangement and transcription. J. Exp. Med. 173:711-720.
10. Morrow, T and Schlissel, M (1992) The purification of B cell precursors from mouse fetal liver. Curr. Topics Micro. Immunol. 182:55-64.
11. Mueller, PR and Wold, B (1989) In vivo footprinting of a muscle specific enhancer by ligation mediated PCR. Science 246:780-786.
12. Shinkai, Y, *et al.* (1992) RAG-2 deficient mice lack mature lymphocytes owing to inability to initiate V(D)J rearrangement. Cell 68:855-867.
13. Hardy, RR, Carmack, CE, Shinton, SA, Kemp, JD, and Hayakawa, K (1991) Resolution and characterization of pro-B and pre-pro-B cell stages in normal mouse bone marrow. J. Exp. Med. 173:1213-1225.
14. Schlissel, M and Baltimore, D (1989) Activation of immunoglobulin kappa gene rearrangement correlates with induction of germline kappa gene transcription. Cell 58:1001-1007.
15. Yancopoulos, G and Alt, F (1985) Developmentally controlled and tissue-specific expression of unrearranged V_H gene segments. Cell 40:271-281.
16. Schatz, DG, Oettinger, MA, and Schlissel, MS (1992) V(D)J Recombination: Molecular Biology and Regulation. Ann. Rev. Immunology 10:359-383.
17. Lee, DY, Hayes, JJ, Pruss, D, and Wolffe, AP (1993) A positive role for histone acetylation in transcription factor access to nucleosomal DNA. Cell 72:73-84.

Mutations in the Coding Region of c-myc Occur Independently of Mutations in the Regulatory Regions and are Predominantly Associated with myc/Ig Translocation

Bhatia. K.[1], Spangler. G.[1], Hamdy.N.[1], Neri. A.[2], Brubaker. G[3].Levin. A.[4], and Magrath.I[1]

[1]Lymphoma Biology Section, NCI, NIH, Bethesda, MD 20892. [2]Servizio di Ematologia, Istituto di Scienze Mediche, Unversidad di Milano, Ospedale Maggiore, I.R.C.C.S., Milan Italy. [3] Shirati Hospital Tanzania [4]Research Triangle Institute Unit, Department of Virology, The Medical College of St Bartholomew's Hospital London U.K.

Abstract

Constitutive expression of c-myc resulting from a chromosomal translocation, which juxtaposes c-myc to an immunoglobulin gene, is a pivotal lesion in Burkitt's lymphomas. This deregulated expression of c-myc is associated with mutations in the regulatory regions, i.e. the first exon and the first intron of c-myc in tumors where the chromosomal breakpoint is not itself within the regulatory region. Until recently it was widely believed that the c-myc protein in these tumors is wild type. We have demonstrated that in a fraction of Burkitt's lymphomas from Africa and from the continental USA, and in mouse plasmacytomas, the c-myc gene carries mutations in the coding region. We now show that, occasionally, such mutations are also present in multiple myelomas - tumors which do not carry translocations or amplifications of c-myc. We also show that the frequency of the c-myc coding region mutations in BL is independent of the frequency of mutations in the regulatory region. These results suggest that the mechanisms that induce missense mutations involving the coding region of c-myc may be different from those that lead to mutations in the regulatory regions.

Introduction:

Proliferation in eukaryotic cells is generally dependent upon the expression of the c-myc protooncogene. Deregulation of the c-myc gene in lymphoid cells is associated with lymphomagenesis (Adams et al 1985). In Burkitt's lymphomas, c-myc deregulation results from a translocation that brings the c-myc gene under the transcriptional influence of immunoglobulin sequences (Magrath. I. 1990). It has been proposed that point mutations within the regulatory regions of c-myc, which occur when the chromosomal breakpoint is further upstream, may also contribute to the deregulation of c-myc (Ceaserman et al 1987, Zajac-Kaye et al 1988, Show and Croce, 1987., Spencer and Groudine, 1990 Yu et al 1993, Pellici et al 1986). Inspite of some reports of sporadic mutations in the protein coding region of the translocated myc allele, the general consensus, has been that Burkitt's lymphoma results from the deregulation of a normal c-myc protein. (Rabbits et al 1983 and 1984). We have recently reported that the integrity of the c-myc protein in BL is frequently compromised by mutations that cluster within regions of c-myc that mediate it's transactivating function and that contain phosphorylation sites. Some of these mutations do not appear to be heterozygous (Bhatia et al 1993), implying that either, the mutational events precede the translocation or there is an imbalance in the gene dosage for the translocated allele. In contrast to the mutations that affect the coding sequences, mutations that affect regulatory regions are invariably heterozygous. The opposing functions of c-myc i.e. proliferation and apoptosis (Evans et al 1992 and Shi et al 1992), and the frequent loss of heterozygosity of wild type c-myc - both functional (since only the translocated c-myc is usually expressed) and physical - in BL make it tempting to hypothesize that the mutations serve to abrogate a function of wild type c-myc. Gu et al (1994) have recently demonstrated that one of the effects of coding region mutations is to inhibit the p107 mediated supression of the c-myc transcriptional activity. .

In this present study, we wished to address the question whether the coding region mutations occur concomitantly with mutations in the regulatory region. We have also tried to determine whether the coding region mutations are specific for BL and thus are "translocation dependent", or whether they also occur in other B-cell neoplasms, and can sometimes be "translocation independent". Since we have previously described coding region mutations in murine plasmacytomas (Bhatia et al 1993), we extended our study to multiple myelomas (MM). None of the MMs included in this study possess a rearranged c-myc gene (Neri et al 1989). We also expanded the information on BL by including in this study seven SNCLs obtained from Tanzania.

Materials and Methods:

The strategy for the SSCP analysis of the coding regions within the c-myc gene has been described before (Bhatia et al 1993 and 1994). MM from nineteen patients were analyzed, Bone marrow aspirates from the patients were collected for routine diagnostic procedures. Five of these tumors were clinically indolent, another five patients were classified as having chronic disease and the remaining 9 were classified as having acute disease or plasma cell leukemia. SNCL biopsies were obtained from Tanzania and Southern blot analysis, using the c-myc probes spanning the first and third exons, was performed as previously described (Guttierez et al 1992).

DNA was obtained from the tumor samples using established protocols. For the PCR generated SSCP, the initial PCR was performed using 250ng of DNA and 10pmoles of each flanking primer. The conditions of the PCR reaction and the SSCP analysis and the sequences of the primers used for the SSCP have been described previously (Bhatia et al 1993). BL DNA samples that were found to contain mutations in exons 2 and 3 from a previous analysis (Bhatia et al 1993) were used as positive controls in the SSCP analysis. Additionally a negative control and a blank PCR (reaction without template) was always included in the analysis as a test for the absence of exogenous DNA contamination. PCR products with aberrant migration and a randomly chosen PCR product with normal migration were sequenced. Direct sequencing of the PCR amplified product was carried out using the sequenase kit (USB).

Results and Discussion:

The results of SSCP analysis for coding region mutations are shown in Fig 1. All but one MM demonstrated wild type c-myc. PCR generated SSCP for regions 2.2 and 2.4. showed abnormal migrating bands for the DNA from Tumor 402. A mutation detected by SSCP analysis was found in 1 (in the 2.1 region) of the 7 SNCL samples obtained from Tanzania (Fig 1).

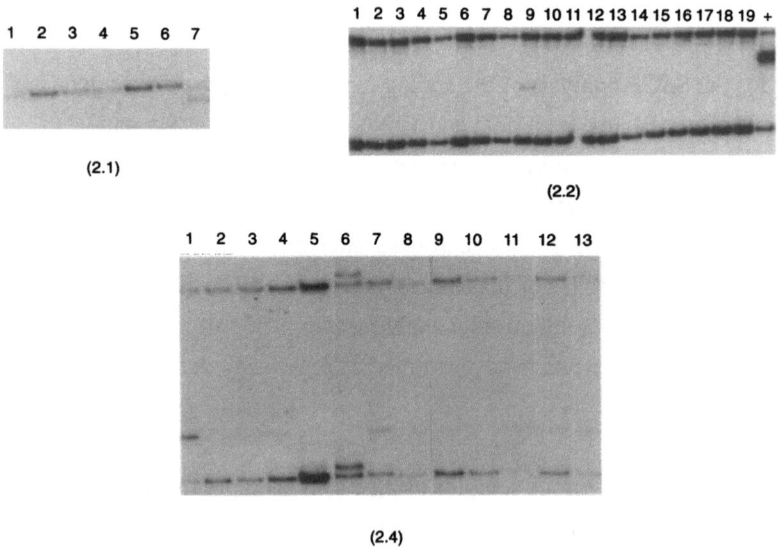

Fig 1: SSCP analysis of the coding region of the c-myc gene. Migration patterns for regions designated 2.1, 2.2 and 2.4 are shown for a representative set of DNA samples. Abnormal migration of PCR product from DNA of a BL (T79) is evident in lane 7. Abnormal migration is also seen for the SSCP analysis of the 2.2 and the 2.4 regions for the DNA from MM 402 lane 9

and lane 6 respectively. Positive control is depicted in the rightmost lane in the panel for the 2.2 region.

Mutations implied by the abnormal migration of coding region PCR fragments 2.2 and 2.4 from one MM were confirmed by sequencing (Fig 2). In this tumor a misense mutation (ATC-GTC) in the 2.2 region in MM 402, which changes isoleucine to valine was present at amino acid 130. A mutation at amino acid 231, was confirmed by sequencing the 2.4 region. The latter mutation changed the third base of the codon CCG to CCA, retaining its coding potential for proline. We have previously described mutations in the 2.2 region in BL which occurred predominantly between amino acids 86-92. We have also observed, however, misense mutations at amino acids 115 and 129.

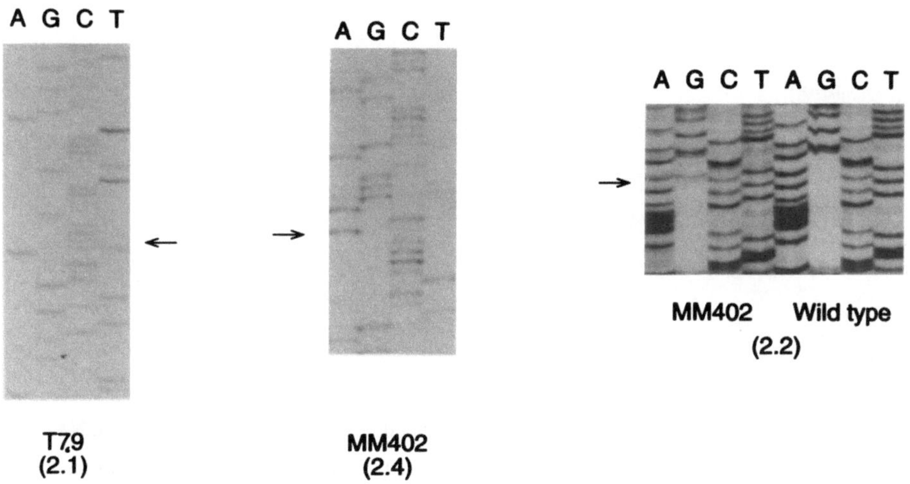

Fig 2: Direct DNA sequencing of PCR amplified products. Sequences for samples with variant SSCP patterns are shown for the 2.1 region for T79 and 2.2 region for MM402. Sequence comparision of a wild type 2.4 product and MM402 is also shown.

Sequencing of the 2.1 variant PCR fragment from the Tanzanian SNCL (T79) revealed the presence of a misense mutation in codon 58 (Fig 2). The mutation , a transition, resulted in threonine being replaced by isoleucine (ACC-ATC). Although the location of the mutation is consistent with our earlier studies, the frequency of the coding region mutations appears to be lower in Tanzanian tumors than in samples from Ghana or from North and South America. Whether the differences in frequency signify epidemiologic differences is difficult to assess at the present time.

Southern blot analysis of the c-myc gene in the Tanzanian SNCL revealed no rearrangements of c-myc, consistent with our data from other equatorial African tumors. EBV analysis of these tumors demonstrated the presence of monoclonal EBV in each tumor (Fig 3).

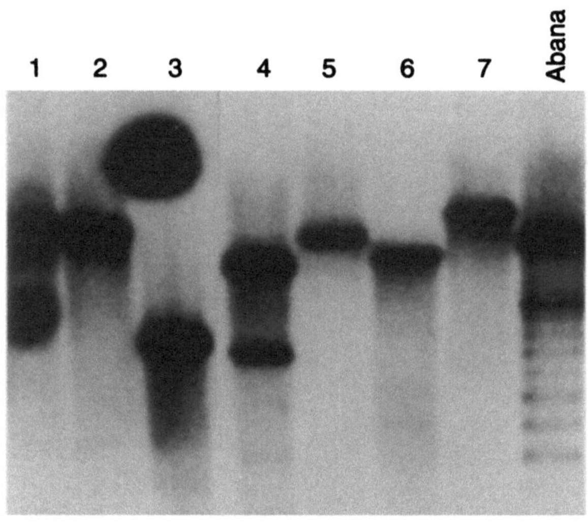

Fig 3: EBV analysis of the BL from Tanzania. The blot depicts monoclonality of EBV. DNA from the tumors was digested with BamHI and electrophoresed in a 1% agarose gel. Following Southern transfer to a nylon membrane, the blot was hybridised with a p32 labelled EBV termini probe.

Tumors from Africa, generally have translocations that break chromosome 8 far upstream 5' of the c-myc gene (Pellici et al 1986. Magrath 1990). The absence of detectable c-myc gene rearrangements in Southern blot analysis is consistent with far 5' breakpoints or variant translocations where the breakpoints are associated far 3'of the c-myc gene. Sixty percent of tumors with far 5' breakpoints (when analyzed by restricting the DNA with PvuII and subsequent hybridization with a first exon c-myc probe) demonstrate a rearranged PvuII band on Southern blotting (Guttierez et al 1992) resulting from the loss of a PvuII site as a consequence of mutations in the 3' end of the first exon. We therefore examined the Tanzanian tumors for the presence of mutations in the regulatory region of the intron both by PvuII restriction analysis and by SSCP analysis. A regulatory region in intron one, mif, was chosen for SSCP analysis because it has been shown to be frequently mutated in BL. Primers were designed that flanked this region and SSCP analysis performed. Fig4, shows the results of SSCP analysis of the SNCL from Tanzania. Six of the seven tumor samples demonstrated a mutation in the mif region, PvuII restriction digests showed the presence of first exon mutations in four of the seven tumors , including the one that did not show a mutation in the mif region . None of the myelomas carried a mutation in the regulatory region (Fig 4).

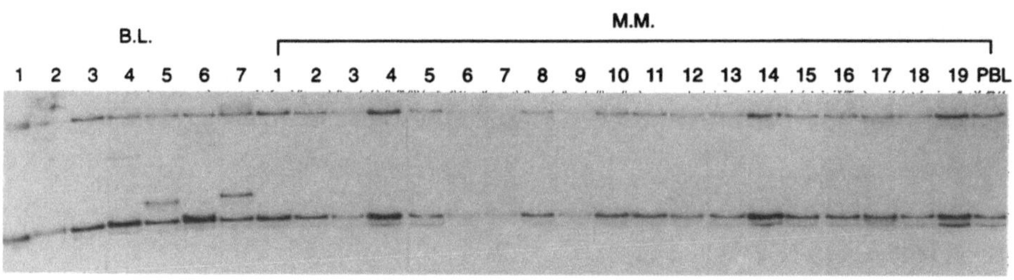

Fig 4: SSCP analysis of DNA from tumor biopsies of BL and MM. SSCP analysis was targeted

to the intron region of the c-myc gene that encompasses the mif 1 through mif3 binding sites. Abnormal migration of the bands is evident in BL samples 1 through 7. All MM samples including MM 402 (lane 12) show a wild type migration pattern similar to that obtained from normal peripheral blood cells (lane 20, PBL).

It is apparent from this data, that although all of the SNCL(7/7) carry mutations in regulatory regions, such mutations are not necessarily coding region mutations. The only tumor with mutations in both the regulatory region and the coding region was T79. Conversely, as seen from the data from MM, coding region mutations can be present in the absence of c-myc rearrangements or regulatory mutations albeit rarely. MM402 which showed two mutations in the coding regions did not have any mutations in the regulatory region of c-myc.

It would thus appear that the hypermutational mechanisms (Klobeck et al 1987) that contribute to the mutations in the first exon and intron do not necessarily encompass the coding exons. However, the low frequency of both types of mutations in myelomas could be a result of the absence of c-myc translocations, suggesting that the mutations are generally translocation dependent or make a contribution to the functional consequences arising from the translocation. The frequent presence of the coding region mutations in HIV associated lymphomas that also carry a rearranged c-myc gene, but only infrequently in lymphomas without rearranged c-myc, further supports this notion (Bhatia et al 1994). The presence of coding region mutations albeit at a low frequency in MM indicates that the mutations in the coding region can occasionally occur in the absence of a translocation.

Acknowledgement: Part of this work was supported by a contract from the NCI to Dr A.Levin

References:

Adams JM, Harris AW, Pinkert CA, Corcoran LM, Alexander WS, Cory S, Palmiter RD, Brinster RL. (1985). The c-myc oncogene driven by immunoglobulin enhancers induces lymphoid malignancy in transgenic mice.

Bhatia K, Spangler G, Gaidano G, Hamdy N, Dalla-Favera R and Magrath I. (1994). Mutations in the coding region of c-myc occur frequently in AIDS associatedlymphomas. Blood in press.

Bhatia K., Huppi, K., Spangler, G., Siwarski, D., Iyer, R and Magrath, I. (1993) Point mutations in the c-myc transactivation domain are common in Burkitt's lymphoma and mouse plasmacytomas.Nature Genetics, 5, 56-61

Ceaserman, E., Dalla-Favera, R., Bentley, D and Groudine, M. (1987), Mutations in the first exon are associated with altered transcription of c-myc in Burkitt lymphoma Science.238, 1272-1275.

Evan, G.I., Wyllie AH, Gilbert CS, Littlewood TD, Land H, Brooks M, Watters CM, Penn LZ, Hancock DC. (1992). Induction of apoptosis in fibroblasts by c-myc protein. Cell 69, 119-128.

Gu W, Bhatia K, Magrath I, Dang CV and Dalla-Favera R. (1994) Binding and supression of the myc transcriptional domain by p107. Science 264, 251-254.

Gutierrez, M.I., Bhatia, K., Barriga, F., Diez, B., Sackmann-Muriel, F., de Andreas, M.L., Eppelman, S., Risueno, C. and Magrath, I.T. (1992) Molecular epidemiology of Burkitt lymphoma from South America: differences in breakpoint location and Epstein-Barr virus association from tumors in other world regions. Blood 79, 3261-3266.

Klobeck, H., Combriato, G., and Zachau, H. (1987) N segment insertion and region directed somatic hypermutation in a kappa gene of a t(2:8) chromosomal translocation. Nucl Acid Res. 15, 4877-88

Magrath, I.(1990).The pathogenesis of Burkitt's lymphoma.Adv.Cancer Res.,55,133-270.

Neri, A., Murphy, J.P., Cro, L., Ferrero, D., Tarella, C,. Baldini, L. and Dalla-Favera, R. (1989) Ras Oncogene Mutation in MultipleMyeloma.J.Exp.Med 170, 1715-1725.

Pellici, P. G., Knowles, D.M., Magrath, I. and Dalla-Favera, R. (1986) Chromosomal breakpoints and structural alterations of the c-myc loci differ in endemic and sporadic forms of Burkitt lymphoma. Proc. Natl. Acad. Sci. U.S.A. 83, 2984-2988

Rabbits, T.H., Forster, A., Hamlyn, P. and Baer, R. (1984). Effect of somatic mutations within translocated c-myc genes in Burkitt lymphoma.Nature 309. 592-597.

Rabbits, T.H., Hamlyn, R.H. and Baer, R. (1983).Altered nucleotide sequence of a translocated c-myc gene in Burkitt lymphoma. Nature, 306, 760-765.

Shi Y, Glynn JM, Guilbert LJ, Cotter TG, Bissonnette RP, Green DR.(1992). Role of c-myc in activation induced apoptic death in T cell hybridomas. Science 257, 212-214.

Showe, L.C. and Croce, C.M. (1987). The role of chromosomal translocations in B and T cell neoplasia. Ann. Rev. Immunol., 5, 253-277.

Spencer, C.A. and Groudine. M. (1990). Control of c-myc regulation in normal and neoplastic cells. Adv Cancer Res.56, 1-48.

Yu, B.W., Ichinose, I., Bonham, M.A., and Zajac-Kaye, M. (1993) Somatic mutations in the c-myc intron 1 cluster in discrete domains that define protein binding sequences. J Biol Chem. 268, 19586-19592.

Zajac-Kaye, M., Gelmann, E.P., and Levens, D., (1988), A point mutation in the c-myc locus of a Burkitt lymphoma abolishes the binding of a nuclear protein. Science. 240, 1776-1780.

The Generation of Pvt-1/Ck Chimeric Transcripts as an Assay for Chromosomal Translocations in Mouse Plasmacytomas

K. Huppi
Mol. Genetics Section, Lab. of Genetics, NCI/NIH, Bethesda, MD 20892

Introduction

A region of mouse chromosome 15 termed the c-Myc/Pvt-1 mega-gene locus is frequently associated with chromosomal breakpoints in a number of B-cell tumors including the mouse plasmacytoma and Burkitt's lymphoma (for review see Potter 1990). In mouse plasmacytomas, one of two distinct types of chromosomal translocations can be found in nearly 100% of all tumors: 1). a t(12;15) translocation involving the Ig-H locus and c-Myc or 2). a t(6;15) translocation involving the Ig-k locus and Pvt-1. While the focus of such translocation events appears to be c-Myc, it is not yet clear whether Pvt-1 plays an active role in this process.
A major step in our understanding of Pvt-1 was achieved when cDNAs corresponding to transcripts of Pvt-1 were found in both humans (Shtivelman et al. 1990) and mouse (Huppi et al. 1990a). Over the past several years, we have succeeded in resolving much of the exon structure of the mouse Pvt-1 gene as determined from extensive cDNA cloning and sequencing studies (Huppi et al. 1992). Interestingly, the four established exons of Pvt-1 encompass precisely the same region wherein a majority of chromosomal breakpoints seem to cluster. Since most of the breakpoints lie within the transcript associated with Pvt-1, abundant and truncated transcripts of Pvt-1 are predictably found in such tumors. As part of our interest in determining what role Pvt-1 plays in tumorigenesis, I have designed and implemented an assay based on the observation that the presence of truncated transcripts of Pvt-1 could be a useful diagnostic indicator of Pvt-1 associated translocations.

Results and Discussion

A compilation of chromosome 6 and chromosome 15 breakpoints from mouse plasmacytomas reveals some important guidelines in the generation of a useful assay for

Pvt-1 translocations (Fig. 1). On mouse chromosome 6, most of the breakpoints fall within a 3-4-kb segment that spans the Jk or IVS regions of Ig-k. Similarly, the Pvt-1 breakpoints cluster to a 4-kb region that encompasses four of the exons determined for Pvt-1. Two translocations have also been shown to reside 14-18-kb upstream of the Pvt-1 region on mouse chromosome 15 (Shapiro et al. 1987; Shaughnessy et al. 1994). However, in both of these cases, a succession of chromosomal breaks have occurred creating a complex translocation. Thus, nearly all reciprocal t(6;15) translocations identified to date have a chromosomal breakpoint that resides within the Ig-k and/or Pvt-1 transcriptional domains. Like the situation with Ig-H/c-Myc translocations in t(12;15) tumors, the precise location of the breakpoint may occur over a region of several kilobases, making the rapid identification of such translocations tenuous at the Southern hybridization level. Similarly, a PCR-based assay using a series of oligonucleotide primers from the breakpoint regions of Ig-k and Pvt-1 appears to be inefficient in that extraordinarily large PCR products must be generated.

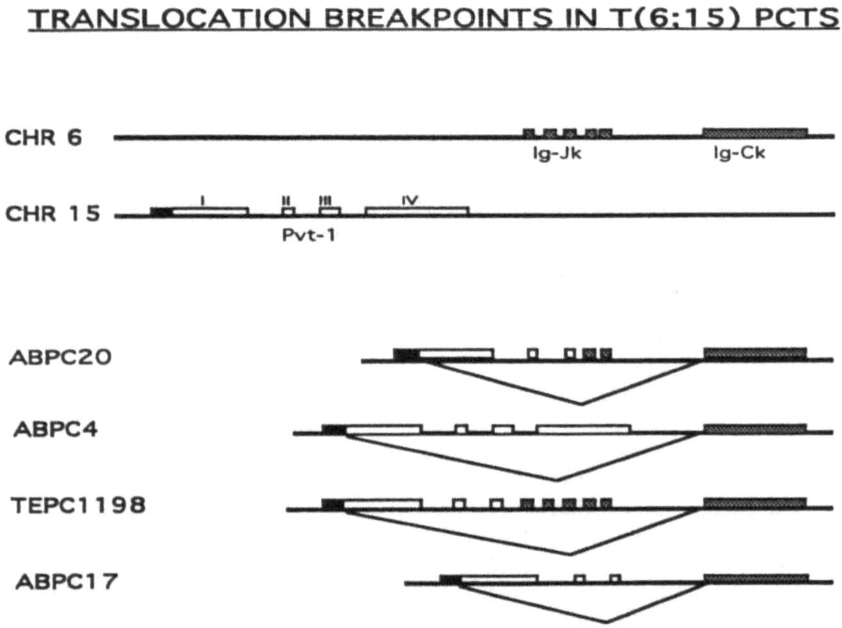

Fig. 1. Organization of chromosome 6 and chromosome 15 associated breakpoints in mouse plasmacytomas.
(Above) A schematic organization of exons for Ig-k (chr. 6) and Pvt-1 (chr.15).
(Below) The molecular organization of chromsomal translocations found in the mouse plasmacytomas ABPC20, ABPC4, TEPC1198 and ABPC17 (Webb et al. 1984, Cory et al. 1985). The Pvt-1 exons have recently been determined (Huppi et al. 1992) and have been placed accordingly. The filled-in box corresponds to the Pvt-1a segment in exon 1. RNA splicing leads to a Pvt-1a/Ig-Ck product in each of the tumors as depicted.

It is observed that both the Pvt-1 and Ig-k breakpoint sites are located within transcriptionally active regions such that an alteration in the transcriptional pattern may accompany t(6;15) translocations. Such a change in transcription, may for example, be found as a loss of expression and/or modified RNA splicing. Normally, a mouse Pvt-1 cDNA probe hybridizes to a large 14-kb transcript that is readily detectable in mRNAs from most mouse tissues (Huppi et al. 1990a). However, more abundant, yet shorter transcripts have consistently been observed in t(6;15) plasmacytomas. Therefore, it seems probable that some form of alternative RNA splicing may be a reliable indicator of chromosomal breakage within Pvt-1. Through extensive cloning and sequencing efforts, we have determined the structure of these transcripts in t(6;15) plasmacytomas (Huppi et al manuscript in prep.). The most common sequence obtained from these tumors is composed of a short 57-bp segment of Pvt-1 juxtaposed to Ig-Ck (Fig. 1). This structure is particularly intriguing in that there is a continuous open reading frame (ORF) from the

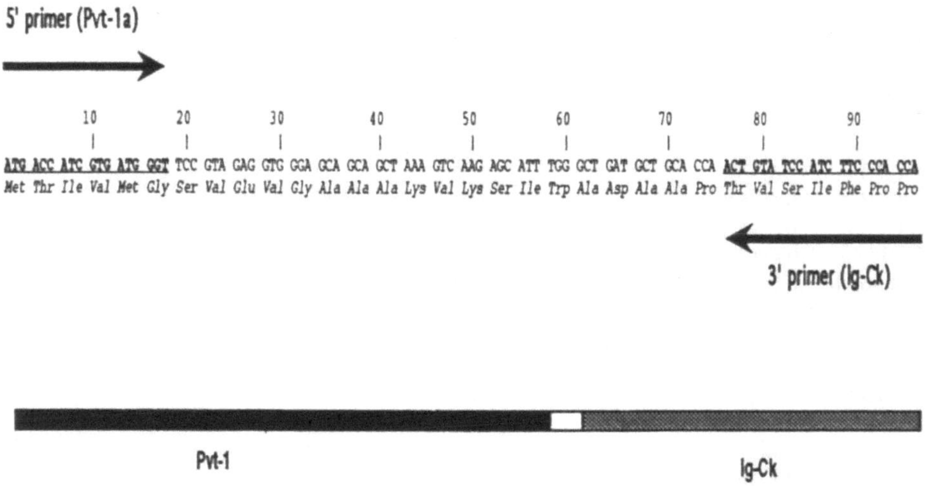

Fig. 2 Sequence of Pvt-1a/Ig-Ck PCR products.
The primers used to generate the 97-bp RT-PCR product from t(6;15) plasmacytomas are highlighted in bold with directional arrows. The deduced amino acid sequence for Pvt-1a and Ck is shown below the DNA sequence. The bar represents regions of Pvt-1a (filled-in), the new trytophan residue generated by splicing (open) and Ig-Ck (partially filled-in). The Pvt-1 RT-PCR assay consists of the generation of a cDNA strand from poly (A)+ mRNA of tumors with the 3' Ig-Ck primer. Following synthesis of the cDNA, 30 cycles of PCR amplification are performed in the presence of the 3' Ig-Ck and 5' Pvt-1a primers. PCR conditions are as follows: 94°- denaturation, 55° reannealing (30 sec) and 72° extension (60 sec).

beginning of the 57-bp segment (referred to as Pvt-1a) to the end of the Ig-Ck segment, suggesting that the Pvt-1a/Ck chimera could actually be a functional transcript. RNA splicing has also removed the Jk segment from the chimeric Pvt-1a/Ig-Ck transcript and inserted a unique tryptophan codon between Pvt-1a and Ig-Ck, preserving the ORF (Fig. 2).

This result is most exciting in view of the fact that no ORF of significance has thus far been documented for normal Pvt-1 transcripts. To date, 10-kb of the full-length 14-kb mouse Pvt-1 transcript have been sequenced and the largest ORF determined is 140 amino acids in length (and includes Pvt-1a, Huppi et al. 1993).

To determine whether this product is really common to all t(6;15) plasmacytomas, I designed an RT-PCR assay using oligonucleotide primers from the amino-terminal side of Pvt-1a and the Ig-Ck region (Fig. 2). In all t(6;15) plasmacytomas examined, regardless of the translocation breakpoint, the same Pvt-1a/Ck product is obtained from tumor mRNA (Table 1). While this result seems to be consistent with the presence of a chromosomal breakpoint in the Pvt-1 region, it is physiologically relevent as well in that an ORF seems to be maintained as a result of RNA editing in vivo (Huppi et al. manuscript in prep.).

Table 1-Mouse plasmacytomas tested with the Pvt-1 RT-PCR assay

Tumor	Translocation	c-Myc	Pvt-1	Pvt-1/Ck
ABPC4	t(6;15)	NR	R	+
ABPC20	t(6;15)	NR	R	+
ABPC17	t(6;15)	NR	R	+
ABPC105	t(6;15)	NR	R	+
TEPC1198	t(6;15)	NR	R	+
SiPCT5634	t(6;15)	NR	R	+
TEPC1033	t(12;15)	R	NR	-
TEPC1017	t(12;15)	R	NR	-
ABPC18	t(12;15)	R	NR	-
TEPC1173	t(12;15)	R	NR	-
TEPC2251	t(12;15)	R	NR	-
TEPC1165	t(12;15)	R	NR	-

NR=non-rearranged, R=rearranged

I next asked if this particular primer pair could successfully predict a majority of Pvt-1 associated translocations. To date, five t(6;15) plasmacytomas with known translocations in the Pvt-1 region have scored positive using the RT-PCR assay with Pvt-1a/Ig-Ck primers (Table I). On the contrary, no cases have been identified with positive Pvt-1a/Ck products from plasmacytomas with non-Pvt-1/ Ig-Ck rearrangements [i.e., t(12;15)plasmacytomas]. Recently, a silicone-induced plasmacytoma, see Potter et al. this issue) was successfully identified as a t(6;15) plasmacytoma by this method. In total, 6/6 t(6;15) plasmacytomas have scored positive with the Pvt-1a/Ck RT-PCR assay. Unfortunately, those few plasmacytomas that contain complex translocations upstream of Pvt-1a have not been included in this assay due to the absence of available mRNA. Predictably, these PCTs would not score positive with the assay, and would be missed in a random search for Pvt-1 translocations. Upstream regions of Pvt-1 may also be transcriptionally active and may provide reasonable splice sites for the development of additional primer pairs if such an assay seems necessary (Fig. 3). Based on the low frequency with which upstream Pvt-1 (upstream of Pvt-1a) translocations occur, however, the development of such an assay seems to be of a lower priority.

RT-PCR ASSAY FOR THE PVT-1 BREAKPOINT REGION

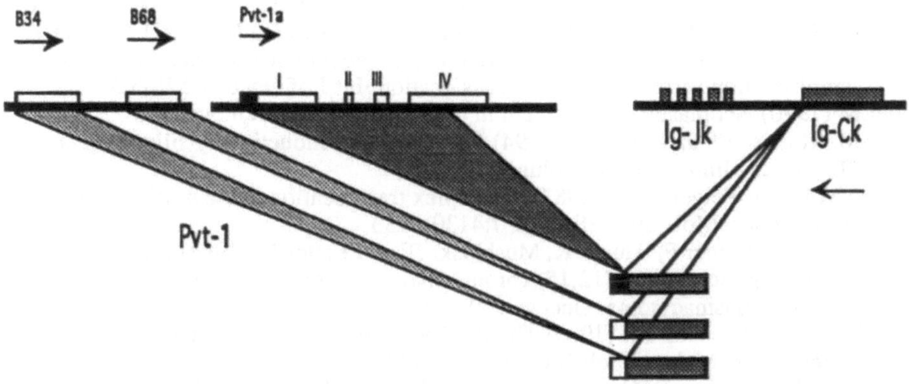

Fig. 3 A schematic RT-PCR assay for breakpoints in the Pvt-1 region. The exons corresponding to Pvt-1 and Ig-Ck are shown on the left and right, respectively. Shown below are shaded regions corresponding to potential breakpoint locations and the RT-PCR primers used (labeled as B34, B68 and Pvt-1a specific primer assays). Since the Pvt-1a region represents the majority of the chromosomal breakpoints, this region is the focus of the RT-PCR assay (darker shade).

Of greater interest is the design and implemention of a similar Pvt-1 assay for the variant translocations of T(2;8) and T(8;22) found in 20% of Burkitt's lymphoma. Analogous to the situation in mouse, PVT transcripts have been identified and PVT/IGL chimeric transcripts have also been observed in variant translocations (Shtivelman et al. 1990). Unlike the mouse, the human PVT transcripts appear to reside much closer to MYC (60-120-kb). Despite this difference, the chromosomal breakpoints occur at 120-kb (Ly61) and 240-kb (JBL2) 3' of c-MYC, respectively. Thus, RNA editing effectively removes more than 100-kb of intervening sequence. A human-based assay for PVT associated translocations would be extremely useful if one or a few PVT primer pairs could be efficiently designed as described here for the mouse.

References

Axelson H, Panda CK, Silva S, Sugiyama H, Wiener F, Klein G and Sumegi J (1991) A new variant 15;16 translocation in mouse plasmacytoma leads to the juxtaposition of c-myc and immunoglobulin lambda. Oncogene 6:2263-2270
Cory S, Graham M, Webb E, Corcoran L and Adams JM (1985) Variant (6;15) translocations in murine plasmacytomas involve a chromosome 15 locus at least 72 kb from the c-myc oncogene. EMBO J 4:675-681
Huppi K, Siwarski D, Skurla R, Klinman D and Mushinski JF (1990a) Pvt-1 transcripts are found in normal tissues and are altered by reciprocal (6;15) translocations in mouse plasmacytomas. Proc Natl Acad Sci (USA) 87:6964-6968
Huppi K, Siwarski D, Skurla RM Jr, Goodnight J and Mushinski JF (1990b) Isolation of normal and tumor-specific Pvt-1 clones. Curr Top Micro Immuol 166:233-241
Huppi K, Siwarski D, Goodnight J, Skurla RM Jr and Mushinski JF (1992) Alternative splicing of Pvt-1 transcripts in murine B-lymphocytic neoplasms accompanies amplification and chromosomal translocation. Int J of Oncol 1:525-532
Potter M (1990) Neoplastic development in B-lymphocytes. Carcinogenesis 11:1-13
Potter M, Morrison S and Miller F (1994) Plasmacytoma induction by silicone gels. Curr Top Micro Immunol (this volume)
Shapiro MA and Weigert M.A (1987) A complex translocation at the murine k light-chain locus. Mol. and Cell. Biol. 7:4130-4133
Shaughnessy J, Wiener F, Huppi K, Mushinski JF and Potter M.(1994) A novel c-myc-activating reciprocal T(12;15) chromsomal translocation juxtaposes Sa to Pvt-1 in a mouse plasmacytoma. Oncogene 9:247-253
Shtivelman E and Bishop JM (1989) The PVT gene frequently amplifies with MYC in tumor cells. Mol and Cell Biol. 9:1148-1154
Shtivelman E, Henglein B, Groitl P, Lipp M and Bishop JM (1989) Identification of a human transcription unit affected by the variant chromosomal translocations 2;8 and 8;22 of Burkitt lymphoma. Proc Natl Acad Sci USA 86:3257-3260
Shtivelman E and Bishop JM (1990) Effects of translocations on transcription from PVT. Mol and Cell Biol 10:1835-1839
Webb E, Adams JM and Cory S (1984) Variant (6;15) translocation in a murine plasmacytoma occurs near an immunoglobulin k gene but far from the myc oncogene. Nature 312:777-779

Interacisternal A-Particle (IAP) Genes show Similar Patterns of Hypomethylation in Established and Primary Mouse Plasmacytomas

K.K. Lueders and E.L. Kuff
Laboratory of Biochemistry, National Cancer Institute, Bethesda, MD 20892.

Introduction

Mouse plasmacytomas generally express higher levels of RNA transcripts from endogenous IAP proviral elements than do lipopolysaccharide-stimulated normal lymphocytes. Lymphocytes express a limited and highly characteristic set of IAP elements (LS elements) (Mietz et al. 1992) that share a unique 218 bp U3 regulatory region in their long terminal repeats (LTRs). The LS elements are also expressed in plasmacytomas. In addition, plasmacytomas express IAP elements (PC elements) which have different regulatory sequences in their LTRs than those in the LS element LTRs (Lueders et al. 1993). Differences were found in recognized nuclear factor binding sites such as ATF/CRE, Sp1, and TATA sites.

IAP proviral elements are 2,000-fold reiterated and widely, and apparently randomly, distributed in the mouse genome (Kuff and Lueders 1988). Activation of particular IAP elements is thought to be determined by the methylation state of their LTRs as well as availability of appropriate transcription factors. In normal mouse cells the IAP LTRs are highly methylated (Mietz and Kuff 1990), and proviral expression is generally low and tissue specific (Kuff and Lueders 1988). In contrast, IAP elements in the DNA of transformed cells are extensively hypomethylated in their 5' LTRs, and are expressed at high levels (Feenstra et al. 1986, Morgan and Huang 1984). Treatment of IAP-negative or low level-expressing cells with the methyltransferase inhibitor, 5-azacytidine, results in a rapid increase in IAP expression (Hojman-Montes de Oca et al. 1984), suggesting that expression can be limited by the number of demethylated copies. These data, as well as the results of in vitro transcription experiments, indicate that hypomethylation of the 5' LTR is a prerequisite for IAP expression.

The methylation state of many IAP proviruses is likely to be determined by their position in the genome. Distinctive patterns of IAP proviral hypomethylation common to several normal tissues have been detected using a two-dimensional electrophoretic technique and a 500 bp general IAP probe (Mietz and Kuff 1990). Strain-specific patterns of hypomethylation in sequences contiguous to the IAP elements were also observed. Development of the oligonucleotide probes, which detect a restricted number of IAP elements, has made it possible to carry out a similar analysis of the hypomethylation state of individual proviral loci on conventional one-dimensional gels. Many individual IAP elements have already been mapped on the mouse chromosome complement using these oligonucleotide probes (Lueders et al. 1993; Lueders and Frankel 1994). Mapping of hypomethylated IAP elements may permit identification and isolation of regions of cellular gene activity that are not detected by other means.

Here we show that many of the same IAP loci are hypomethylated in three independently established mouse plasmacytoma cell lines, and that some of these loci are hypomethylated in primary plasmacytomas as well, suggesting that these tumors have a common pattern of early hypomethylation.

Methods

Individual IAP proviruses were identified as 5'-junction fragments generated by restriction endonuclease digestion of genomic DNA (Mietz and Kuff 1992). The junction fragments were created by HindIII sites in the flanking DNA and at a conserved position immediately downstream of the probe sequences in the IAP 5' LTR (Fig. 1). The corresponding HindIII site in the 3' LTR eliminated detection of 3' flanks. The methylation status of each provirus was assessed using a conserved methylation sensitive HaeII site in the 5' LTR.

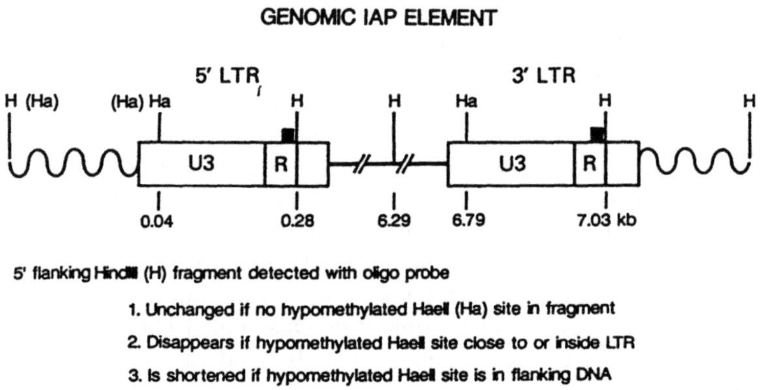

Fig. 1. Partial restriction map of an IAP element showing the location of HindIII (H) and HaeII (Ha) sites. The open boxes represent the LTRs; U3 and R are functional regions of the LTR. The heavy solid line represents IAP internal sequences, and the wavy lines represent flanking sequences. Numbers refer to nucleotide positions on the map of IAP element MIA14 (Mietz et al. 1987). Small solid boxes show the locations of the oligonucleotide probe sequences LS2, LS3, and T1. The possible effects of digestion with the methylation sensitive enzyme HaeII on the 5' HindIII fragments detected with the oligonucleotide probes are listed.

Fragments were detected by hybridization of genomic DNA to oligonucleotide probes (23 nucleotides in length) derived from expressed IAP elements. LS2 and LS3 probes detect two subfamilies of the LS elements expressed in normal B-cells (Mietz et al. 1992). The T1 probe detects a subfamily of the PC elements expressed in plasmacytomas (Lueders et al. 1993). Location of the probes is indicated in Fig. 1. DNAs were electrophoresed on 0.9% agarose gels with a pulse controller and hybridization was carried out in dried gels as previously described (Mietz and Kuff 1992; Lueders et al. 1993). Specificity of hybridization was achieved by washing the gels in 3.2 M tetramethylammonium chloride, which eliminates dependence of hybrid stability on base composition.

Established plasmacytomas MPC11, MOPC104E, and MOPC21 were independently induced in BALB/cAn mice by intraperitoneal mineral oil, and have been grown for many generations as cell lines (MPC11 and MOPC21) or as solid tumors (MOPC104E). Primary plasmacytomas 416, 518, and 3 were induced with pristane, and 294, 295, and 297 were induced with a viral vector (RIM) carrying c-myc and v-Ha-ras genes (Clynes et al. 1988). Primary tumors were isolated from mesenteric surfaces as diffuse nodular implants.

Results

The high level of IAP proviral expression in plasmacytoma cells is associated with increased hypomethylation of genomic IAP elements. The methylation status of individual IAP proviruses can be determined by comparing the restriction fragment patterns obtained from DNA cut with HindIII (H) versus the patterns obtained from DNA cut with HindIII plus methylation-sensitive HaeII (H/Ha). The predicted results are described in Fig. 1. Fragments are marked with the following symbols in all the figures:

- • 5' -junction fragments cut by HaeII in or near the IAP LTR
- ○ Newly appearing fragments from a distant cut HaeII in the flanking sequence
- ◂ Fragments in tumor DNA that are not in germline DNA (transpositions)

The results obtained from analysis of DNAs from established plasmacytomas with LS2 probe are shown in Fig. 2. Liver DNA digested with HindIII only is shown to establish the pattern for normal cells; none of the fragments was hypomethylated. The LS2 probe detected 2 hypomethylated IAP proviral locus (marked with asterisks) in normal LPS-stimulated B cells (not shown). In contrast to normal B-cells the established plasmacytomas had 7 to 11 LS2 proviral loci that were hypomethylated. Seven of these were common to all three tumors. These results indicate that IAP provirus hypomethylation in the established plasmacytomas does not occur entirely randomly (Lueders et al. 1993).

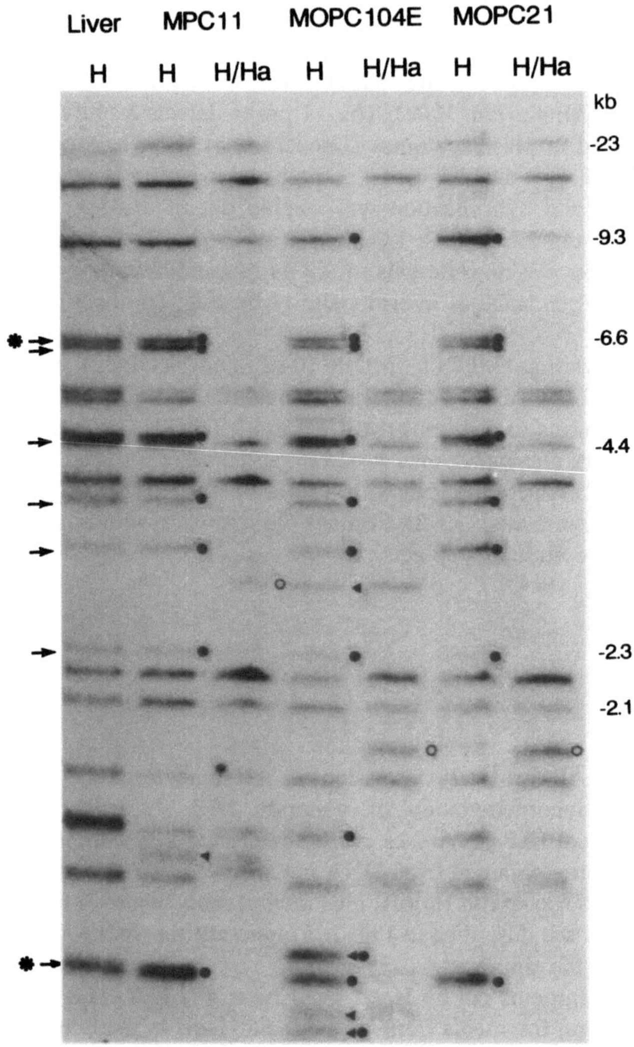

Fig. 2. Hybridization of the LS2 probe to HindIII (H) and HindIII plus HaeII (H/Ha) digested genomic DNAs from established plasmacytomas. Asterisks indicate fragments that were hypomethylated in normal B-cells. →, fragments hypomethylated in all tumors.

Not all IAP elements were hypomethylated, however, even when their LTR sequences were very similar. Four additional hypomethylated loci were unique to 1 or 2 of the tumors. Each tumor had only one new fragment in the HindIII plus HaeII digest, indicating that the rest of the hypomethylated sites were in the LTR or close to it in the flanking DNA. The patterns also showed one newly inserted provirus in MPC11 and 4 newly inserted proviruses in MOPC104E. Two of the latter were also hypomethylated. LS2 type IAP transcripts are commonly found in established plasmacytomas (Lueders et al. 1993).

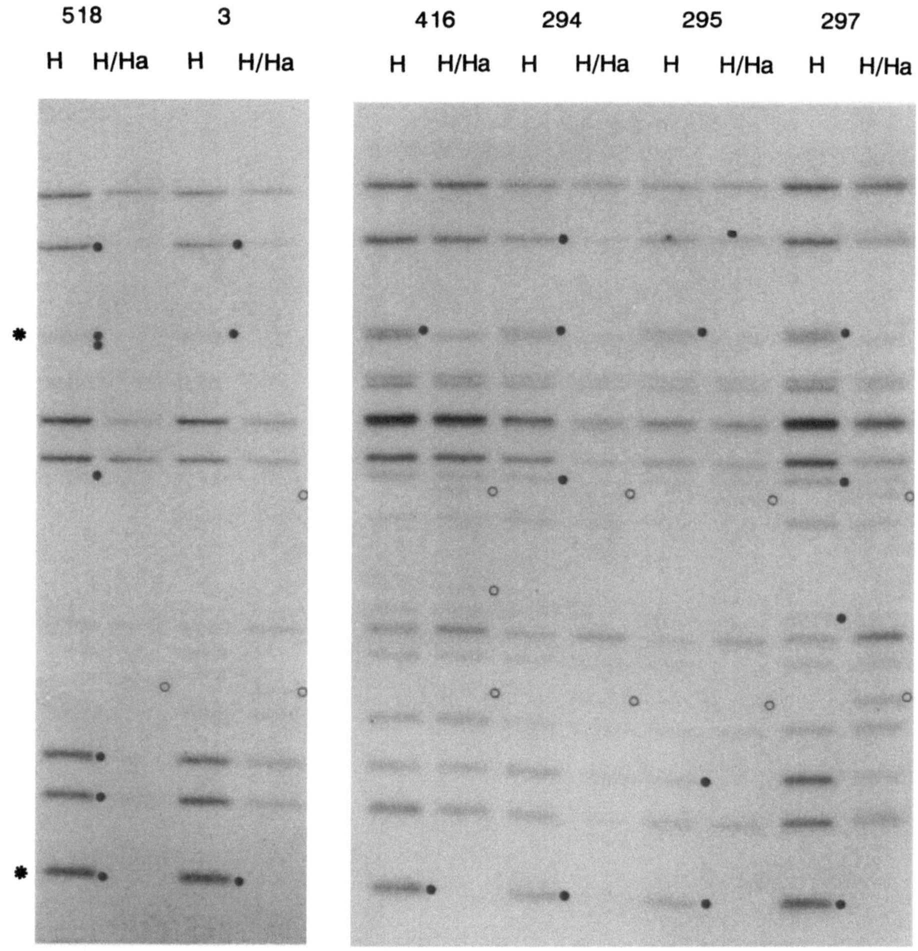

Fig. 3. Hybridization of the LS2 probe to genomic DNAs from primary plasmacytomas.

A similar analysis of DNAs from primary plasmacytomas was carried out to determine whether the increased hypomethylation of IAP elements is an early property of the developing tumors. Results with the LS2 probe are shown in Fig. 3. The only hypomethylated LS2 loci that were common to all 6 primary tumors analyzed were those that were also hypomethylated in normal stimulated B-cells (marked with asterisks). Three of the tumors each had an additional hypomethylated LS2 locus that was also hypomethylated in the established plasmacytomas. The appearance of several new fragments after HaeII digestion (marked with open circles) indicated that the hypomethylated sites for these loci were in the flanking DNA rather than in the IAP LTR. Similar sizes for these fragments in all the tumors suggested the same flanking HaeII site was cut in each case.

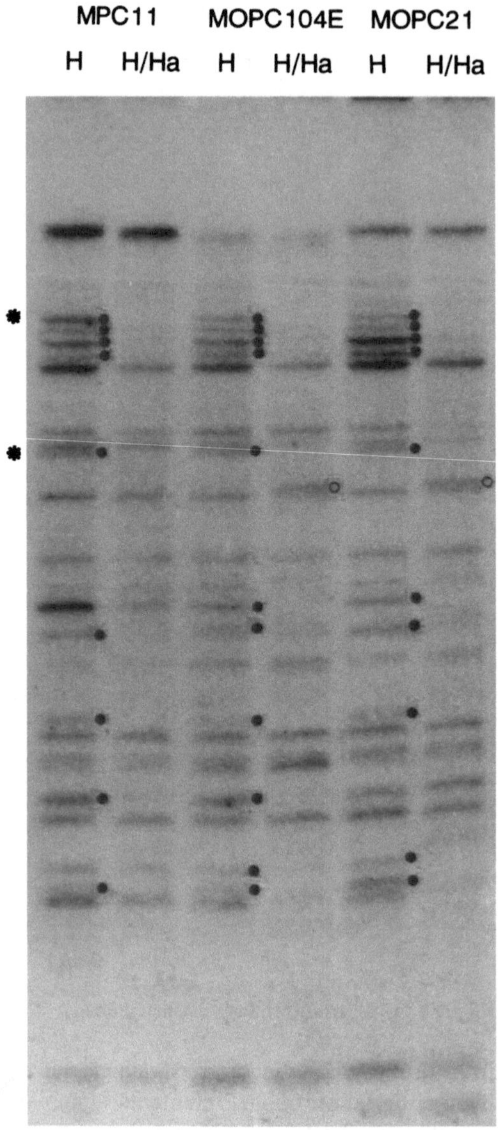

Fig. 4. Hybridization of the LS3 probe to genomic DNAs from established plasmacytomas.

Analysis of DNA from established plasmacytomas with LS3 probe is shown in Fig. 4. This probe detected 2 hypomethylated loci in DNA from normal cells (marked with asterisks). All three of these tumors again had multiple, common fragments that were hypomethylated. The small number of new fragments in the HindIII plus HaeII lanes indicated that the hypomethylated sites were primarily in the LTR or nearby in the flanking DNA, as was the case for the LS2 loci in the established tumors.

Analysis of DNAs from primary plasmacytomas with the LS3 probe is shown in Fig. 5. Many of the loci hypomethylated in the established tumors were also hypomethylated in the primary tumors. The number of LS3 loci hypomethylated

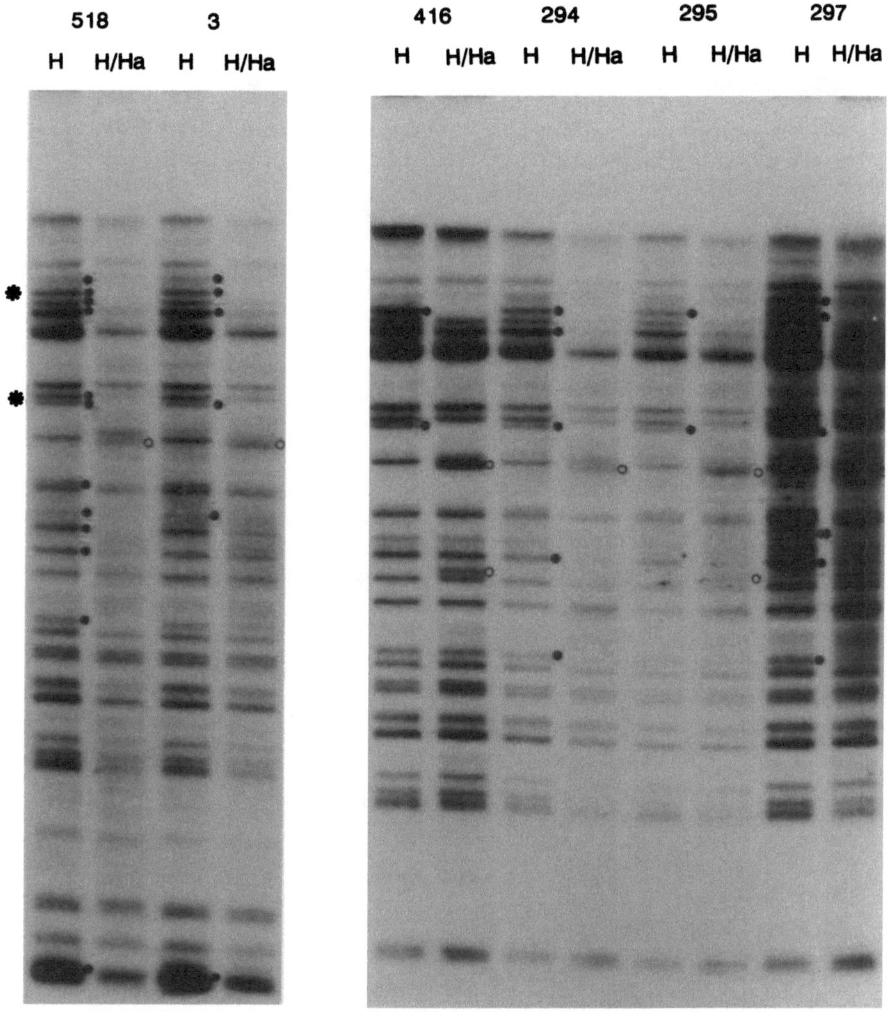

Fig. 5. Hybridization of the LS3 probe to genomic DNAs from primary plasmacytomas.

in each tumor was not the same; tumor 518 had a pattern most closely resembling that of the established tumors, while 416 was indistinguishable from LPS-stimulated B-cells with the LS3 probe. Tumor 297 had an IAP fragment that was not present in normal genomic DNA, and this fragment was also hypomethylated.

Fig. 6 shows analysis of DNA from primary plasmacytomas with the T1 probe. IAP elements carrying the T1 sequence are highly expressed in many

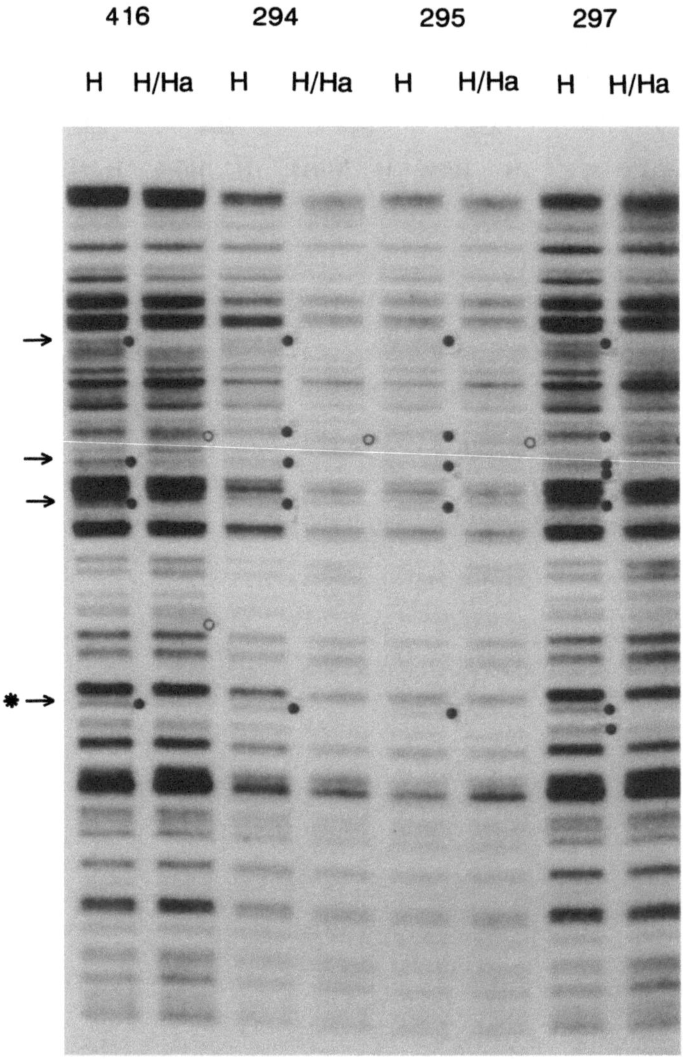

Fig. 6. Hybridization of the T1 probe to genomic DNAs from primary plasmacytomas. →, fragments hypomethylated in all tumors.

established plasmacytomas (Lueders et al. 1993). Multiple common T1 loci were hypomethylated in all 4 of the primary tumors analyzed. The same loci were also hypomethylated in established plasmacytomas (Lueders et al. 1993, not shown). One T1 locus (marked with an asterisk) was also hypomethylated in normal B-cells, although T1 sequences were not detected in the RNA of these cells. The new fragments resulting from hypomethylation of a HaeII site in the flanking DNA, which were present in these primary plasmacytomas, were not seen in the established tumors.

Summary

Alterations in cell programming associated with neoplastic transformation may involve widespread changes in patterns of DNA methylation. Increased expression of IAP elements in plasmacytomas compared with LPS-stimulated normal B-cells is accompanied by extensive hypomethylation of IAP sequences (Mietz and Kuff 1990), subsets of which are revealed with the LS2, LS3 and T1 probes. Multiple common LS- and PC-specific IAP loci are hypomethylated in established plasmacytomas, showing that hypomethylation does not occur entirely randomly. Many of the same IAP loci are hypomethylated in primary plasmacytomas induced by two different methods, as soon as recognizable tumor tissue can be isolated. In primary tumors hypomethylation frequently appears to occur in DNA flanking the IAP elements. In the established tumors the hypomethylated sites occur primarily in the IAP LTR, suggesting that for these loci hypomethylation begins in the flanking DNA and is extended into the IAP LTRs during progression of the tumors.

The newly hypomethylated IAP LTRs in primary plasmacytomas (as compared to normal B cells) may provide a set of reporter genes for chromosomal regions that are characteristically hypomethylated in these transformed cells and that may contain cellular genes whose activation is related to the transformation process.

References

Clynes R, Wax J, Stanton, LW, Smith-Gill S, Potter M, Marcu KB (1988) Rapid induction of IgM-secreting murine plasmacytomas by pristane and an immunoglobulin heavy-chain promoter/enhancer-driven c-myc/v-ras retrovirus. Proc. Natl. Acad. Sci USA 85:6067-6071

Feenstra A, Fewell J, Kuff EL, Lueders KK (1986) In vitro methylation inhibits the promoter activity of an intracisternal A-particle LTR. Nucleic Acids Res 14:4343-4352

Hojman-Montes de Oca F, Lasneret J, Dianoux L, Canivet M, Ravicovitch- Ravier R., Peries J (1984) Regulation of intracisternal A-particles in mouse teratocarcinoma cells; involvement of DNA methylation in transcriptional control. Biol. Cell 52:199-204

Kuff EL, Lueders KK (1988) The intracisternal A-particle gene family: structure and functional aspects. Adv. Cancer Res. 51:183-276

Lueders KK, Fewell JW, Morozov, VE, Kuff EL (1993) Selective expression of intracisternal A-particle genes in established mouse plasmacytomas. Mol. Cell. Biol. 13:7439-7446

Lueders KK, Frankel WN, Mietz, JA, Kuff EL (1993) Genomic mapping of intracisternal A-particle proviral elements. Mammalian Genome 4:69-77

Lueders KK, Frankel WN (1994) Mapping of mouse intracisternal A-particle proviral markers in an interspecific backcross. Mammalian Genome, in press

Mietz JA, Kuff EL (1990) Tissue and strain-specific patterns of endogenous proviral hypomethylation analyzed by two-dimensional gel electrophoresis. Proc. Natl. Acad. Sci. USA 87:2269-2273

Mietz JA, Kuff EL (1992) IAP-specific oligonucleotides provide multilocus probes for genetic linkage studies in the mouse. Mammalian Genome 3:447-451

Mietz JA, Fewell JW, Kuff EL (1992) Selective activation of a discrete family of endogenous proviral elements in normal BALB/c lymphocytes. Mol. Cell. Biol. 12:220-228

Mietz JA, Grossman Z, Lueders KK, Kuff EL (1987) Nucleotide sequence of a complete mouse intracisternal A-particle genome: relationship to known aspects of particle assembly and function. J. Virol. 61:3020-3029

Morgan RA, Huang RCC (1984) Correlation of undermethylation of intracisternal A-particle genes with expression in murine plasmacytomas but not in NIH/3T3 embryo fibroblasts. Cancer Res. 44:5234-5241

Regulatory Elements in the Immunoglobulin Kappa Locus Induce c-myc Activation in Burkitt's Lymphoma Cells

K.HÖRTNAGEL, A.POLACK, J.MAUTNER, R.FEEDERLE and G.W.BORNKAMM

Institut für Klinische Molekularbiologie und Tumorgenetik, GSF, D 81377 München

Introduction

The characteristic reciprocal translocations in Burkitt's lymphomas (BL) always involve the *c-myc* proto-oncogene on chromosome 8 and one of the immunoglobulin (Ig) loci on chromosomes 2, 14, or 22. The breakpoints relative to both *c-myc* and the immunoglobulin genes vary considerably and may be located either within the *c-myc* transcription unit or up to several hundred kilobases 5' or 3' of *c-myc*. The translocated *c-myc* allele in BL cells displays several characteristic features: (i) *c-myc* is predominantly expressed from the translocation chromosome, whereas the normal allele is transcriptionally silent or expressed at low level only, (ii) structural alterations occur consistently in and around *c-myc* exon 1, (iii) the block to RNA elongation is functionally missing, and (iv) the P1 promoter is the preferential site of transcriptional initiation, in contrast to the normal *c-myc* gene where 80 - 90 % of total *c-myc* RNA is derived from the P2 promoter (promoter shift) (Taub *et al.*, 1984a,b; Yang *et al.*, 1985; Bornkamm *et al.*, 1988; Nishikura and Murray, 1988; Spencer and Groudine, 1991)

The demonstration of normal expression of a highly mutated *c-myc* gene after stable transfection ruled out that the functional changes in *c-myc* activation were a consequence of the structural alterations only (Richman and Hayday, 1989a,b; Spencer *et al.*, 1990; Polack *et al.*, 1991). It is therefore assumed that the deregulation of the translocated *c-myc* allele in BL cells is brought about by juxtaposition of an active Ig locus (Cory, 1986; Bornkamm *et al.*, 1988; Magrath, 1990) At present however remained elusive, which elements of the Ig locus are involved in *c-myc* activation and how the long distance from the Ig locus to the *c-myc* gene is overcome.

We have chosen the variant t(2;8) translocation as a model to study the interplay between *c-myc* and regulatory elements in the Ig loci. In this type of translocation the breakpoints fall into a small region on chromosome 2 in the vicinity of the kappa joining region (Henglein *et al.*, 1989), leading to the colocalisation of the *c-myc* gene and the kappa matrix attachment region (MAR), the intron enhancer (Ei), the kappa constant region (Ck) and the 3' enhancer (E3') (Cockerill and Garrard, 1986; Atchison and Perry, 1987; Pongubala and Atchison, 1991). The intron enhancer is among the best studied transcription control elements and its up to 10-fold enhancing activity has been reported for various promoters (Picard and Schaffner, 1984; Atchison and Perry, 1988; Sen and Baltimore, 1989). However, the intron enhancer alone can activate a juxtaposed *c-myc* gene only slightly (Polack *et al.*, 1991). The existence of another enhancer located 3' had been postulated after the description of cell lines capable of transcribing Ig genes in the apparent absence of an active intron enhancer (Meyer and Neuberger, 1989). Functional analysis of the region downstream of the kappa constant region revealed a strong B-cell specific enhancer, which stimulates transcription from the V kappa promoter over a distance of 14 kb (Blasquez *et al.*, 1989; Judde and Max, 1992).

To assess the role of the described Ig elements in *c-myc* activation in Burkitt's lymphoma we reconstructed the translocation on a minichromosome consisting of the *c-myc* gene, a t(2;8) chromosomal breakpoint and the adjacent Ig kappa locus. Functional analysis of this construct and deletional constructs thereof led to the delineation of important Ig regions for transcriptional *c-myc* activation and induction of the promoter shift.

Episomal vectors as tools to reconstruct *c-myc* activation

It was our aim to set up a system in which the features of *c-myc* activation and differential promoter usage of Burkitt's lymphoma could be reconstructed. Therefore a series of plasmids encompassing the *c-myc* gene and various parts of the Ig kappa locus was constructed in the Epstein-Barr virus (EBV) derived eukaryotic vector pHEBO (Sudgen *et al.*, 1985). This vector was chosen since it replicates as an episome and permits to study gene regulation without interfering position effects. The structure of the different constructs is shown in Figure 1. Two versions of each plasmid were constructed, using a normal or a mutated *c-myc* gene, respectively. All following experiments were carried out using the two versions in parallel to further evaluate a possible role of the mutations in the translocated *c-myc* gene. The constructs were transfected into Raji cells and four stable transfectants were obtained for each construct after selection with hygromycin. The integrity of the constructs in the transfectants was tested by Southern blot analysis and the copy number ranging between 20 and 50 was determined by densitometry (Fig. 2A). Importantly, the *c-myc* gene on the episomal vectors exhibited its cognate chromatin configuration as revealed by the formation of DNAse I hypersensitive sites (Fig. 2B). In the corresponding Northern blot (Fig. 2C), *c-myc* RNA of the correct size with a level of expression considerably above that of untransfected cells was detected. No difference between constructs carrying a normal *c-myc* gene (pKH99-6) and those carrying a highly mutated *c-myc* gene (pKH100-6)could be observed.

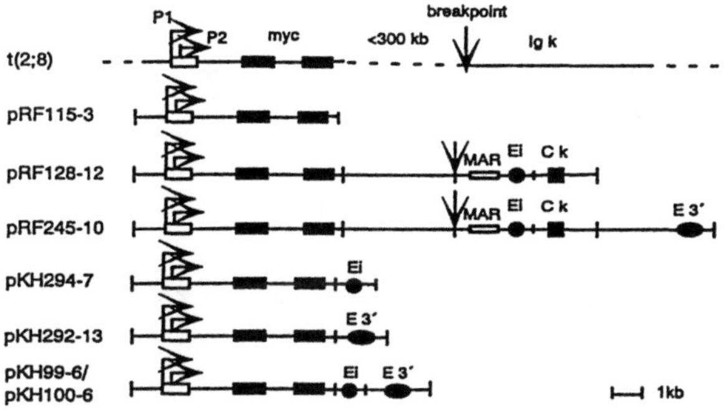

Fig. 1. Schematic representation of the t(2;8) chromosomal translocation and the constructs used for stable transfection. Coding exons are shown as solid boxes, the *c-myc* promoters P1 and P2 are indicated as horizontal arrows. The breakpoint region derived from LY91 is marked by a vertical arrow. The matrix attachment region is represented by a shaded box, the kappa intron and 3' enhancers are shown as filled circles and ellipses, respectively. The constructs pKH99-6 and PKH100-6 only differ in carrying a normal or mutated *c-myc* gene, respectively. Only the names of the clones carrying the mutated c-myc are presented for the other constructs.

Induction of the promoter shift

The BL cell line Raji carries a t(8;14) translocation and harbors a deletion on the translocation chromosome removing the 3' part of the first *c-myc* exon. S1 experiments using a exon 1 specific probe therefore allow differentiation between transcripts derived from the endogenous translocated allele and transcripts from the transfected constructs. As a second advantage Raji cells display a prominent promoter shift in S1 analysis, indicating the presence of all necessary transacting factors and serving as an internal control (Rabbitts *et al.*, 1983). An S1 analysis of the different transfectants is presented in Figure 3. As described previously , mutations in the first exon and intron of the *c-myc* gene (pRF115-3 in Figure 1) were incapable of inducing *c-myc* expression, whereas the construct consisting of *c-myc*, the LY91 derived t(2;8) breakpoint and kappa sequences extending from the joining region J5 to 2.7 kb 3' of the constant kappa gene (pRF128-12 in Figure 1) displayed slight activation of P2 (Polack *et al.*, 1991). This indicated that additional or other elements apart from the kappa intron enhancer are involved in the BL specific *c-myc* activation. Consequently a fragment encompassing the kappa 3' enhancer was added to the construct (pRF245-10, Figure 1). S1 analysis revealed a dramatic increase in P1 and P2 specific transcripts from the transfected construct as compared to the endogenous translocated *c-myc* allele. Densitometric evaluation of the autoradiogram (not shown) resulted in a ratio of P1 to P2 derived transcripts > 1 indicating the formation of a promoter shift. In comparison with construct pRF128-12 addition of the 3' enhancer (pRF245-10) increased the activity of the P1 and P2 promoters by a factor of 30 and 8, respectively.

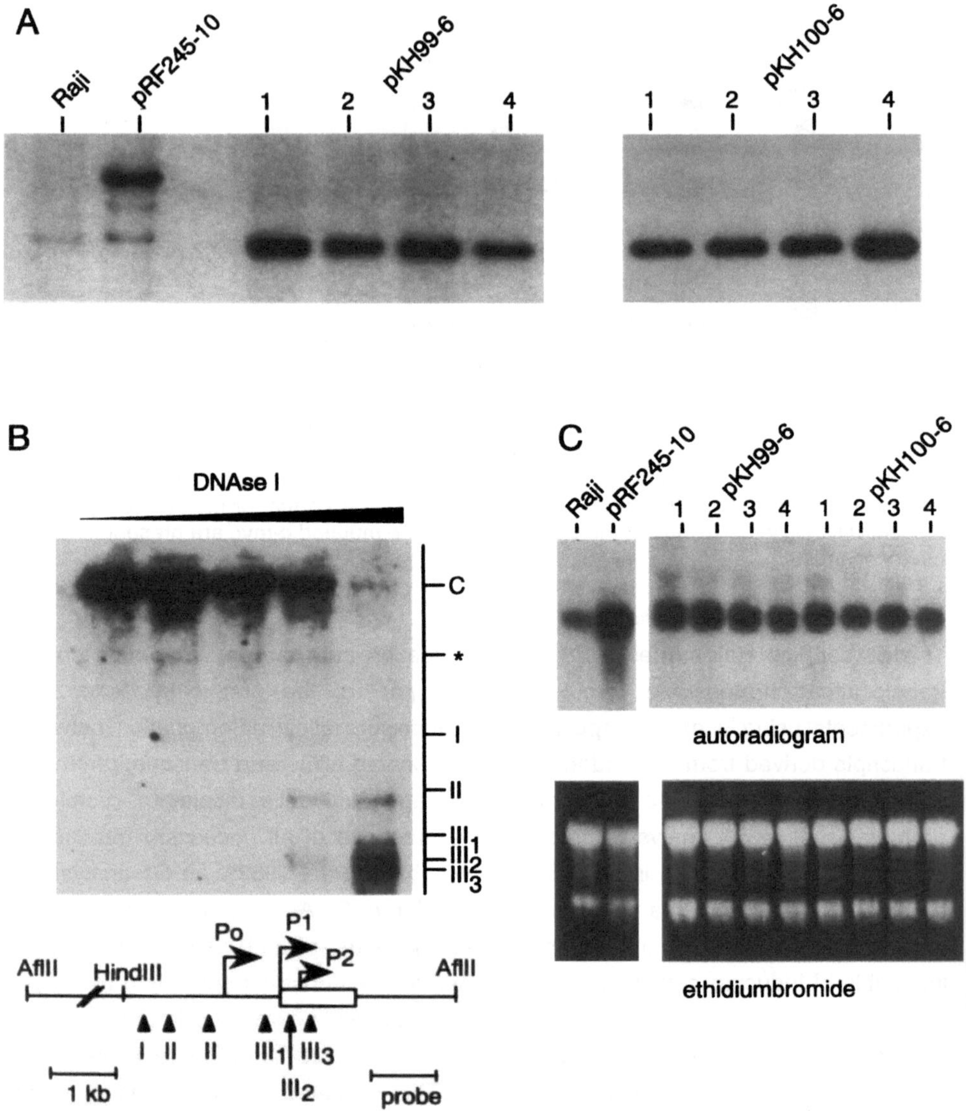

Fig. 2. (**A**) Southern blot analysis of Raji cells transfected with the indicated constructs. The blot was hybridized with a ^{32}P-labeled probe of *c-myc* exon 3. Only a selection of transfectants is presented to demonstrate the integrity of the episomal constructs and their copy number in comparison to the signal in the first lane which is derived from the endogenous *c-myc* of untransfected Raji cells. (**B**) DNAse I-hypersensitive sites of the c-myc promoter region in stable transfectants carrying pKH100-6. Fragments corresponding to DNAse I-hypersensitive sites are designated by roman numerals according to Siebenlist *et al.* (1984, 1988). The site marked by an asterisk corresponds to a hypersensitive site in the vector. (**C**) The Northern blot revealed strong *c-myc* expression from all transfectants carrying pKH99-6 or pKH100-6, respectively. Even stronger expression was detected from the transfectant with pRF245-10. RNA from untransfected Raji cells served as a control.

The kappa 3' enhancer alone is not sufficient to activate *c-myc*

The enhancing activity of the human kappa 3' enhancer was described to be 5- to 7-fold higher than that of the kappa intron enhancer (Judde and Max, 1992). Moreover, the fact that a kappa gene can be expressed in a cell which lacks the kappa intron enhancer, indicates that the kappa 3' enhancer is sufficient for the activity of the V promoter (Atchinson and Perry, 1987, 1988; Blasquez *et al.*, 1989). With the construct pKH292-13 we addressed the question whether this element was sufficient to activate *c-myc* and to induce a promoter shift. Again, stable transfectants were created and S1 experiments were performed. Transcripts derived from the P1 promoter on the construct were hardly visible and the P2 promoter became activated only slightly (Figure 3).

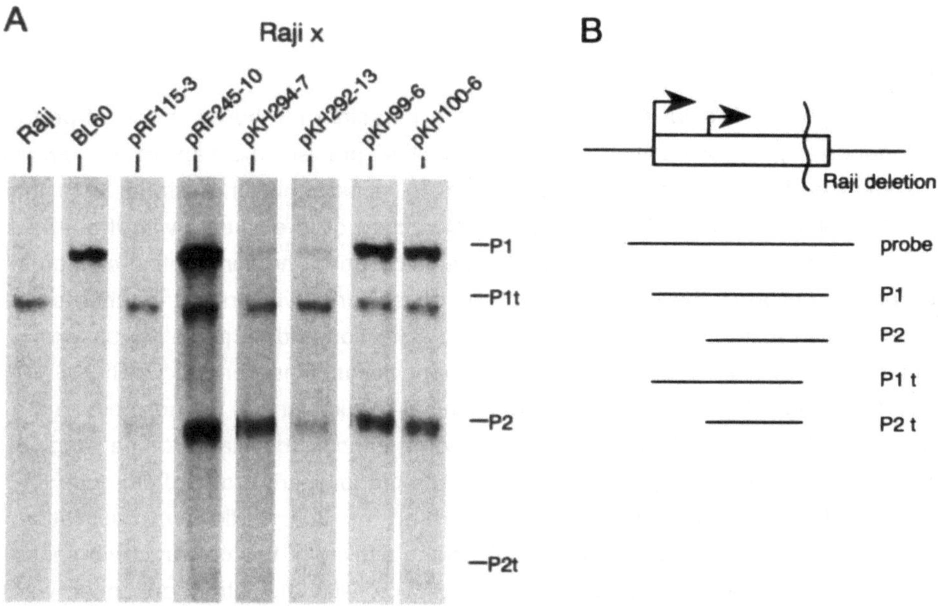

Fig. 3. (**A**) Induction of strong *c-myc* activation (pRF245-10) and promotershift (pRF245-10, pKH99-6,pKH100-6) by the Ig kappa locus. Nuclease S1 experiments analysing the c-myc promoter usage of Raji cells stably transfected with the indicated constructs. Total cellular RNA of the different transfectants and control RNA from untransfected Raji and BL60 cells were analysed with the probe shown in (**B**). Signals corresponding to the endogenous translocated allele are marked with P1t and P2t whereas signals derived from the transfected constructs are marked with P1 and P2.

The two kappa enhancers induce a promoter shift, but only moderately activate *c-myc*

In the mouse system the two kappa enhancers have been shown to synergistically activate gene expression (Blasquez *et al.*, 1992). Our speculation that the two kappa enhancers together might accomplish the BL specific *c-myc* activation was supported by transient transfection assays, which demonstrated strong synergy between the enhancers in the activation of the *c-myc* P1 promoter (unpublished data). The plasmid pKH100-6 consisting of *c-myc* and the core-fragments of both enhancers was constructed and the RNAs of the corresponding transfectants were analysed. A preferential usage of P1 over P2 was observed, but the overall *c-myc* activation was reduced as compared to the transfectants with pRF245-10. The densitometrically scanned signals revealed a P1 to P2 ratio of > 1, the transcriptional activity of pKH100-6, however was reduced by a factor of 5 compared to pRF245-10. For that reason additional element(s) present on pRF245-10 must be involved in the BL specific *c-myc* activation.

Conclusion

Even though Burkitt's lymphoma represents a paradigm for the activation of proto-oncogenes by chromosomal translocations the precise mechanism of *c-myc* activation by the translocationally juxtaposed Ig locus is not understood. With the long term intention to study the molecuar interaction of Ig elements with the *c-myc* gene in BL cells we first focussed on the identification of the contributing elements. By reconstruction of the t(2;8) translocation on episomal vectors we established a system which allowed the detection of *c-myc* activation and deregulation under the influence of different elements of the Ig kappa locus. The construct pRF245-10, encompassing the *c-myc* gene, the chromosomal breakpoint and the adjacent kappa elements (matrix attachment region, intron enhancer, constant kappa and 3' enhancer) displayed strong *c-myc* activation with predominant usage of promoter P1, indicating the presence of all elements necessary for BL specific *c-myc* activation. Since the 3' enhancer added dramatically to the activity of this construct we asked whether the 3' enhancer alone or in conjunction with the intron enhancer was sufficient for *c-myc* activation in BL cells. Analysis of the respective construct (pKH292-13) clearly revealed only slight *c-myc* activation by the 3' enhancer alone. However, the two enhancers in concert (pKH99-6/pKH100-6) activated *c-myc* to a larger extent and induced a promoter shift. Apparently *c-myc* activation requires the interaction of both enhancers, but at least one additional element which promotes full activation of *c-myc* must be present in the kappa locus. The matrix attachment region (MAR) described by Cockerill and Garrard (1986) and reported by Blasquez *et al.* (1989) to have a stimulating effect on V-gene transcription represents the prime candidate for this missing element. Experiments to study the role of MAR for *c-myc* activation in BL cells are currently under way.

References

Atchison, M.L. and Perry, R.P. (1987). The role of the kappa enhancer and its binding factor NF-kappa B in the developmental regulation of kappa gene transcription. Cell *48*, 121-128.

Atchison, M.L. and Perry, R.P. (1988). Complementation between two cell lines lacking kappa enhancer activity: implications for the developmental control of immunoglobulin transcription. EMBO J. *7*, 4213-4220.

Blasquez, V.C., Xu, M., Moses, S.C., and Garrard, W.T. (1989). Immunoglobulin kappa gene expression after stable integration. I. Role of the intronic MAR and enhancer in plasmacytoma cells. J. Biol. Chem. *264*, 21183-21189.

Blasquez, V.C., Hale, M.A., Trevorrow, K.W., and Garrard, W.T. (1992). Immunoglobulin kappa gene enhancers synergistically activate gene expression but independently determine chromatin structure. J. Biol. Chem. *267*, 23888-23893.

Bornkamm, G.W., Polack, A., and Eick, D. (1988). *c-myc* deregulation by chromosomal translocation in Burkitt's lymphoma. In Cellular oncogene activation. G. Klein, ed. (New York, Basel: Dekker,M.,Inc.), pp. 223-273.

Cockerill, P.N. and Garrard, W.T. (1986). Chromosomal loop anchorage of the kappa immunoglobulin gene occurs next to the enhancer in a region containing topoisomerase II sites. Cell *44*, 273-282.

Cory, S. (1986). Activation of cellular oncogenes in Hemopoietic cells by chromosome translocation. Adv. Cancer Res. *47*, 189-234.

Henglein, B., Synovzik, H., Groitl, P., Bornkamm, G.W., Hartl, P., and Lipp, M. (1989). Three breakpoints of variant t(2;8) translocations in Burkitt's lymphoma cells fall within a region 140 kilobases distal from *c-myc*. Mol. Cell. Biol. *9*, 2105-2113.

Judde, J.-G. and Max, E.E. (1992). Characterization of the human immunoglobulin kappa gene 3' enhancer: Functional importance of three motifs that demonstrate B-cell-specific in vivo footprints. Mol. Cell. Biol. *12*, 5206-5216.

Magrath, I. (1990). The pathogenesis of Burkitt's lymphoma. Adv. Cancer Res. *55*, 134-270.

Meyer, K.B. and Neuberger, M.S. (1989). The immunoglobulin kappa locus contains a second, stronger B-cell-specific enhancer which is located downstream of the constant region. EMBO J. *8*, 1959-1964.

Nishikura, K. and Murray, J.M. (1988). The mechanism of inactivation of the normal *c-myc* gene locus in human Burkitt lymphoma cells. Oncogene *2*, 493-498.

Picard, D. and Schaffner, W. (1984). A lymphocyte-specific enhancer in the mouse immunoglobulin kappa gene. Nature *307*, 80-82.

Polack, A., Strobl, L., Feederle, R., Schweizer, M., Koch, E., Eick, D., Wiegand, H., and Bornkamm, G.W. (1991). The intron enhancer of the immunoglobulin kappa gene activates *c-myc* but does not induce the Burkitt specific promoter shift. Oncogene *6*, 2033-2040.

Pongubala, J.M. and Atchison, M.L. (1991). Functional characterization of the developmentally controlled immunoglobulin kappa 3' enhancer: regulation by Id, a repressor of helix-loop-helix transcription factors. Mol. Cell Biol. *11*, 1040-1047.

Richman, A. and Hayday, A. (1989). Normal expression of a rearranged and mutated *c-myc* oncogene after transfection into fibroblasts. Science *246*, 494-497.

Richman, A. and Hayday, A. (1989). Serum-inducible expression of transfected human *c-myc* genes. Mol. Cell. Biol. *9*, 4962-4969.

Sen, R. and Baltimore, D. (1989). Factors regulating immunoglobulin-gene transcription. In Imunoglobulin Genes. T. Honjo, F.W. Alt, and T.H. Rabbitts, eds. (London: Academic Press), pp. 327-342.

Siebenlist, U., Hennighausen, L., Battey, J., and Leder, P. (1984). Chromatin structure and protein binding in the putative regulatory region of the c-myc gene in Burkitt lymphoma. Cell *37*, 381-391.

Siebenlist, U., Bressler, P., and Kelly, K. (1988). Two distinct mechanisms of transcriptional control operate on c-myc during differentiation of HL60 cells. Mol. Cell. Biol. *8*, 867-874.

Spencer, C.A., LeStrange, R.C., Novak, U., Hayward, W.S., and Groudine, M. (1990). The block to transcription elongation is promoter dependent in normal and Burkitt's lymphoma *c-myc* alleles. Genes Dev. *4*, 75-88.

Taub, R., Moulding, C., Battey, J., Murphy, W., Vasicek, T., Lenoir, G.M., and Leder, P. (1984). Activation and somatic mutation of the translocated *c-myc* gene in Burkitt lymphoma cells. Cell *36*, 339-348.

Taub, R., Kelly, K., Battey, J., Latt, S., Lenoir, G.M., Tantravahi, U.T., and Leder, P. (1984). A novel alteration in the structure of an activated *c-myc* gene in a variant t(2;8) Burkitt lymphoma. Cell *37*, 511-520.

Yang, J.-Q., Bauer, S., Mushinski, J.F., and Marcu, K.B. (1985). Chromosome translocations clusterd 5` of the murine *c-myc* qualitatively affect promoter usage: implications for the site of normal *c-myc* regulation. EMBO J. *4*, 1441-1447.

Genomic Instability: Heavy Chain Switch-Related Problems

Illegitimate Recombinations Between c-myc and Immunoglobulin Loci are Remodeled by Deletions in Mouse Plasmacytomas but not in Burkitt's Lymphomas

J.R. Müller, S. Janz, and M. Potter
Laboratory of Genetics, NCI, NIH

Abstract

Recombinations between c-*myc* and immunoglobulin loci are a hallmark of Burkitt's lymphomas and mouse plasmacytomas. Analyzing the fine structure of these illegitimate rearrangements has revealed differences between the recombinations in these two tumors. Recombinations are nearly reciprocal in Burkitt's lymphomas, whereas in most BALB/c plasmacytomas large stretches of c-*myc* sequences have been deleted. The recombinations detected during preneoplastic development of plasmacytomas, in contrast, have most of the c-*myc* sequences retained on either one of the translocated chromosomes. We conclude that initial recombination structures are subjected to secondary changes in mouse plasmacytomas. In Burkitt's lymphomas the primary recombination sequence is preserved in tumor cells.

Introduction

Illegitimate recombinations between the proto-oncogene c-*myc* and immunoglobulin sequences play a major role in B cell lymphomagenesis in mice and man [1-3]. In the majority of mouse plasmacytomas, c-*myc* is juxtaposed to the immunoglobulin heavy chain α locus (IgHα). In Burkitt's lymphomas, most rearrangements join c-*myc* sequences with the immunoglobulin heavy chain μ locus (IgHμ). To study the fine structures of these recombinations PCR techniques were designed to target the most common rearrangements between c-*myc* and IgHμ in sporadic Burkitt's lymphomas and between c-*myc* and IgHα in pristane-induced mouse plasmacytomas.

Results and Discussion

Recombinations in Plasmacytomas and Burkitt's Lymphomas Show Differences
The analysis of amplified recombination products in 15 plasmacytomas and in 9 Burkitt's lymphomas revealed several differences (Fig. 1). Comparing the breaksites in c-*myc* in both translocated chromosomes 8- and 14+ in Burkitt's lymphomas and 12+ and 15- in plasmacytomas reveals that these sites are closely located with no more than 39 bp distance in the Burkitt's lymphomas, whereas in plasmacytomas the distance is significantly larger (Fig. 1a). In other words, recombinations are nearly reciprocal in Burkitt's lymphomas. In contrast, in most plasmacytomas stretches of c-*myc* as large as 750 bp are missing on both translocated

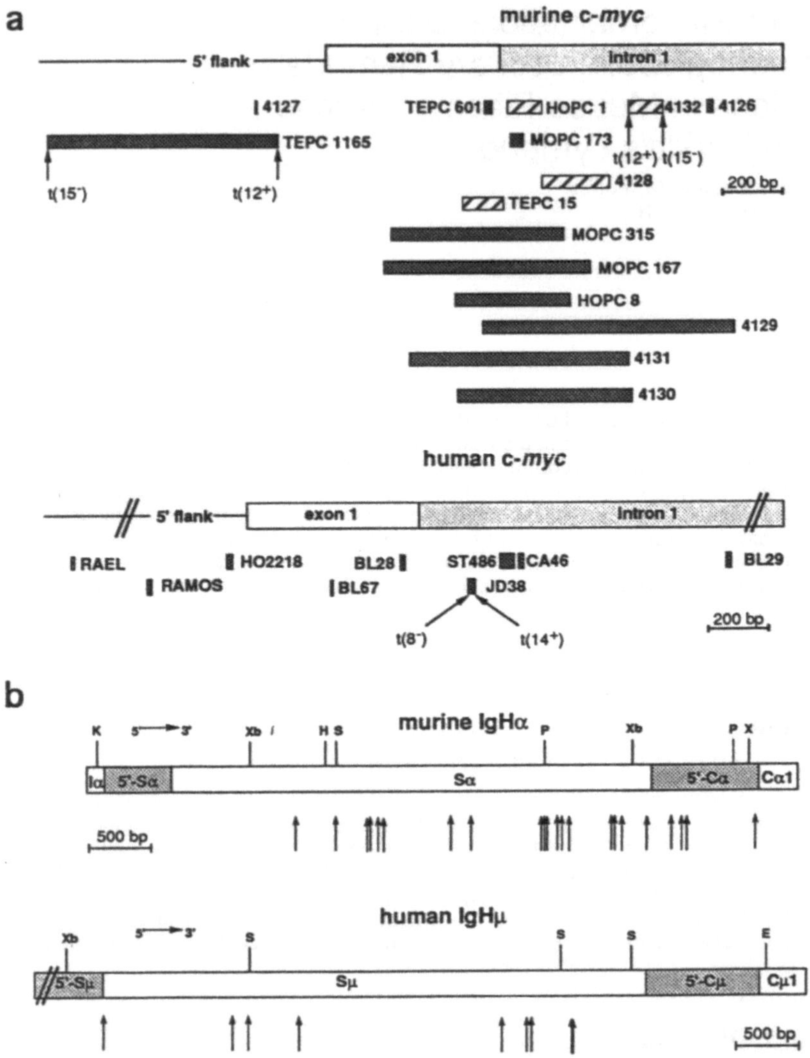

Fig. 1. Differences between IgH/c-*myc* recombinations in mouse plasmacytomas and in Burkitt's lymphomas. a, The comparison of the breakpoints in both translocated chromosomes in plasmacytomas reveals regions in c-*myc* that have been deleted (grey bars) or duplicated (striped bars). In Burkitt's lymphoma, in contrast, break points are always closely located. b, IgH recombination sites in BALB/c plasmacytomas and in Burkitt's lymphomas on the c-*myc* activating chromosomes.

chromosomes and in some cases c-*myc* DNA is duplicated. In addition, the recombination sites in the immunoglobulin locus on the c-*myc*-activating chromosome in Burkitt's lymphomas are exclusively found within the repetitive switch region, whereas some of the plasmacytomas have their breaksites downstream of the switch α region (Fig. 1b).

Preneoplastic Recombinations in Plasmacytomagenesis Are Nearly Reciprocal
To determine whether the differences between plasmacytomas and Burkitt's lymphomas are caused by different recombination mechanisms or by secondary alterations of originally nearly reciprocal exchanges in mouse plasmacytomas, we studied preneoplastic stages of pristane-induced plasmacytoma development. We analysed preneoplastic B cells residing in the oil granuloma one month after treatment with pristane which is about 3 to 4 months before the appearence of tumors. Recombination structures found in preneoplastic B cells revealed features similiar to those in Burkitt's lymphomas. The length of missing c-*myc* sequences appears to be significantly smaller than in plasmacytomas and is limited to less than 110 bp (Fig. 2). Two of the illustrated cases (OG10 and OG12) show almost perfect reciprocal exchanges. In addition, recombination sites in the immunoglobulin locus are found as in Burkitt's lymphomas in switch regions only.

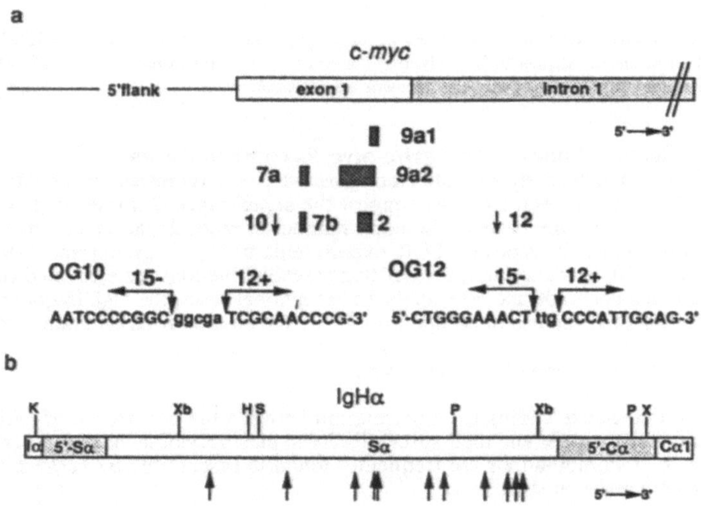

Fig. 2. IgH/c-*myc* recombinations in preneoplastic lesions in pristane-treated BALB/c mice. a, Length of the deletions in the c-*myc* locus. b, Breaksites in the IgHα locus

These data suggest that original recombinations in plasmacytomas are remodeled by secondary deletions. The remodeling hypothesis allows one to predict the occurrence of multiple clones in the same mesentery. In the case of an early tumor (Fig.3) the two amplified chromosome 12+ structures are characterized by an identical Sμ/c-*myc* recombination. However, a 158-bp deletion of switch sequences was found in only one of these indicating that one recombination structure was remodeled by a deletion confined to the Sμ region.

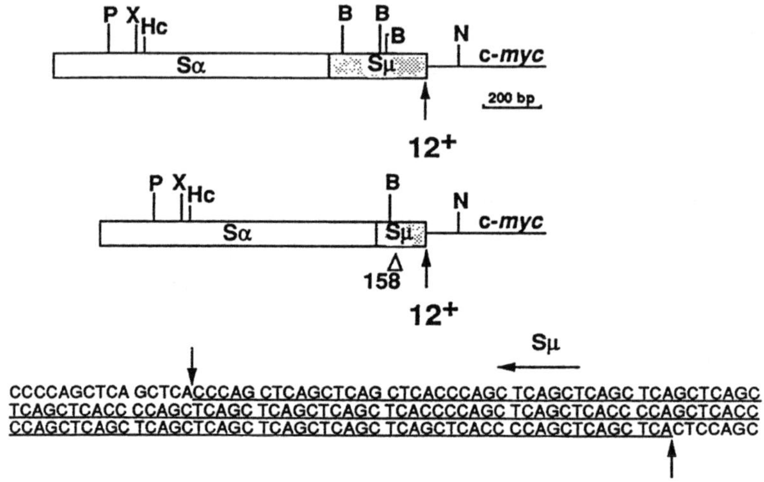

Fig. 3. Related Sμ/c-*myc* recombination-positive subclones in an early primary plasmacytoma. The 158-bp deletion (between the vertical arrows in the sequence shown below) leaves the c-*myc*/Sμ recombination unaffected and allows the unambigious assignment of both fragments to the same translocation event.

Significance of Preneoplastic Clones with IgH/c-*myc* Recombinations

We do not have direct evidence that the early recombination-positive clones represent the precursor cells of plasmacytomas. However, several findings suggest the significance of these cells. We estimated the size of these clones on the basis of the copy number of recombination structures. Based on the amplification success in repeated PCR experiments using oil granuloma DNA compared to the success rate with low copy numbers of tumor recombinations we estimated the majority of clones to contain between 200 and 2000 cells. In exceptional cases, the high frequency of recombination structures allowed the use of competitive PCR. We co-amplified oil granuloma DNA with known numbers of tumor recombinations at different ratios resulting in estimations of up to several thousand cells per mesentery (data not shown).

We have also studied different mouse strains for a correlation between the occurrence of early IgH/c-*myc* recombination-positive clones and their susceptibility to plasmacytoma induction (Tbl. 1). Using conventional PCR, recombinations are frequently found in susceptible BALB/c mice and occur only exceptionally in resistant strains.

Table 1. Comparison of the frequency of preneoplastic B cell clones positive for IgH/c-*myc* recombinations, as detected by PCR in mesenteric oil granulomas 30 days post-pristane with the susceptibility to plasmacytoma development in different strains of mice

Strain	Mice positive for IgH/c-*myc* recombinations at day 30	Incidence of plasmacytomas
BALB/c	37 % (24/64)	30-60 %
C.D2-MIA	10 % (2/20)	<10 %
CDF1	5 % (1/20)	<2 %
DBA	0 %	0 %

In summary, we conclude that IgH/c-*myc* recombination bearing cells are expanded at an early stage of plasmacytomagenesis in susceptible mice strains. The initial IgH/c-*myc* recombinations in mouse plasmacytoma precursor cells are subject to secondary deletions. In contrast, during the development of Burkitt's lymphomas, the original translocation structure is preserved in tumor cells.

References

1. Potter M, Wiener F (1992) Plasmacytomagenesis in mice: model of neoplastic development dependent upon chromosomal translocations. Carcinogenesis 13:1681-1697
2. Janz S, Müller J, Shaughnessy J, Potter M (1993) Detection of recombinations between c-*myc* and immunoglobulin switch α in murine plasma cell tumors and preneoplastic lesions by polymerase chain reaction. Proc Natl Acad Sci USA 90: 7361-7365
3. Dalla-Favera R, Bregni M, Erikson J, Patterson D, Gallo RC, Croce CM (1982) Assignment of the human c-*myc* oncogene to the region of chromosome 8 which is translocated in Burkitt's lymphoma cells. Proc Natl Acad Sci USA 79: 7824-7827

Sγ3 SNIP and SNAP Binding Motifs are Occupied in vivo in Mitogen-Activated I.29μ + Cells

Amy L. Kenter and Robert Wuerffel

Department of Microbiology and Immunology
University of Illinois College of Medicine at Chicago
835 South Wolcott Avenue, Chicago, IL. 60680
Telephone (312) 996-5293
FAX (312) 996-6415

1. Introduction

The Ig heavy (H) chain class switch, permits the expression of a variable region with new constant regions associated with various effector functions (for review see [5,10,14]). The IgH chain class switch occurs by a DNA rearrangement which brings one of seven downstream C_H genes [5] near to the mature V gene replacing the Cμ with another of the C_H genes [5]. The recombination event focuses on the switch (S) DNA, regions of repetitive sequence upstream of each C_H gene, (with the exception of Cδ) and produces a new hybrid DNA combination with the concomitant deletion of the intervening genomic material [5,10,14]. The looping-out and deletion model for switch recombination predicts that the intervening DNA between S regions will be excised as a circle. Circular excision products of Ig switch recombination have recently been isolated from LPS stimulated spleen cells [7,11,20].

Since switch recombination is clearly focused on switch regions we hypothesized that some DNA-binding protein factor(s) might be involved in specifically recognizing and facilitating the alignment of switch regions prior to recombination. We have identified a DNA-binding protein complex which specifically recognizes the Sμ tandem repeat and have termed the binding protein SNUP, for switch nuclear μ protein [23]. The kinetics of expression of SNUP in splenic B cells treated with LPS/DxS parallels the induction of recombinational activity at Sμ in these cells ([23] and unpublished data). Two DNA-binding proteins which specifically and individually interact with two discrete regions of the Sγ3 tandem repeat have been identified in crude and partially purified nuclear extracts derived from LPS/DxS activated splenic B cells [21]. One binding complex is defined as unique and specific for the Sγ3 A site by methylation interference assays and competition-binding analyses and is refered to as switch nuclear A protein or SNAP [21]. The second binding protein has been found indistinguishable from NF-κB p50 homodimer by mobility shift assays, methylation interference, competition-binding studies and supershift analysis using an antiserum specific for the p50 component [1,21]. NF-κB is found to bind to the B site in the Sγ3 tandem repeat and is refered to as switch nuclear protein or SNIP/NF-κB [21]. We have observed that the SNIP and SNAP binding sites are conserved in Sγ2b and Sγ1 DNA. SNAP was found to specifically bind to Sγ2b and Sγ1 by mobility shift and competition binding studies [9]. SNIP/NF-κB was found to interact with its cognate sites at Sγ2b and Sγ1 B-sites as determined by mobility shift assays, competition-binding studies and supershift analysis using an antiserum specific for the p50 component [9].

It is important to ascertain the functional significance of SNIP/NF-κB and SNAP in switch recombination. To approach this question the known switch recombination breakpoints for Sγ3 were examined. Interestingly, 19/22 recombination breakpoints fell within the region spanning the SNIP and SNAP binding sites [21,22]. This analysis was then extended to Sγ2b and Sγ1. Distinctions in switch DNA substrate usage were found when primary ($\mu \rightarrow \gamma$) and successive ($\gamma \rightarrow x$) switch recombination events were considered. In primary ($\mu \rightarrow \gamma$) switch events, all three Sγ switch regions showed focusing of breakpoints generally to the region spanning the SNIP/SNAP binding sites (region I) whereas the long spacer (region II) displayed a relative absence of breakpoints. When the secondary events were subdivided into secondary donor and acceptor recombination substrates a distinctive pattern emerged. The secondary donor breakpoints clustered in region I in proportions identical with those found for primary ($\mu \rightarrow \gamma$) recombination breakpoints. In marked contrast, the secondary acceptor breakpoints are located in

Sγ region II, the long spacer. The use of reciprocal domains to contribute DNA substrates to the recombination transaction suggests that switch recombination is an orderly process.

Our *in vitro* analysis of Sγ3 switch DNA suggests that this switch region may assume an orderly structure dictated by occupancy of an array of multiply interdigitated elements that function as binding sites for SNIP and SNAP. The task of determining the physiological relevance of these recognition motifs is complicated. The functional contribution of SNIP and SNAP binding sites to switch recombination cannot be directly assessed since there is no *in vitro* assay for this event. Generally, functional protein:DNA binding sites are not easily discriminated from nonfunctional sites by primary sequence alone. This problem is especially acute in comparison of individual SNIP and SNAP binding sites throughout Sγ3 since there is approximately a 12% sequence variation in these sites among the tandem repeats. *In vivo* footprinting studies can circumvent some of these problems by providing information as to whether SNIP and SNAP binding sites are engaged and then when and how these recognition motifs are occupied.

2. Experimental Results

The Sγ switch regions, Sγ3, Sγ1, Sγ2b and Sγ2a are composed of regular 49 to 52bp unit repeats which are organized in a head to tail fashion [5]. The Sγ3 switch region contains the basic 49bp unit which is tandemly repeated 44 times [19]. Using the *in vivo* footprinting method there are two questions we propose to address. Are the SNIP and SNAP binding motifs occupied by protein? Since there is some sequence variation from unit repeat to repeat, are all tandem repeats equally engaged?

I.29μ+ lymphoma cells initially express only IgM but will upon LPS induction undergo switch recombination *in vitro* to express IgA [16,17]. When the tumor cells are passaged *in vivo* (in the mouse) switching to IgG3 is also observed [16,17]. In I.29μ+ cells the Sγ3 region is hypomethylated and when stimulated with LPS the Sγ3 germline transcript is expressed [15-17]. We have found that in I.29μ+ cells SNIP is constitutively expressed while SNAP is induced by LPS (unpublished data, A.L.K.). This suggests that it is reasonable to study the Sγ3 region in I.29μ+ cells by *in vivo* footprinting since SNIP is continuously expressed while SNAP and Sγ3 germline transcripts are mitogen inducible.

To determine the extent of occupancy of SNIP and SNAP binding sites in Sγ3 DNA we used the ligation-mediated polymerase chain reaction (LMPCR) protocol developed by Mueller and Wold [13]. This procedure results in the exponential amplification of all the fragments represented in the footprint sequence ladder with conservation of single nucleotide resolution. First, the cells under investigation are treated with dimethyl sulfate (DMS). Since DMS readily permeates cellular membranes and has a high reactivity with DNA without introducing drastic structural changes, it is an excellent reagent for true *in vivo* footprinting. DMS methylates the N7 of guanines making them susceptible to subsequent cleavage with piperidine [12]. Nuclear proteins bound at or near individual guanines can either enhance or reduce the frequency of DMS methylation relative to the same residues in naked DNA [4]. Following DMS treatment the genomic DNA is harvested, purified and then heated in the presence of base to cause strand cleavage at all modified G residues [3, 12, 13]. Guanines that were involved in protein binding will not be methylated, thus, not cleaved and consequently not amplified. This will result in a hole in the sequencing ladder.

The DMS footprint of several Sγ3 tandem repeats in I.29μ+ cells grown in the presence or absence of LPS is shown as a representative example (Fig.1). Pairwise comparison of *in vitro* DMS-treated DNAs with *in vivo* DMS-treated samples (from cells grown in the presence or absence of LPS) reveals a series of changes in the profile of DMS reactive G residues, as will be discussed in the sections which follow. DNA preparations from these DMS treatments were assayed in multiple independent experiments. All interactions detected and reported here were highly reproducible.

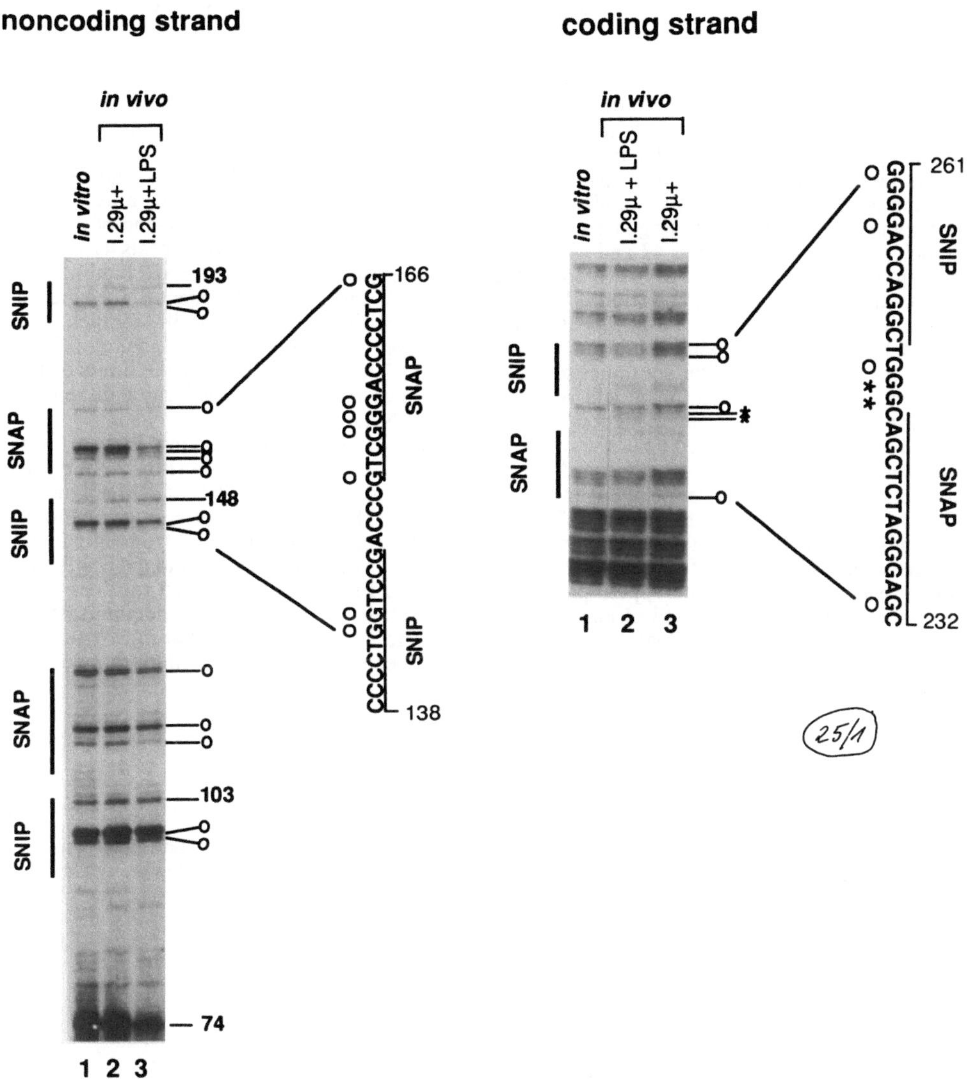

Fig. 1. *In vivo* footprinting of the Sγ3 coding and noncoding strands. SNIP and SNAP binding sites are indicated for individual tandem repeats. (Lanes 1) LMPCR of control (naked) DNA that was methylated with DMS for 2 minutes *in vitro*. (Lanes 2) LMPCR of DNA from I.29μ+ cells treated with DMS for 2 minutes *in vivo*. (Lanes 3) LMPCR of DNA from I.29μ+ cells activated with LPS for 24 hours then treated with DMS for 2 minutes *in vivo*. The DNA sequence of selected tandem repeats is shown in detail. Protected guanines are indicated by open circles while enhanced residues are shown by asterisks. The numbering of the genomic sequence ladders starts with the first residue polymerized onto the labeling primer. On the noncoding strand residues 103, 148, and 193 are located in tandem repeats 2, 3, and 4 respectively. On the coding strand the sequence between positions 232 and 261 is located in tandem repeat 38.

The oligonucleotides used as primers in the LMPCR are complementary to residues on the coding strand at the extreme 3'-end and residues on the noncoding strand at the extreme 5'-end of the Sγ3 region. The sequences of the primers will be reported elsewhere. Using these primers we can visualize tandem repeats 1-6 at the 5'-end of Sγ3 and tandem repeats 36-44 at the 3'-end. The sequence analyzed in these studies is shown in Figures 2 and 3 where the SNIP/SNAP recognition motifs are boxed and the sequences aligned to highlight the reiterative nature of the repeat units. It is noteworthy that tandem repeat 1 contains an unrecognizable SNIP binding site while tandem repeats 2,3 and 4 contain perfect SNIP binding motifs. Tandem repeats 1-4 contain imperfect but clearly recognizable SNAP binding sites while repeat 5 contains a slight permutation in the SNIP binding site and a perfect SNAP binding site. At the 3'-end of the Sγ3 switch region tandem repeats 43 and 44 are essentially unrecognizable while repeats 41 and 42 contain imperfect SNIP and SNAP binding motifs and repeat 40 has a good SNIP binding site but no apparent SNAP binding motif. Tandem repeats 38 and 39 contain intact SNIP binding sites. The SNAP binding site in repeat 39 is imperfect while that in repeat 38 is intact. We routinely obtain high quality footprints up to about 300bp from the end of the labeling primer while the sequencing ladder continues to around 500bp. These are essentially the limits of the technique as described by Mueller and Wold [13]. For these reasons we have focused on repeats 2-5 at the 5'-end and repeats 37-40 at the 3'-end of Sγ3 DNA.

On the coding strand, comparison of the *in vitro* and *in vivo* guanine ladders derived from uninduced I.29μ+ cells shows enhanced G residues at positions 247 and 248 (Fig.1). We find that guanines in symmetrical positions in the short spacer separating the SNIP and SNAP binding sites are enhanced in tandem repeats 37 and 38 (Fig.2). This suggests that a conformational difference exists between the naked and *in vivo* DMS treated samples derived from uninduced I.29μ+ cells. When DNA from LPS induced I.29μ+ cells is analyzed, residues in the SNIP site, short spacer, and SNAP site appear protected (Fig.1 and Fig.2). No protected residues were observed in the long spacer (Fig.1 and Fig.2). Similar results were found in tandem repeats 37, 38 and 40 with some individual differences (Fig.2). These results suggest that in I.29μ+ cells there is a conformational change centered in the short spacer and that following mitogen induction SNIP and SNAP motifs are occupied.

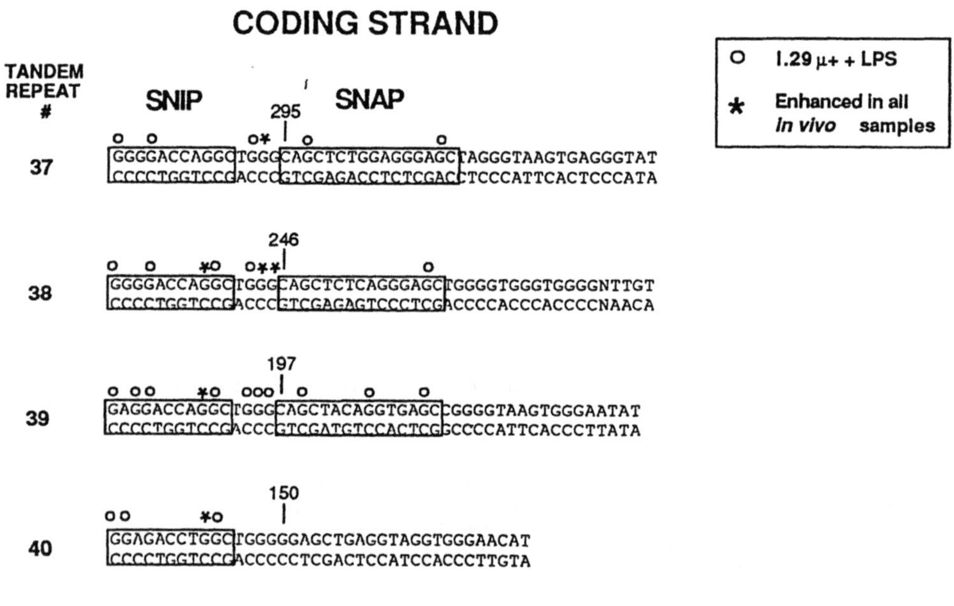

Fig. 2. Summary of the *in vivo* DMS footprints observed on the coding strand at the 3'-end of the Sγ3 region. SNIP and SNAP binding motifs, as previously defined (21), are boxed.

On the noncoding strand, the naked and *in vivo* DMS treated samples from induced and uninduced I.29μ+ cells were analyzed (Fig.1 and Fig.3). Protected guanines were observed only in samples derived from LPS activated I.29μ+ cells. LMPCR reveals strong protections over the three SNIP and two SNAP motifs in this region of Sγ3 DNA. Since all the guanine residues of the noncoding strand are located within SNIP and SNAP sites it is important to confirm the integrity of the genomic sequencing ladder. It is notable that the G residues at the ends of the SNIP sites, at positions 103, 148 and 193, show undiminished intensity. This verifies the protection of the adjacent G residues. The extent of SNIP and SNAP site occupancy appears to increase in the more internally located tandem repeats.

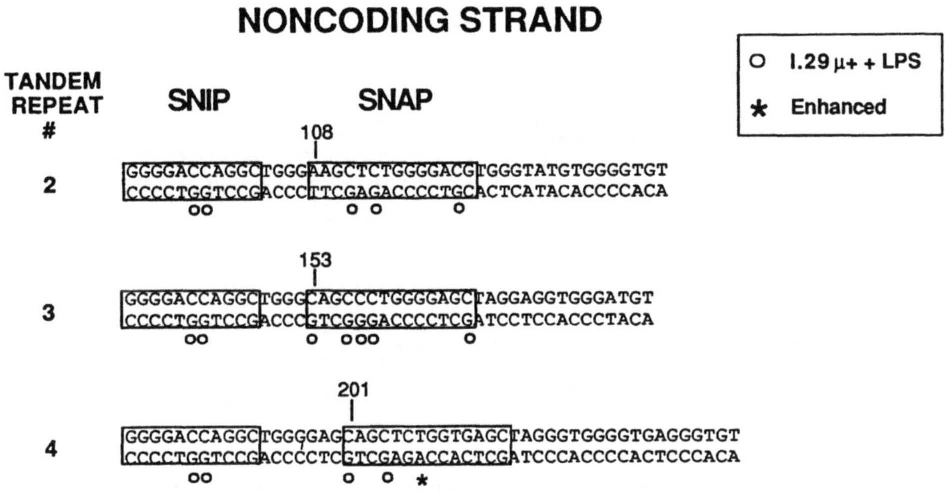

Fig. 3. Summary of the *in vivo* DMS footprints observed on the noncoding strand at the 5'-end of the Sγ3 region. SNIP and SNAP binding motifs, as previously defined (21), are boxed.

The degree of SNIP/SNAP site occupancy appears significantly greater on the noncoding strand as compared to the coding strand. Complete DMS protection will be observed when all copies of a site are occupied continuously. When guanines are partially protected in an *in vivo* footprint a portion of the sites may be completely occupied during the DMS treatment or alternatively, all sites may be partially occupied during the DMS treatment. The LMPCR method requires that primers anneal specifically to a single site in genomic DNA. Analysis of switch regions, containing long stretches of tandem repeats, limits application of LMPCR to the upstream and downstream ends of Sγ3 DNA. The difference in the degree of guanine protection on the coding and noncoding strands is difficult to evaluate since we are analyzing tandem repeats from different ends of the switch region. It may be significant that recombination breakpoints preferentially occur in the first third of Sγ3 [6,19,22]. It appears from our results that greater G residue protection is correlated with biased distribution of Sγ3 recombination joins. This observation may imply a higher order structure of the Sγ3 switch region *in vivo*.

We do not know the identity of the proteins which occupy the SNIP and SNAP binding sites *in vivo*. To begin to approach this question the *in vitro* methylation interference data has been compared with the *in vivo* footprinting results reported here (Fig.4). A summary of affected guanines found in symmetrical positions in tandem repeats is shown (Fig.4). To qualify for inclusion as an affected site a guanine must be found protected or enhanced in at least two tandem repeats and in at least one perfectly representative SNIP and SNAP binding motif. A protected G residue found in two tandem repeats but derived from an imperfect SNIP site, for example, would not be included in the summary. There are fundamental differences in the way the *in vitro* and *in vivo* data sets were obtained which make direct comparison complex. *In vivo* footprinting relies on methylation protection rather than methylation interference and may produce a somewhat different set of protein:DNA contacts. Nonetheless, the DNA contacts generated by these methods should be largely overlapping if the proteins involved are the same in the living cell as those analyzed *in vitro*.

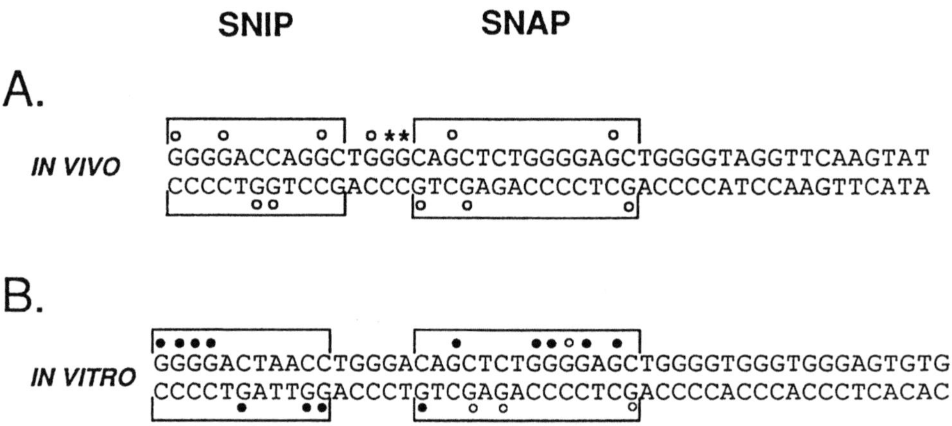

Fig. 4. Comparison of *in vitro* and *in vivo* Sγ3 DMS footprints. A. The *in vivo* (methylation protection) footprint pattern over Sγ3 is a composite where the coding strand information was taken from the extreme 3'-end of Sγ3 while the noncoding strand data was derived from the extreme 5'-end of Sγ3. Guanines which are protected (o) from methylation or for which methylation is enhanced (*) are shown. B. The *in vitro* (methylation interference) footprint over the SNIP and SNAP sites in Sγ3 DNA (26). Guanines for which methylation either strongly (•) or moderately (o) interferes with binding are indicated.

We find that the protein:DNA contacts observed by the *in vivo* footprinting method represent a subset of contacts found by the methylation interference studies with several exceptions (Fig.4). Our *in vitro* studies found no contacts in the short spacer nor on the coding strand at the 3'-end of the SNIP site. The methylation interference studies were performed using partially purified SNIP and SNAP separately. This may account for why no effect was found in the short spacer. The sequence of the genomic Sγ3 DNA probe used for the *in vitro* methylation interference assays differed from the most prevalent SNIP sequence in that it contained an AC on the coding strand at the 3'-end of the SNIP motif where, most often, GG is seen (compare the sequences in Fig.4, A and B). Thus the protection and enhancement scored at these positions *in vivo* would not have been seen in *in vitro* experiments. The results obtained using the *in vivo* footprinting method are consistent with our *in vitro* studies but do not rule out the possibility that other or additional proteins not identified *in vitro* occupy the region spanning the SNIP/SNAP binding sites. It is striking that no protein:DNA contacts were observed in the long spacer.

3. CONCLUSION

The analysis of DNA-protein interactions and chromatin configuration *in vivo* can provide insights into the molecular mechanisms which regulate tissue specific and stage specific gene expression. Several recent studies have demonstrated differences in DNA-protein interactions *in vivo* as compared to protein binding *in vitro* [2,3,8,13,18]. *In vitro* analysis showed that factors which specifically bind to the E boxes of the immunoglobulin heavy chain enhancer are present in many cell types [3]. However, *in vivo* footprinting revealed that the E box motifs were occupied only in B cells [3]. A similar observation has been made for a regulatory element of the human Igκ gene 3' enhancer which is engaged *in vivo* only in B cells even though nuclear extracts from T cells have been shown to contain the specific binding protein [8]. In these cases, cell type-specific binding of ubiquitous factors was apparent only by *in vivo* footprinting. These results suggest that in living cells, the binding of trans-acting factors to cis-regulatory elements is controlled not only by DNA sequence specificity but also by complex epigenetic factors, such as local chromatin structure, which are not easily reconstituted *in vitro*.

The expression of different IgH isotypes during B cell differentiation is mediated by the process of switch recombination. By *in vitro* binding and footprinting assays we have shown that switch specific binding proteins are induced in splenic B cells by stimuli that cause isotype switching to occur [9,21-23]. We have also found that the positions of switch recombination breakpoints are strongly correlated with the recognition motifs for switch DNA binding proteins [9,21,22]. In the absence of a functional *in vitro* assay for switch recombination it is difficult to assess the physiological relevance of these proteins and their binding sites. To determine whether SNIP and SNAP binding motifs are important during switch recombination we have used the LMPCR protocol to ask whether SNIP and SNAP binding sites are occupied by protein in I.29μ+ cells. In these cells, SNIP and SNAP binding sites were found occupied by protein following mitogen induction but were not occupied in unstimulated I.29μ+ cells. These studies confirm the importance of the SNIP and SNAP binding motifs. It is likely that Sγ3 DNA assumes an orderly structure during switch recombination since protein occupancy occurs at multiple interdigitated binding sites which form an array along the Sγ3 region.

ACKNOWLEDGMENTS

This work is supported by the National Institutes of Health grant GM39231, Council for Tobacco Research Award 3175 and ACS IM-729 to A.L.K.

REFERENCES

1. Baeuerle PA, Baltimore D (1990) The physiology of the NF-κB transcription factor. In: Cohen P, Foulkes JG (eds) Hormonal regulation of transcription Elsevier-Biomedical, Amsterdam (Molecular Aspects of Cellular Regulation, vol 6)
2. Becker PB, Ruppert S, Schutz G (1987) Genomic footprinting reveals cell-type specific DNA binding of ubiquitous factors. Cell 51:435-443
3. Ephrussi A, Church GM, Tonegawa S, Gilbert W (1985) B-lineage specific interactions of an immunoglobulin enhancer with cellular factors *in vivo*. Science 227:134-140
4. Gilbert W, Maxam A, Mirzabekov A (1976) Contacts between the lac repressor and DNA revealed by methylation. In: Kjeldgaard NO, Maaloe OM (eds) Control of ribosome synthesis. Academic Press NY pp 139-148 (Alfred Benzon Symposium IX)
5. Gritzmacher CA (1989) Molecular aspects of heavy-chain class switching. CRC Crit Rev Immunol 9:173-200
6. Iwasato T, Arakawa H, Shimizu A, Honjo T, Yamagishi H (1992) Biased distribution of recombination sites within S regions upon immunoglobulin class switch recombination induced by transforming growth factor β and lipopolysaccharide. J Exp Med 175:1539-1546

7. Iwasato T, Shimizu A, Honjo T, Yamagishi H (1990) Circular DNA is excised by immunoglobulin class switch recombination. Cell 62:143-149
8. Judde J-G, Max EE (1992) Characterization of the human immunoglobulin kappa gene 3' enhancer: Functional importance of three motifs that demonstrate B-cell specific *in vivo* footprinting. Mol Cell Biol 12:5206-5216
9. Kenter A, Wuerffel R, Sen R, Jamieson CE, Merkulov G (1993) Switch recombination breakpoints occur at nonrandom positions in the Sγ tandem repeat. J Immunol 151:4718-4731
10. Marcu KB (1982) Immunoglobulin heavy-chain constant-region genes. Cell 29:719-721
11. Matsuoka M, Yoshida K, Maeda T, Usuda S, Sakano H (1990) Switch circular DNA formed in cytokine-treated mouse splenocytes: evidence for intramolecular DNA deletion in immunoglobulin class switching. Cell 62:135-142
12. Maxam AM, Gilbert W (1980) Sequencing end-labeled DNA with base specific chemical cleavages. In: Grossman L, Moldave K (eds) Academic Press N Y pp 499-560 (Methods in enzymology, vol 65)
13. Mueller PR, Wold B (1989) *In vivo* footprinting of a muscle specific enhancer by ligation mediated PCR. Science 246:780-786
14. Shimizu A, Honjo T (1984) Immunoglobulin class switching. Cell 36:801-803
15. Stavnezer J, Radcliffe G, Lin Y-C, Nietupski J, Berggren L, Sitia R, Severinson E (1988) Immunoglobulin heavy-chain switching may be directed by prior induction of transcripts from constant region genes Proc Natl Acad Sci 85:7704-7708
16. Stavnezer J, Sirlin S, Abbott J (1985) Induction of immunoglobulin isotype switching in cultured I.29 B lymphoma cells: Characterization of the accompanying rearrangements of heavy chain genes. J Exp Med 161:577-601
17. Stavnezer-Nordgren J, Sirlin S (1986) Specificity of immunoglobulin heavy-chain switch correlates with activity of germline heavy chain genes prior to switching. EMBO J 5:95-102
18. Strauss EC, Andrews NC, Higgs DR, Orkin SH (1992) *In vivo* footprinting of the human α-globin locus upstream regulatory element by guanine and adenine ligation-mediated polymerase chain reaction. Mol Cell Biol 12:2153-2142
19. Szurek P, Petrini J, Dunnick W (1985) Complete nucleotide sequence of the murine γ3 switch region and analysis of switch recombination in two γ3 expressing hybridomas. J Immunol 135:620-626
20. vonSchwedler U, Jack H-M, Wabl M (1990) Circular DNA is a product of immunoglobulin class switch rearrangement. Nature 345:452-455
21. Wuerffel R, Jamieson CE, Morgan L, Merkulov GV, Sen R, Kenter AL (1992) Switch recombination breakpoints are strictly correlated with DNA recognition motifs for immunoglobulin Sγ3 DNA-binding proteins. J Exp Med 176:339-349
22. Wuerffel R, Kenter AL (1992) Protein recognition motifs of Sγ3 DNA are statistically correlated with switch recombination breakpoints. Curr Top in Micro and Immunol 182:149-156
23. Wuerffel RA, Nathan AT, Kenter AL (1990) Detection of an immunoglobulin switch region-specific DNA-binding protein in mitogen stimulated mouse splenic B cells. Mol Cell Biol 10:1714-1718

Chromosomally Integrated Retroviral Substrates are Sensitive Indicators of an Antibody Class Switch Recombinase-Like Activity

J. Ballantyne[1], L. Ozsvath[2], K. Bondarchuk[2,*] and K. B. Marcu[1-4]
[1]Genetics Graduate Program and Departments of [2]Biochemistry and Cell Biology, [3]Microbiology and [4]Pathology, State University of New York, Stony Brook, NY11794, USA

*Present address: New York State Department of Health, Albany, NY 12143, USA

Introduction

The main function of B lymphoid cells is to produce antibodies against an almost infinite variety of antigens which might be harmful to the organism. B cells comprise a clonally diverse population in which each cell has a mono-specific antibody receptor bound to its surface. A foreign antigen encounters its cognate antibody receptor bound to a B cell and, with the cooperation of helper T cells, causes the clonal proliferation and subsequent differentiation of this sub-set of B cells into antibody-secreting plasma cells [1].

Both the heavy (H) and light (L) immunoglobulin polypeptide chains possess a variable (V) region, which is responsible for the binding specificity of the antibody, and a constant (C) region, which is responsible for effector functions facilitating antigen-antibody complex inactivation and/or removal from the body. A unique feature of B cell ontogeny is the creation of a functional antibody molecule by somatic recombination of multiple non-contiguous germ-line gene segments at both the V and C gene loci. In the case of the H chain, two distinct types of somatic DNA rearrangements take place. The first wave is a developmentally regulated, site-specific recombination which fuses V (variable), D (diversity) and J (joining) segments to create a functional VDJ gene [2]. The second type of somatic rearrangement is the principal mediator of antibody class (or isotype) switching, by which antibodies of a given specificity acquire different effector functions (for review see [3]). This typically occurs during the secondary immune response which effectively results in the transfer of effector function from IgM to IgG, IgA or IgE antibodies. An illegitimate non-homologous DNA rearrangement event juxtaposes a functional VDJ gene (initially 5' of the Cμ gene in an IgM-bearing B cell) in close proximity to a downstream CH gene. This occurs by a deletional mechanism resulting in the loss of all the intervening CH genes [4, 5, 6].

The murine CH locus consists of eight functional genes which are arranged as 5'-Cμ-Cδ-Cγ3-Cγ1-Cγ2b-Cγ2a-Cε-Cα-3' at the telomeric end of chromosome 12 [7]. Recombinagenic switch (S) sequences are positioned 1-4kb upstream of each of these functional CH gene segments with the exception of C_δ [3]. Switch recombinations within or nearby these S sequences have been documented in a large number of hybridomas, plasmacytomas and A-MuLV-transformed pre-B lines [8-11]. S sequences are composed of direct tandem repeats of a basic core structure, which often contains the pentamers GAGCT and GGGGT although other repeats such as TGAGC, GAGCTG (Sμ, Sε and Sα) and GCAGC, ACCAG (Sγ) are also prevalent [3]. S regions vary in length and repeat unit sequence and display different degrees of homology with one another [3]. The hierarchy of homology is Sμ > Sε > Sα > Sγ3 > Sγ1 > Sγ2b > Sγ2a with Sμ showing a considerable degree of homology with Sε and Sα and being least homologous with Sγ2b and Sγ2a [3]. Varying degrees of divergence from the consensus sequences occurs at the 5' and 3' ends of each switch sequence [3].

Switching always originates within the Sμ region which consists primarily of dense clusters of two pentameric units $(GAGCT)_n$.GGGGT, where n=3 but can range from 1 to 7 [8-11]. A heptameric sequence YAGGTTG is found clustered 5' to murine Sμ and may also represent a switch recombination signal sequence [3]. The Sγ regions are moderately homologous to one another and comprise direct repeats of a 49bp (Sγ3, Sγ1, Sγ2b) or 52bp (Sγ2a) consensus sequence [3]. These direct repeats themselves comprise two homologous 24bp or 26bp sequences. Higher order structures involving long direct repeats unrelated to the 49bp structure are found in the Sγ1 and Sγ2b regions [3].

Switching can be regulated *in vivo* to yield a specific or limited set of isotypes [12]. Several T cell-derived cytokines have been demonstrated to be important regulators of class switching [12]. Directed class switching may be manifested by the differential transcriptional competence of the CH genes. Prior to class switch recombination 'sterile' or so-called germline transcripts encompassing a short I exon (residing 5' to the S sequences) spliced to an adjacent CH region are expressed in a regulated inducible fashion [13]. This transcribed status is believed to engender a particular CH gene with the 'accessibility' to undergo switch recombination [14]. The lack of sterile CH transcripts is correlated, both *in vitro* and *in vivo*, with the absence of class switch recombination to that CH gene [15].

We have previously provided direct evidence that S sequences themselves are a necessary and sufficient substrate for mediating switch recombination in B lineage cells, provided that the switch sequences are transcriptionally competent and therefore presumably accessible [16-18]. Using a selectable retroviral delivery system, a switch deletion substrate was introduced into a limited number of murine pre-B, mature B and fibroblast cell lines to assay for switch recombinase activity. The S retrovector comprised a switch substrate cassette consisting of a *Herpes simplex* thymidine kinase gene (*HSVtk*) flanked by genomic switch fragments from the murine Sμ and Sγ2b regions as well as a neomycin (*neo*) resistance gene to score for viral integration. The Sμ genomic fragment comprised about 265bp and 890bp of non-repetitive and repetitive S sequences, respectively, whereas the Sγ2b fragment consisted of 670bp of 49bp tandem repeats and 580bp of 3' flanking sequence. Sμ was chosen since it normally acts as the donor sequence in switch recombination whereas Sγ2b was chosen because of its general lack of Sμ homology which should reduce background homologous recombination. Transcriptional competence, and hence accessibility, of the switch substrate would be provided by an Mo-MuLV LTR. Stable integration of the S retrovector provirus was obtained by selection for neomycin resistance in geneticin. Subsequent selection in bromodeoxyuridine (BUdR) media with endogenous *tk-* cells, yielded neo^R clones which had lost their *HSVtk* phenotype. Several pre-B cell lines exhibited *HSVtk* loss by virtue of Sμ-->Sγ2b switch deletion events. This switch-recombinase like activity was not found in fibroblasts nor in a terminally differentiated antibody secreting hybridoma [16-18]. Both Sμ and Sγ2b sequences were required since Sμ alone would not suffice. In some pre-B cells, but not all, inter-S segment recombination within the retrovector was not accompanied by endogenous Sμ to Sγ2b switching indicating that the CH loci were subject to some type of negative control possibly related to their 'accessibility' status. Sequence analysis of several S substrate retrovector recombinants revealed S region pentamers nearby the breakpoints but there was no evidence of a site-specific consensus sequence.

It is unclear how the tandem repetitive S regions mediate switching and how switching is regulated at the molecular level. It may be that these sequences represent a structural motif which is directly recognized by the DNA binding domain of a general switch recombinase or there may even be a class of isotype-specific switch recombinases whose substrate activities are under independent control [19]. Alternatively, or in addition, S region binding proteins may act as accessory molecules to recruit the recombination machinery by specific protein-protein interactions. Several nuclear proteins have been identified which bind to S regions but their precise function(s) are unknown [20-22].

The present work extends and refines the use of chromosomally integrated retroviral switch substrates as a means to detect switch recombinase-like activity. We have extended the host range of the original retrovector by means of an amphotropic packaging line, prepared a new series of retrovectors with novel features and developed a sensitive direct PCR assay for the detection of recombination. The results indicate that this putative switch recombinase activity appears to be B cell type specific and is present to varying degrees throughout B cell differentiation from the pre-B to the plasma cell stages. Moreover murine switch sequences apparently serve as targets for a human switch recombinase activity. We also demonstrate the efficacy of the assay for the detection of the unrearranged parental retrovector in transient assays which should facilitate the detection of recombination in primary cells.

Results

Retroviral Vector Switch Substrates and Recombination Assays

Three retrovectors were used (see Fig.1) : ZN(Sμ/Sγ2b)tk1 was first described by Ott et al. [16]. LNL(Sμ/Sγ2b)Hytk and LNSL(Sμ/Sγ2b)Hytk were derivatized from the LNSL series of retrovectors [23] and have several improvements over the first substrate vector including: (i) higher titre, (ii) a chimeric *Hytk* gene [24] and (iii) more Sμ tandem repeat sequences (1540bp compared to 890bp).

Fig.1 Retroviral Vector Substrates

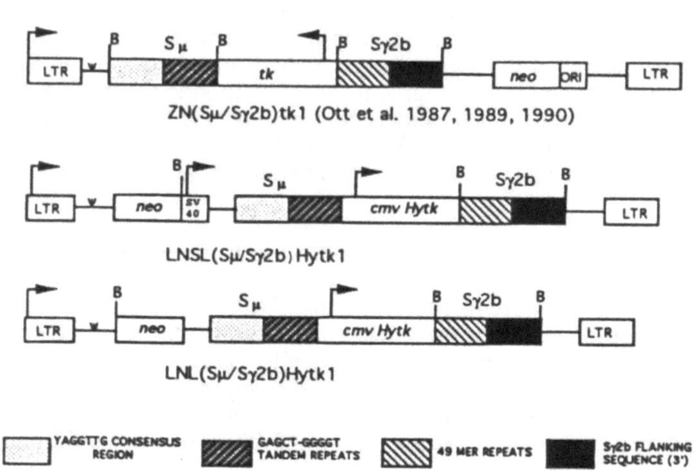

To measure the recombination status of the chromosomally integrated retrovector a sensitive direct PCR analysis scheme was developed (see Fig. 2).

Fig.2 Direct PCR Analysis For S-S Mediated Recombination Status

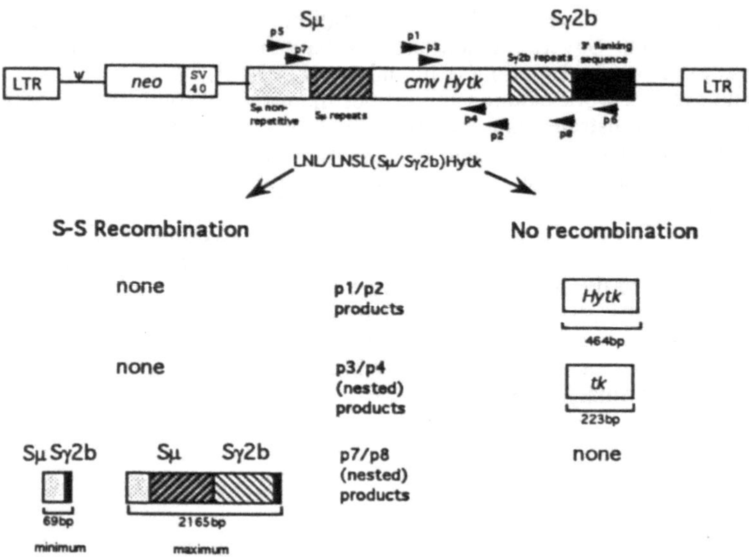

In the absence of S-S recombination *Hytk* remains intact, whereas recombination results in its loss. The presence of *Hytk* is detected by a 464bp PCR product formed as a result of using two primers, p1 and p2, which span the *hygro* and *HSVtk* junction of the chimeric *Hytk* gene. If necessary, increased sensitivity is obtained by the use of an additional round of PCR using primers p3 and p4 which produce a 223bp nested *HSVtk* fragment.

If S-S mediated recombination occurs between the Sμ and Sγ2b tandem repeats of the retrovector a range of possible recombinants can be formed depending upon the location of the recombination breakpoints within each switch sequence. The use of a nested PCR analysis employing primers p5/p6 and then p7/p8 (which have been designed to span only the tandem repeats of Sμ and Sγ2b) enables the detection of recombinant fragments ranging in size from 69bp to 2165bp. The unrearranged parental retrovector would be expected to produce a 5kb PCR product but is undetectable with the particular PCR conditions employed. The *Hytk* and Sμ/Sγ2b PCR products are all detected in ethidium bromide-stained polyacrylamide gels.

Analysis of Individual Clones of Switch Retrovector-Infected Cells

Cell lines representing different developmental stages of the T and B lymphoid lineages were infected with the ZN(Sμ/Sγ2b)tk1 retrovector and grown in the presence of geneticin to select for proviral integrants. Clones of geneticin resistant cells were expanded to ~2x107 cells, high molecular weight genomic DNA was isolated and digested with Bam H1, separated in 0.8% agarose gels and transferred to a nylon membrane. The DNA was sequentially probed for *HSVtk*, Sμ and Sγ2b sequences. A specific Sμ-Sγ2b deletion would be detected as an

Sμ/Sγ2b positive band even if only 10% of the integrated proviruses had undergone an inter-S segment recombination.

Results obtained with ten representative clones of EL-4 (mature T) (ATCC), *bcl-1* B1 (mature B) [25] and five clones of KK125 (human Burkitt lymphoma) (ATCC) are shown in Fig.3. The monoclonality of each isolate was confirmed by novel *neo* bands in Southern hybridizations (data not shown).

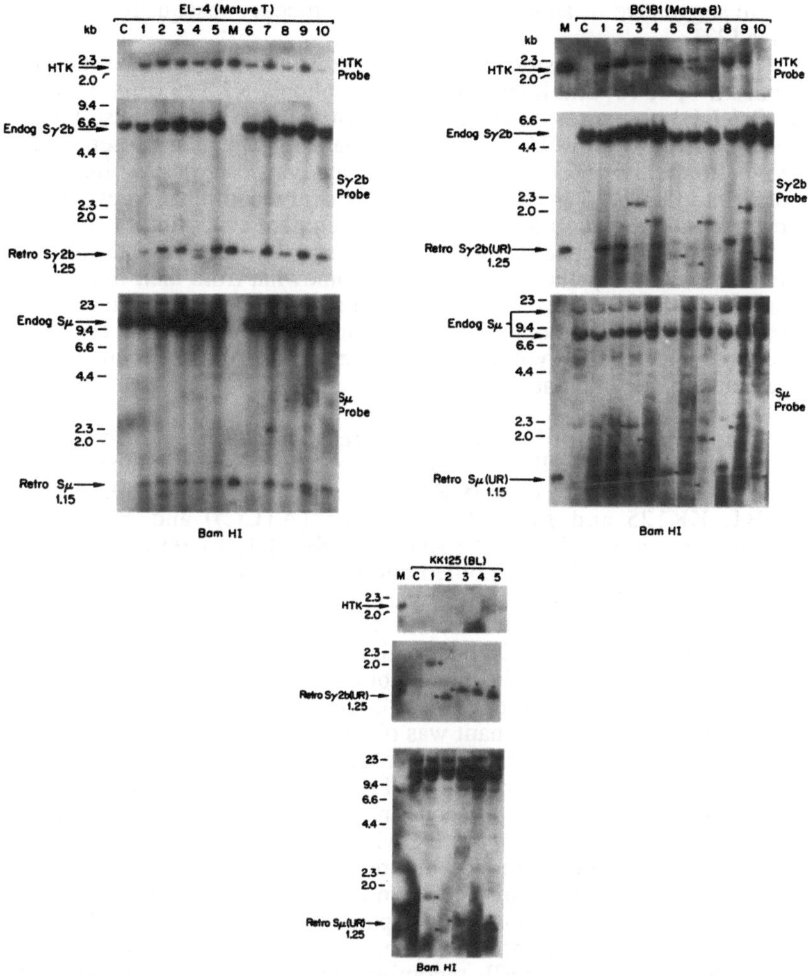

Fig.3. Analysis of Individual Clones of ZN(Sμ/Sγ2b)tk1 Infected Cells.
(a) EL-4 (b) *bcl-1* B1 (c) KK125. Southern blots of BamH 1 digested DNA with, in each case, *HSVtk* (HTK) (upper panel), Sγ2b (middle panel) and Sμ (lower panel) DNA probes. The migration positions of endogenous (cellular) and exogenous (proviral) switch segments are indicated. Unrearranged (UR) proviral (Retro) Sγ2b and Sμ BamH 1 fragments are 1.25kb and 1.15kb. Rearranged S segment bands corresponding to Sμ/Sγ2b fusions are indicated by small arrowheads. M and C lanes contain BamH 1 digested DNA from the retrovector plasmid and the uninfected cell line respectively.

With respect to the EL-4 T cell line no instance of Sμ-Sγ2b mediated recombination was observed in any of the clones. However several instances of intra-S segment rearrangements were observed with both Sμ (Fig. 3a, clones 2,9) and Sγ2b (Fig. 3a, clone 4) sequences undergoing deletional rearrangement. Similar results were observed in the M14T pre-T line [26] (data not shown). These findings suggest the presence of a homologous recombination-like activity in T cells. However, internal S deletions were not observed in fibroblasts [16]. Intra-S segment increases in length were also occasionally observed (data not shown) which are suggestive of an error prone repair process subsequent to intra-S strand breakage [27].

Unique Sμ-Sγ2b fusion events were detected in the *bcl-1* B1 mature B (Fig. 3b, clones 3-7, 9, 10) and human Burkitt lymphoma KK125 (Fig. 3c, clones 1-5) lines. *Bcl-1* B1 co-expresses cell surface IgM and IgG1 and the isotype switch is believed to occur by an RNA mediated mechanism [25]. The presence of Sμ/Sγ2b recombination activity (confirmed by PCR analysis, see below) implies that the commitment to undergo an endogenous S segment recombination event could have molecular requirements in addition to gene segment accessibility (transcriptional competence in this instance) and recombinase activity. The KK125 results indicate that murine switch sequences, which differ somewhat from their human analogues [28], appear to be recognized by the human recombination machinery. Human S sequences have been demonstrated to be functional in a murine context *in vivo*. [29]. This implies an evolutionary conservation of S sequence recognition requirements for recombination.

PCR Analysis of Populations of Switch Retrovector-Infected Cells

A very sensitive indicator of the recombination status of the switch substrate retrovectors can be achieved by PCR analysis. B lineage cells (300-18 pre-B [16], *bcl-1* B1, KK125 and J558 plasmacytoma (ATCC)) and NIH3T3 (ATCC) fibroblasts were infected with the LNSL(Sμ/Sγ2b)Hytk1 retrovector. Clones of geneticin resistant cells were pooled and expanded to ~2x107 cells , DNA was isolated and 100ng subjected to PCR analysis. DNA from the non-infected parental lines served as negative controls. The PCR assays were utilized to detect the presence of *Hytk* and any Sμ/Sγ2b recombinant hybrids .

Multiple different Sμ/Sγ2b recombination hybrids were detected in the 300-18, *bcl-1* B1 and KK125 cell lines (Fig.4, lanes 10, 8 and 6 respectively) whereas only a single Sμ/Sγ2b recombinant was obtained from the J558 plasmacytoma line (Fig. 4, lane 4). No recombination was observed in 3T3 fibroblasts (Fig. 4, lane4) nor in any of the parental uninfected cell line controls (Fig.4, lanes 3, 5, 7, 9, 11). Unlike the other cell lines tested, recombination (albeit with a smaller number of S-S hybrids) was observed in *bcl-1* B1 without recourse to a second nested PCR reaction (data not shown). Indeed this result agreed with the Southern blot data which revealed a predominant S-S recombinant band in the *bcl-1* B1 population accompanied by complete loss of the parental retrovector Sγ2b band (data not shown). In contrast to the pre-B and mature B cells wherein a large number of multiple, independent Sμ/Sγ2b recombination events were observed, the J558 plasma cell tumor line exhibited a single predominant band indicating a lower level of S-S recombinase-like activity in this more differentiated B cell. The intense fluorescent bands at the origins of the KK125, *bcl-1* B1 and 300-18 represent high molecular weight DNA (>23kb), and are probably due to incompletely extended, partially single-stranded Sμ/Sγ2b hybrids.

As expected there was a reciprocal relationship between the extent of recombination and the presence of the *Hytk* gene. The lines exhibiting the most recombination activity, namely 300-18, *bcl-1* B1 and KK125, demonstrated low (300-18) or barely visible (*bcl-1* B1, KK125) *Hytk* signals (Fig.5, lanes 11, 9 and 7, respectively). 3T3 possessed a relativly strong *Hytk* signal (Fig.5, lane3) which is consistent with the absence of detectable recombination in the fibroblast. J558

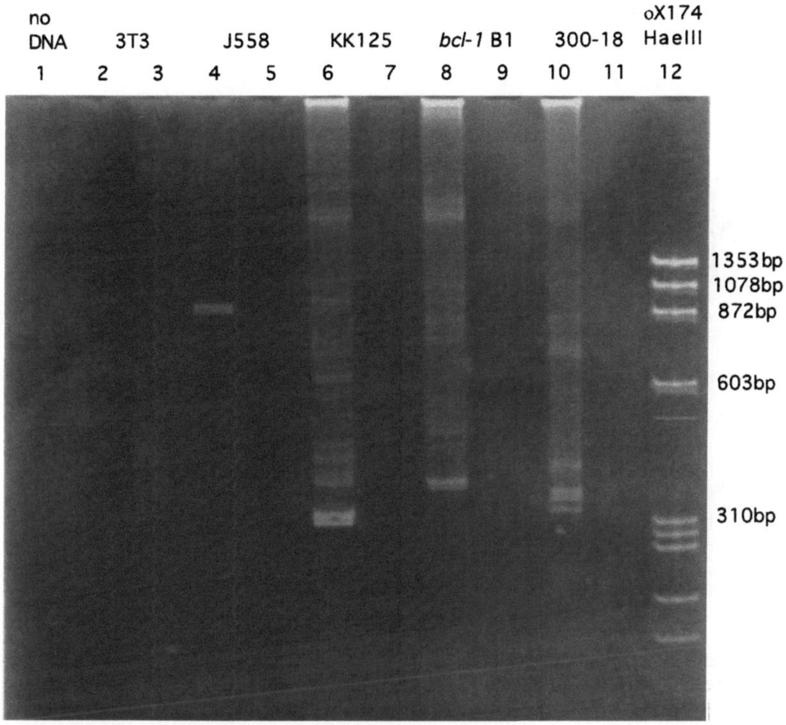

Fig.4. The detection of Sμ/Sγ2b recombination by direct PCR.
Cells were infected with the LNSL(Sμ/Sγ2b)Hytk1 retrovirus. DNA obtained from pooled populations of G418[r] cells were analyzed by PCR for the presence of Sμ/Sγ2b recombinants, as described in the text, and 15% of the reaction products separated on a 5% acrylamide gel. Results are shown for NIH3T3 (lane 2), J558 (lane 4), KK125 (lane6), *bcl-1* B1 (lane8) and 300-18 (lane 10). Amplified DNA from uninfected parental lines are shown in lanes 3, 5, 7, 9 and 11 respectively. Lane 1 is a PCR reagent blank (no DNA added).

demonstrated an intermediate intensity (Fig.5, lane5). As with the Sμ/Sγ2b assay, DNA from each non-infected parental cell line was utilized as a negative control (Fig.5, lanes 4,6,8,10,12).

Transient Assay

The selection and expansion of stable clones of geneticin resistant cells over several weeks precludes the possibility of using this approach for the detection of recombinants in non-transformed, primary cells. Normal splenic lymphocytes can be induced to proliferate and terminally differentiate in response to mitogens and cytokines over a 1-2 week period [30]. The development of a transient assay for the detection of S-S recombination activity within days of retrovector infection would allow the evaluation of the switch recombination potential of a number of primary cells of both the lymphoid and non-lymphoid lineages.

Bcl-1 B1 and 300-18 cells (~5x10^5) were separately infected at a multiplicity of infection (moi) of 1:1 with two different isolates of the LNL(Sμ/Sγ2b)Hytk1 retrovirus. Sixty hours after infection high molecular weight DNA was isolated and 100ng subjected to PCR analysis using nested primers for the *HSVtk* component of the *Hytk* chimeric gene. Fig.6 demonstrates that it was relatively straightforward to detect the presence of the unrearranged retrovectors' *Hytk* gene in both *bcl-1* B1 (lanes 2,3) and 300-18 (lanes 5,6).

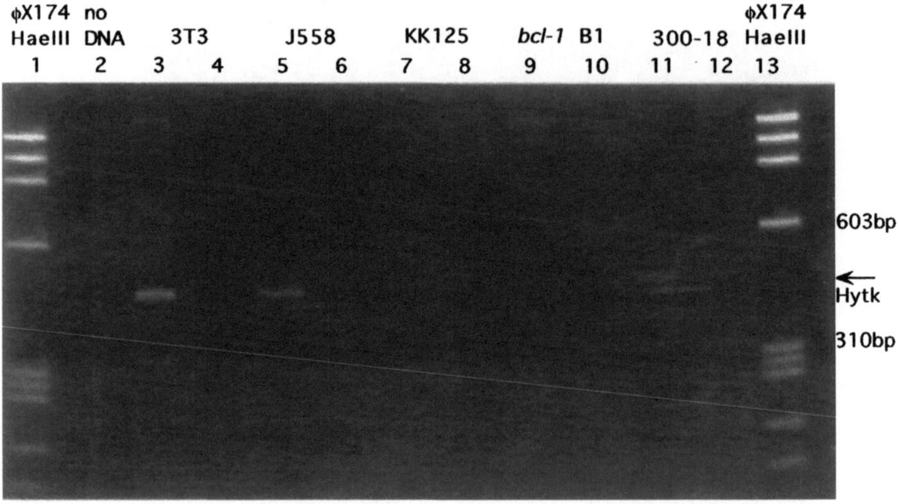

Fig.5. *Hytk* status determination by direct PCR.
Cells were infected with the LNSL(Sμ/Sγ2b)Hytk1 retrovirus. DNA obtained from the same pooled populations of G418[r] cells, as described in the text and Fig.4, were analyzed by PCR for the presence of *Hytk* and 15% of the reaction products separated on a 5% acrylamide gel. The expected position of the *Hytk* PCR product is indicated. Results are shown for NIH3T3 (lane 3), J558 (lane 5), KK125 (lane 7), *bcl-1* B1 (lane 9) and 300-18 (lane 11). Amplified DNA from the uninfected parental lines are shown in lanes 4, 6, 8, 10 and 12 respectively. Lane 2 is a PCR reagent blank (no DNA added).

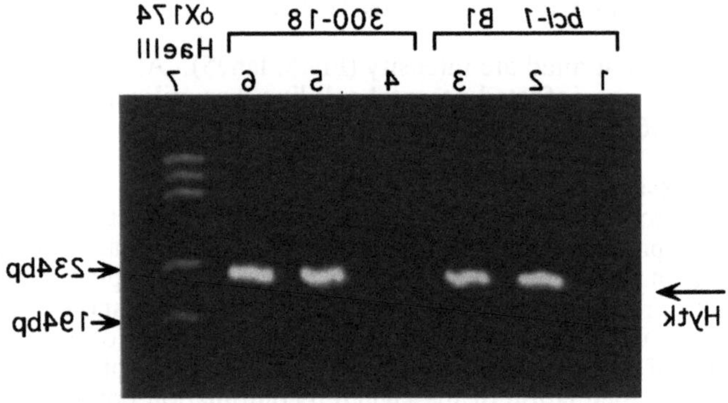

Fig. 6. Sensitive detection of the switch retrovector in transiently infected cells.
Cells were infected with two different isolates of the LNL(Sμ/Sγ2b)Hytk1 retrovirus. 60 hours after infection, DNA was prepared, subjected to nested PCR analysis and 15% of the reaction products separated on a 5% acrylamide gel. The expected position of the nested *Hytk* PCR product is indicated. Results are shown for *bcl-1* B1 (lanes 2, 3) and 300-18 (lanes 5, 6). Amplified DNA from the uninfected parental lines are shown in lanes 1 and 4 respectively.

Conclusions

Portions of the Sμ and Sγ2b regions in a selectable retroviral vector delivery system appear to serve as efficacious substrates for a B cell specific activity with the properties of a class switch recombinase.

Retroviral switch substrates allow for the assessment of switch-recombinase activity on chromosomally integrated molecules in a transient fashion. This should greatly facilitate switch recombinase assays in normal proliferating and differentiating B cells in the future.

Acknowledgements

This work was supported by NIH grant GM26939 awarded to KBM.

References

1. Paul WE (1989) The immune system: an introduction. In: Paul WE (ed) Fundamental Immunology, Second Edition, Raven Press Ltd., New York, pp 3-19
2. Alt FW, Oltz EM, Young F, Gorman J, Taccioli G and Chen J (1992) VDJ recombination. Immunol. Today 13:306-314
3. Gritzmacher CA (1989) Molecular aspects of heavy-chain class switching. Crit. Rev. Immunol. 9:173-200
4. Schwedler U-V, Jack H-M and Wabl M (1990) Circular DNA is a product of the immunoglobulin class switch rearrangement. Nature 345:452-455
5. Matsuoka M, Yoshida K, Maeda T, Usuda S and Sakano H (1990) Switch circular DNA formed in cytokine-treated mouse splenocytes: evidence for intramolecular DNA deletion in immunoglobulin class switching. Cell 62:135-142
6. Iwasato T, Shimizu A, Honjo T and Yamagishi H (1990) Circular DNA is excised by immunoglobulin class switch recombination. Cell 62:143-149
7. Shimizu A, Takahashi N, Yaoita Y and Honjo T (1982) Organization of the constant-region gene family of the mouse immunoglobulin heavy chain Cell 28: 499-506
8. Honjo T (1983) Immunoglobulin genes. Ann. Rev. Immunol. 1:499-528
9. Marcu KB (1982) Immunoglobulin heavy-chain constant-region genes Cell 29: 719-721
10. Radbruch A, Bruger C, Klein S and Muller W (1986) Control of immunoglobulin class-switch recombination. Immunol. Rev. 89:69-83
11. Shimizu A and Honjo T (1984) Immunoglobulin class switching. Cell 36:801-803
12. Cebra JJ, Komisar JL and Schweitzer PA (1984) CH isotype 'switching' during normal B-lymphocyte development. Ann. Rev. Immunol. 2:493-548
13. Coffman RL, Lebman DA and Rothman P (1993) The mechanism and regulation of immunoglobulin isotype switching. Adv. Immunol. 54:229-270
14. Alt FW, Blackwell TK, DePinho RA, Reth MG and Yancopolous GD (1986) Regulation of genome rearrangement events during lymphocyte development. Immunol. Rev. 89:5-30
15. Jung S, Rajewsky K and Radbruch A (1993) Shutdown of class switch recombination by deletion of a switch region control element. Science 259:984-987
16. Ott DE, Alt FW and Marcu KB (1987) Immunoglobulin heavy chain switch region recombination within a retroviral vector in murine pre-B cells. EMBO J. 6:577-584
17. Ott DE and Marcu KB (1989) Molecular requirements for immunoglobulin heavy chain constant region gene switch-recombination revealed with swich substrate retroviruses. Intern. Immunol. 1:582-591

18. Ott DE, Kim M-G and Marcu KB (1990) Immunoglobulin heavy chain class switching: molecular requirements for constant-region gene switch recombination. Cytokines 3:61-84
19. Winter E, Krawinkel U and Radbruch A (1987) Directed Ig class switch recombination in activated murine B cells. EMBO J. 6:1663-1671
20. Adams B, Dorfler P, Aguzzi A, Kozmik Z, Urbanek P, Maurer-Fogy I and Busslinger M (1992) Pax-5 encodes the transcription factor BSAP and is expressed in B lymphocytes, the developing CNS and adult testes. Genes and Dev. 6:1589-1607
21. Schultz CL, Elenich LA and Dunnick WA (1991) Nuclear protein binding to octamer motifs in the immunoglobulin γ1 switch region. Intern. Immunol. 3:109-116
22. Wuerffel R, Jamieson E, Morgan L, Merkulov GV, Sen R and Kenter A (1992) Switch region breakpoints are strictly correlated with DNA recognition motifs for immunoglobulin Sγ3 DNA-binding proteins. J. Exp. Med. 176:339-349
23. Osborne WRA and Miller AD (1988) Design of vectors for efficient expression of human purine nucleoside phosphorylase in skin fibroblasts for enzyme-deficient humans. Proc. Natl. Acad. Sci. Usa 85:6851-6855
24. Lupton SD, Brunton LL, Kalberg VA and Overell RW (1991) Dominant positive and negative selection using a hygromycin phosphotransferase-thymidine kinase fusion gene. Mol. Cell. Biol. 11:3374-3378
25. Nolan-Willard M, Berton MT and Tucker PW (1992) Coexpression of μ and γ1 heavy chains can occur by a discontinuous transcription mechanism from the same unrearranged chromosome. Proc. Natl. Acad. Sci. USA 89:1234-1238
26. Marolleau J-P, Fondell JD, Malissen M, Trucy J, Barbier E and Marcu KB, Cazenave P-A and Primi D (1988) The joining of germ-line Va-Ja complexes in a T cell receptor a,b positive T cell line. Cell 55:291300
27. Dunnick W, Wilson M and Stavnezer J (1989) Mutations, duplication and deletion of recombined switch regions suggest a role for DNA replication in the immunoglobulin heavy-chain switch. Mol. Cell. Biol. 9:1850-1856
28. Mills FC, Brooker JS and Camerini-Otero RD (1990) Sequences of immunoglobulin switch regions:implications for recombination and transcription. Nucl. Acids Res. 18:7305-7316
29. Taylor LD, Condie EC, Huszar D, Higgins KM, Mashayekh R, Sequar G, Schramm SR, Kuo C-C, O'Donnell SL, Kay RM, Woodhouse CS and Lonberg N (1994) Human immunoglobulin transgenes undergo rearrangement, somatic mutation and class switching in mice that lack endogenous IgM. Intern. Immunol. 6:579-591
30. Finkelman FD, Holmes J, Katona IM, Urban Jr. JF, Beckmann MP, Park LS, Schooley KA, Coffman RL, Mosmann TR and Paul WE (1990) Lymphokine control of in vivo immunoglobulin isotype selection. Annu. Rév. Immunol. 8:303-333

The Role of BSAP in Immunoglobulin Isotype Switching and B-Cell Proliferation

Edward E. Max[1], Yoshio Wakatsuki[2], Markus F. Neurath[2] and Warren Strober[2].
[1] Laboratory of Cell and Viral Regulation, FDA-CBER
[2] Mucosal Immunity Section, Laboratory of Clinical Investigation, NIAID

Abstract

A role for the transcription factor B cell-specific activator protein (BSAP) in switch recombination has been proposed because binding sites for this protein have been found near switch regions of several isotypes. We have attempted to assess BSAP's role by altering the expression of this protein in B cells switching in culture to IgG1. We found that a phosphorothioate oligonucleotide antisense to the BSAP translation initiation site was able, when incubated with B cells, to decrease BSAP activity in nuclear extracts, and that IgG1 expression was reduced in such cells compared to cells incubated with control oligonucleotides. However, it is not clear whether this apparent reduction in switch recombination was mediated by the known BSAP binding sites in the immunoglobulin heavy chain locus because the antisense experiments revealed an additional activity of this protein: it is a rate-limiting regulator of cell proliferation. Down-regulation of BSAP was associated with decreased proliferation, while increasing BSAP (by transfection with a BSAP expression plasmid) increased proliferation. Thus because switch recombination apparently requires cell division, the effect of BSAP down-regulation on switching might have resulted from decreased proliferation. The role of BSAP in B cell proliferation suggests that dysregulation of this protein could contribute to neoplastic transformation of B cells. Because of BSAP's many activities, experiments to elucidate the mechanisms of its effects on switching and proliferation will be challenging.

Introduction

The DNA recombination that underlies the immunoglobulin isotype switch marks an important event in B cell development, but little is known about its mechanism. Comparisons between DNA structures before and after isotype switching have revealed that the switch is accompanied by deletion of a DNA segment whose endpoints lie in or near the "switch regions" of the two isotype loci involved in the switch -- generally μ and the newly expressed isotype. This recombination replaces the Cμ gene with the constant region gene for the newly expressed isotype, which comes to lie downstream of the V region gene in roughly the position formerly occupied by Cμ. The switch regions are DNA segments characterized by many internal repeats of sequence motifs over several kb of DNA. Switch regions lie upstream of most immunoglobulin heavy chain constant region genes.

To investigate the mechanism of the switch recombination event, several laboratories have pursued an approach which has met with some success in analyzing the mechanism of the V(D)J recombination: namely, the characterization of nuclear proteins that bind to DNA motifs in or near the recombining loci. Recently several labs have independently reported (Waters et al., 1989; Liao et al., 1992; Marcu et al., 1992; Xu et al., 1992) examples of such binding motifs that all appear to interact with the same protein, which we here call BSAP (B cell-specific activator protein) (See Fig. 1). The gene encoding this

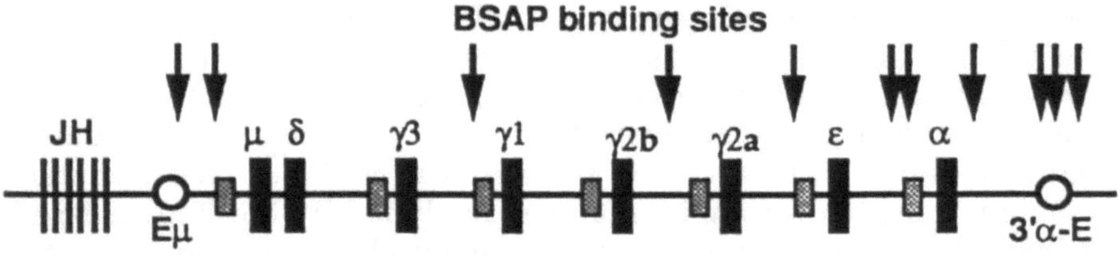

Fig. 1. Location of reported BSAP binding sites in the immunoglobulin heavy chain locus, based on data from (Waters et al., 1989; Liao et al., 1992; Marcu et al., 1992; Xu et al., 1992)

protein was simultaneously cloned by still another laboratory (Barberis et al., 1990; Adams et al., 1992), which was pursuing murine proteins capable of interacting with a known regulatory motif from the promoter of the sea urchin histone gene H2A-2.2. BSAP turns out to be a member of a family of transcription factors (reviewed in Walther et al., 1991; Gruss and Walther, 1992) widely conserved in evolution and encoded by "paired box" or PAX genes (the name deriving from one of the first members identified, the *paired* gene of drosophila). BSAP is encoded by the PAX-5 gene. All PAX proteins contain a 128-residue segment -- the "paired domain" -- which contacts a complex DNA motif (Czerny et al., 1993). In addition, most contain a characteristic octapeptide; and some (including BSAP) contain a segment homologous to a classic homeodomain.

BSAP is now known to be a transcription factor, participating in the regulation of several genes expressed in the B lymphocyte lineage. BSAP itself is expressed in pre-B and B cells but not in secreting plasma cells. The genes it regulates include CD19 (Kozmik et al., 1992), CD20 (J. Kehrl, personal communication), λ5 and VpreB (Okabe et al., 1992); in these examples BSAP apparently stimulates transcription by binding to motifs in the promoter. BSAP can also act as a negative regulator; two BSAP binding sites in the enhancer 3' of the immunoglobulin Cα gene inhibit transcription (Singh and Birshtein, 1993; Neurath et al., 1994). The existence of BSAP sites near immunoglobulin switch regions has led to the hypothesis that this protein might play a role in switching by up-regulating the synthesis of "sterile transcripts." These are RNAs that are known to be transcribed from a particular immunoglobulin constant region which has not yet undergone switch rearrangement, but which is about to do so. It is believed that such transcription may be necessary for switch recombination, either because it makes the switch region "accessible" to switch recombination enzymes or because the resultant transcript participates in the recombination reaction. Recently, the proposed role for BSAP in stimulating sterile transcription received support from the report that a BSAP site in the promoter of the murine ε heavy chain sterile transcript is required for optimal activity of that promoter (Liao et al., 1994).

In evaluating the generality of such a role for BSAP in isotype switching, it is important to note that BSAP is already known to regulate many genes in the B cell lineage and undoubtedly regulates more that are presently unknown; furthermore switch recombination requires more than production of a sterile transcript, so other roles for BSAP in switch recombination should be considered. In particular, B cells incubated under conditions that stimulate sterile transcript production do not undergo DNA switch recombination unless agents are added that stimulate cell proliferation. Thus it seems possible that the recombination requires not only transcription of a switch target gene but DNA replication as well. Thus we have examined the relationship of BSAP to proliferation (Wakatsuki et al., 1994).

Results

BSAP down-regulation reduces surface IgG1 expression

We have studied the role of BSAP in switching to IgG1 in splenic B cells. In this system resting B cells are prepared by depleting splenic cells of other cell types using complement lysis and Percoll centrifugation. The resulting cells are incubated with LPS and IL-4, which together promote -- in a percentage of the cells -- surface expression of IgG1 (Radbruch et al., 1986; Kepron et al., 1989) associated with switch recombination (Chu et al., 1993).

We attempted to suppress BSAP in these B cells by coincubation with a synthetic antisense (vs control "sense") phosphorothioate oligonucleotide overlapping the translation start site of BSAP mRNA.

```
             M  D  L  E  K  N  Y
AATATCGAAATGGATTTAGAGAAAAATTAC   BSAP cDNA
      ||||||||||||||||||||
   5'-CGAAATGGATTTAGAGAA    sense oligo
      GCTTTACCTAAATCTCTT-5' antisense oligo
```

An additional control ("nonsense" or "scrambled") oligonucleotide was a random mixture of oligos having the same average base composition as the antisense oligo. To evaluate the effectiveness of the antisense oligonucleotide in lowering BSAP levels, an electrophoretic mobility shift assay (EMSA) was used to visualize BSAP in nuclear extracts from oligo-treated B cells; the probe for these assays was the double stranded H2A-2.2 oligonucleotide originally used to characterize BSAP. As shown in Fig. 2, nuclear extracts from cells treated with control oligonucleotides generated a band which comigrated with the

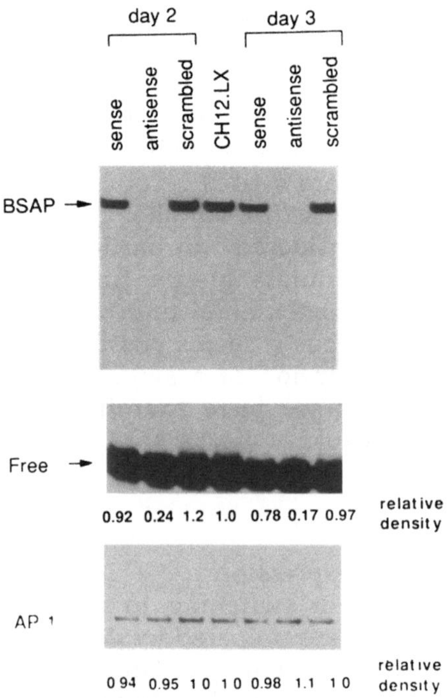

Fig. 2 Reduction of BSAP by antisense oligonucleotide. Splenic B cells stimulated by LPS were incubated in 30 μM of the indicated phosporothioate oligonucleotide. Nuclear extracts were assayed in EMSAs for BSAP activity using the H2A-2.2 probe (Barberis, et al., 1989) and for AP-1 activity (Lee, et al., 1987).

BSAP band from the cell line CH12.LX; the CH12LX band in earlier experiments demonstrated sequence-specific competition with appropriate unlabelled oligonucleotides (data not shown). In comparison with control oligonucleotides (sense and scrambled), the antisense oligonucleotide reduced detectable BSAP activity substantially by day 2 and almost completely by day 3. Levels of another transcription factor, AP-1, were not altered. To test whether the antisense oligo was effective in reducing functional *in vivo* BSAP activity, we transfected CH12LX cells with a construct in which transcription is inhibited by BSAP sites derived from the enhancer downstream of Cα; the antisense oligonucleotide caused increased expression of the construct, consistent with a release from BSAP-induced inhibition (data not shown).

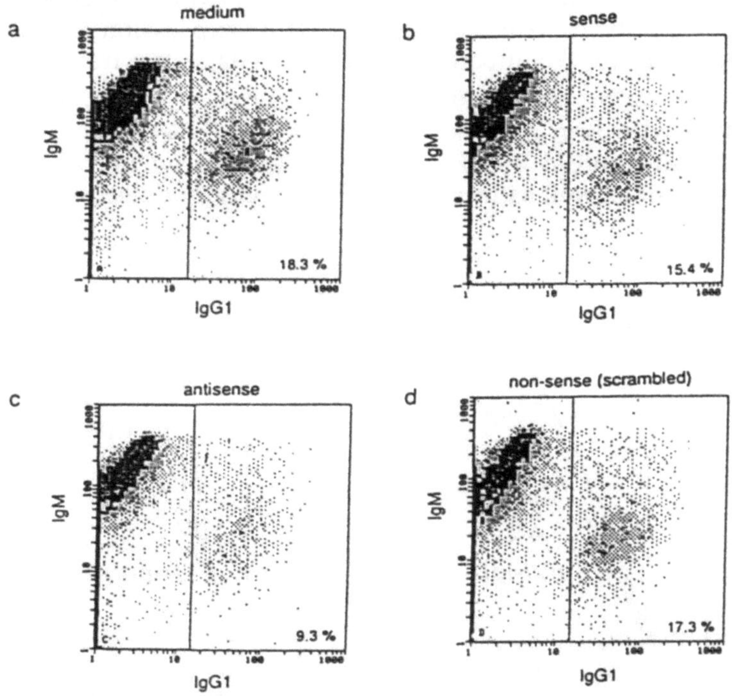

Fig 3. Effect of BSAP antisense oligo on surface expression of a"switched" isotype. Splenic B cells were incubated for four days in LPS and IL-4 plus the indicated oligonucleotide (30μM) and surface IgG1 and IgM were assessed by two-color flow cytometry.

Since the antisense oligonucleotide appeared to substantially reduce BSAP levels in B cells, we could ask whether this reduction affected the switch to surface IgG1 expression that occurs in culture in response to IL-4 and LPS. As shown in Fig. 3, number of sIgG1+ cells detected by flow cytometry was significantly reduced in cells treated with BSAP antisense oligo, in comparison with that observed with control oligos. A BSAP binding site reported 5' of the initiation site of sterile transcripts from the γ1 locus (Waters et al., 1989) represents a possible mediator of this effect, although this BSAP site did not appear to regulate promoter activity in conventional transient transfection assays (J. Stavnezer, personnel communication and Xu and Stavnezer, 1992).

BSAP's role in cell proliferation

Because cell proliferation is an important prerequisite for isotype switching, we examined the relationship between proliferation and BSAP activity. Splenic B cells were incubated in the presence of three proliferative stimuli: (1) CD40 ligand (CD40L), expressed on the surface of L cells stably transfected with a CD40L expression construct, (2) monoclonal anti-IgD antibody bound to L cells stably transfected with an FcγII receptor expression construct (L_{CDw32} cells), and (3) LPS (incubated with L_{CDw32} cells but no antibody). As shown in Fig. 4, all three stimuli induced B cell proliferation (measured by ^{3}H-thymidine incorporation) as well as increases in BSAP activity detected by EMSA.

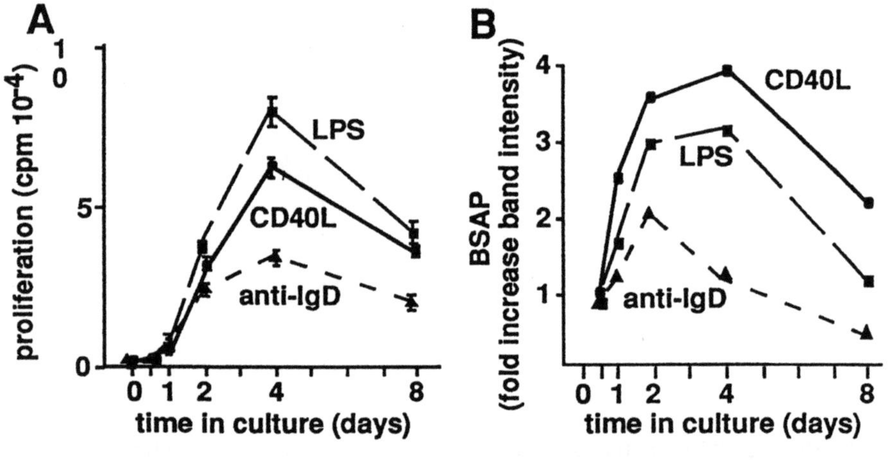

Fig. 4. Proliferation and BSAP responses to B cell activation. B cells were incubated under the three activation signals shown. (A) Cell proliferation, meared by ^{3}H-thymidine incorporation. (B) BSAP, measured by densitometry of EMSA bands.

The correlation of BSAP activity with proliferation suggested the possibility that BSAP plays a regulatory role in B cell proliferation. To explore this possibility, splenic B cells stimulated by LPS were incubated with varying concentrations of BSAP antisense oligo (or control oligos) and ^{3}H-thymidine incorporation was measured after 4 days. At the highest concentration tested (40μM) the antisense oligo almost completely abolished BSAP activity and caused about a 10-fold reduction in proliferation, while control oligos had minimal effect on proliferation (see Fig.5).

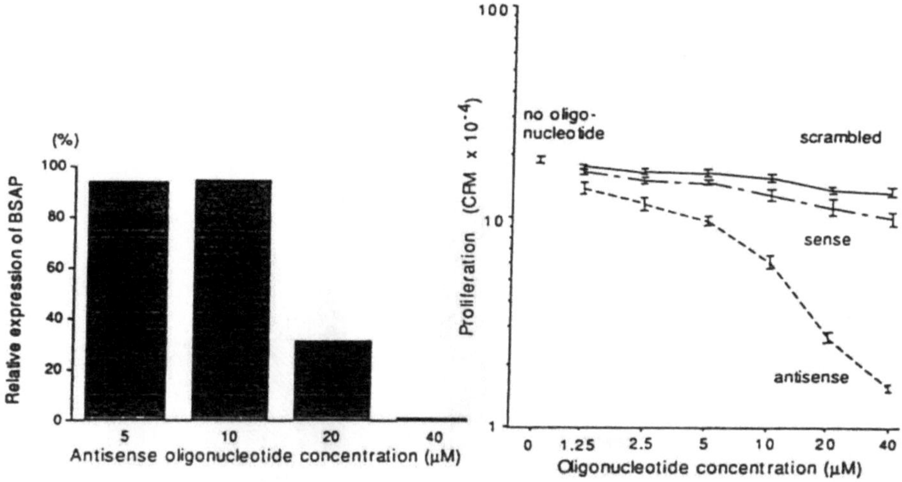

Fig. 5. Effects of BSAP oligos on splenic B cells. Resting spleen B cells were incubated with LPS and the indicated oligonucleotides and harvested after 4 days. (A) BSAP activity (EMSA band intensity). (B) proliferation (^{3}H-thymidine incorporation).

If the antiproliferative effect of the antisense oligo is mediated by down-regulation of BSAP, then this oligo should have no effect on the proliferation of cells lacking endogenous BSAP activity. This point was tested by incubating the antisense oligo (or control sense oligo) with the BSAP-negative lines MOPC315 (a plasmacytoma) and EL-4 (T cell) as well as the BSAP-positive line CH12.LX. At 40μM, the antisense oligo inhibited proliferation of CH12.LX more than 100-fold, while exerting little effect on the two BSAP-negative lines (data not shown). This observation supports the interpretation that the antisense oligo inhibits proliferation by suppression of BSAP and not by non-specific toxicity.

BSAP up-regulation stimulates proliferation

To further explore the relationship between BSAP and proliferation, we transfected B cells with a BSAP expression plasmid to assess whether up-regulation of BSAP could stimulate proliferation. The expression plasmid was constructed by PCR amplification of BSAP mRNA, followed by cloning into the expression vector BCMGSNeo (Karasuyama et al., 1989). To verify that the final expression construct encoded functional BSAP, the plasmid was transfected into the BSAP-negative MOPC315 line and into splenic B cells, and BSAP activity in nuclear extracts was visualized by EMSA; the transfected MOPC315 cells demonstrated a clear BSAP band, and the endogenous BSAP band of the B cells showed increased intensity (data not shown). Fig. 6 shows that transfection of the BSAP-expressing plasmid into splenic B cells

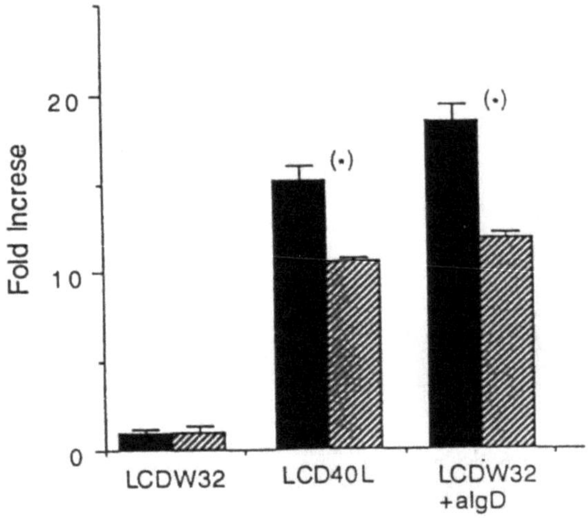

Fig. 6. Transfection of B cells with a BSAP expression plasmid increases proliferation in response to CD40 ligand and to anti-IgD. Splenic B cells were transfected by electroporation with either the BSAP expression plasmid (filled bars) or a control vector (hatched bars) and then incubated with the indicated stimuli.

caused a significant increase in proliferation, as compared with transfection of a control vector lacking the BSAP cDNA. This stimulation, though modest in magnitude, is impressive considering that only a minority (no more than 30%) of the cells were effectively transfected, so the average proliferation rate measured in this experiment must underestimate the proliferation in the minority of cells with transfection-boosted BSAP levels. Transfection of BSAP into MOPC315 also induced an increase in proliferation (data not shown), indicating that this plasmacytoma is capable of responding to BSAP even though its baseline proliferative rate appears to be maintained by other factors in the absence of BSAP.

Discussion

Our results indicate that BSAP is required for optimal sIgG1 expression in splenic B cells stimulated with LPS and IL-4. The modest inhibition of sIgG1 expression that we observed under conditions that dramatically suppress BSAP activity suggests either that only small amounts of BSAP are required to support sIgG1 expression or that some switching to sIgG1 production can occur independently of BSAP. Presumably the decrease in sIgG1 expression that we observed reflects a decrease in Sμ-Sγ1 switch rearrangement, but this point has not been explicitly addressed in our experiments to date. Whether the BSAP binding sites near the immunoglobulin μ or γ1 loci play a role in the decreased sIgG1 expression we have observed is not clear. As mentioned above, the BSAP binding motif upstream of the initiation site for sterile γ1 transcription has not been found to play a role in the associated promoter. Regardless of the role of these BSAP sites, our data suggest another potential role for BSAP in isotype switching: maintenance of cell proliferation, which is believed to be essential for switch recombination to occur. Our results indicate that BSAP is a rate-limiting controller of proliferation, in that experimental manipulations that up- or down-regulate BSAP levels cause corresponding changes in proliferation rates. The mechanism of BSAP's role in proliferation is unknown. Although this protein is known to regulate several genes in the B cell lineage, none of the known genes would be expected to directly influence cell division. It seems likely that additional genes will be found to be regulated by BSAP, including some that may explain its role in proliferation. One speculative possibility is that dysregulation of BSAP could be involved in some lymphoproliferative disorders.

References

Adams, B, Dorfler, P, Aguzzi, A, Kozmik, Z, Urbanek, P, Maurer, FI, Busslinger, M (1992). Pax-5 encodes the transcription factor BSAP and is expressed in B lymphocytes, the developing CNS, and adult testis. Genes Dev 6:1589-607.

Barberis, A, Widenhorn, K, Vitelli, L, Busslinger, M (1990). A novel B-cell lineage-specific transcription factor present at early but not late stages of differentiation. Genes Dev 4:849-59.

Chu, CC, Max, EE, Paul, WE (1993). DNA rearrangement can account for in vitro switching to IgG1. J Exp Med 178:1381-90.

Czerny, T, Schaffner, G, Busslinger, M (1993). DNA sequence recognition by Pax proteins: bipartite structure of the paired domain and its binding site. Genes Dev 7:2048-61.

Gruss, P, Walther, C (1992). Pax in development. Cell 69:719-22.

Karasuyama, H, Tohyama, N, Tada, T (1989). Autocrine growth and tumorigenicity of interleukin 2-dependent helper T cells transfected with IL-2 gene. J Exp Med 169:13-25.

Kepron, MR, Chen, YW, Uhr, JW, Vitetta, ES (1989). IL-4 induces the specific rearrangement of gamma 1 genes on the expressed and unexpressed chromosomes of lipopolysaccharide-activated normal murine B cells. J Immunol 143:334-9.

Kozmik, Z, Wang, S, Dorfler, P, Adams, B, Busslinger, M (1992). The promoter of the CD19 gene is a target for the B-cell-specific transcription factor BSAP. Mol Cell Biol 12:2662-72.

Liao, F, Birshtein, BF, Busslinger, M, Rothman, P (1994). The transcription factor BSAP (NF-HB) is essential for immunoglobulin germline epsilon transcription. J Immunol 152:2904-2911.

Liao, F, Giannini, SL, Birshtein, BK (1992). A nuclear DNA-binding protein expressed during early stages of B cell differentiation interacts with diverse segments within and 3' of the Ig H chain gene cluster. J Immunol 148:2909-17.

Marcu, KB, Xu, L, Kim, MG (1992). S alpha BP/BSAP/NF-S mu B1, a murine and human B cell stage specific nuclear factor with DNA binding specificity implying roles in switch-recombination and transcription. Curr Top Microbiol Immunol 182:167-74.

Neurath, M, Strober, W, Wakatsuki, Y (1994). The murine immunoglobulin 3' alpha enhancer is a target site with repressor function for the B-cell lineage-specific transcription factor BSAP (NF-HB, S-alpha-BP). J Immunol in press.

Okabe, T, Watanabe, T, Kudo, A (1992). A pre-B- and B cell-specific DNA-binding protein, EBB-1, which binds to the promoter of the VpreB1 gene. Eur J Immunol 22:37-43.

Radbruch, A, Muller, W, Rajewsky, K (1986). Class switch recombination is IgG1 specific on active and inactive IgH loci of IgG1-secreting B-cell blasts. Proc Natl Acad Sci U S A 83:3954-7.

Singh, M, Birshtein, BK (1993). NF-HB (BSAP) is a repressor of the murine immunoglobulin heavy-chain 3' alpha enhancer at early stages of B-cell differentiation. Mol Cell Biol 13:3611-22.

Wakatsuki, Y, Neurath, MF, Max, EE, Strober, W (1994). The B cell-specific transcription factor BSAP regulates B cell proliferation. J Exp Med 179:1099-1108.

Walther, C, Guenet, JL, Simon, D, Deutsch, U, Jostes, B, Goulding, MD, Plachov, D, et al. (1991). Pax: a murine multigene family of paired box-containing genes. Genomics 11:424-34.

Waters, SH, Saikh, KU, Stavnezer, J (1989). A B-cell-specific nuclear protein that binds to DNA sites 5' to immunoglobulin S alpha tandem repeats is regulated during differentiation. Mol Cell Biol 9:5594-601.

Xu, L, Kim, MG, Marcu, KB (1992). Properties of B cell stage specific and ubiquitous nuclear factors binding to immunoglobulin heavy chain gene switch regions. Int Immunol 4:875-87.

Xu, MZ, Stavnezer, J (1992). Regulation of transcription of immunoglobulin germ-line gamma 1 RNA: analysis of the promoter/enhancer. Embo J 11:145-55.

MIX
Papier aus verantwortungsvollen Quellen
Paper from responsible sources
FSC® C105338

If you have any concerns about our products,
you can contact us on
ProductSafety@springernature.com

In case Publisher is established outside the EU,
the EU authorized representative is:
Springer Nature Customer Service Center GmbH
Europaplatz 3, 69115 Heidelberg, Germany

Printed by Libri Plureos GmbH
in Hamburg, Germany